T0329635

Smart Grid and Enabling Technologies

Smart Grid and Enabling Technologies

Shady S. Refaat
Texas A&M University at Qatar, Doha, Qatar

Omar Ellabban
CSA Catapult Innovation Centre, Newport, UK

Sertac Bayhan
Qatar Environment and Energy Research Institute,
Hamad bin Khalifa University, Doha, Qatar

Haitham Abu-Rub
Texas A&M University at Qatar, Doha, Qatar

Frede Blaabjerg
Aalborg University, Aalborg, Denmark

Miroslav M. Begovic
Texas A&M University, College Station, USA

Registered Office(s)
John Wiley & Sons, Inc., 111 River Street, Hoboken, NJ 07030, USA
John Wiley & Sons Ltd, The Atrium, Southern Gate, Chichester, West Sussex, PO19 8SQ, UK

Editorial Office
The Atrium, Southern Gate, Chichester, West Sussex, PO19 8SQ, UK

For details of our global editorial offices, customer services, and more information about Wiley products visit us at www.wiley.com.

Wiley also publishes its books in a variety of electronic formats and by print-on-demand. Some content that appears in standard print versions of this book may not be available in other formats.

Library of Congress Cataloging-in-Publication Data

Names: Refaat, Shady S., author. | Ellabban, Omar, author. | Bayhan, Sertac, author. | Abu-Rub, Haithem, author. | Blaabjerg, Frede, author. | Begovic, Miroslav M., 1956- author.
Title: Smart grid and enabling technologies / Shady S. Refaat, Texas A&M University at Qatar, Doha, Qatar, Omar Ellabban, CSA Catapult Innovation Centre, Newport, UK, Sertac Bayhan, Qatar Environment and Energy Research Institute, Hamad bin Khalifa University, Doha, Qatar, Haitham Abu-Rub, Texas A&M University at Qatar, Doha, Qatar, Frede Blaabjerg, Aalborg University, Aalborg, Denmark, Miroslav M. Begovic, Texas A&M University, College Station, USA.
Description: First edition. | Hoboken, NJ : Wiley, 2021. | Includes bibliographical references and index.
Identifiers: LCCN 2021012116 (print) | LCCN 2021012117 (ebook) | ISBN 9781119422310 (hardback) | ISBN 9781119422433 (adobe pdf) | ISBN 9781119422457 (epub)
Subjects: LCSH: Smart power grids.
Classification: LCC TK3105 .R44 2021 (print) | LCC TK3105 (ebook) | DDC 621.31–dc23
LC record available at https://lccn.loc.gov/2021012116
LC ebook record available at https://lccn.loc.gov/2021012117

Cover Design: Wiley
Cover Image: © NicoElNino/Getty Images

Set in 9.5/12.5pt STIXTwoText by Straive, Pondicherry, India
Printed and bound by CPI Group (UK) Ltd, Croydon, CR0 4YY

C9781119422310 _150721

Contents

About the Authors

Shady S. Refaat received the BASc, MASc, and PhD degrees in Electrical Engineering in 2002, 2007, and 2013, respectively, all from Cairo University, Giza, Egypt. He has worked in the industry for more than 12 years as Engineering Team Leader, Senior Electrical Engineer, and Electrical Design Engineer on various electrical engineering projects. Currently, he is an associate research scientist in the Department of Electrical and Computer Engineering, Texas A&M University at Qatar. He is a senior member of the Institute of Electrical and Electronics Engineers (IEEE), a member of The Institution of Engineering and Technology (IET), a member of the Smart Grid Center – Extension in Qatar (SGC-Q). He has published more than 105 journal and conference articles. His principal work area focuses on electrical machines, power systems, smart grid, Big Data, energy management systems, reliability of power grids and electric machinery, fault detection, and condition monitoring and development of fault-tolerant systems. Also, he has participated and led several scientific projects over the last eight years. He has successfully realized many potential research projects.

Omar Ellabban (S'10–M'12–SM'13) is a senior researcher and creative manager with more than 20 years of combined experiences (teaching, research, industrial experience, consulting services and project management) between academia, research institutes, industry and power utility companies in various fields.

Dr. Ellabban is conducting and leading many research projects in different areas, such as: power electronics, electric vehicles, automatic control, motor drive, energy management, grid control, renewable energy, energy storage devices, distributed energy systems and their integration into the smart grid. Dr. Ellabban received his BS (Hons) degree in electrical machines and power engineering from

Helwan University, Egypt; his MS degree in electrical machines and power engineering from Cairo University, Egypt; and his PhD (Hons.) degree in electrical engineering from Free University of Brussels, Belgium, in 1998, 2005, and 2011, respectively. He joined the Research and Development Department, Punch Powertrain, Sint-Truiden, Belgium, in 2011, where he and his team developed a next-generation, high-performance hybrid powertrain. In 2012, he joined Texas A&M University in Qatar as a postdoctoral research associate and became an assistant research scientist in 2013, where he is involved in different renewable energy integration projects. In 2016, he joined Iberdrola Innovation Middle East as the Research and Development Director to lead a number of research, development and innovation projects under various topics focusing on transforming the current electric grid into a smart grid and integrating renewable energies and energy storage systems interfaced by power electronics converters as microgrids penetrating the distribution networks. In 2020, he joined CSA Catapult as Principal Power Electronics Engineer to lead different projects focusing on Compound Semiconductors applications across different sectors.

Dr. Ellabban has authored more than 70 journal and conference papers, one book chapter, two books entitled, "Impedance Source Power Electronic Converters, 2016" and "Smart Grid Enabling Technologies, 2020" and many international conference tutorials. His current research interests include renewable energies, grid control, smart grid, automatic control, motor drives, power electronics, and electric vehicles. He is a Senior Member of the IEEE, IET member and currently serves as an Associate Editor of the IEEE Transactions on Industrial Electronics.

Sertac Bayhan received the MS and PhD degrees in electrical engineering from Gazi University, Ankara, Turkey, in 2008 and 2012, respectively. In 2008, he joined the Electronics and Automation Department, Gazi University, as a Lecturer, where he was promoted to Associate Professor in 2017. From 2014 to 2018, he worked at Texas A&M University at Qatar as a Postdoctoral Fellow and Research Scientist. Dr. Bayhan is currently working in the Qatar Environment and Energy Research Institute (QEERI) as a Senior Scientist and he is a faculty member with the rank of Associate Professor in the Sustainable Division of the College of Science and Engineering at Hamad Bin Khalifa University.

Dr. Bayhan is the recipient of many prestigious international awards, such as the Research Fellow Excellence Award in recognition of his research achievements and exceptional contributions to the Texas A&M University at Qatar in 2018, the Best Paper Presentation Recognition at the 41st and 42nd Annual Conference of the IEEE Industrial Electronics Society in 2015 and 2016, Research Excellence Travel Awards in 2014 and 2015 (Texas A&M University at Qatar), and Researcher Support Awards from the Scientific and Technological Research Council of Turkey (TUBITAK). He has acquired $13 M in research funding and published more than 150 papers in mostly prestigious IEEE journals and conferences. He is also the coauthor of two books and four book chapters.

Dr. Bayhan has been an active Senior Member of IEEE. Because of the visibility of his research, he has recently been elected as Chair of IES Power Electronics Technical Committee and selected as a Co-Chair of IEEE-IES Student and Young Professional Activity Program. He currently serves as Associate Editor for IEEE Transactions on Industrial Electronics, IEEE Journal of Emerging and Selected Topics in Industrial Electronics, and IEEE Industrial Electronics Technology News, and Guest Editor for the IEEE Transactions on Industrial Informatics.

Haitham Abu-Rub is a full professor holding two PhDs from Gdansk University of Technology (1995) and from Gdansk University (2004). Dr. Abu Rub has much teaching and research experience at many universities in a number of countries including Qatar, Poland, Palestine, the USA, and Germany.

Since 2006, Dr. Abu-Rub has been associated with Texas A&M University at Qatar, where he has served for five years as chair of Electrical and Computer Engineering Program and has been serving as the Managing Director of the Smart Grid Center at the same university.

His main research interests are energy conversion systems, smart grid, renewable energy systems, electric drives, and power electronic converters.

Dr. Abu-Rub is the recipient of many prestigious international awards and recognitions, such as the American Fulbright Scholarship and the German Alexander von Humboldt Fellowship. He has co-authored around 400 journal and conference papers, five books, and five book chapters. Dr. Abu-Rub is an IEEE Fellow and Co-Editor in Chief of the IEEE Transactions on Industrial Electronics.

Frede Blaabjerg (S'86–M'88–SM'97–F'03) was with ABB-Scandia, Randers, Denmark, from 1987 to 1988. He became an Assistant Professor in 1992, an Associate Professor in 1996, and a Full Professor of power electronics and drives in 1998. From 2017, he became a Villum Investigator. He is *honoris causa* at University Politehnica Timisoara (UPT), Romania and Tallinn Technical University (TTU) in Estonia. His current research interests include power electronics and its applications such as in wind turbines, PV systems, reliability, harmonics and adjustable speed drives. He has published more than 600 journal papers in the fields of power electronics and its applications. He is the co-author of four monographs and editor of 10 books in power electronics and its applications. He has received 32 IEEE Prize Paper Awards, the IEEE PELS Distinguished Service Award in 2009, the EPE-PEMC Council Award in 2010, the IEEE William E. Newell Power Electronics Award 2014, the Villum Kann Rasmussen Research Award 2014, the Global Energy Prize in 2019, and the 2020 IEEE Edison Medal. He was the Editor-in-Chief of the IEEE Transactions on Power Electronics from 2006 to 2012. He has been Distinguished Lecturer for the IEEE Power Electronics Society from 2005 to 2007 and for the IEEE Industry Applications Society from 2010 to 2011 as well as 2017 to 2018. In 2019–2020 he serves as President of IEEE Power Electronics Society. He is also Vice-President of the Danish Academy of Technical Sciences. He was nominated in 2014–2019 by Thomson Reuters to be among the most 250 cited researchers in Engineering across the world.

 **Miroslav M. Begovic** (FIEEE'04) is Department Head of Electrical and Computer Engineering and Carolyn S. & Tommie E. Lohman '59 Professor at Texas A&M University. Prior to that, he was Professor and Chair of the Electric Energy Research Group in the School of Electrical and Computer Engineering, and an affiliated faculty member of the Brook Byers Institute for Sustainable Systems and University Center of Excellence in Photovoltaic Research at Georgia Tech. Dr. Begovic obtained his PhD from Virginia Tech University. His research interests are in monitoring, analysis, and control of power systems, as well as development and applications of renewable and sustainable energy systems. For the Centennial Olympic Games in 1996 in Atlanta, he designed with Professor Ajeet Rohatgi, a 340 kW photovoltaic system on the roof of Aquatic Center at Georgia Tech, which at that time was the largest roof-mounted PV system in the world. He has been a member of the IEEE PES Power System Relaying Committee for two decades and chaired a number of its working groups. Professor Begovic was Editor of the section on Transmission Systems and Smart Grids in the *Springer Encyclopedia on Sustainability* (published in 2012), coordinated by an Editorial Board consisting of five Nobel Prize Laureats, has also served as guest editor of the IET Generation, Transmission & Distribution Special Issue on Wide Area Monitoring and Control in 2010, authored one section of a book, nearly 200 journal and conference papers, two IEEE special publications, and delivered more than 100 keynote and invited presentations. He authored invited papers in three Special issues of IEEE Proceedings: on Future Energy Systems (2010), on Critical Infrastructures (2005) and on Renewable Energy (2001).

Dr. Begovic is a Fellow of IEEE and member of Sigma Xi, Tau Beta Pi, Phi Kappa Phi and Eta Kappa Nu. Dr. Begovic is a former Chair of the Emerging Technologies Coordinating Committee of IEEE PES, IEEE PES Treasurer (2010–2011), IEEE PES Distinguished Lecturer, and serves as President of the IEEE Power and Energy Society.

Acknowledgments

We would like to take this opportunity to express our sincere appreciation to all the people who were directly or indirectly helpful in making this book a reality.

We are grateful to the Qatar National Research Fund (a member of Qatar Foundation) for funding many of the research projects, whose outcomes helped us in preparing a major part of this book chapters. Chapters 1, 8, 9, and 17 for NPRP grant [NPRP12S-0226-190158], chapter five for NPRP grant [NPRP9-310-2-134], Chapters 10, 14 and 15 for NPRP grant [NPRP10-0101-170082], and Chapter 6 for NPRP grant [NPRP12S-0214-190083]. The statements made herein are solely the responsibility of the authors.

Also, we appreciate the help from many colleagues and students for providing constructive feedback on the material and for help with the editing. Particular appreciation goes to Mohammad Saleh, Amira Mohammed, and Mohamed Massoudi.

We are indebted to our family members for their continuous support, patience, and encouragement without which this book would not have been completed.

Preface

Smart grid (SG) is an emerging area of engineering and technology which integrates electricity, communication, and information infrastructures to ensure an efficient, clean, and reliable electric energy supply. This is an extremely complex field with different disciplines and engineering areas pooled together. This book aims to cover SG technologies and their applications in a systematic and comprehensive way. Different areas of SGs have been included in this book, such as architectural aspects of the SG, renewable energy integration, power electronics domination in the SG, energy storage technologies for SG applications, smart transportation, communication and security aspects, the pivotal role of artificial intelligence toward the evolution of SGs, SG challenges and barriers, standardization, and future vision. For this reason, the book has been written by experienced individuals who specialize in various areas of SGs.

The objective of this book is to equip readers with up-to-date knowledge of the fundamentals, emerging grid structure, current research status, and future vision in the development and deployment of SGs. The concepts presented in this book include four main areas of SGs and its applications: **Advanced SG Architecture** which includes smart power systems, communication systems, information technology, security, and the advancement of microgrids. **Renewables energies,** entail technologies of both energy storages, and power electronics suitable for renewable energy systems and SG applications. **SG applications** are divided into fundamental and emerging applications. The fundamental applications refer to energy management strategies, reliability models, security, and privacy, in addition to promoting demand-side management (DSM). Emerging applications include the deployment of electric vehicles (EVs) and mobile charging stations. **SG tools** are divided into crucial tools for distribution grids such as Big Data management and analytics, cloud management and monitoring tools, consumer engagement, and artificial intelligence for the SG, the requirements for the simulation tools and the recently adopted standards, in addition to the challenges and future business models of SGs.

The book builds its foundation by introducing the SG architecture and integrating renewable energy sources and energy storage systems in the next generation power grid. The first chapter provides a basic discussion on the infrastructure of SGs followed by the technologies used in the SG. An overview of different renewable energy resources is discussed in Chapter 2 showing their current status, future opportunities, and the challenges of integrating them with the grid. Energy storage systems and power electronics converters as grid integration units are presented in Chapters 3–4. A comprehensive review of microgrids, including their characteristics, challenges, design, control, and operation either in grid-connected or islanded modes are introduced in Chapter 5. Chapter 6 is devoted to one of the most emerging applications of SGs, which is smart transportation. This chapter presents an overview of EVs; their current status and future opportunities, in addition to the challenges of integrating them into the SG. The impact of EVs on SG operation and modeling EV mobility in energy service networks are also exemplified in this chapter. The net-zero energy cost building uses energy efficiency and renewable energy strategies as part of the business model. Chapter 7 describes the zero energy buildings (ZEBs) definition, design, modeling, control, and optimization. This chapter discusses the benefits and barriers of the current state and the future trends of ZEBs as a step to reduce energy consumption in the building sector. The goal of Chapter 8 is focused on the SG features utilizing multi-way communication among energy production, transmission, distribution, and usage facilities. The multi-way communication among energy generation, transmission, distribution, and usage facilities is discussed in Chapter 8. The reliable, efficient, and intelligent management of complex power systems necessitates an employment of high-speed, reliable, and secure data information and communication technology into the SG to manage and control power production and usage; this topic is described in Chapter 9. The electric energy sector is sitting on a data goldmine, the so-called Big Data. The real value of Big Data utilization resides in a good understanding of its analytics technologies, promise, and potential applications in SGs, which is discussed in Chapter 10. Driven by concerns regarding electric sustainability, energy security, and economic growth, it is essential to have a coordination mechanism based on heuristic rules to manage the energy demand and enhance the survivability of the system when failures occur or at peak periods, which is achieved by the principle of DSM systems defined in Chapter 11. It is essential to know the business model concepts, their main components, and how they can be used to analyze the impact of SG technology to create, deliver, and capture values for the utility business. The value chain for both traditional and smart energy industries are needed. The different electricity markets have been described and presented in Chapter 12. Chapter 13 aims to provide energy systems researchers and decision-makers with proper insight into the underlying drivers of consumer acceptance of the SG and the logical steps for their engagement to promote SG technology and make it feasible promptly. The fundamental relationship between SG and cloud computing services is also covered. The architectural principles, characteristics of cloud computing services, and the examination of the advantages and disadvantages of those characteristics for SG are defined. Furthermore, the opportunities and challenges of using cloud computing in SGs, and the major categories of data security challenges of cloud computing are described in Chapter 14.

In Chapter 15, the latest taxonomy of Artificial Intelligence (AI) applications in SGs is discussed, including load and renewable energy forecasting, power optimization, electricity price forecasting, fault diagnosis, and cyber and physical layers security. Chapter 16 discusses the current state of simulation-based approaches including multi-domain simulation, co-simulation, and real-time simulation and hardware-in-the-loop for SGs. Furthermore, some SG planning and analysis software are summarized with their advantages and disadvantages. Chapter 17 presents an overview of SG standards; new standardization studies, SG policies of some countries, and some important standards for the smart grid. Chapter 18 depicts the concepts of distributed generation, micro-grid, SG, and distributed operation, which all pose more complexity and challenges to the modern power systems. This chapter presents the challenges and barriers that modern SGs face from different perspectives.

This book has the typical attributes of a contemporary book and discusses several aspects that will appeal to students, researchers, professionals, and engineers from various disciplines looking to increase their knowledge, insights, and ideas for the future development of SG as the next energy paradigm. This work perfectly fills the current gap and contributes to the realization and a better understanding of SG and its enabling technologies.

List of Abbreviations

3 GPP	Third Generation Partnership Project
ABC	Ant bee colony
Adaboost–MLP	Adaptive boosting-Multilayer perceptron
AER	all-electric-range
AMI	Advanced metering infrastructure
AMR	Automated Meter Reading
AMS	Automatic metering services
AM	analytical methods
ANSI	American National Standards Institute
AC	Alternative Current
AVR	Automatic Voltage Regulator
AWS	Amazon Web Services
API	Application programming interface
ANN	artificial neural net-work
ARIMA-XGBoost	Autoregressive integrated moving Average-Extreme gradient boosting
ARMA-TDNN	Autoregressive and moving average-Time delay neural network
ARMA	Autoregressive and Moving Average
ARIMA	Autoregressive integrated moving average
Adaboost–MLP	Adaptive boosting-Multilayer perceptron
ARIMA-ANFIS	Autoregressive integrated moving average-*adaptive neuro-fuzzy inference system*
ANFIS	*Adaptive neuro-Fuzzy inference system*
AODE	Aggregating One-Dependence Estimators classifier
ARMA	Autoregressive moving average
ACO	Ant colony optimization
Ant colony	Ant colony
ABC	Ant bee colony
ARIMA Neurofuzzy-	Artificial neural networks-fuzzy logic-Autoregressive integrated moving average
ARIMA Mixed	Mixed autoregressive integrate moving average
Improved ARIMAX	Improved Autoregressive integrated moving average process with exogenous inputs
BAC	Building Automation and Control
BM	business model
BPE	Building Produced Energy

BANs	Building/Business Area Network
BPL	Broadband over Power Line
Bayesian	Wavelet-Extreme learning machine
Boosting additive quantile regression	Boosting additive quantile regression
BNN	Bayesian neural network
BAC	Bayesian actor-Critic algorithms
BBN	Bayesian belief network
Bio	Biological swarm chasing algorithm
bioenergy	Biomass energy
biofuels	liquid fuels
BEVs	battery electric vehicles
CC	Cloud Computing
CIM	Common Information Model
CPP	critical peak pricing
CHIL	controller HIL
CHP	Combined heat and power systems
CIS	Customer Information System
CEPRI	China Electric Power Research Institute
COAG	Council of Australian Governments
CAES	compressed air energy storage
CHB	cascaded H-bridge converter
CMC	central management controller
Faster R-CNN	Faster Region-based Convolutional neural network
CNN-WT	Convolutional neural network-Wavelet transform
CVAELM	Complementary ensemble empirical mode decomposition with adaptive noise- Variational mode decomposition – Adaptive boosting-Extreme learning machine
CRfs	Conditional random fields
CRO-SL	Coral reefs optimization algorithm
CSP	concentrating solar power
c-Si	silicon
CO_2	carbon dioxide
CAES	Compressed Air Energy Storage
CICEVs	conventional internal-combustion-engine vehicles
DG	Distributed Generation
DR	Demand Response
DMS	Distribution Management System
DSI	Demand-Side Integration
DSM	Demand Side Management
DSL	Digital Subscriber Line
DSO	distribution network operator
DER	Distributed Energy Resources
DRMS	Demand Response Management System
DFIG	doubly fed induction generator
DC	Direct Current
DPR	digital protective relay
DFR	digital fault recorder

DGs	distributed generators
DES	Distributed Energy Storage
DBN	Deep belief networks
DCNN	Deep convolutional neural network
DRN-DWWC	Deep residual networks - Dynamically weighted wavelet coefficients
DQL	Deep Q-learning
DNI	direct component
DoD	*Depth of Discharge*
E2E	End-to-End
EMS	Energy management system
EU	European Union
EPBD	Energy Performance of Building Directive
ETL	extract, transform and load
ETAP	Electrical Transient Analyzer Program
EPRI	Electric Power Research Institute
EEGI	European Electricity Grid Initiative
ESSs	energy storage systems
EMI	electromagnetic interference
ERP	Enterprise Resource Planning
EV	Electric Vehicles
ESN	Echo state networks
EKF-based NN	Extended Kalman filter method Neural Network
Extra tree	Extra tree
ETC	evacuated tube solar collectors
ESOI	energy stored on energy invested
EVSE	Electric Vehicle Supply Equipment
EPS	Electric power system
FACTS	Flexible AC transmission systems
FERC	Federal Energy Regulatory Commission
FMI	Functional Mockup Interface
FMU	Functional Mock-up Units
FAN	Field Area Network
FC	flying capacitor converter
FL	Fuzzy logic
FCRBM	Factored conditional restricted Boltzmann Machine
Faster R-CNN	Faster Region-based Convolutional neural network
FIS-LSE	Fuzzy inference system-least-squares estimation
FLC-FA	Fuzzy logic controller-Firefly algorithm
FR	Fresnel Reflector
FPC	flat-plate solar collectors
FES	Flywheel Energy Storage
FCEVs	fuel-cell electric vehicles
GOOSE	generic object-oriented system event
GSM	Global System for Mobile Communications
GIS	Geographic Information System
GPS	Global Positioning System
GWAC	Grid Wise Architecture Council

GASVM	Genetic functionality support vector machine
GRU NN	Gated recurrent unit neural network
GP	Gaussian process
GANs	Generative adversarial networks
GBM	Gradient boosting machine
GARCH	Generalized Autoregressive Conditional Heteroskedastic
GA-NN	Genetic algorithm-Neural network
GBRT	Gradient boosted regression tree
Glow-worm	Glow-worm optimization
GWO	grey wolf optimization
GBTD	Gradient boosting theft detector
GHP	Geothermal Heat Pump
GWEC	The Global Wind Energy Council
GHG	greenhouse gases
HANs	Home area networks
HEMs	home energy management
HIL	Hardware-in-the-Loop
HPCC	HomePlug Command and Control
HVDC	High-Voltage Direct Current
HCS	hill climbing searching
HESSs	hybrid energy storage systems
HPF	high pass filter
HDFS	Hadoop Distributed File System
HNN	Hybrid neural network
HMM	Hidden Markov models
HPPs	Hydropower plants
HEVs	hybrid electric vehicles
ICT	Information and Communication Technologies
IEDs	Intelligent Electronic Devices
IHD	In-Home Display
IANs	Industrial Area Networks and
IEC	International Electro Technical Commission
IEEE	Institute of Electrical and Electronics Engineers
ITL	Information Technology Laboratory
ISO	International Organization for Standardization
ISA	International Society of Automation
IC	internal combustion
IaaS	Infrastructure as a Service
IGCC	Integrated Gasification Combined Cycle
ITU	International Telecommunication Union
ISACA	Information Systems Audit and Control Association
Ipso-ANN	Improved article Swarm Optimization Algorithm-Artificial neural network
Improved ARIMAX	Improved Autoregressive integrated moving average process with exogenous inputs
IRENA	International Renewable Energy Agency
IEA	International Energy Agency
KPIs	Key Performance Indicators

KATS	Korean Agency for Technology and Standards
kbps	kilobits per second
KNN	K-nearest neighbors
KNN-ANN	K-nearest neighbor-Artificial neural network
KAIST	Korea Advanced Institute of Science and Technology
LPF	low pass filter
LAN	local area network
LVDC	low-voltage direct current
LVAC	low-voltage alternating current
LSTM NN	Long short-term memory Neural network
LES	Linear exponential smoothing
LightGBM	Light gradient boosting method
LCOE	levelized cost of electricity
LCOS	levelized cost of storage
Li-ion	*Lithium-Ion*
Li-Po	*Lithium-Polymer*
MG	Microgrid
MDMS	Meter Data Management System
MPPT	maximum power point tracking
MMC	modular multilevel converter
MVDC	Medium Voltage DC
MPC	Model Predictive Control
MOPSO	Multi-Objective Particle Swarm Optimization
MOGA	Multi-Objective Genetic
MDMS	Meter Data Management System
MVAC	medium voltage AC
MAS	multi agent-based control system
MV	medium voltage
MSS-Ada	MSS-Adaptive boosting
Adaboost–MLP	Adaptive boosting-Multilayer perceptron
Mixed ARIMA	Mixed autoregressive integrate moving average
MNB	Multinomial naïve bayes
MS	Molten Salt
MP	mathematical programming
NANs	neighbored area networks
NIST	National Institute of Standards and Technology
NPC	neutral-point clamped converter
nZEB	net Zero Energy Buildings
nZEC	net Zero Energy Community
NAN	Neighborhood Area Network
NN	Neural network
Neuro-Fuzzy	Artificial neural networks-Fuzzy logic
Neurofuzzy-ARIMA	Artificial neural networks-fuzzy logic-Autoregressive integrated moving average
NARX	Nonlinear autoregressive exogenous
NB	Naïve bayes
Ni-Cd	*Nickel-Cadmium*

Ni-MH	*Nickel-Metal Hydride*
OMS	Outage Management System
OPFR	Optimal Power Flow Reserves
OpenADR	Open Automated Demand Response
OPF	Optimal Power Flow Tool
OTEC	*Ocean thermal energy conversion*
OTEGs	ocean thermo-electric generators
OLEV	on-line electric vehicle
ORNL	Oak Ridge National Laboratory
PHIL	power HIL
PHEV	Plug-in Hybrid Electric Vehicle
PSCAD	Power Systems Computer Aided Design
PLC	Power Line Communication
PV	photovoltaic
P2P	peer-to-peer
PLL	Phase Locked Loop
PVQV	Voltage Adequacy and Stability Tool
PES	Power and Energy Society
PSF	power signal feedback
PI	proportional integral
PID	proportional–integral–derivative
PMU	phasor measurement units
PaaS	Platform as a Service
PSO	Particle swarm optimization
PCC	point of common coupling
PCA-LSSVM	Principal component analysis- Least squares support vector machines
PDRNN	Pooling-based deep recurrent neural network
PT	Parabolic Trough
PHS	Pumped Hydroelectric Storage
PEVs	plug-in electric vehicles
PHEVs	plug-in hybrid electric vehicles
Pb-acid	*Lead-Acid*
PATH	Partners for Advance Transit and Highways
QoS	Quality of Service
RTS	power grid real-time
RERs	renewable energy resources
ROCOF	rate of change of frequency
RTP	Real-Time Pricing
RVM-XGBoost Ensemble	Ensemble of relevance vector machines and boosted trees
RF-XGBoost	Random forest-extreme gradient boosting
RBFNN	Radial basis neural networks
RF	Random forest
RFL	Reinforcement learning
RE	renewable energy
RC	Resistance to Change
RF	Radio Frequency
SMs	smart meters

SCADA	Substation supervisory control and data acquisition
STATCOM	Static Synchronous Compensator
SG	Smart Grid
SCADA	Supervisory Control and Data Acquisition
SCOPF	Security Constrained Optimal Power Flow Tool
SGCC	State Grid Corporation of China
SC	supercapacitors
SMES	superconducting magnetic energy storage
SRF-PLL	synchronous reference frame PLL
SGs	synchronous generators
SMs	submodules
SSR	Sub-synchronous resonance
SER	sequence of event recorder
SQL	Structured Query Language
SaaS	Software as a Service
SVM	Support vector machines
SD-EMD-LSTM	Similar days-selection, empirical mode decomposition-long short-term memory neural networks
SDAs	Stacked denoising Autoencoders
SARSA	State–action–reward–state–action
STARMA	Spatiotemporal auto-regressive moving average model
SELM	Stacked extreme learning machine
SMLE	Stacking heterogeneous ensemble learning model
SMO	spider monkey optimization
SA	Simulated annealing
SARIMAX	Seasonal autoregressive integrated moving average process with exogenous inputs
STPAR	Smooth transition periodic autoregressive
STARMA	Spatiotemporal models
SD	Solar Dish
ST	Solar Tower
STS	Solar thermal systems
SF	Solar Fuels
SNG	Synthetic Natural Gas
SLI	starting, lighting, and ignition
SoC	*State of Charge*
SoH	*State of Health*
SGCC	Smart Grid Consumer Collaborative
TSO	transmission system operator
ToU	Time of use pricing
TEPCO	Tokyo Electric Power Co.
TSR	tip speed ratio
TD learning	Temporal difference learning
TF	thin-films
TES	Thermal Energy Storage
TRA	Theory of Reasoned Action
TPB	Theory of Planned Behavior
TAM	Technology Acceptance Model

TTM	Transtheoretical Model
UNFCCC	United Nations Framework Convention on Climate Change
VSC	Voltage Source Converters
VSG	Virtual synchronous generator
VIC	virtual impedance control
VPN	virtual private network
VSI	voltage source inverter
VMD-CNN	Variational mode decomposition -Convolutional neural network
VARMAX	Vector autoregressive moving Average with exogenous inputs
V2G	vehicle-to-grid
G2V	grid-to-vehicle
V2H	vehicle to home
V2V	vehicle to vehicle
VAM	Value-based Adoption Model
WASA	wide-area situational awareness
WAN	Wide Area Network
WAMS	Wide Area Monitoring Systems
WLAN	wireless local area network
WPT-RF	Wavelet packet transform-Random forest
WESN	Wavelet echo state networks
WT-TES-WNN	Wavelet transform combined with Holt-Winters- Weighted nearest-neighbor model
WNN	Weighted nearest neighbors
WCD	World Commission on Dams
WECs	wave energy converters
WPT	wireless power transfer
XML	Extensible Markup Language
XGboost-DWT	Extreme gradient boosting- Discrete wavelet transform
XGBoost	Extreme gradient boosting
ZEBs	zero energy buildings

About the Companion Website

Smart Grid and Enabling Technologies is accompanied by a companion website:

www.wiley.com/go/ellabban/smartgrid

The website includes:

- PowerPoint Slides for Lecturers

Scan this QR code to visit the companion website.

1

Smart Grid Architecture Overview

The electric power system is the largest and best engineering invention and achievement in human history. However, this grid paradigm faces serious challenges with regard to the increasing demand for electricity, the expanding penetration of intermittent renewable energies, and the need to respond to emerging needs such as wide usage of electric vehicles. The newly faced and expected challenges and expectations from the grid are forcing drivers to transform the current power system into a smarter grid. Smart grid (SG) is a new paradigm shift that combines the electricity, information, and communication infrastructures to create a more reliable, stable, accessible, flexible, clean, and efficient electric energy system. The SG comprises two main parts, SG infrastructure, and smart applications and operation. SG infrastructure entails a smart power system, information technology (IT), and communication system, while SG applications and operation are categorized into fundamental and emerging areas. The fundamental ones refer to energy management strategies, reliability models, security, privacy, and demand-side management (DSM). Emerging applications include the wide deployment of electric vehicles and mobile charging and storage stations. All this indicates that SGs are characterized by automated energy generation, delivery, monitoring, and consumption with stakeholders from smart utilities, markets, and customers.

Initially in this chapter, the principles of current electrical power systems will be briefly discussed. After that, the implications of the transformation trend toward SG architecture will be investigated. Following this, SGs are addressed in greater depth, covering fundamentally diverse concepts and classifications. Lastly, some SG architectures will be highlighted and the future challenges and directions will be addressed.

1.1 Introduction

Today, power grids are being challenged to meet the ever-growing energy demands of the twenty-first century. Energy usage is projected to rise by 50% by 2050, according to the Energy Information Administration [1]. Today's grid is an aging infrastructure, combined with the growth of distributed energy resources (DERs), it is therefore more prone to outages and disturbances leading to poor reliability and power quality. These factors present a significant challenge for distributed renewable energy integration to the grid with unidirectional power flow [2]. The current grid is characterized by one-directional electricity flow, lack of information exchange, centralized bulk generation, lack of flexibility to directly trade in electricity markets, inefficient monitoring and control of the power distribution networks, lack of flexibility and accessibility to new innovative solutions such as flexible loads, accommodating large scale of fluctuating energy resources, electric vehicles wide usage, etc. The SG is designed to tackle all these challenges by integrating and smartly utilizing the electricity, information, and communication infrastructures.

The SG is the solution to overcome the aforementioned challenges while also responding to the current and future humanity energy expectations. SG's implementation will not only have environmental benefits through high penetration of renewable sources, but will also have significant regional, national, and global impacts related to achieving a more reliable, efficient, and economic energy system. The SG paradigm integrates a variety of modern advanced technologies such as smart sensors and measurements, advanced communication and information, edge computing and control. This paradigm allows a flexible and reliable electricity system with bi-directional power and information flows [3]. The structures of the SG anticipate and respond to electric system disturbances, optimize asset utilization, and operate efficiently. SG houses all generation and storage options, which hinders the dependency on peak demand back-up power stations – thus, cutting significant costs related to the generation, transmission, and distribution. Furthermore, SG enables active participation of customers, new products, services, and markets – thus can support the uptake of new industries. SG functions resiliently against attacks and natural disasters, delivers power quality for the digital economy- thus, creating new jobs and regenerating the economy at a time of financial crisis. The SG is a power network that contains distributed nodes, which operate under the pervasive control of smart subsystems, so-called smart microgrids. A microgrid is a small-scale version of the electric grid, however possessing distributed generation and potentially energy storage (ES). Microgrids can operate in a grid-connected mode, islanded mode, or in both modes which improve the grid's reliability, controllability, and efficiency. Widespread installations of microgrids enable a faster transformation to the SG paradigm from the current grid infrastructure [4].

1.2 Fundamentals of a Current Electric Power System

The two main characteristics of conventional electrical power systems are: centralized energy generation and unidirectional power delivery systems. This means that electric power is first produced by bulk power generation and then transmitted across the electricity grid to the distribution layer, and finally to the end users. The flow of electricity in the grid is from top (high-voltage network) to bottom (low-voltage network). Figure 1.1 shows the three main stages of generation, transmission, and distribution [5]. Those grid elements are briefly explained in the next sub-sections.

1.2.1 Electrical Power Generation

Traditional power plants burn the fossil fuel to generate electricity therefore they contribute in significant amount of greenhouse gas emission to the environment. The crucial need

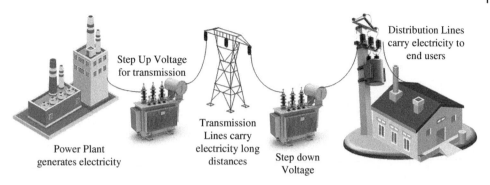

Figure 1.1 The fundamentals of electric power system. Adapted from Ref Num [5].

to adopt power-generation approaches that have fewer environmental impacts is an essential requirement for modern power grids, which means moving toward more renewable energy systems. Among renewable energies, sunlight (photovoltaics) is converted into electricity, wind kinetic energy is converted into electricity, water gravitational and kinetic energy is converted to electricity (hydro) [6]. Continuous technology development is used to convert renewable energies into electricity at increased efficiency and lower cost. Therefore, it is essential for the current power grid to efficiently accommodate a constant increase of fluctuating renewable energy sources.

1.2.2 Electric Power Transmission

The transmission system has the highest voltage rating and is used to transmit the energy from the bulk generation plant to the distribution networks through interconnecting substations. The transmission system may include overhead lines, underground and under water cables. This transmission system could be high voltage alternating current (HVAC) or high voltage direct current (HVDC). Traditionally, an HVAC system is mostly used, however, the HVDC is rapidly gaining popularity due to reduced losses and cost particularly over large distances. The electric power could also be transported and distributed via underground cables. Construction of an underground transmission system normally costs 4 to 10 times that of an equal distance overhead line [7]. The overall amount of power and distance that the power needs to be transported, overdetermine the essential design of the transmission and distribution systems. Hence, the greater the distance and the more power to be transferred, the higher the rated voltage is, and the higher the cost of the system will be as shown in Figure 1.2 [8]. This shows that distributed renewable energy generation allows for the reduction of the cost not only of the central power generation plants, but also of the transmission and distribution infrastructures. This is one of the drivers for the SG energy paradigm.

1.2.3 Electric Power Distribution

The distribution network is the last stage in electric power delivery responsible for carrying electricity from the transmission and sub-transmission systems to end users. There are four main arrangements used in electric power distribution: radial, parallel feeders, ring main, and interconnected (mesh) systems as shown in Figure 1.3 [9]. Distribution networks

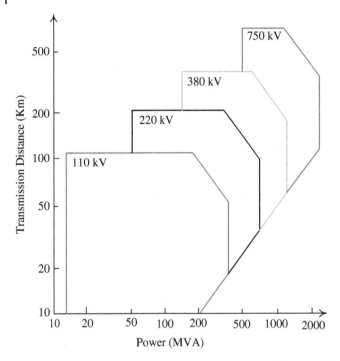

Figure 1.2 Selection of rated voltage for three-phase AC transmission line. Ref [8]. Reproduced with permission from John Wiley & Sons.

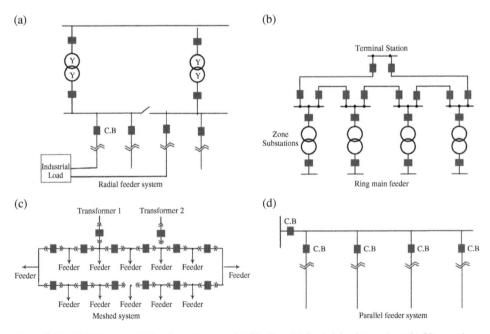

Figure 1.3 Main types used in electric power distribution, (a) Redial feeder system. (b) Ring main feeder. (c) Meshed system. (d) Parallel feeder system. Adapted from Ref Num [9].

typically have a radial topology "star network," with merely a single power flow path between the distribution substation and a certain load (unidirectional power flow). Distribution networks rarely implement a ring or loop topology, with two power flow paths between the distribution substation and the load [10]. Moving toward renewable energy systems and introducing distributed generators (DGs) enforces the use of bidirectional power flows which causes many challenges for traditional distribution systems. A transformation to the SG paradigm includes smart elements to allow for bidirectional power flow and ES technologies. ES systems may include pumped hydro, compressed air, electrochemical batteries, flow batteries, compressed air, superconducting magnetic ES, supercapacitors, and flywheels.

1.3 Limitations of the Traditional Power Grid

The first AC generating system was built more than 130 years ago as a centralized and unidirectional system [11]. The traditional power network consists of various elements such as conductors, transformers, switches, relays, etc. for safely delivering electricity to end consumers. The traditional power grid is a centralized control and management system that uses a supervisory control and data acquisition (SCADA) as shown in Figure 1.4.

Developing real-time control, supervision, and monitoring systems with a smart protection system is essential to optimize the production and consumption of electricity, to improve the overall efficiency, and to ensure the grid's reliability. The challenge is that several new generation sources must randomly connect and disconnect seamlessly with the distribution grid. Moreover, controlling a large number of different sources with different characteristics is of the utmost importance due to the possibility of conflicting requirements and limited communication resources [12]. Those are essential challenges for transforming the current grid into an SG. Other challenges before implementing the SG could be as follows:

1.3.1 Lack of Circuit Capacity and Aging Assets

In many locations worldwide, the power system has extended widely since the 1950s and the distribution and transmission equipment are now beyond the expected lifetime and require replacement. The capital costs of like-for-like change are very high. A typical large-scale utility may require hundreds of billions of dollars to fully modernize the whole grid's infrastructure, which is an overrated cost and should decrease gradually over time. The practical solution is to transform this overstrained infrastructure into an SG, which may even take 20 years or more.

1.3.2 Operation Constraints

Every power system should operate within predefined voltage and frequency limits. The capacity of traditional distribution circuits is restricted by power limits and the changes in voltage and frequency that exist between maximum and minimum loads. Different actions should be taken to vary the output of generations according to the demand, as well as to reconfigure and optimize the network to prevent system collapse. Increasing the output of

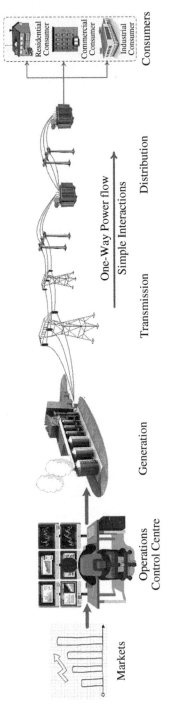

Figure 1.4 Traditional power grid.

generation as per demand at different points in the network is an acceptable solution in case the overloads are small and infrequent [13]. If constraints continue to exist, other solutions should be considered such as advancing the capacity of current power lines, including new lines, or producing separate electricity "highways" to avoid the affected parts, which is still of high cost. SG paradigm can achieve the energy management in a cheaper way by engaging end users in managing their consumption levels which protects the grid without the need to add new expensive infrastructure.

1.3.3   Self-Healing Grid

Nowadays, society needs a growing reliable electricity supply as crucial loads are increasingly connected. The conventional approach is to add more redundant systems, at a considerable environmental impact and capital cost. No action has been needed to maintain supply after a fault other than disconnecting the faulty circuit. The SG paradigm allows for distributed renewable energy generation and intelligent post-fault reconfiguration and self-healing for sustained electricity supply at significantly reduced cost [14]. Better utilization of assets can be achieved with SG at fewer redundant circuits, lower cost, and higher efficiency.

1.3.4   Respond to National Initiatives

Many agencies and policy makers are promoting SG initiatives as cost-effective solutions for modernizing their power systems while activating the deployment of low-carbon energy resources. Advancements of the SG can be noticed in a number of countries. Governments are aware that the SG principle is able to mitigate various blackouts and contingencies at lower cost. Many of SG technologies are already deployed, whereas others are in the demonstration and planning phases. Lastly, the increasing concerns over the natural disasters and terrorist attacks in many countries have motivated the governments to call for a more resilient grid that is less dependent on centralized power stations. SG solutions respond to those challenges and expectations.

1.4   Smart Grid Definition

The Department of Energy (DOE) defines the SG as "the electricity delivery system, from point of generation to point of consumption, integrated with communications and information technology for enhanced grid operations, customer services, and environmental benefits" [15]. The SG implements electricity, information, and communication infrastructures to generate power more efficiently and reliably, and as cleanly and safely as possible for preserving the environment [16]. The European Technology Platform defined the SGs as "an electricity network that can intelligently integrate the actions of all users connected to it generators, consumers, to efficiently deliver sustainable, economic and secure electricity supply" [17]. From the previous definitions, it is evident that the SG is an electrical grid that entails a variety of smart technologies, operations, and measurements such as smart meters, smart appliances, renewable energy resources, electric vehicle, flexible loads, smart

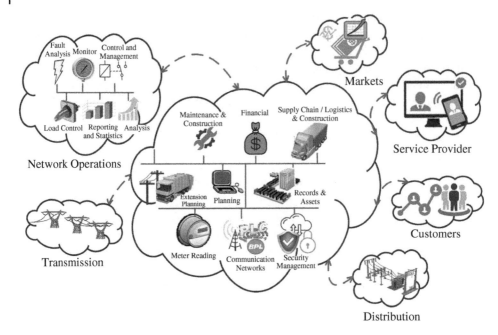

Figure 1.5 The conceptual model of SG framework. Ref [18]. Reproduced with permission from Walter de Gruyter GmbH.

markers, various energy-efficient programs, and smart end users. The SG includes the benefits of advanced communications and information technologies that provide real-time information which can intelligently and cost-effectively integrate the behaviors and actions of all users connected to it, i.e. generators, operators and consumers. This will ensure reliable, efficient, and economically viable solutions for the continuous delivery of clean and affordable energy. One important difference between present grids and the SG is the two-way exchange of power and information within the grid. The conceptual model of a SG is shown in Figure 1.5 [18]. SG implements innovative products and services along with intelligent monitoring, control, communication, and processing to:

- Improve facilitation between the grid elements of all sizes.
- Permit customers to play an important role in improving the operation of the system.
- Offer customers more information and options to participate in the energy market.
- Significantly decrease the environmental impact of electricity generation.
- Improve the electric system efficiency, reliability, quality, and security.
- Improve service quality and reduce electricity cost.

1.5 Smart Grid Elements

The SG architecture consists of three main systems: power, communication, and information. Proper architecture is necessary to ensure SG functionality. The design and analysis of future SGs require fundamental insight into the impact of power network topology and integrated network control with Big Data utilization. Furthermore, it is essential to have an insight into the complex interaction between the physical layer and cyber layer that includes the supporting communication, information, and computational systems. SG architecture can be represented as a layered structure including the following five main layers as shown in Figure 1.6: System architecture, Distribution Control, Applications, Standards, and Cybersecurity measures.

The grid modernization is expected to make the grid more flexible, accessible, and manageable with interconnected networks consisting of a number of smaller-sized subsystems integrated with a large number of renewable energy sources. Making the grid more accessible is possible by having grid resources available and considering the access to the loads. The SG serves the needs of multiple stakeholders in the electricity industry. Devices and systems developed individually by different vendors and for different electric utilities are employed by various customers, so they must work together in harmony; these systems must achieve interoperability requirement. The upcoming technology in the SG interoperability framework is the real-time dynamic control and management systems. The components of SG are the combination of intelligent appliances and equipment that play an important role in the production, delivery, and consumption of electricity. SG elements can be grouped into seven key technology areas [19]. These areas are distributed generation, electric storage system, smart meters, advanced control, integrated communications, sensing and measurement, improved interfaces, and decision support using customer engagement and demand response (DR) as shown in Figure 1.7.

1.5.1 Distributed Generation

DER are defined as small-scale decentralized power generation sites as shown in Figure 1.8 [20]. The systems with DER are modular, and flexible usually located in the vicinity of the load. DER systems utilize renewable energy sources such as small hydro, biomass, biogas, solar,

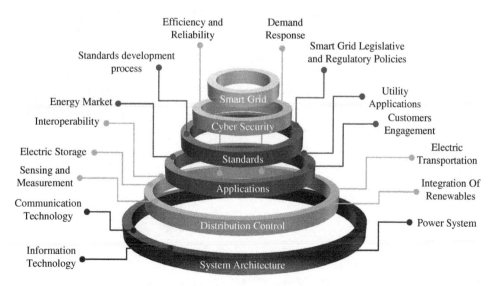

Figure 1.6 SG components.

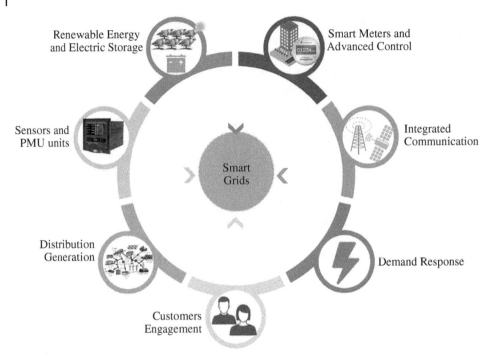

Figure 1.7 Main key technology areas of smart grid.

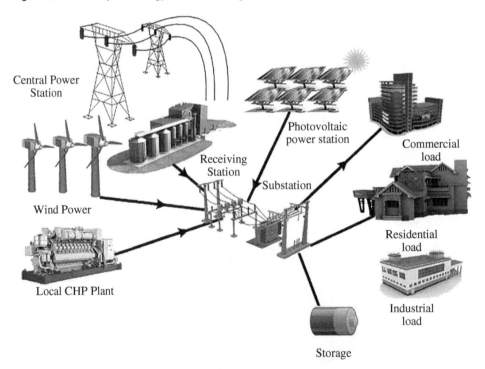

Figure 1.8 Distributed energy resources paradigm in smart grid. Ref [20]. Reproduced with permission from IEEE.

wind, and geothermal. Such systems can be controlled and coordinated within an SG. Using the distributed generation makes the grid active of bidirectional flow of power [21, 22].

DERs can offer potential benefits to the electric grid such as improving energy efficiency, enhancing energy security, and ensuring faster recovery of electricity services. DERs may serve in a single structure connected to an isolated grid, become part of a microgrid, or be connected to the distribution system. Distributed generation can support the delivery of clean, reliable power which supports reducing electricity losses over transmission and distribution lines. The impact of the DERs depends on different factors [23]; such as:

- Size and penetration level of the DGs.
- Type of the DGs, unit ratings, unit impedance, and used transformers used, etc.
- Mode of operation and the interconnection methods with the grid or with the local loads.

On other side, the penetration of DGs increases the complexity of power grids and presents significant stability and control challenges, which may cause greater voltage and frequency deviations and coordinating problems. To overcome these challenges, a coordinated control and managing system must be used to provide the continuity of service, while still meeting customer demands and ensuring the vulnerability of the power system.

1.5.2 Energy Storage

ES is an essential technology for obtaining effective utilization of renewable energy while ensuring continuous energy supply and grid support. Therefore, the storage provides a way to settle the peaks and valleys of supply and end disruptive electricity supply. Storage technologies needs more advances to accomplishing higher power and energy capacities to enabling large-scale deployment. ES systems are combined with advanced power electronics as the interface with the electrical grid. The technical benefits of ES are various forms of grid support [22]. Distributed storage systems may have enormous potential to provide various services to the grid such as supporting the grid's voltage and frequency, providing spinning reserves, enhancing national grid security, and improving grid resiliency. Distributed ES systems are installed at a number of locations on and off the grid. Such systems have two main elements for charging and bi-directional energy flow as shown in Figure 1.9. Most of the existing large- and utility-scale storage resources are hydro and pumped storage. ES can also provide many financial benefits [24].

Effective storage relies on storing and discharging electricity at the required time, and in a way that relies on clear and automatic pricing signals transmitted to smart storage systems. Such storage can give a solution to some challenges, for example, the power congestion at the distribution level, hence avoid/defer potential upgrades in grid infrastructure. However, there are many storage related challenges that must be taken into consideration such as [25]:

- Policies enhancement on net metering, DR, grid reliability standards, generation-based incentives vs hybrid solutions, and the need to consider energy efficiency policies at equipment level vs efficiency at the systemic level.
- Distortion of price signals due to subsidies or lack of real-time pricing signals for consumers.
- Need to consider life-cycle vs capital costs for the selection of government-funded projects.

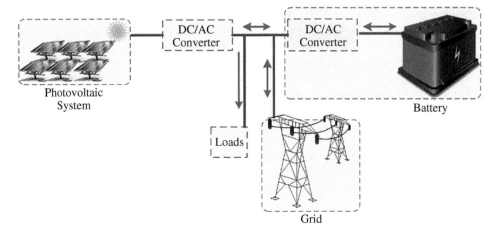

Figure 1.9 The distributed energy storage system.

- Awareness of available technologies and opportunities in various sectors.
- Cost of technology role for the localization and system integration.
- Need for innovative business models.
- Need for financing mechanisms.

1.5.3 Demand Response

The concept of DR has emerged as a solution to demand-side control in the microgrid [26]. Using smart meters and the bidirectional way of communication in the SG opens the door to the technology to participate in the electricity market improvement [27]. The DR programs can be defined as the most successful solution to solve the peak-load burden on the grid and engage customers in the wholesale market operations. A greater number of active consumers can change the profile of the load by minimizing or maximizing the demand as per the generation instructions. This ensures that the load will follow the generation, rather than the generation following the load as in the current energy paradigm. However, these types of programs attract a number of customers' schemes, but designing it remains a major challenge. Most DR programs and the challenges faced when implementing these programs will be presented in detail in Chapter 11.

1.5.4 Integrated Communications

Communication infrastructure is essential for the effective functioning of SGs. The implementation of communication technologies guarantees the decrease of energy consumption, ensures best implementation of the SG, and provides coordination among all SGs' components from generation to the end-users. Examples of existing communication network technologies used for SG are fiber optics, WLAN, cellular communication, WiMAX, and power pine communication (PLC). Detailed discussion of the integrated communication in the SG and a comparison of communication infrastructure between the legacy grid and the SG communication standards and research challenges and future trends are presented in Chapter 8.

1.5.4.1 Communication Networks

The communication system connects various components of SG architecture for real-time control, monitoring, and data utilization. Integrated communication is the connector for all SG technologies. The communication infrastructure of the SG is predicated upon three types of networks: Home Area Network (HAN), Neighborhood Area Network (NAN), and Wide Area Network (WAN). Figure 1.10 shows the diagram of the SG communication infrastructure [28]. **HAN** is installed and operated in a small area (tens of meters) and has a lower transmission data rate of hundreds of bits per second. HAN consists of a broadband internet connection used to communicate and share the data between devices over a network connection and smart meters. HAN offers more efficient home energy management [28]. **NAN** is installed and operated in an area over hundreds of meters. A number of HANs can be connected to one NAN to transmit the data of other NAN networks and to local data centers for storage and further analytics. The NAN has a 2 Kbps transmission data rate. Different technologies can be used to implement the NAN network such as PLC, Wi-Fi, and Cellular [29].

WAN is installed and operated in an area of tens of kilometers and it contains several NANs and LDCs. The communication between SG components such as renewable energy generation, transmission, distribution, and the operator control center are predicated upon a WAN network [30]. SG communication infrastructures share the same main challenge, which is how to be merged effectively. A number of technologies can be employed to the SG to achieve an effective merge between communication infrastructure. These technologies are ZigBee, WLAN, Cellular networks, and Power Line Communication (PLC).

ZigBee is utilized in applications requiring a small data rate, prolonged battery life, low price, and safe networking. Applications also include wireless light switches, traffic control systems, meters for in-home-displays, and extra consumer and industrial devices that require a short-range of wireless data transmission at relatively low rates. The benefits of ZigBee application in the SG are low cost, decreased size, and relatively decreased bandwidth. The drawbacks of the ZigBee are the small battery which suggests a short lifetime, small memory, limited data rate, and low processing capability [31].

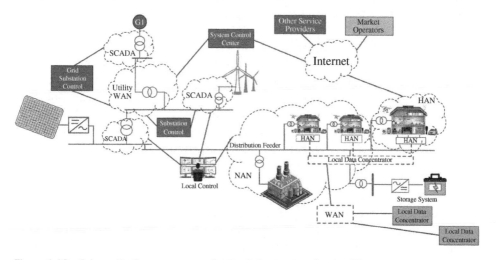

Figure 1.10 Schematic diagram communication infrastructure for the SG.

WLAN is a wireless local area network (WLAN) that links two or more devices through the use of spread-spectrum or Orthogonal Frequency Division Multiplexing (OFDM) [32] and generally delivering a connection through an access point to the internet. This provides customers with the chance to roam around in a local coverage area and at the same time maintain connection with the network. The benefits of WLAN are low price, huge installations worldwide, and plug and play (PnP) devices. The main drawback of WLAN is possible interferences with other devices that communicate on similar frequencies.

Cellular networks are vastly employed in the majority of countries and possess a well-recognized infrastructure. Cellular networks could be utilized for communication among a number of components and devices in the SG. There are a number of current technologies for cellular communication including GSM, GPRS, 3G, 4G, 5G, and WiMAX [33]. The benefits of the cellular networks are presently available infrastructure across a vast area of implementation, elevated rates of data transmission, existing security systems implemented in cellular communication. The main drawback is that cellular networks are shared with other customers and are not fully devoted to SG communications.

1.5.4.2 Power Line Communication (PLC)

PLC permits data exchange among devices by electrical power lines. PLC is employed by including a modulated carrier signal to power cables. The benefit of the PLC is the currently recognized infrastructure that decreases deployment costs. Drawbacks include the existence of higher harmonics in the power lines that interfere with communication signals and the limited frequency of communication. Installing a number of smart meters and communication infrastructure should be implemented according to certain standards that are acknowledged by all companies and utilities taking part in building the SG. Different organizations are working on SG standardization such as the European Committee for Standardization, American National Standards Institute (ANSI), International Telecommunication Union (ITU), Institute of Electrical and Electronics Engineers (IEEE), and others. The well-known standards for SG communications are IEEE P2030, which provides guidelines for interoperability of the electric power system with end-use applications and loads; IEEE P1901 used by all classes of Broadband over Power Line (BPL) devices, entailing devices utilized for the connection to internet access services and devices utilized in buildings for LANs, smart energy applications, transportation platforms (vehicle), and other data distribution applications; IEC62351, which deals with cybersecurity issues of the SG; IEC62056 for electricity metering data exchange; PLC G3 to allow data and control messages to be transferred to generation, transmission, and distribution systems. There are more standards for SG communication including 802.15.4, ISO 1802, IPv4, DNP3, IEC61970, etc. [34]. More details on the standardization will be discussed in Chapter 17.

1.5.5 Customer Engagement

In the traditional grid, consumers have a passive role while occupying a marginal position in the energy market. Customer engagement has been negligible in the aspect of energy monitoring, controlling, management, generation, storing, and trading. However, with the introduction of the SG paradigm, the consumers have a crucial and active role in all aspects above. The SG technologies open the opportunity for active engagement through real-time insight in their energy consumption patterns, price changes, local renewable energy production, storing, and energy sharing. Customers engagement is the language of energy DSM in the SG which is used to achieve supply–demand balancing, load shifting, and increased reliability, high efficiency, and resiliency in the electric system. Electric utilities are putting more focus on energy demand management to realize three main tasks: enhancing energy efficiency, direct load control, and to meet a dynamic DR [35].

Energy management on the demand-side acts on the consumers for controlling electrical energy usage. There are a number of solutions to attain demand management and flexible energy consumption. Direct load control and dynamic DR programs are addressing the biggest priorities and challenges for the successful implementation of demand management in SGs. The participant customers can rely on their generation sources to meet their demand (typical usage requirements) whenever possible. Customer engagement, through (DSM) market prices change, increases with time. Figure 1.11 shows the global market of customer engagement. The global energy management systems (EMS) (industrial, home, and building) market size was USD 9.8 billion in 2017 and is expected to increase to USD 72.73 billion in 2024 [36, 37]. The right-side column of the Figure 1.11 indicates the spending per region. The EMS is becoming a crucial tool for both the utility and the customer to monitor, analyze, shift, optimize, and control energy and assets in real-time.

1.5.6 Sensors and PMU Units

Sensing and measurement technologies collect data to evaluate and monitor the state operation and equipment health status which support the grid's functionality and higher reliability. Also, this serves customers in improving their electrical usage by giving them information regarding their daily demands. Sensing and measurement technologies include sensors, phasor measurement units, and advanced metering infrastructures (AMI). All this supports a wide-area monitoring system, time-of-use, assets functionality, real-time pricing, and the

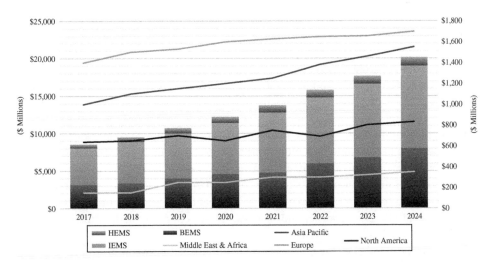

Figure 1.11 Customer engagement demand side management spending by region, 2017–2024 (USD Million).

system proper operation. A phasor measurement unit is a high-speed sensor integrated with the power grid to monitor power quality by allowing data to be obtained at certain instants of time. Phasor measurement units can be considered as a health-meter of the grid as they collect different measurements of voltage, phase, and current to be analyzed. This will help to reduce blackouts and provide a wide-area situational awareness.

1.5.7 Smart Meters and Advanced Metering Infrastructure

Smart meters are a two-way communicator that helps create a bridge between the utilities and the end consumer. In comparison to existing meters, smart meters have included functionalities by using real-time sensors, power outage notification, and power quality monitoring. Smart meters function digitally and permit automatic and complex transmissions of data between utilities and customers. Sharing information through smart meters can be linked to a Home EMS, which allows the consumers to see it in a comprehensible format which helps them to control their energy usage. To have a safe and reliable grid, various devices and algorithms that allow for rapid diagnosis and analysis should be developed.

AMI includes the implementation of various technologies that allow for a two-way flow of information, providing consumers and utilities with information on electricity cost and use, including the time and amount of electricity used. AMI gives a wide range of functionalities such as [38]:

1) Remote consumer price signals, which can provide time-of-use pricing information.
2) Collect, store, and report users' energy consumption data for any needed periods.
3) Enhance energy diagnostics from detailed load profiles.
4) Obtain location and degree of outages remotely.
5) Provide the possibility for remote connection and remote disconnection.
6) Allow identification of electricity theft and losses.

1.6 Smart Grid Control

The future SG is expected to be a flexible and manageable interconnected network consisting of small-scale and self-contained sub-areas, integrated with the large-scale electric power grid as the backbone. Utilizing micro sources, such as renewable energy sources and combined heat and power plants, into the SG makes them feed their local loads in an economic and environmentally friendly manner [39]. Therefore, the SG control architecture should therefore be dynamic and multilayer to handle real-time operation and provide tradeoff between performance and implementation. Advanced control uses high-speed communication infrastructure, distributed intelligent agents, analytical tools, and operational functionalities. The advanced control systems in the SG monitor the essential components, provide timely response, and enables the detection, prediction, disconnection, and self-healing of faults in the system.

Hierarchical control systems of the SG are distinguished between multilevel systems and multilayer systems. The multilevel system is based on the cooperation of independent controllers which cooperate to control the trading of the power. The multilayer

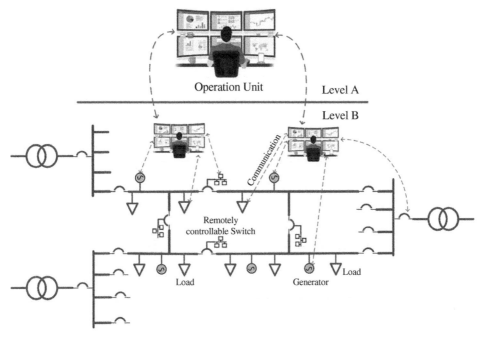

Figure 1.12 Distributed operation architecture with two levels.

system is based on individual actions with each controller having its own objective. A multitude of different architectures of the SG exists to realize such integrated systems. They are known as "distributed," "decentralized," "local," or "central." [40]. If an information exchange exists among the independent controllers, the control architecture is assumed to be distributed as shown in Figure 1.12. The system could be fully or partially distributed, and this is reliant on the condition that the information is shared between all controllers or among a subset of controllers. A decentralized control architecture-independent controller controls distinct subsystems. In particular, no information is exchanged among them as shown in Figure 1.13. The local architecture restricts the control on a single device or a single facility. The input data should be in existence locally, and no external communication exists as shown in Figure 1.14. In the event that a central operation unit manages all other devices in a system and aggregates and processes all the corresponding information, a central operation architecture is the case as presented in Figure 1.15.

1.7 Smart Grid Characteristics

SG implements ground-breaking products and services together with intelligent monitoring, control, communication, and self-healing technologies which can be characterized by the following goals and functionalities:

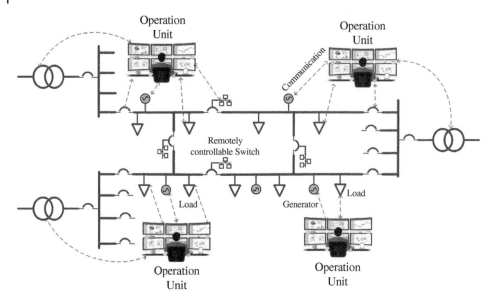

Figure 1.13 Decentralized operation architecture.

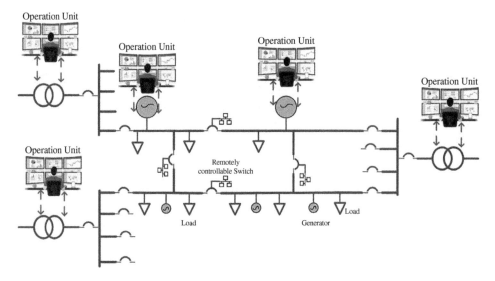

Figure 1.14 Local operation architecture.

1.7.1 Flexibility

SG distribution, transmission, and generation infrastructures allow for bidirectional power flow and are flexible to accommodate various types of generations, storage, loads, and emerging technologies such as electric vehicles (EVs) and mobile storage. SG allows the integration and operation of generators of all sizes and types at different locations [41]. SG accommodates all renewable energy sources and storage options and flexibly responds to differing

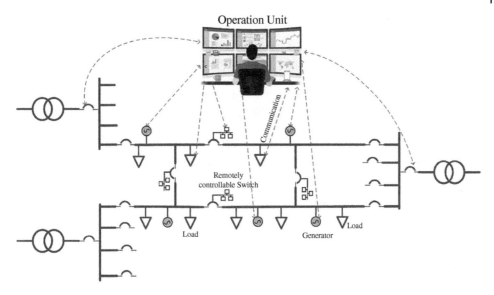

Figure 1.15 Central operation architecture.

humanity expectations and innovations, now and for decades to come. The SG paradigm will also be flexible to customers' active role regardless of the many existing obstacles.

The residential sector is not being targeted by many programs in the traditional grid paradigm as it is hard to deal with due to a multitude of factors such as high acquisition costs and limited access to the individuals. However, currently, new smarter devices can be incorporated with a number of residential appliances to respond locally to the price signals in an automated manner. The flexibility of demand creates values for the grid and customers by minimizing customer bills, shifting consumption to lower prices at off-peak hours, and reducing demand (during peak periods). Flexibility within demand can also help suppliers in some events to defer investments in central generation, distribution, and transmission.

The SG will be flexible to the customers' active participation. Consumers will utilize the grid in a number ways, more consumers will be "prosumers": both producers and consumers of energy and to additionally store the energy. The grid will no longer be merely a "delivery pipe" for electric power. All connected to the grid will be masters, no slave and master roles for them in future SGs [42]. The grid can constantly deliver power against disturbances (in extreme climate conditions and periods of natural disasters) without outages over a large area and could maintain information security against various attacks [43].

1.7.2 Improved Efficiency

Energy efficiency and product innovation programs are coupled together to make industrial and consumer sectors more efficient than they have been for a long time [44]. A SG with distributed energy generation allows for lower transmission and distribution losses which make the whole system more efficient. All parties connected to the grid work smartly for programs that improve the efficient use and delivery of electricity.

1.7.3 Smart Transportation

Electric transportation has been evolving rapidly during the past few years. The installation of smart EVs in the energy market can compensate for the need of major grid's infrastructure expansion. This compensation can be achieved if the EVs battery technologies allow for vehicle to grid (V2G), grid to vehicle (G2V) and vehicle to building (V2B) power flows to perform large-scale mobile storage and are combined with suitable pricing schemes to support the grid performance and economy. EV technology might be one of the most significant accelerators of SG adoption. Also, battery cost, size, and weight declination are considered as some of most important research topics related to EV deployment [45].

1.7.4 Demand Response Support

SG permits generators and loads to interrelate in an automated way in real-time, which allows customers to play a major role in optimizing the operation of the whole grid. Also, giving the consumers timely information enables them to reduce their energy bills by modifying their consumption patterns to overcome some of the constraints in the power system. DR and DSM are essential programs in SGs. DSM is applied for long-term planning such as for shifting the load peak over time. DR ensures short-term load response to improve the energy consumption profile over time. This could be realized by creating a dynamic electricity price. Consumers have the choice and authority over their consumption patterns. Various financial incentives could be created to adjust the level of demand and generation at strategic periods of the day. Figure 1.16 presents a classification of DR programs. Implementing DR programs into the operational aspects of the system has different benefits such as reducing the peak load and avoiding the need for new power plants and infrastructure oversizing [46].

DR programs are categorized into two main programs, time-based programs, and incentive-based programs. In the time-based DR programs, the change in electricity prices can vary automatically at different times based on customer electricity consumption as per the

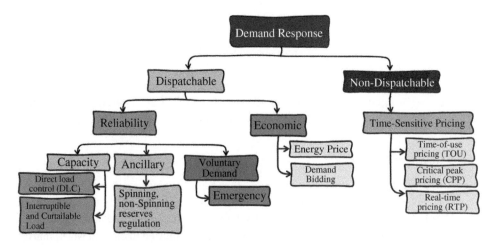

Figure 1.16 Classification of DR.

contracts signed with the operator. While, with the incentive-based DR programs, the customer is offered some incentives to participate in a fixed or varying period. The benefits will increase by triggering an incentive price to affect customer behavior by decreasing demand consumption.

Currently, DR is only used by large commercial consumers, and its operation is based on informal signals such as phone calls by the utility or by the DR provider asking the consumer to lower their energy consumption during peak times for energy demand [47]. There are four main obstacles in the face of the uptake of DR:

1) The shortfall of market integration.
2) Improving incentive-based DR programs.
3) The need for increased adoption of enabling technologies.
4) More communication in the power grid which is associated with privacy and security concerns.

1.7.5   Reliability and Power Quality

The SG utilizes technologies such as improved fault detection, state estimation, and enabling self-healing of the network without the need for specialized personnel. This leads to a reliable supply of electricity and minimized vulnerability to attacks or natural disasters. Smart grid operates resiliently in disasters and during physical, or cyber-attacks. Advanced control methods and monitoring oversee essential elements of the grid, enable rapid diagnosis and solutions to events that affect the grid's integrity, power quality, and smooth operation. The grid can monitor both on-line and in real-time as well as assess its current state and predict its future situation. The SG has robust risk warning procedures to employ preventive capabilities, automatic fault diagnosis, self-fault isolation, and self-restoration [48]. With all-new energy resources and entities, optimization and handling the system will become more challenging, even with the availability of new technologies and tools. Interdependencies and interactions between distribution and transmission systems will keep rising. The increase in the grid's complexity will require many technological, computational, and business operation requirements such as [49, 50]:

1) Self-learning systems.
2) Increased coordination between transmission-level balancing areas as well as additional balancing abilities at the distribution level.
3) Balancing abilities using both load-side and supply-side operations.
4) Privacy and security to be applied in all parts of the system, down to end-use devices.
5) PnP capabilities in SG enhanced levels.

1.7.6   Market-Enabling

The SG enables systematic communication between suppliers (their price of energy) and consumers (willingness-to-pay) and allows both the consumers and supplier increasing transmission paths, aggregated supply, DR initiatives, and ancillary service provisions [51].

1.8 Transformation from Traditional Grid to Smart Grid

There is a huge need to transform the traditional grid structure to SG. The current electric grid is on the way to SG at various rates of acceleration. Much has been accomplished to mitigate the possibility of blackouts, especially in utilizing new technologies that can assist electricity grids to be more reliable. Many of these technologies are smart and widely deployed now, whereas others are still in the demonstration and planning stages. Advanced components are already being used to analyze and diagnose the grid state and assist in its healing within a limited period of time. Figure 1.17 shows the transfer process from the traditional grid to a SG which indicates moving from one-way power flow (simple interaction) into two-way power flow (multi-interaction). A detailed comparison between the traditional power grid and SG are presented in Table 1.1. This shows that the majority of SG features are originated from the massive amount of generated information, an uncountable number of internet-connected control, and programmable auto-operated equipment [52, 53].

1.8.1 The Necessity for Paradigm Shift to SG

Maintaining economic growth and improving the quality of human life are reliant on the availability of affordable and reliable electricity. Up to now, conventional grids function in almost the same way as those of 130 years ago, i.e. power flows in a single direction across

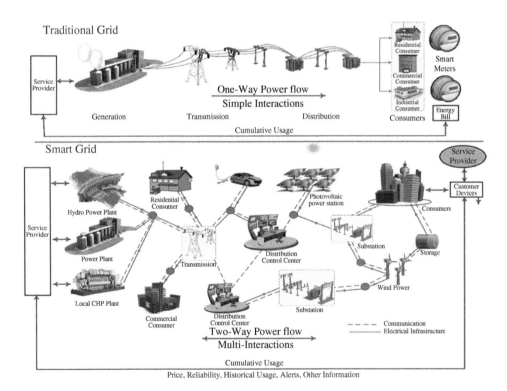

Figure 1.17 The difference between the conventional power grid and smart grid structure.

Table 1.1 A detailed comparison between conventional power grids and smart grids.

Characteristics	Traditional grid	Smart Grid
Technology	• Electromechanical • Mechanical devices electricity operated • No communication between devices • Little internal regulation	• Digital • Digital devices • Increased communication between devices • Remote control and self-regulation
Flow of power and communication	• One way • Power flow starts from the main plant using traditional energy structure to the customer	• Two way • Power flow goes to and from various grid users
Generation	Centralized	Distributed
Fault location	Difficult to determine	Can be determined remotely as well as predicted
Monitoring	Manual	Self- monitoring
Equipment failure	System responds to deal with post failure and blackout incidents	Adaptive and can be isolated and automatically reconnected.
Control	Limited control system	Pervasive control system
Operation and maintenance	Manually equipment checks	Remotely monitor equipment

the grid, from the central power plants to the customers. The reliability is maintained by conserving the excess capacity, which is inefficient, uneconomic, and environmentally unfriendly. Current grid topologies cannot be used with the distributed renewable energy sources and two-directional power flow. The alternating nature of renewable energy sources creates additional challenges. The aging grid also faces new problems due to increased demand, and nonlinear loads. Such a grid experiences an inability when the demand for power delivery and consumption boosts, which has happened frequently worldwide in recent years. The main reason that acts on the decrease of the traditional power grid's reliability is the lack of information exchange [54]. The current grid is of limited ability to react quickly to handle congestion, instability, and power quality challenges. The inflexibility of the existing grid cannot support the high integration of renewable energy. These limitations can lead to blackouts, equipment outages, and unscheduled downtime. Approximately 90% of all power outages and disturbances have their origins in the distribution network, therefore, transforming to the SG paradigm ensures a significant improvement of the grid's reliability. Furthermore, electric utility customers currently have a passive role; they have no access to the real-time consumption and pricing information that allow them actively participate within the power grids and optimize their energy usage and bills during peak and off-peak times. The two-way communication and power flow within the SG allow for effective energy control and energy management. This feature also allows achievement of both environmental and economic sustainability.

In summary, transforming to SG paradigm will help in the wide-scale integration of energy sources, enhancing network reliability, and improving power quality and load profile.

1.8.2 Basic Stages of the Transformation to SG

Successful transformation to SG requires a real, national level roadmap. Many countries have already established their SG roadmaps [55–60]. Each country has its own unique definition of a SG based on their own policies, goals, and objectives. The transformation roadmap should be developed based on different technical and economic realities and challenges facing each country. To develop a complete SG Roadmap that responds to the nation's electric power sector goals, some basic stages should be involved to define the priorities of their energy sectors. Also, there is a need for specific objectives, actions, and tools to fully achieve the roadmap set goals. Examples of steps and targets that countries put to transform traditional grid into SG are:

1) Install smart meters and AMI around the entire country by a specific time.
2) Significant percentage of cars to become electric within a specific time.
3) Specific reduction of country's emissions by a specific year.
4) Significant level of renewable generation by a specific time.
5) New tariff system.
6) Activating a DSM and customers activated role before a specific year.
7) Establish effective information and big data centers with a clear strategy for secure data storing and utilization. Timely publish current and historical data that is not confidential nor sensitive.
8) Building human capacity by adopting specific training programs on SG areas.

1.9 Smart Grid Enabling Technologies

The current technology revolution is leading to unwitnessed shifts in the economy, society, business, and individuals. A great effort is being made worldwide toward clean energy in order to protect the environment and improve operational efficiencies and customer services for better grid availability and reliability. The transfer of the traditional grid to a SG requires modifications and upgrades at various levels of the electric grid [61]. Nowadays, information and communication technologies are being developed very quickly and can significantly support the SG vision. This transformation is pushed by several factors, which include electrification, decentralization, and digitalization as shown in Figure 1.18. Those drivers work in harmony to assist in enabling, scaling up, and reinforcing advancements in SGs. Decentralization enables active elements of the system but requires a high level of coordination. Digitalization supports electrification and decentralization by having better management, which includes automatic control, consumption real-time optimization, and interaction with customers.

1.9.1 Electrification

Electrification is the process of powering by electricity and, within the context of the SG principle [62]. As the electricity generation in the SG moves toward more renewable

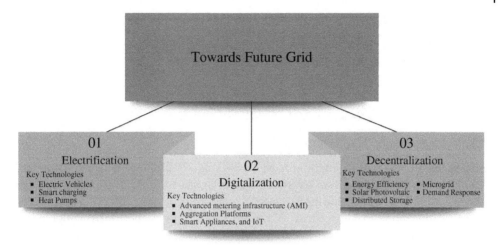

Figure 1.18 Three trends of the grid edge transformation.

sources, this trend enables more environmental gains and adding to it more end sectors such as heating, transportation, away from fossil fuel resources. This trend also increases the efficiency of energy utilization. An example of electrification is using EVs in transportation. This technology has been evolving rapidly over the past five years. The EV range was improved to exceed 480 km for particular models. Batteries' costs which were above $1100 per kilowatt-hour in 2010, have fallen 87% in real terms to $156/kWh in 2019 [63]. These reductions have reduced the cost gap with traditional internal-combustion engine vehicles. It's expected to have economically competitive EVs, although multiple electrification challenges within the infrastructure can limit the successful adaptation rate of EVs. The deployment of EVs boosts electricity consumption and offers a huge opportunity to maximize the utilization of the grid. This can be achieved if the charging/recharging technology is combined with suitable pricing and flexible usage. Electrification and economic growth are highly correlated [64].

1.9.2 Decentralization

Decentralization is the transformation of the bulk, "one-way direction" of energy into a distributed, multi-directional flows, known as multi-lane highways [65]. A decentralized electrical grid has several environmental and security benefits. Microgrids coordinated with distributed energy generation give systems significantly enhanced reliability and grid efficiency. Distributed power generation and microgrids are independent of the grid which means this power is provided to the local loads even when the main grid is not available. Decentralized generation allows the reduction of power outages for critical facilities, for instance, hospitals or police stations or any facility that may need continuous power. Hence, decentralized grids are more energy-efficient than centralized electricity grids [47]. Consequently, implementing renewable energy sources in the current power grid does not automatically imply that the current power grid is decentralized. The transformation from a centralized to a decentralized electricity grid requires a number of technologies with various implications that need to be considered to become a reality such as [66]:

1) Distributed generation (from renewable energy resources).
2) Distributed storage.
3) DR.

1.9.3 Digitalization and Technologies

Advancements in digitalization enhance the grid's utilization and management. Technologies that are digitalized have been increasingly used to allow devices across the grid to exchange beneficial data for the customers and grid operators. There is a huge interest in utilizing innovative Internet of Things (IoT) sensors, smart meters, network remote automation, distributed and centralized control systems, digital platforms, to spot the light on the grid's proper optimization and aggregation, enable real-time operation of the network, and enhance grid's situational awareness and utility services [67]. The rise in the deployment of AMI shows clear opportunities for enhancing the quality of service, the observability of the network, and data gathering for short- and long-term utilization. The grid's various digital data provides opportunities for fault detection, energy consumption and generation, as well as prediction of faults and outages. Digitalization of the network is an obvious opportunity for cost-effective development and management of the electricity system with high returns in cost and quality to serve. On the consumer side, chances for the use of smarter customer technologies are being increased. The installation of digital technologies in the network should not be slowed down by outdated regulations. As more digital devices are deployed, the communication between these devices will be increased. Broadband communication infrastructure will support a wide range of services (both network and consumer services). Not having updated policies and standards can slow down the development of this infrastructure and hinder innovation in this space. There are more challenges for digitalization including available infrastructure and replacement cycles [68]. Technologies related to the SG have evolved from previous attempts at utilizing electronic control, metering, and monitoring. Back in the 1980s, automatic meter reading was implemented to keep track of loads from heavy consumers and emerged into the AMI of the 1990s [69]. These meters can record and store how electricity is utilized at various times of the day. Smart meters give continuous information so that monitoring can be achieved in real-time and could be utilized as a gateway for "smart sockets" in the premises and for DR-aware equipment [70]. The main technologies in the SG consist of three main parts: electric energy technologies, operation technologies, and information and communication technologies as described in Figure 1.19. Energy technologies include renewables energy, distribution generation, DR, EVs, ES, integrated gasification combined cycle, etc. The operation technologies refer to a variety of hardware and software that detect or cause a change through direct monitoring and/or control of physical devices, processes, and events in the SG [71]. It includes the EMS, distribution management system (DMS), DSM, SCADA, outage management system (OMS), asset management system (AMS), flexible alternating current transmission system (FACTS), wide area monitoring systems (WAMS), process control application (PCA), etc. The information technologies include enterprise resource planning,

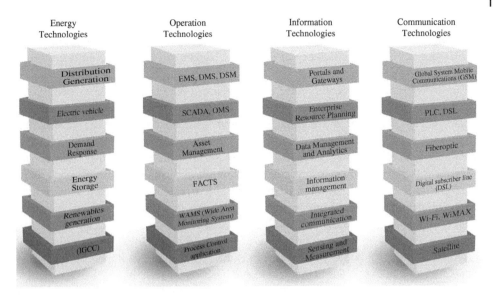

| Energy Technologies | Operation Technologies | Information Technologies | Communication Technologies |

Figure 1.19 Technologies for the evolution of the SG.

portals/gateways, information management, sensing and measurements, asset management, data collection, storing, mining, and analytics. Communication technologies, which include Wireline and Wireless technologies. The Wireline technologies include PLC, fiberoptic, Digital subscriber line (DSL). Wireless technologies include Wi-Fi, Worldwide interoperability for microwave access (WiMAX), Global System for Mobile Communications (GSM), and Satellite. The technology in SGs can be divided into different areas, each consisting of sets of assets, starting from generation through transmission, distribution, and different consumers [72, 73].

1.10 Actions for Shifting toward Smart Grid Paradigm

The electric grid's modernization should begin with a careful assessment of existing technologies within the power grid. Proper utilization of the existing power plants, transmission, and distribution lines and other facilities are the key factor in accomplishing grid modernization within the available financial and technical resources. The country should determine its current level of technology deployment followed up with a focused financial study and analysis of missing grid modernization aspects and associate them with timeline and cost. There are specific analyses that should be conducted for shifting to the SG paradigm. Each utility should conduct its own different analyses and implement specific strategies within a defined timeline. Examples of required analyses are [74, 75]:

Gap analysis: refers to identifying the incomplete actions between current status and future targets to achieve the desired outcomes. Also, it identifies the gaps in a particular grid's technology evolution and developed pillars. This will be used in the development of any shortcomings, as well as the implementation plans. The next step is to assess the current status, concerns, degree of technology deployment, and success rate.

The simple form of gap analysis is comparing the current situation to the objective statement for each focus technology and pillar. An example of the gap analysis on increased reliability and efficiency of electricity transmission and distribution system is shown in Figure 1.20.

Cost–benefit analysis: refers to the analysis that should be conducted to identify the associated cost of the actions in the various pillars that are necessary to mitigate the gaps which are identified in the gap analysis. This should be done to measure the physical performance of the application of the SG in a way that is quantifiable and with high certainty and accuracy. The cost-effectiveness analysis takes a step further and provides scientifically based measurements and protocols to identify the physical impacts and monetary costs that come with the SG projects. The payback time for all sectors would then be identified which will help in promoting and adopting the SG paradigm at a very wide scale and get support from policymakers and market players.

Risk analysis: refers to the analysis conducted in order to identify the risks associated with taking actions in the various pillars. The risk assessment has an important role in the adopted SG road map in order to timely mitigate the associated risks.

Barrier analysis: refers to the conducted analysis to identify any potential barriers that must be overcome in adopting and achieving the SG era. These may relate to policies, regulations, technology, human resources, finances, etc.

1.10.1 Stages for Grid Modernization

The process of grid modernization can be divided into four main stages as shown in Figure 1.21. The requirements, benefits, and risks associated with each stage are summarized below. Also, individual countries and grids operators should have updated policies and regulations supporting the transformation to a SG.

Stage 0: Includes manual control and local automation with little local and remote automation.

Stage 1: Includes the substations automation and remote control to be built on stage 0 by adding intelligent electrical devices (IED), remote terminal units (RTU), streaming

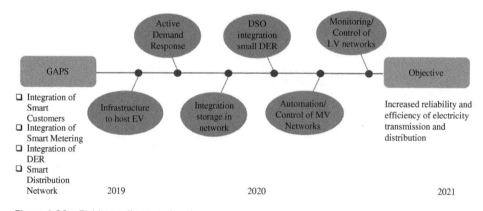

Figure 1.20 Fishbone diagram showing gaps.

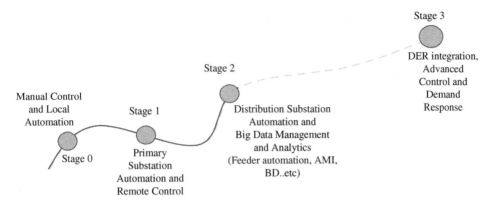

Figure 1.21 The main stages for achieving grid modernization.

sensors, and data communication facilities to achieve local and remote monitoring and control capabilities at HV/MV/LV substations.

Stage 2: Includes feeder automation and remote control built on Stage 1 by extending remote monitoring and advanced control to the outgoing feeders. This stage includes information from communicating meters to large industrial customers for improved control, management, and decision making.

Stage 3: Includes large-scale penetrations and integration for DER, control, and management with big data management analytics and distributed data centers, DR, and self-healing application. Also, adding ES, static VAR sources, and developed communication and control facilities to successfully implement high penetrations of DERs on the distribution feeders. This stage of grid modernization also entails the use of AMI to achieve on-demand reading of consumer meters along with DR capabilities. Also, adding the intelligent energy management system (IEMS) and home energy management system (HEMS) will help with improved control, management, and decision-making.

SG implementation requires special support and mechanisms including economic incentives, technology-specific actions, new policies, and feed-in tariffs. Without these particular requirements, the grid will not fully satisfy the national energy sector goals and future generations' needs.

1.10.2 When a Grid Becomes Smart Grid

Seven principal characteristics are considered a must for any network to be declared as a "Smart Grid":

1) Be self-healing from power disturbance events (Self-Healing).
2) Enable active participation by consumers in DR (Demand Response).
3) Operate resiliently against physical and cyber-attacks (Resiliency and Immunity).
4) Provide power quality for twenty-first century needs (21st Power Quality).
5) Accommodate any level and type of renewable generation and storage options.
6) Enable new technologies, products, services, and markets (Modern Services).
7) Optimize assets and operating efficiently (Optimal Asset Management).

1.11   Highlights on Smart Grid Benefits

Smart grid plays a critical role to enable the overall electric power sector objectives. The role of SG within the energy sector policies and regulations is shown in Figure 1.22. The general goals of SG are to ensure delivery of electric energy in an efficient, sustainable, resilient, and environmentally friendly manner. This allows for ambitious power sector targets that trigger new investment needs and call for new ways to control, manage and operate network infrastructures.

SG is therefore a conceptual goal whose achievement will require continuous grid modernization with the use of conventional and advanced technologies and operations. Through its pillars, SG provides great benefits to different parties connected to it. Utilities will gain reduced maintenance cost and lower distribution cost, while grid operators will ensure better monitoring and control capabilities which lead to greater efficiency and resiliency. Finally, consumers will gain better supply quality and continuity and will have control over their profile of consumed power and reduced electricity consumption. SG adaptation will benefit the nation and serve the environment by reducing electrical energy consumption, achieving continuous electricity supply, and ensuring clean and sustainable energy. The SG vision presents a power system that is more intelligent, more decentralized, and more controllable than today's grid. Integrating numerous intelligent SG technologies and operations in power generation, transmission and distribution are considered as the required trend for the energy paradigm shift.

The benefits that could be obtained from adopting the SG are divided into two types, direct benefits, and indirect benefits. Direct benefits include the following:

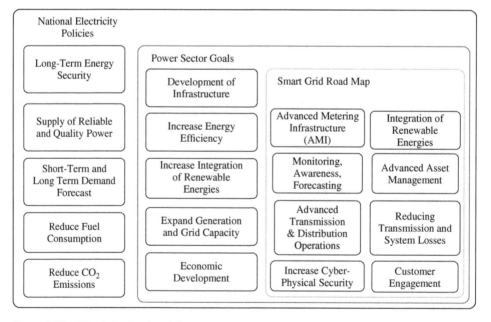

Figure 1.22   SG role in the electricity power sector.

- Enabling active participation of consumers, which transforms the centralized grid control to less centralized and more consumer interactive based.
- Improving energy system resiliency, flexibility, and load management.
- Increasing operating efficiency.
- Reducing transmission and distribution losses.
- Anticipating and responding to system disturbances (achieving self-healing and resilient system).
- Providing power quality for the digital economy.
- Allowing for a high level of renewable energy penetration.
- Accommodating the needs for a high level of EV integration.

Indirect benefits of the SG include the reduction of the overall expenses by reducing both long-term capital expense and operating expense. As demand increases, utilities must provide the required power to meet the peak loads which results in extremely high-cost infrastructure. Smart management of energy supply and demand will help reduce the need to build more power plants in addition avoiding oversize of transmission and distribution infrastructures which, in turn, decreases the long-term capital cost. In addition, SG adaptation enables the utility to reduce power outages, decrease the risk of premature failure and, in turn, increase resiliency of the overall grid. Furthermore, new economic growth and job creation are important indirect benefits of the SG.

1.12 Smart Grid Challenges

The power system is migrating from the conventional grid to the SG which faces many obstacles and challenges. With all-new energy resources, smart devices, and entities, the complexity of the system will significantly increase. Optimal design, operation and handling of such systems will become more challenging, even with the availability of new technologies and tools. Chapter 18 identifies the most important challenges facing the development of the SG. Here we briefly touch upon specific points as the main challenges facing the transformation to the SG energy paradigm.

1.12.1 Accessibility and Acceptability

Customers' acceptability, privacy, costs, cybersecurity, and regulatory considerations possess a substantial influence on the advancement of SGs. Undeniably, customers play a vital part in SG expansion: they are needed to transform their part as customers from passive to active, having full knowledge of their consumption, and capable of managing it depending on the availability of energy which is considered a challenge [76]. The SG technology should be accessed, accepted, and adopted by various grid operators and end-users. Success of this transformation depends on all levels of acceptance, without exception.

1.12.2 Accountability

Accountability is essential to enhance the security of the SG including privacy, integrity, and confidentiality. Accountability refers to a system that is recordable and traceable, consequently, including various parties responsible for those communication principles for its actions. Accountability logic is considered as an essential process to examine the accountability of a secure system [77]. All grid players hold responsibility for its success and utilization, but also include various levels of accountability. This should be identified and made clear to all grid players.

1.12.3 Controllability

SG technologies contain a large number of different controllable elements, systems, and subsystems which require coordination and proper management. As an example, DERs are connected to the grid directly or through a power electronic interface. The voltage source inverter (VSI) is connected to the grid as an interface to support the suitable modification of the grid voltage and frequency [61]. The entire grid is becoming more power electronics dominated with very low inertia. Controllability of such a system becomes a major challenge for stable and efficient operators.

1.12.4 Interoperability

The IEEE definition for interoperability is "the ability of two or more systems or components to exchange information and to use the information that has been exchanged" [78]. The SG is an interoperable system that should be capable of exchanging meaningful, actionable information, addressing safe, secure, efficient, and reliable operations of power systems. The systems will distribute a pre-defined purpose of the exchanged information, and this information can expect suitable types of responses. The reliability, fidelity, and security of information exchanges among SG subsystems and components should achieve required performance levels [79]. The GWAC (Grid Wise Architecture Council) [80] has looked at, interoperability between components of the same system, or between different systems and suggests several implications to be derived;

1) The infrastructure should permit exchange of data and its transfer from senders to receivers; thus, network connectivity should be assured.
2) The implementations of the participating solutions should be able to make sense of the data given. Therefore, common symbols, protocols, and implementation-specific interpretations thereof are required.

In the context of the SG, interoperability includes seamless, end-to-end connectivity of hardware and software from end-use devices to the power source, improving the coordination of energy flows with real-time information and analysis. Interoperability is an essential part of the success of SGs. According to the National Institute of Standards and Technology (NIST) [81] "Once appropriate levels of interoperability are achieved, policymakers, investors, engineers, and other stakeholders can turn their attention to solving a broad set of challenges: improving the efficiency of power delivery, transitioning to cleaner energy sources, and enabling new markets that surround electricity delivery." The NIST-coordinated interoperability effort is organized to include a wide set of stakeholders among the industry, and it is examining the purpose of standards across a wide range of SG interoperability areas [79].

1.12.5 Interchangeability

Interchangeability defines the process of two or more components interchanged by joint replacement without affecting system performance [82]. Interchangeability needs devices to aid in the present functional behavior on their communication interfaces or permit alterations in functionality to be reinforced by the corresponding communication protocol [83]. Thus, interchangeability deals with additional requirements with regard to the functional behavior of devices at their communication interfaces.

1.12.6 Maintainability

It is essential to ensure the grid's ability to maintain reliable operation and undergo timely modifications and repairs to ensure high-quality power regardless of external factors variations [84]. The SG sub-systems and components should be able to perform their functions for the pre-defined period of time. Maintainability is an essential part of SG reliability.

1.12.7 Optimality

SG is characterized with the variations in power sources that are produced from conventional and various renewable sources. Also, the load capability for peak demand decreases will the increase of power network complexity. This requires highly distributed and optimal schemes and elements that ensure the grid's reliability and economic operation. Economic, size, and technical optimality should be ensured at the generation and demand sides [85]. Optimal placement and sizing of the distributed generations, charging stations, system modularizing, measurement systems, etc. are essential for creating SG energy paradigm and its scalability.

1.12.8 Security

It is essential to create a secure SG at various levels, control, communication, and physical. The SG should have measures to protect its massive amount of data and to secure consumers' data privacy. Security needs a system-wide solution for the various anomalies that could hinder physical and cyber levels of the grid [86]. The SG should be resilient against various coordinated and non-coordinated attacks.

1.12.9 Upgradability

Upgradability is related to smart-grid equipment adaptation criteria and substation equipment service life. Designers go through complex procedures related to substation equipment requirements. The equipment should implement long life cycles that consider reliability, upgradability, and interchangeability [87]. SG areas consist mostly of a long-life lasting equipment as opposed to typical IT systems. Test and replacement of these devices usually requires hard work and should consider the high cost due to their large-scale implementation and high importance usage. Furthermore, utilizing cryptographic strategies that surpass current security conditions is considered delaying the probable requirement of further upgrades [88].

1.13 Smart Grid Cost

Grid operators are required to entirely assess the estimated costs, benefits, and potential risks of implementing the SG applications in order to define a reasonable investment plan for grid modernization. The investment plan should include a list of practical projects and applications to be implemented, their cost, and realization timeframes. Such a plan should mainly rely on the available technical and financial resources in order to achieve the maximum guaranteed return on investment (ROI) with minimum risk. SG will benefit both grid operators and their customers through new technologies development and new applications. The investment on SG is influenced by the targeted power grid reliability, security, efficiency, and resilience. However, disturbances, faults, blackouts, equipment damages, outages, customer interruption, loss of data, and losses of resources over time can result in investment delays in this sector which will also negatively affect a nation's economic growth. SG achieves a significant level of ROI rate and may deliver the highest and long-term returns to the electric operators and customers, as shown in Figure 1.23 [89].

The SG implementation is a continuous process that includes a set of technologies and additional features that can be added gradually to reach the most effective supply and demand balance in addition to reliable and clean electricity. A market study by Electric Power Research Institute (EPRI), indicates that the investment level at utility-scale in the power grid is between $17 and $24 billion per year over the next 20 years [36]. Figure 1.24 and Table 1.2 list the major components of the SG total cost [90].

Low refers to an EPRI low estimate of $ total SG costs; HIGH refers to EPRI high estimate of $ total SG costs. The wide variety in these estimates of the investment that is needed to realize the grid modernization reflects the uncertainty of the current industry modernization stage [91]. Again, these costs are modest when compared with the yield fruitful benefits from SG implementation.

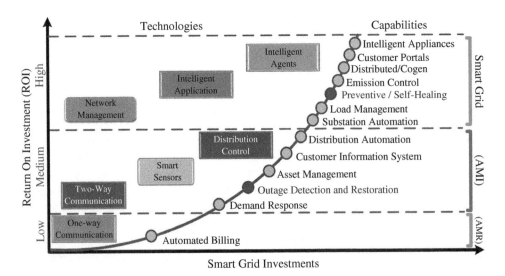

Figure 1.23 SG investment. Adapted from [89].

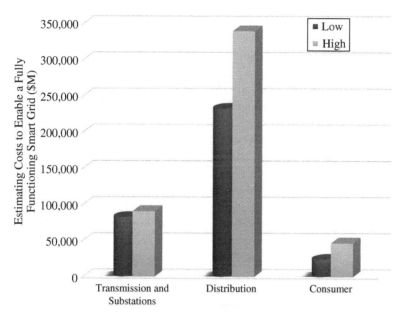

Figure 1.24 SG costs Ref [90]. Reproduced with permission from EPRI (Electric Power Research Institute).

Table 1.2 Investment costs of a fully functioning SG ($ M) [90]. Reproduced with permission from EPRI (Electric Power Research Institute)

	Low	High
Transmission and substations	82046	90413
Distribution system	231960	339409
Consumer Engagement	23672	46368
Total	337678	476190

1.14 Organization of the Book

This book is comprised of 18 different chapters dealing with different SG related issues. Chapter 1 provides an elementary discussion on the fundamentals of the SG; its concept and definition, characteristics, and challenges. The chapter provides also the benefits of moving toward SG

Chapter 2 presents an overview of different renewable energy resources; their current status and also future opportunities, as well as the challenges of integrating them into the electricity grid and the operation in distributed mode as part of the SG.

Chapter 3 describes the power electronics technology for distributed generation integrated into the SG. An introduction to typical distributed generation systems with the power electronics is presented. Power electronics converters in grid-connected AC systems and their control technologies are introduced. Then power electronics enabled autonomous AC power systems are discussed with the coordination and power management schemes. This chapter presents the basic control of power converters. Then autonomous DC power systems are illustrated. Finally, conclusions are drawn with future works.

Chapter 4 is dedicated to the impact of the Energy Storage Systems (ESS) on the future grid and presents how these technologies have the potential to shift energy utilization away from peak demand periods and increased costs, enhance the reliability and resilience of the power grid and considering the large integration of intermittent renewable resources. A detailed presentation of different ES technologies and the development of the current technologies and future status are introduced. One of the topics presented in this chapter is the use of ESS in SG applications, technical and financial benefits for the deployment of these technologies in the future SG.

A comprehensive review of microgrid including their characteristics, challenges, design, control, and operation either in grid-connected or islanded modes are introduced in Chapter 5. This chapter presents a detailed study of communications issues between microgrids.

Chapter 6 is devoted to one of the most important applications of the SG which is smart transportation. This chapter presents an overview of electric vehicles; their current status and also future opportunities, in addition to the challenges of integrating them into the SG. The impact of EVs on SG operation and Modeling EV mobility in energy service networks are also depicted in this chapter.

Chapter 7 describes the zero energy buildings (ZEBs) definition, design, modeling, control, and optimization. Furthermore, generalizing its concept into the SG community. This chapter discusses the benefits and barriers of the current state and the future trends of (ZEBs) as a step to reduce the energy consumption in the building sector.

The goal of Chapter 8 is to shed light on the SG features multi-way communication among energy production, transmission, distribution, and usage facilities. The reliable, efficient, and intelligent management of complex power systems necessitates employment of high-speed, reliable, and secure data information and communication technology into the SG to manage and control power production and usage is described in detail in Chapter 8.

Chapter 9 presents two main parts when studying the SG, SG infrastructure, and SG applications. SG infrastructure entails three main layers: power system, information, and communication layers and these are discussed in this chapter. Although the cyber system made the grid more energy-efficient, it has introduced threats of cyber-attack such as operational failures, loss of synchronization, damage of power components, and loss of system stability. Because of this, information security is a major element for information and communication infrastructure in the SG to improve the grid efficiency and reliability as well as considering privacy which is described in detail in Chapter 9.

Chapter 10 elaborates on the evolution and benefits of moving the energy grid to a SG. The main obstacle that faces this transfer is the difficult management of an unprecedented deluge of data. Unfortunately, utilities still do not make full use of this huge volume of data. To achieve high performance in SGs, several techniques and approaches must be used to manage all the data in order to generate viable values from this big data which can improve the utility's chances of reaping optimal long-term returns from its SG investment as presented in detail in Chapter 10.

The SG principle transfers the future generation electricity network to a smarter and intelligent grid by enabling bi-directional information and active participation from all parties connected to it. Coordination and communication between both sides, generation, and consumption is an important topic for research which is discussed in detail in Chapter 11. Driven by concerns regarding electric sustainability, energy security, and economic growth, it is essential to have a coordination mechanism based on heuristic rules to manage energy demand and enhance the survivability of the system when failures occur or at peak periods achieved by the principle of DMSs clearly defined in Chapter 11.

Chapter 12 presents the business model concept, its main components, and how they can be used to analyze the impact of SG technology to create, deliver, and capture value for the utility business. Then, the value chain for both the traditional and smart energy industry are discussed. After that, different electricity markets have been described. This is followed by a review of previously proposed SG business models with its future levers. Finally, the chapter highlights the potential of applying blockchain technology in the electricity market.

Chapter 13 sheds light on fully motivating the residential space in the SG which is still an unresolved problem. This chapter mentions the importance of offering power systems researchers and decision-makers suitable knowledge about the fundamental drivers of consumer acceptance of the SG and the methods to be followed for their engagement in order to implement SG technology and make it feasible earlier.

Cloud computing is considered the next-generation computing paradigm because of its advantages in network access, massive computation services, storage capacities, and various application opportunities including the SG. Chapter 14 defines the fundamental relationship between SG and cloud computing services. The architectural principles, characteristics of cloud-computing services as well as the advantages and disadvantages of those characteristics for the SG are discussed. Furthermore, opportunities and challenges of using cloud computing in SG, and the major categories of data security challenges of cloud computing are also touched upon.

Chapter 15 discusses the latest taxonomy of Artificial Intelligence (AI) applications in SGs is discussed, including load and renewable energy forecasting, power optimization, electricity price forecasting, fault diagnosis, and cyber and physical layers security.

Chapter 16 discusses the current state of simulation-based approaches including multi-domain simulation, co-simulation, and real-time simulation, and hardware-in-the-loop for the SG. Furthermore, some SG planning and analysis software are summarized with their advantages and disadvantages.

Many issues require to be handled before the SG becomes a major player of the main utility grid. One of the important issues with the SG is standards. These standards include the generation sources, the smart home appliances, and the EMS that need to communicate with each component of the SG to activate the energy trade between customers and producers. Chapter 17 presents an overview of SG standards; new standardization studies, SG policies of some countries, and some important standards for the SG.

Chapter 18 depicts the concepts of distributed generation, micro-grid, smart-grid, and distributed operation pose more complexity and challenges to the modern power systems. This chapter presents the challenges and barriers that modern smart-grids face from different perspectives.

References

1 U.S. Energy Information Administration (2019). International Energy Outlook. https://www.eia.gov/outlooks/ieo (accessed 31 January 2021).

2 Alshahrani, A., Omer, S., Su, Y. et al. (2019). The technical challenges facing the integration of small-scale and large-scale PV systems into the grid: a critical review. *Electronics* 8 (12): 1–28.

3 Sarkar, S., Chakrabarti, U., Bhattacharyya, S., and Chakrabarti, A. (2020). A comprehensive assessment of the need and availability of smart grid technologies in an electricity distribution grid network. *Journal of the Institution of Engineers (India)*: 1–9.

4 Mohammed, A., Refaat, S.S., Bayhan, S., and Abu-Rub, H. (2019). Ac microgrid control and management strategies: evaluation and review. *IEEE Power Electronics Magazine* 6 (2): 18–31.

5 Sayed, A., Magdy, A., Badr, A. and Eldebeiky, S. (2019). Optimal Management of Distribution Networks Regarding Reactive Power Generation. *21st International Middle East Power Systems Conference (MEPCON)*, Cairo, Egypt (17–19 December 2019). IEEE.

6 Ellabban, O., Abu-Rub, H., and Blaabjerg, F. (2014). Renewable energy resources: current status, future prospects and their enabling technology. *Renewable and Sustainable Energy Reviews* 39: 748–764.

7 Bascom, E.C.R., Muriel, K.M., Nyambega, M. et al. (2014). Utility's strategic application of short underground transmission cable segments enhances power system. *IEEE PES T&D Conference and Exposition*, Chicago, IL, USA (14–17 April 2014). IEEE.

8 Schavemaker, P. and Van der Sluis, L. (2017). *Electrical Power System Essentials*. Wiley.

9 Prakash, K., Lallu, A., Islam, F.R. and Mamun, K.A. (2016). Review of power system distribution network architecture. *3rd Asia-Pacific World Congress on Computer Science and Engineering (APWC on CSE)*, Nadi, Fiji (5–6 December 2016). IEEE.

10 Bastiao, F., Cruz, P. and Fiteiro, R. (2008). Impact of distributed generation on distribution networks. *5th International Conference on the European Electricity Market*, Lisboa, Portugal (29–30 May 2008).

11 Blalock, T.J. (2012). In the Berkshires, part 1: William Stanley started something [history]. *IEEE Power and Energy Magazine* 10 (4): 85–94.

12 Veldman, E., Geldtmeijer, D.A.M., Knigge, J.D., and Slootweg, J.G. (2010). Smart grids put into practice: technological and regulatory aspects. *Competition and Regulation in Network Industries* 11 (3): 287–307.

13 Network Development Roadmap Consultation (2018). https://www.nationalgrideso.com/document/113896/download

14 Refaat, S.S., Mohamed, A. and Kakosimos, P. (2018). Self-Healing control strategy; challenges and opportunities for distribution systems in smart grid. *12th International Conference on Compatibility, Power Electronics and Power Engineering*, Doha, Qatar (10–12 April 2018). IEEE.

15 US Department of Energy (2003). Grid 2030 – A Vision for Electricity's Second 100 Years. https://www.energy.gov/oe/downloads/grid-2030-national-vision-electricity-s-second-100-years (accessed 31 January 2021).

16 Webb, M. (2008). Smart 2020: enabling the low carbon economy in the information age. *The Climate Group. London* 1 (1): 1–87. https://www.compromisorse.com/upload/estudios/000/36/smart2020.pdf.

17 Hertzog, C. (2009). *Smart Grid Dictionary*. GreenSpring Marketing LLC.

18 Mohamed, A., Refaat, S.S., and Abu-Rub, H. (2019). A review on big data management and decision-making in smart grid. *Power Electronics and Drives* 4 (1): 1–13.

19 Grunwald, A. and Orwat, C. (2019). Technology Assessment of Information and Communication Technologies. In: *Advanced Methodologies and Technologies in Artificial Intelligence, Computer Simulation, and Human-Computer Interaction* (ed. M. Khosrow-Pour), 600–611. Hershey, PA: IGI Global.

20 Refaat, S.S., Abu-Rub, H., Trabelsi, M. and Mohamed, A. (2018). Reliability evaluation of smart grid system with large penetration of distributed energy resources. *IEEE International Conference on Industrial Technology,* Lyon, France (20–22 February 2018). IEEE.

21 Eskandari, M., Li, L., Moradi, M.H. et al. (Oct. 2020). Optimal voltage regulator for inverter interfaced distributed generation units part I: control system. *IEEE Transactions on Sustainable Energy* 11 (4): 2813–2824. https://doi.org/10.1109/TSTE.2020.2977330.

22 Eskandari, M., Blaabjerg, F., Li, L. et al. (Oct. 2020). Optimal voltage regulator for inverter interfaced distributed generation units part II: application. *IEEE Transactions on Sustainable Energy* 11 (4): 2825–2835. https://doi.org/10.1109/TSTE.2020.2977357.

23 Parhizi, S., Lotfi, H., Khodaei, A., and Bahramirad, S. (2015). State of the art in research on microgrids: a review. *IEEE Access* 3: 890–925. https://doi.org/10.1109/ACCESS.2015.2443119.

24 Ela, E., Kirby, B., Botterud, A. et al. (2013). *Role of Pumped Storage Hydro Resources in Electricity Markets and System Operation: Preprint*. United States https://www.osti.gov/servlets/purl/1080132.

25 Das, C.K., Bass, O., Kothapalli, G. et al. (2018). Overview of energy storage systems in distribution networks: placement, sizing, operation, and power quality. *Renewable and Sustainable Energy Reviews* 91: 1205–1230.

26 Vahid-Ghavidel, M., Javadi, M.S., Gough, M. et al. (2020). Demand response programs in multi-energy systems: a review. *Energies* 13 (17): 4332.

27 Liu, Y., Yuen, C., Ul Hassan, N. et al. (April 2015). Electricity cost minimization for a microgrid with distributed energy resource under different information availability. *IEEE Transactions on Industrial Electronics* 62 (4): 2571–2583. https://doi.org/10.1109/TIE.2014.2371780.

28 Ekanayake, J.B., Jenkins, N., Liyanage, K. et al. (2012). *Smart Grid: Technology and Applications*. Wiley.

29 Rafique, Z., Khalid, H.M., and Muyeen, S.M. (2020). Communication systems in distributed generation: a bibliographical review and frameworks. *IEEE Access* 8: 207226–207239. https://doi.org/10.1109/ACCESS.2020.3037196.

30 Elyengui, S., Bouhouchi, R. and Ezzedine, T. (2014). The enhancement of communication technologies and networks for smart grid applications. arXiv preprint arXiv:1403.0530, Cornell University.

31 Jayashree, L.S. and Selvakumar, G. (2020). The internet of things: connectivity standards. In: *Getting Started with Enterprise Internet of Things: Design Approaches and Software Architecture Models*, 1–30. Cham: Springer.

32 van Nee, R. and Prasad, R. (2000). *OFDM for Wireless Multimedia Communications*. Artech House, Inc.

33 Nuaymi, L. (2007). *WiMAX Technology for Broadband Wireless Access*. Chichester: Wiley.

34 Baimel, D., Tapuchi, S., and Baimel, N. (2016). Smart grid communication technologies. *Journal of Power and Energy Engineering* 04 (08): 1–8.

35 Khan, I. (2019). Energy-saving behaviour as a demand-side management strategy in the developing world: the case of Bangladesh. *International Journal of Energy and Environmental Engineering* 10 (4): 493–510.

36 US Department of energy (2018). Smart Grid System Report. 2018 Report to Congress. https://www.energy.gov/sites/prod/files/2019/02/f59/Smart%20Grid%20System%20Report%20November%202018_1.pdf (accessed 1 February 2021).

37 Kelly, M. and Elberg, R. (2017). Customer Management and Experience Technologies. Global Analysis and Market Forecasts, Navigant research.

38 Rashed Mohassel, R., Fung, A., Mohammadi, F., and Raahemifar, K. (2014). A survey on advanced metering infrastructure. *International Journal of Electrical Power & Energy Systems* 63: 473–484.

39 Henderson, M.I., Novosel, D., and Crow, M.L. (2017). Electric power grid modernization trends, challenges, and opportunities. *IEEE Power and Energy*: 1–17. https://www.ieee.org/content/dam/ieee-org/ieee/web/org/about/corporate/ieee-industry-advisory-board/electric-power-grid-modernization.pdf.

40 Wenderoth, F., Drayer, E., Schmoll, R. et al. (2019). Architectural and functional classification of smart grid solutions. *Energy Informatics* 2 (1), article number: 33: 1–13.

41 Al-Badi, A.H., Ahshan, R., Hosseinzadeh, N. et al. (2020). Survey of smart grid concepts and technological demonstrations worldwide emphasizing on the Oman perspective. *Applied System Innovation* 3 (1): 1–27.

42 Xinghuo, Y., Cecati, C., Dillon, T., and Godoy Simoes, M. (2011). The new frontier of smart grids. *IEEE Industrial Electronics Magazine* 5 (3): 49–63.

43 Cunjiang, Y., Huaxun, Z., and Lei, Z. (2012). Architecture design for smart grid. *Energy Procedia* 17: 1524–1528.

44 Loschi, H.J., Leon, J., Iano, Y. et al. (2015). Energy efficiency in smart grid: a prospective study on energy management systems. *Smart Grid and Renewable Energy* 6 (08): 250.

45 Veldman, E., Gibescu, M., Postma, A. et al. (2009). Unlocking the hidden potential of electricity distribution grids. *20th International Conference and Exhibition on Electricity Distribution, Prague*, Czech Republic (8–11 June 2009).

46 Siano, P. (2014). Demand response and smart grids – a survey. *Renewable and Sustainable Energy Reviews* 30: 461–478.

47 Jain, A. and Mishra, R. (2016). Changes & challenges in smart grid towards smarter grid. *2016 International Conference on Electrical Power and Energy Systems,* Bhopal, India (14–16 December 2016). IEEE.

48 Electricity Advisory Committee (2008). Smart Grid: Enabler of the New Energy Economy. A Report by the Electricity Advisory Committee. https://www.energy.gov/sites/prod/files/oeprod/DocumentsandMedia/final-smart-grid-report.pdf (accessed 1 February 2021).

49 European Technology Platform Smart Grids (2008). Strategic Deployment Document for European's Electricity Networks of the Future. Draft for 3rd General Assembly, Belgium. http://kigeit.org.pl/FTP/PRCIP/Literatura/020_SmartGrids_ETP_SDD_FINAL_APRIL2010.pdf (accessed 1 February 2021).

50 Platt, G. (2007). The Decentralised Control of Electricity Networks – Intelligent and Self-Healing Systems. Grid Interop 2007 Forum Proceedings. Richland, WA.

51 DOE (2008). Metrics for Measuring Progress toward Implementation of the Smart Grid. Washington, DC: DOE, Office of Electricity Delivery and Energy Reliability, Washington, DC.

52 Bagdadee, A.H. and Zhang, L. (2019). A review of the smart grid concept for electrical power system. *International Journal of Energy Optimization and Engineering (IJEOE)* 8 (4): 105–126.

53 Md, R.H., Amanullah, M.T.O., and Shawkat Ali, A.B.M. (2013). Smart grid. In: *Smart Grids* (ed. A.B.M.S. Ali), 23–44. London: Springer https://www.springer.com/gp/book/9781447152095.

54 Liu, X., Chen, B., Chen, C., and Dong, J. (2019). Electric power grid resilience with interdependencies between power and communication networks–a review. *IET Smart Grid* 3 (2): 182–193.

55 Madrigal, M., Uluski, R., and Gaba, K.M. (2017). *Practical Guidance for Defining a Smart Grid Modernization Strategy: The Case of Distribution (Revised Edition)*. The World Bank.

56 International Electrotechnical Commission (2010). IEC smart grid standardization roadmap. SMB Smart Grid Strategic Group.

57 ETIP, SNET (2016). Final 10-year ETIP SNET R&I roadmap covering 2017–26. Brussels: European Technology & Innovation Platforms Smart Networks for Energy. https://www.etip-snet.eu/wp-content/uploads/2017/03/Final_10_Year_ETIP-SNET_RI_Roadmap.pdf (accessed 1 February 2021).

58 Rebec, G., Moisan, F. and Gioria, M. (2009). Road-map for smart grids and electricity systems integrating renewable energy sources. https://inis.iaea.org/search/search.aspx?orig_q=RN:45087571 (accessed 1 February 2021).

59 Scarsella, B. (2009). Flexibility Roadmap Future Smart, A smart grid for all: Our transition to Distribution System Operator. UK Power Networks. https://innovation.ukpowernetworks.co.uk/wp-content/uploads/2019/07/futuresmart-flexibility-roadmap.pdf (accessed 1 February 2021).

60 Farhangi, H. (2014). A road map to integration. *IEEE Power and Energy Magazine*.

61 The U.S. Department of Energy by Litos Strategic Communication (2004). The SMART GRID: an introduction. https://www.energy.gov/sites/prod/files/oeprod/DocumentsandMedia/DOE_SG_Book_Single_Pages%281%29.pdf (accessed 1 February 2021).

62 SEFI, UNEP (2010). Global Trends in Sustainable Energy Investment, Analysis of Trends and Issues in the Financing of Renewable Energy and energy effeciency. https://energy-base.org/app/uploads/2020/03/13.SEFI-Global-Trends-in-Sustainable-Energy-Investment-2010.pdf (accessed 1 February 2021).

63 Naimoli, S., and Ladislaw, S. (2020). Climate Solutions Series. Decarbonizing the Electric Power Sector, *Center for Strategic & International Studies* (12 May), p. 1–7.

64 Hirsh, R.F. and Koomey, J.G. (2015). Electricity consumption and economic growth: a new relationship with significant consequences? *The Electricity Journal* 28 (9): 72–84.

65 Ansari, N., and Lo, C-H. (2017). Decentralized controls and communications for autonomous distribution networks in smart grid. US Patent 9, 804,623, filed 10 October 2012 and issued October 31 2017.

66 Cárdenas, A.A. and Safavi-Naini, R. (2012). Security and privacy in the smart grid. In: *Handbook on Securing Cyber-Physical Critical Infrastructure* (eds. S.K. Das, K. Kant and N. Zhang), 637–654. Elsevier.

67 Saleem, Y., Crespi, N., Rehmani, M.H., and Copeland, R. (2019). Internet of things-aided smart grid: technologies, architectures, applications, prototypes, and future research directions. *IEEE Access* 7: 62962–63003. https://doi.org/10.1109/ACCESS.2019.2913984.

68 Astarios, B., Kaakeh, A., Lombardi, M. and Scalise, J. (2017). The future of electricity: New technologies transforming the grid edge. World Economic Forum. http://www3.weforum. org/docs/WEF_Future_of_Electricity_2017.pdf (accessed 1 February 2021).

69 World Bank (2011). Applications of Advanced Metering Infrastructure in Electricity Distribution. https://openknowledge.worldbank.org/handle/10986/12948 (accessed 1 Feburyary 2021).

70 Mohassel, R.R., Fung, A., Mohammadi, F., and Raahemifar, K. (2014). A survey on advanced metering infrastructure. *International Journal of Electrical Power & Energy Systems* 63: 473–484.

71 Glenn, C., Sterbentz, D., and Wright, A. (2016). Cyber Threat and Vulnerability Analysis of the U.S. Electric Sector. https://www.osti.gov/servlets/purl/1337873. (accessed 1 February 2021).

72 Wilson, E., Stephens, J., and Peterson, T. (2015). *Smart Grid (R)Evolution: Electric Power Struggles*. Cambridge: Cambridge University Press.

73 Jiang, T., Yu, L., and Cao, Y. (2015). *Energy Management of Internet Data Centers in Smart Grid*. Springer Berlin Heidelberg.

74 Farmanbar, M., Parham, K., Arild, Ø., and Rong, C. (2019). A widespread review of smart grids towards smart cities. *Energies* 12 (23): 1–18.

75 SAIC Smart Grid Team (2006). San Diego Smart Grid Study Final Report. https://www. sandiego.edu/law/documents/centers/epic/061017_SDSGStudyES_FINAL.pdf (accessed 1 February 2021).

76 Back, A-K., Evens, C., Hukki, K. et al. (2011). Consumer acceptability and adoption of Smart Grid, SGEM Research Report Helsinki. http://sgemfinalreport.fi/files/SGEM%20 Research%20Report%20D1.2%202011-04-04.pdf (accessed 1 February 2021).

77 Liu, J., Xiao, Y. and Gao, J. (2011). Accountability in smart grids. *Consumer Communications and Networking Conference*, Las Vegas, NV, USA (9–12 January 2011). IEEE.

78 Geraci, A. (1990). *IEEE Standard Computer Dictionary: A Compilation of IEEE Standard Computer Glossaries*. New York, NY: Institute of Electrical and Electronics Engineers Inc.

79 Authorship Team (2011). A Smart Grid Policy Center White Paper. https://www.smartgrid. gov/files/documents/Paths_Smart_Grid_Interoperability.pdf (accessed 1 Februry 2021).

80 The GridWise Architecture Council (2008). GridWise Interoperability Context-Setting Framework. https://www.gridwiseac.org/pdfs/interopframework_v1_1.pdf (accessed 1 February 2021).

81 FitzPatrick, G.J. and Wollman, D.A. (2010). NIST interoperability framework and action plans. *IEEE PES General Meeting*, Providence, RI, USA (25–29 July 2010). IEEE.

82 Greer, C., Wollman, D.A., Prochaska, D.E. et al. NIST framework and roadmap for smart grid interoperability standards, release 3.0. No. Special Publication (NIST SP)-1108r3.

83 Strabbing, W. (2017). Smart meter interoperability and interchangeability in Europe. https://esmig.eu/news/smart-meter-interoperability-and.

84 Alves, G., Marques, D., Silva, I. et al. (2019). A methodology for dependability evaluation of smart grids. *Energies* 12 (9): 1817.

85 Lestas, I., Kasis, A., Monshizadeh, N., and Devane, E. (2017). Stability and optimality of distributed secondary frequency control schemes in power networks. *IEEE Transactions on Smart Grid* 10 (2): 1747–1761. https://doi.org/10.1109/TSG.2017.2777146.

86 Altera Corporation (2013). Overcoming Smart Grid Equipment Design Challenges with FPGAs. https://www.intel.com/content/dam/www/programmable/us/en/pdfs/literature/wp/wp-01191-smart-grid-design.pdf (accessed 1 February 2021).

87 Jokar, P., Arianpoo, N., and Leung, V.C.M. (2016). A survey on security issues in smart grids. *Security and Communication Networks* 9 (3): 262–273.

88 Mustafa, M.A. (2015). Smart grid security: protecting users'privacy in smart grid applications. Doctoral Thesis. University of Manchester. https://www.escholar.manchester.ac.uk/api/datastream?publicationPid=uk-ac-man-scw:276339&datastreamId=FULL-TEXT.PDF (accessed 1 February 2021).

89 Hossain, M.R., Oo, A.M.T. and Shawkat Ali, A.B.M. (2010). Evolution of smart grid and some pertinent issues. *20th Australasian Universities Power Engineering Conference*, Christchurch, New Zealand (5–8 December 2010). IEEE.

90 Electric Power Research Institute. (2011). Estimating the Costs and Benefits of the Smart Grid: A Preliminary Estimate of the Investment Requirements and the Resultant Benefits of a Fully Functioning Smart Grid. https://smartgrid.gov/files/documents/Estimating_Costs_Benefits_Smart_Grid_Preliminary_Estimate_In_201103.pdf (accessed 28 January 2013).

91 Abu-Rub, H., Refaat, S.S., Bayhan, S. et al. (2019). Optimizing KAHRAMAA's Smart Grid Capabilities and Setting its Future Roadmap. TASK FORCE ON Qatar's Smart Grid Road Map, (not published).

2

Renewable Energy

Overview, Opportunities and Challenges

The path toward achieving sustainable development goals necessarily pass by the integration of Renewable Energy Sources (RES) as the key factor for socio-economic growth and improved public health. Contrary to traditional sources, i.e. fossil fuel and coal, the energy inexhaustibility and fast replenishment of RES gathered the attention of research community and stakeholders to promote the use of RSE to meet the ever-growing demand for electricity. With the growing interest in RES, the incorporation of RES in the bulk power system has led to an inherent dynamic characteristic evolution in energy systems. This chapter provides a systematic review of the actual state of RES implementation, the challenging problems and the direction of future research. Furthermore, the operational integration of RES in the smart grid (SG) environment is also extensively discussed and included in this chapter.

2.1 Introduction

The increasing damage and rapid depletion of traditional energy sources compel the worldwide population to achieve the necessary transition toward RES. It is vital that RES are included in the energy mix, especially, with an average growing rate of 1.8% energy consumption per year [1]. Existing electric power systems rely on fuel and coal to generate energy. The escalating permeation of RES aims to satisfy the expected global energy demand increase and the global energy demand to a large extant in order to meet the world's energy demand growth [2].

As a result of some environmental issues, a number of related organizations have engaged in research to increase efficiency and green power plants using developed technology. Concerns regarding environmental protection are rising, RES is therefore sought and examined. Fossil fuel and renewable energy costs, social and environmental prices are going in different directions and the economic and policy plans required to aid the extensive spreading of sustainable markets for renewable energy systems are developing quickly. Future growth in the energy sector will mainly be in the new regime of renewables. Hence, the transition to renewable energy can support us in meeting the challenges of decreasing

Smart Grid and Enabling Technologies, First Edition. Shady S. Refaat, Omar Ellabban, Sertac Bayhan, Haitham Abu-Rub, Frede Blaabjerg, and Miroslav M. Begovic.
© 2021 John Wiley & Sons Ltd. Published 2021 by John Wiley & Sons Ltd.
Companion website: www.wiley.com/go/ellabban/smartgrid

greenhouse gas emissions, hindering future extreme weather and climate effects, and maintaining a reliable, timely, and cost-efficient delivery of energy. The integration of renewable energy may result in substantial dividends for the future of energy security.

Renewables, with nuclear and hydroelectric power, deliver 50% of the extra energy needed out to 2035. Furthermore, renewable energy is the fastest growing source of energy as a result of decreasing capital costs coupled with rising penetration and due to the present state and federal policies investing in its employment, with its share in the primary energy rising to 10% by 2035, up from 3% in 2015. In addition, renewables account for 40% of the increase in power generation, making their share of global power rise from 7% in 2015 to approximately 20% by 2035 [3, 4].

RES is helped by nature and produce energy straight from the sun (thermal, photo-chemical, and photo-electric), indirectly from the sun (wind, hydropower, and biomass), or from other natural phenomena of the environment (geothermal and tidal energy). Renewable energy does not include energy resources originating from fossil fuels, waste products from fossil sources, or waste products from inorganic sources [5]. Renewable resources are gained from solar energy, wind, falling water, the heat of the earth (geothermal), plant materials (biomass), waves, ocean currents, temperature differences in the oceans and the energy of the tides. Renewable energy technologies turn these natural energy sources into practical forms of energy – usually electricity, heat, chemicals, or mechanical energy. Figure 2.1 illustrates an outline of renewables utilized across the globe and Figure 2.2 illustrates the theoretical potential of the RES which are able to provide over 3000 times the current energy consumption around the world [6]. In 24 hours, the sunlight that reaches the earth generates sufficient energy to meet the present energy requirements for 8 years [7–9].

The renewable energy markets – electricity, heating and transportation – have been rising over the previous five years. The integration of well-known technologies, for example, hydro and additional advanced technologies including wind and solar photovoltaic, has increased rapidly, which gave confidence in the technologies, decreased prices and increased new opportunities [10]. Currently, renewable energy delivers approximately 18.3% of the final energy consumption, 50% of this percentage consists of advanced renewables, equally divided between electricity and direct heat applications, and the other 50% involves traditional biomass utilized for heating and cooking. The percentage of renewable energy in the total final energy will merely increase by 2030 from 18.3 to 21% [11]. Renewable energy generating capacity experienced its greatest annual rise ever in 2016, with approximately 161 Gigawatts (GW) of capacity added making the total global capacity almost 2017 GW, as illustrated in Figure 2.3.

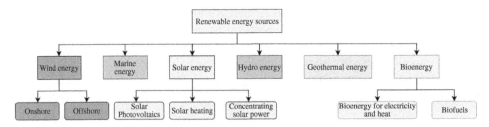

Figure 2.1 Flowchart of the common renewable energy sources.

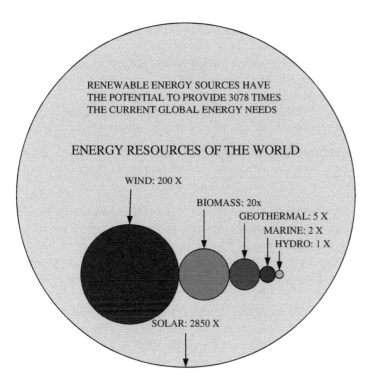

Figure 2.2 Renewable energy resources theoretical potential.

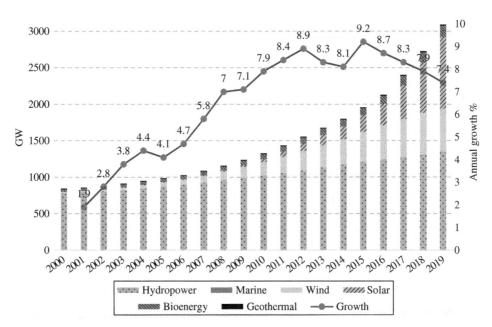

Figure 2.3 Total renewable power installed capacity (GW), including its annual growth rate, 2000–2019. Adapted from [12].

Furthermore, in 2019, renewables were responsible for approximately 7% of net additions to global power generating capacity [12].

This chapter summarizes the benefits, growth, investment and deployment. Furthermore, challenges of integrating them into the electricity grid will be addressed. The content of this chapter is an updated and extension of earlier authors' publication [9].

2.2 Description of Renewable Energy Sources

2.2.1 Bioenergy Energy

Biomass includes all organic materials originating from plants and trees and entails the use and storage of the sun's energy by photosynthesis. Biomass energy (bioenergy) is the transformation of biomass into practical forms of energy including heat, electricity, and liquid fuels (biofuels). Biomass for bioenergy can originate from lands, for example, from dedicated energy crops and from residues produced in the processing of crops for food or different products [13–15].

Biomass energy is renewable and sustainable, but is comparable to fossil fuels. Even although biomass can be burned to acquire energy, it may additionally come as a feedstock to be transformed to numerous liquids or gas fuels (biofuels). Biofuels can be transported and stored, and permit heat and power production when needed, which is crucial in an energy mix with a high dependence on intermittent sources such as wind. These similarities are responsible for the essential contribution biomass is projected to offer in future energy usage [16]. Consequently, a plan to enhance biorefinery and biotransformation technologies to transform biomass feedstock into clean energy fuels is presently being developed. Interconversion of many biomass and energy forms in the carbon cycle is shown in Figure 2.4, [17]. Biomass feedstock can be transformed into bioenergy by thermo-chemical and bio-chemical transformation processes. These processes entail combustion, pyrolysis, gasification, and anaerobic digestion, as illustrated in Figure 2.5. Furthermore, the use of biomass-derived fuels is to substantially counteract current energy security and trade balance problems, and adopt new socio-economic improvements for many nations, as shown in Table 2.1 [18].

Biomass has the ability to reliably deliver baseload power, making it more favorable than other RES including wind and solar, however, the big disadvantage of biomass fuel is the lack of efficiency it possesses. Even although biomass could be utilized to generate energy to meet customer demand, biomass has huge amounts of water per unit of weight, which implies that it lacks energy potential as fossil fuels. Furthermore, transportation costs for biomass are greater per unit of energy than fossil fuels due to its small energy density.

The supply of biomass for energy has been growing at around 2.5% per year since 2010. The global Installed cumulative biopower capacity increased significantly from 39 GW in 2004 to 112.6 GW in 2016, Figure 2.6 shows the global biomass cumulative installed capacity from 2000 to 2013, [19]. Future projections suggest that biomass and waste energy production may rise from 62 GW in 2010 to 270 GW in 2030, as shown by Figure 2.7 [20].

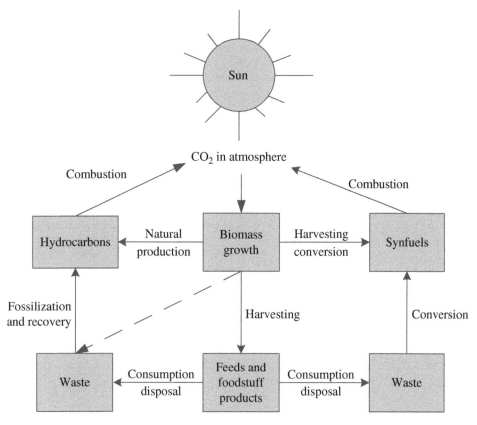

Figure 2.4 Main features of the bioenergy energy technology. Adapted from [17].

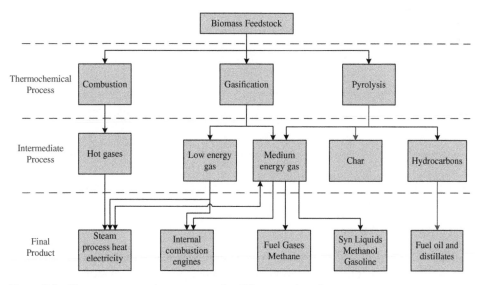

Figure 2.5 Bioenergy conversion processes for different end products.

Table 2.1 Potential benefits and technical limitations of biomass energy. Adapted from Ref [18].

Potential benefits	Technical limitations
Environmental benefits	**Environmental threats**
• Reduced reliance on ecologically harming fossil fuels	• Use of protected soil for the production of biomass
• A decline in greenhouse gas emissions	• Drainage of municipal sources of water
• Reduced brown haze and poisonous chemical emissions;	• Strong demand for fertilizer, herbicides and pesticides, resulting in increasing emissions of air and soil
• Use of squander materials diminishing the requirement for landfill sites	• Potential climate change globally by increased CO_2 production in the atmosphere
Economic gains	• The use of GM crops and microorganisms could theoretically impact ecosystems
• Relatively reasonable resources	• Decreased biodiversity from soil contamination and/or preferred crop agricultural agriculture
• Locally disseminated vitality sources give consistency and reliability	• Increased emissions of wood-burning particulate carbon
• More broadly disseminated which helps achieve energy security	**Associated technologies**
• Generation of work openings in country communities	• Collection capacity of feedstock
• Biomass and bioenergy innovation send out opportunities	• Pre-treatment of biomass
	• Enzyme generation
• Utilizing the full potential of biomass as a renewable and boundless fuel source	• Cost of innovation technology and maintenance

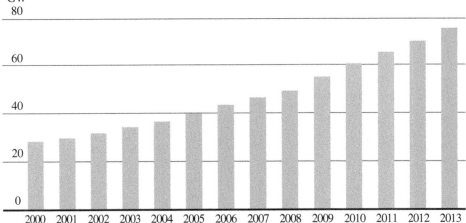

Figure 2.6 Global biomass cumulative installed capacity, 2000–2013. Ref Num [19].

2.2.2 Geothermal Energy

Geothermal energy is an alternative clean energy which gained substantial attention because of its abundant reserves. From heating supply to power generation, geothermal energy is widely employed in buildings' heating, energy generation and hydrogen

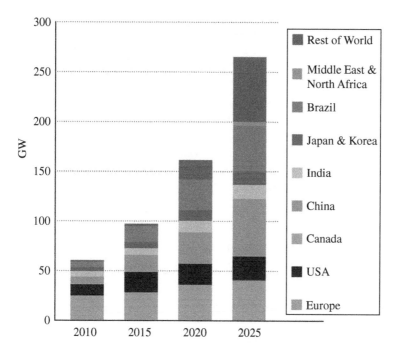

Figure 2.7 Biomass installed capacity for energy systems (2010–2025). Ref Num [20].

production. This could be implemented on a small scale to deliver heat for a residential unit by utilizing a geothermal heat pump (GHP), or on a large scale for energy generation by a geothermal power plant. Geothermal energy is known to be cost-effective, reliable, and an environmentally friendly energy source [21].

Geothermal energy resources involve thermal energy from the Earth's interior stored in rock and trapped steam or liquid water. Geothermal systems arise in a number of geological environments where temperatures and depths of the reservoirs change accordingly. A number of high-temperature hydrothermal systems (greater than 180 °C) are related to present volcanic activities. Intermediate-temperature (between 100 and 180 °C) and low-temperature (less than 100°C) systems exist in continental settings, where above-normal heat production by radioactive isotope decay causes a rise in terrestrial heat flow or where aquifers are charged by water heated by circulation along deeply penetrating fault zones. Under suitable conditions, high-, intermediate-, and low temperature geothermal areas could be utilized for energy production and the direct utilization of heat [22, 23].

Geothermal energy sources are characterized as hydrothermal systems, conductive systems and deep aquifers. Hydrothermal systems involve liquid- and vapor-dominated types. Conductive systems entail hot rock and magma over a wide range of temperatures, and deep aquifers consisting of circulating fluids in porous media or fracture zones at depths usually more than 3 km, but they lack a localized magmatic heat source.

Geothermal energy resource utilization technologies may be divided into different types for electrical energy production, direct use of heat, or combined heat and power in cogeneration applications. GHP technologies are a subset of direct use. Presently,

Table 2.2 Types of geothermal resources, temperatures and their applications. Adapted from Ref [24].

Type	In-situ fluids	Subtype	Temperature Range	Utilization	
				Current	Future
Convective systems (hydrothermal)	Yes	Continental	H, I & L	Power, direct use	
		Submarine	H	None	Power
Conductive systems	No	Shallow (<400 m)	L	Direct use (GHP)	
		Hot rock (EGS)	H, I	Prototypes	Power, direct use
		Magma bodies	H	None	Power, direct use
Deep aquifer systems	Yes	Hydrostatic aquifers	H, I & L	Direct use	Power, direct use
		Geo-pressured		Direct use	Power, direct use

H: High, I: Intermediate, L: Low (temperature range).

the only commercially exploited geothermal systems for energy production and direct use are hydrothermal. Table 2.2 summarizes the resources and utilization technologies [24, 25].

Types of traditional geothermal power technologies are as follows: dry steam, flash and binary. In dry steam plants, a high-pressure steam shoots up from the dry steam reservoir and is used to make the turbines function which then turns on the generator. In flash plants, steam is separated from the high-pressure and high-temperature geothermal fluids which includes water and with a high temperature. The steam is guided to a turbine that then turns on the generator. The liquid (condensed from the steam after going through the turbine) and the water are sent into the reservoir. In binary or ORC (i.e. Organic Rankine Cycle) plants, heat is transferred from the high-temperature water to an organic working fluid that possesses a lower boiling point than water [26].

Approximately 0.4 GW of new geothermal power generating capacity came online in 2016, culminating in total capacities of approximately13.5 GW. Indonesia and Turkey invested in new installations. Kenya, Mexico and Japan also invested in projects during the year, and many other countries had projects under development. The US possesses the biggest geothermal capacity, now just under 3.6 GW, followed by the Philippines (1.9 GW), Indonesia (1.6 GW) and Mexico (1 GW), Figure 2.8 illustrates the installed geothermal electric capacity as of 2019 in the top 10 countries [26]. The United Nations and the International Renewable Energy Agency (IRENA) pledged a fivefold growth in the installed-d capacity for geothermal energy production and twofold growth or more for geothermal heating by 2030 relative to 2014 levels. Therefore, there are many (short, medium and long) targets for geothermal power generation global installed capacity as: 21 GW (by 2020), 65 GW (by 2030) and 140 GW (by 2050), as shown in Figure 2.9 [12, 26, 27].

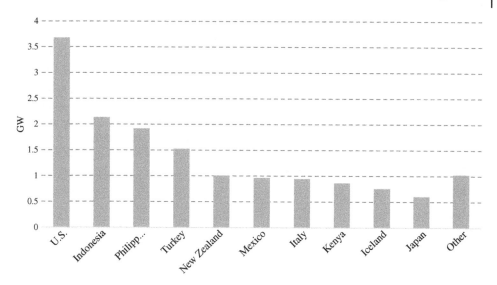

Figure 2.8 Cumulative installed geothermal generating capacity by top 10 countries in 2019.

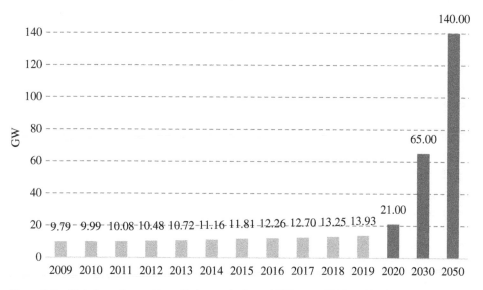

Figure 2.9 Global geothermal installed capacity from 1950 up to 2019 and its forecasting for 2020, 2030 and 2050. Adapted from [12], [26], [27].

2.2.3 Hydropower Energy

Hydropower has a wide resource availability coverage, producing about 17.2% of the nation's electricity, in 2019, [11]. Hydropower is recognized for being the highest density of energy resources; it is ranked first with regards to installed production capacity from renewables. The main goal of hydropower in the global energy supply is in delivering centralized power generation, moreover, hydropower plants (HPPs) has the ability to function in isolation and supply independent systems, usually in rural and remote regions [28].

Hydropower describes the energy obtained due to the movement of water. Flowing water generates energy that can be transformed into electrical power with the use of turbines. The most widespread form of hydropower is dams, even although advanced forms of utilizing wave and tidal power are becoming more common. Hydropower is produced from water flowing in the hydrological cycle, propelled by solar radiation. It is the movement of water propelled by the force of gravity to move from higher to lower elevations that could be utilized to produce hydropower. HPPs span a wide range of scales, from a few watts to several GW. Four broad hydropower typologies are present today [29, 30]:

- *Run-of-river hydropower*: a facility that channels flowing water from a river to spin a turbine. Normally, a run-of-river case will have minimum or no storage possibility. Run-of-river delivers a constant generation of electrical power with limited flexibility of operation for daily changes in demand by water flow management.
- *Storage hydropower*: a large system that utilizes a dam to store water in a reservoir. Electrical power is generated by releasing water from the reservoir by a turbine, which operates a generator. Storage hydropower delivers the base load and can be shut down and operated at short notice depending on the demands of the system. It could deliver sufficient storage capacity to function without the need for the hydrological inflow for several weeks or possibly months.
- *Pumped-storage hydropower*: delivers peak-load supply, taking advantage of water which is circulated between a lower and upper reservoir by pumps, that use the additional energy from the system at periods of low demand. When energy demand is high, water is released back to the lower reservoir by turbines to generate electrical power.
- *Offshore hydropower*: less recognized relative to the others but an increasing group of technologies that utilize tidal currents or the energy of waves to produce electrical power from seawater.

An estimated 21.8 GW of hydropower capacity was employed, with pumped storage, amounting to an installed capacity of 1292 GW worldwide. The total installed capacity has risen by 39% from 2005 to 2015, with an average rate of approximately 2.4% each year. World leaders in producing hydro power are China, Brazil, the United States, Canada, the Russian Federation, India and Norway, which was responsible for almost 68% of installed capacity at the end of 2019, as illustrated in Figure 2.10, [12]. Hydropower is a proven and well-advanced technology based on more than 100 years of experience. Presently, hydropower is a relatively flexible power technology with one of the best conversion efficiencies relative to other energy sources because of its direct transformation of hydraulic energy to electrical energy. Yet, more improvements can be made by refining operations, reducing environmental impacts, adapting to new social and environmental needs and advancing more robust and cost-effective technological solutions. Figure 2.11 shows evolution of world hydropower generation since 1980, [31]. From 1999 through 2005, hydropower advancement was relatively halted globally, showing the influence of the World Commission on Dams (WCD), which was organized to review the improvement of large dam performance and come up with guidelines for the integration of new dams. From 2005 onwards

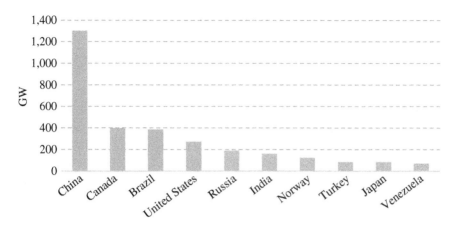

Figure 2.10 Hydropower generation by top 10 countries in 2019. Adapted from Ref Num [12].

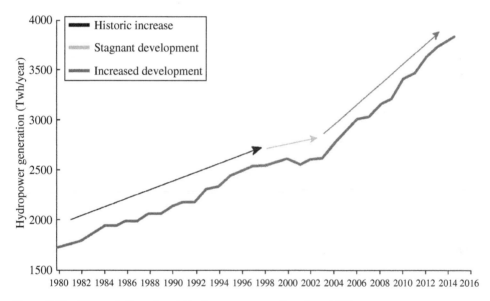

Figure 2.11 The evolution of world hydropower generation since 1980.

(green arrow), hydropower development has seen an upswing in development due to the expansion and use of the Hydropower Sustainability Assessment Protocol [32].

2.2.4 Marine Energy

Ocean and marine energy denote the diverse forms of renewable energy acquired from the ocean. Two kinds of ocean energy exist: mechanical and thermal. The movement of the earth and the moon's gravitational force leads to the mechanical forces experienced. The movements of the earth create wind in the oceans that ultimately produces waves and the gravitational pull of the moon causes the initiation of coastal tides and currents. Thermal energy is acquired from the sun, which increases the temperature of the ocean yet the depths are still of a lower temperature. Therefore, the temperature difference experienced allows for the energy to become trapped and converted to useful energy, typically, electrical energy. Five kinds of ocean energy transformation take place: wave energy, tidal energy, marine current energy, and ocean thermal energy conversion [33–37].

- *Wave energy* is produced by the motion of a device either floating on the surface of the ocean or fixed to the ocean floor and there are several methods for transforming wave energy to electric energy. Wave energy is recognized as the most commercially developed of the ocean energy technologies yet is still far from where it could be practically.
- *Tidal energy*, the tidal cycle takes place every 12 hours as a result of the gravitational force of the moon. The difference in water height from low and high tide is potential energy. Comparable to traditional hydropower produced from dams, tidal water can be trapped in a barrage across an estuary at periods of high tide and forced by a hydro-turbine at periods of low tide.
- *Current energy,* marine current is ocean water moving in one direction. Kinetic energy of the marine current can be trapped with submerged turbines that are relatively comparable to wind turbines, where the marine current forces the rotor blades to move to produce electrical energy.
- *Ocean thermal energy conversion* (OTEC), utilizes ocean temperature variations from the surface to depths lower than 1000 m, to obtain energy. Research focuses on two types of OTEC technologies to obtain thermal energy and transform it to electrical energy: closed and open cycles.
- *Salinity gradient* power is the energy generated from the variation in salt concentration between two fluids, usually fresh and saltwater, e.g. when a river flows into the sea. Collision of fresh and saltwater delivers large amounts of energy, which this technology strives to capture.

Tidal power stations generate tens to hundreds of MW similar to hydropower stations, wave energy converters (WECs) from some kW to MW, salinity gradient power stations from some kW to MW, ocean thermal energy converters (OTECs) from kW to MW and ocean thermo-electric generators (OTEGs) from some watts to kW.

Ocean energy is still in the process of development, and intensive research is required, progress and demonstration efforts needed for learning and cost reduction before it can contribute to the energy supply. Therefore, the ocean energy market is still in its infancy, and the sector must address many issues to confirm the reliability and affordability of its technologies. A number of barriers are present in ocean energy technologies, which include obtaining site permits, the environmental influence of technology implementation, and grid connectivity for transporting the energy generated.

In 2019, the global installed capacity of ocean energies was 536 MW, a rise of 32 MW relative to 2014. Ocean energies amount to 0.03% of the global renewable energies installed capacity. However, there is huge potential to increase the use of ocean energies. Theoretical ocean energy resources might be enough to meet present and future global electricity demand, with a range of 20 000–80 000 TWh/year, culminating up to 100–400% of the present global demand. Moreover, a global potential of 337 GW could be achieved by 2050, one third of this would be in Europe. For the 2020 milestone, approximations suggest an installed capacity ranging from some hundred MW to 2 GW. Figure 2.12 illustrates the global ocean energy forecast based on device technology and infrastructure available, [38, 39].

2.2.5 Solar Energy

Solar energy production includes the sun's energy to deliver hot water by solar thermal systems (STS) or electricity by solar photovoltaic (PV) and concentrating solar power (CSP) systems. These technologies are technically recognized with many systems employed worldwide over the previous few decades.

2.2.5.1 Photovoltaic

Solar PV systems directly transforms solar energy into electrical energy. The basic foundation of a PV system is the PV cell, which is a semiconductor device that transforms solar energy into DC current. PV cells are then connected to form a PV module, normally in the range of 50–300 W. The PV system consists of modules, inverters, batteries, components, mounting systems, etc. PV systems are usually modular, i.e. modules could be connected together to deliver electrical power in the range of some Watts to hundreds of MW.

There are many different PV cell technologies on the market, utilizing different types of materials, and a greater number will be present in the future. PV cell technologies are divided into three categories, pointed out as first, second, and third generation: (i) wafer-based crystalline silicon (c-Si); (ii) thin-films (TF); and (iii) emerging and novel PV technologies, including concentrating PV, organic/polymer material cells and dye-sensitized solar cell, advanced thin films and other novel concepts. During the last two decades, PV technologies have substantially enhanced their performance (i.e. efficiency, lifetime, energy pay-back time) and decreased their costs, and this is projected to continue in the

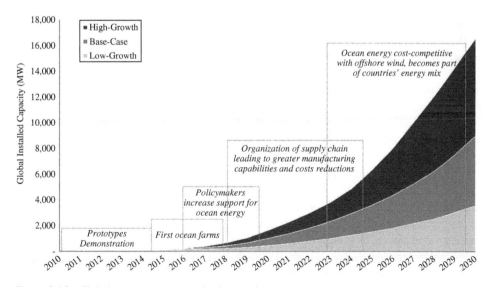

Figure 2.12 Global ocean power capacity forecasting.

future. Research aims to enhance the efficiency and lifetime, and decrease the investment costs to decrease the electricity production cost. Solar PV has two benefits: first, module manufacturing can be implemented in large plants, which allows for economies of scale; second, PV is a relatively modular technology. In comparison to CSP, PV has the upper hand that it utilizes not only direct sunlight but also the diffuse component of sunlight, i.e. solar PV generates electrical energy even if clouds are present. This ability grants the effective integration in numerous areas around the world relative to CSP [40–42].

PV systems are described by two main types: off-grid and grid-connected applications. Off-grid PV systems have a substantial opportunity for economic application in un-electrified regions of developing countries, and off-grid centralized PV mini-grid systems. Centralized PV mini-grid systems have the potential to be one of the most cost-efficient for a pre-defined level of service, and they could have a diesel generator set as an optional balancing system or to function as a hybrid PV-wind-diesel system. These types of system are applicable for decreasing and refraining from utilizing the diesel generator in remote regions [43].

Grid-tied PV systems utilize an inverter to transform electrical current from DC to AC and, after that, supply the electrical power produced to the grid. Relative to an off-grid installation, system costs are lower due to the fact that energy storage is not needed because the grid is utilized as a buffer. Grid-connected PV systems are described as two types of applications: distributed and centralized. Grid-connected distributed PV systems are employed to deliver electric energy to a grid-connected consumer or to the electric network. These systems have several advantages that include: distribution losses in the electric network are decreased because the system is installed at the point of use; additional land is not needed for the PV system, and prices for mounting the systems can be decreased if the system is mounted on an existing structure; and the PV array itself could be utilized as a cladding or roofing material, as in building-integrated PV. Usual sizes are 1–10 kW for residential systems, and 10 kW to several MWs for rooftops on public and industrial buildings. Grid-connected centralized PV systems implement the functions of centralized power stations. The power generated by this system is not related to a particular electricity consumer, and the system is not positioned to perform certain functions on the electricity network other than to produce bulk power. Usually, centralized systems are installed on the ground, and they are greater than 1 MW. The economic benefits of these systems are the optimization of installation and operating costs by bulk buying and the cost-effectiveness of the PV elements and balance of systems on a large scale. Furthermore, the reliability of centralized PV systems can be better than distributed PV systems as they can implement maintenance systems with monitoring equipment, which could be a smaller section of the total system cost [44].

During 2019, approximately 115 GW of solar PV capacity was installed globally, making the global solar PV capacity arrive at a value of 627 GW. More solar PV capacity was installed in 2019 (up 44% over 2015) than the cumulative world capacity five years earlier [45]. The PV market was multiplied by almost 40 in 10 years, as illustrated in Figure 2.13. The global market expansion is due to the rising competitiveness of solar PV, and to new government programs, higher demand for electrical energy and increasing awareness of solar PV's potential to hinder pollution and carbon dioxide emissions. Figure 2.14 shows the solar PV capacity in the top 10 countries in 2019. Solar PV plays a vital role in electrical energy

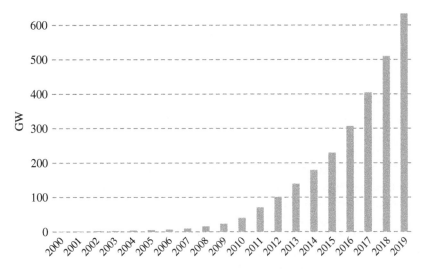

Figure 2.13 Global integrated solar PV capacity from 2000 to 2019.

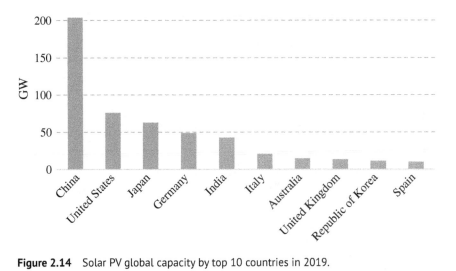

Figure 2.14 Solar PV global capacity by top 10 countries in 2019.

production in many regions around the world. In 2019, solar PV was responsible for 10.7% of net production in Honduras and met 8.6% of the electricity demand in Italy, 8.3% in Greece and 8.2% in Germany. A minimum of 22 countries had sufficient solar PV capacity at the end of 2019 to meet 3% or more of their electrical energy need [12]. Furthermore, the levelized cost of electricity (LCOE) of solar PV fell 58% between 2010 and 2015, making it extremely competitive at the utility scale. While demand is increasing at a fast pace for off-grid solar PV, the capacity of grid-connected systems is increasing much faster. Distributed (residential, commercial and industrial rooftop systems) grid-connected applications have struggled to ensure a stable global market since 2011, while the centralized largescale projects were associated with the increasing share of yearly installations, as illustrated in

Figure 2.15. The IRENA estimates that solar PV capacity could increase to 1760 GW in 2030; reaching this capacity by 2030 needs an average yearly rise of a total capacity of 15%. IRENA approximates that solar PV might be responsible for 7% of global electrical energy generation by 2030 [11].

2.2.5.2 Concentrated Solar Power

Concentrated Solar Power (CSP) is a power generation technology that uses mirrors or lenses to direct sunlight onto a receiver, which gathers and transmits the solar energy to a heat transfer fluid that could be utilized to supply heat for end-use applications or to heat a fluid and generate steam. The steam turns the turbine and produces energy just like traditional power plants. Huge CSP plants could possess a heat storage system to produce electrical power even in the presence of clouds or at sunset. This allows for the generation of dispatchable electricity, which could facilitate grid integration and economic competitiveness. Furthermore, CSP technology can easily be integrated into fossil fuel-based power plants that utilize traditional steam turbines to generate electrical power, whereby the part of the steam generated by the combustion of fossil fuels is replaced by heat from the CSP plant. While CSP plants generated mainly electrical power, they also generate high-temperature heat that could be utilized for industrial processes, space heating (and cooling), and heat-based water desalination processes. Moreover, CSP plants could be utilized to generate steam to inject into mature and dense oil fields for thermal enhanced oil recovery, [46].

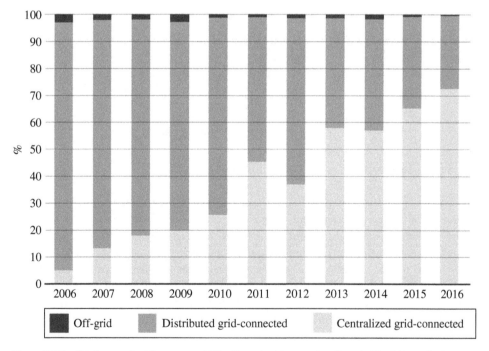

Figure 2.15 PV shares of grid-connected (distributed and centralized) and off-grid installations, 2006–2016.

CSP merely utilizes the direct component (DNI) of sunlight to operate and hence, are an interesting opportunity for implementation in regions with high DNI (i.e. Sun Belt regions like North Africa, the Middle East, the southwestern United States and southern Europe), which is an advantage that solar PV does not possess. CSP applications can be implemented from small distributed systems of tens of kW to large centralized power stations of hundreds of MW. The CSP plant could be constructed in 1–3 years (depending on its size) and could function for over 30 years. Five to six months of full-power utilization are required to reimburse for the energy used for the construction. The CSP technology includes four parts and they are as follows; Parabolic Trough (PT), Fresnel Reflector (FR), Solar Tower (ST) and Solar Dish (SD). In PT and FR plants, mirrors lead the sun's rays on a focal line; the concentration factors range from 60 to 80 with a temperature culminating up to 550 °C. In ST and SD plants, mirrors lead the sun rays on a single focal point with higher concentration factors ranging from 600 to 1000 and temperatures ranging from 800 to 1000 °C, [47, 48].

The commercial implementation of CSP plants began in 1984 in the United States with the Solar Energy Generating Systems (SEGS) project. The last SEGS plant (a 354 MW in California) was finished in 1990. From 1991 to 2005, no CSP plants were constructed globally. Globally installed CSP capacity has risen to almost tenfold since 2004 and increased at an average of 50% per year between 2009 and 2010 and 2013–2014. In 2013, global installed capacity rose by 36% to greater than 3.4 GW and a 310 MW of capacity became active in 2019, increasing globally capacity to a value greater than 6.2 GW, as illustrated in Figure 2.16. Spain and the United States remain the global leaders, yet the number of regions with employed CSP are rising. CSP activity experienced a substantial shift from Spain and the United States to developing regions in 2015, and this trend remained in 2016. There is a significant trend toward developing regions with a high solar radiation, South Africa led the market in new additions in 2016, becoming

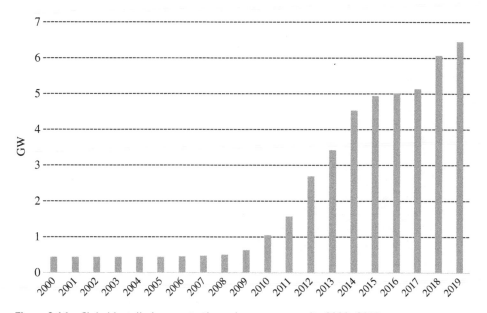

Figure 2.16 Global installed concentrating solar power capacity, 2000–2019.

the second developing country to do so after Morocco in 2015. Figure 2.17 shows the CSP capacity in the top 10 countries in 2019 [12].

Even although CSP costs experienced a substantial decrease, CSP employment has been hindered due to fast and significant declines in the cost of solar PV. In numerous countries, research and industry are working to enhance CSP performance and decrease its costs by focusing on: cost reductions in important CSP components (collectors for instance); other applications of CSP; efficiency of the heat transfer process and increasing the value of CPS by implementing thermal energy storage (TES) systems, which allows CSP facilities to deliver dispatchable power [49].

2.2.5.3 Solar Thermal Heating and Cooling

STS transforms sun-oriented radiation into heat. These frameworks are utilized to raise the temperature of a heat transfer fluid, which can be used to generate energy. The hot liquid can be utilized straightforwardly for hot water requirements or space heating/cooling needs, or a heat exchanger can be utilized to exchange the heat to the ultimate application. The heat produced can, moreover, be put away in a legitimate storage tank for utilization within the hours when the sun is not accessible. The sun powered collector is the key component of an STS.

There are two prevailing plans: flat-plate solar collectors (FPC) and evacuated tube solar collectors (ETC). Both level and emptied tube advances are progressed and each type of innovation includes a wide run of choices. In nations and locales with tall cooling requests, a sun-based heat cooler is a vital innovation. Cost competition and government subsidy, particularly for recently built lodging, are the key drivers. In expansion to comparatively

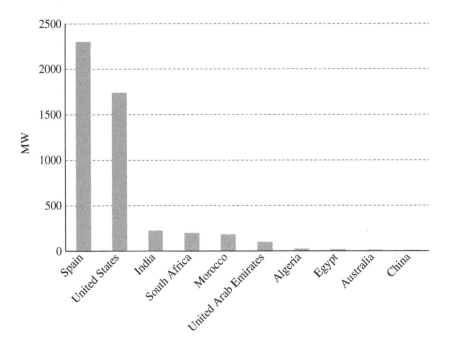

Figure 2.17 Concentrating solar power capacity in the top 10 countries in 2016.

tall forthright costs, there were impediments such as: (i) disappointment to supply adequate administrative instruments to ensure viable working; (ii) hesitance of private shoppers to move from conventional viable warming and cooling frameworks, and (iii) disappointment of the designers and building and vitality businesses to realize the capabilities of STS systems [50].

The installed STS in 127 nations with a worldwide capacity of 456 GWth is promoted by 2016 as seen in Figure 2.18. China (71%), United States (4%), Turkey (3%), Germany (3%), Brazil, India, Australia, Austria, Israel, and Greece were the best 10 nations for full-speed operations, as shown in Figure 2.19.

2.2.6 Wind Energy

Wind power is known to be the transformation of wind energy by wind turbines into a practical form of energy, like utilizing wind turbines to generate electrical energy, wind mills for mechanical power, wind pumps for pumping water or drainage, or sails to drive ships. The first wind turbines for electrical energy production were made at the beginning of the twentieth century. The technology has improved substantially since the early 1970s. By the end of the 1990s, wind energy was well-known as the most crucial sustainable energy resources [51].

Producing electrical power from the wind needs the kinetic energy of wind to be transformed to mechanical and then electrical energy, the wind forces the turbine's rotor to turn, converting the kinetic energy to rotational energy by moving a shaft connected to an electric generator, thereby generating electrical power through electromagnetism, hence challenging the industry to create cost-effective wind turbines and power plants to implement this transformation. Wind power increases with the area of the rotor and to

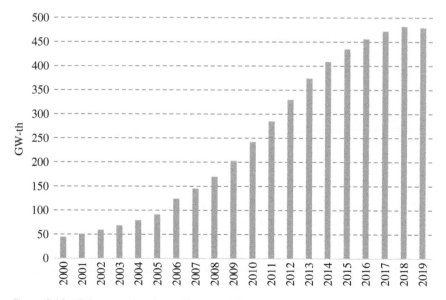

Figure 2.18 Solar water heating collectors' global capacity, 2000–2019.

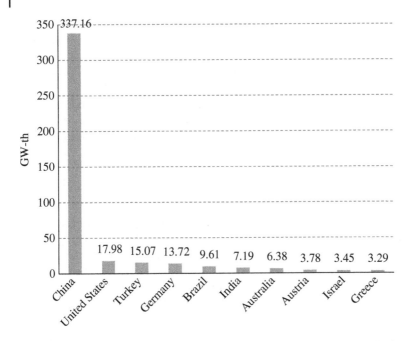

Figure 2.19 Solar water heating collector capacity by top 10 countries in 2016.

the cube of the wind velocity. Theoretically, as the wind speed doubles, the wind power rises eight times. The output power depends on the swept area (related directly to the length of the blades) and the wind speed. During the past few decades, the size of wind turbines has increased constantly, from 0.05 MW in 1985 to 2.0 MW in 2014. The biggest commercially available turbines until now reach 8.0 MW each, with a rotor diameter of 164 m, as illustrated in Figure 2.20. Yet, a turbine merely traps a fraction of that available energy (40–50%), so wind turbine design has focused on increasing energy capture over the range of wind speeds experienced by wind turbines, while striving to decrease the cost of wind energy considering all other parameters. To decrease cost, wind turbine design is also motivated by a desire to decrease materials usage while striving to enlarge turbine size, improve component and system reliability, and to enhance wind power plant functions, [52–54].

The three main components of the wind generation system are: (i) turbine which could be vertical or horizontal-axis, (ii) installation area which could be onshore or offshore, and (iii) application which could be grid connected or stand-alone. The majority of large wind turbines are up-wind horizontal-axis turbines with three blades. The majority of small wind turbines are horizontal-axis. Developed structures of vertical-axis turbines are presently utilized in many regions across the globe. They are associated with an aerodynamic energy loss ranging from 50 to 60% at the blade and rotor, a mechanical loss of 4% at the gear, and a 6% electromechanical loss at the generator. The full production efficiency typically ranges from 30% to 40% at wind energy facilities.

From 1970 to 1980, a number of onshore wind-turbine configurations were examined, considering horizontal and vertical axis designs. The horizontal axis design came to

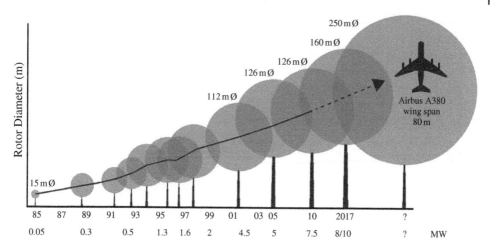

Figure 2.20 Growth in capacity and rotor diameter of wind turbines, 1985–2016 [52]. Reproduced with permission from IEA-ETSAP.

dominate, although configurations changed, particularly the number of blades and blades orientation (upwind or downwind of the tower). Following consolidation, turbine designs centered around the three-blade, upwind rotor; positioning the turbine blades upwind of the tower avoids the tower from blocking wind flow onto the blades, while three-bladed machines usually have lower noise emissions than two-bladed machines. Onshore wind turbines are usually grouped together into wind power plants, also known as wind farms. These wind power plants are usually 5–300 MW in size, although smaller and bigger plants are available. Offshore wind energy technology is less mature than onshore, and has higher investment, operation and maintenance costs. Benefits of implementing offshore wind energy entail: better quality of wind resources located at sea; potential to utilize even larger wind turbines; potential to construct bigger power plants than onshore; and the opportunity to decrease demand for land-based transmission infrastructure [53].

With regard to the reliability of a power system, the power transformation system is of paramount importance when it comes to wind turbines. For large grid-connected turbines, power transformation systems are categorized into three forms. Fixed-speed induction generators are well-known for their stall-regulated and pitch-controlled turbines; in these particular arrangements, wind turbines are net consumers of reactive power that must have been given by the power network. Advanced turbines replaced this with variable-speed machines. Two arrangements are recognized world-wide, doubly-fed induction generators and synchronous generators with a full-power electronic converter, and they come with pitch-controlled rotors. These particular variable speed designs decouple the rotating masses of the turbine from the power system, thus, producing many power quality benefits as opposed to the old turbine configurations. These turbines could transmit real and reactive power, and some fault ride-through capability, which are required by power network operators. To enhance production, development is now going directed toward bigger machines, which means greater dimensions of the blades, more hub heights and greater rotor dimensions. These modifications mean significantly increased capacity factors within given wind resource regimes, enabling more opportunities in both established and new markets [52, 54].

Another strong year for the global wind industry was 2019, with annual installations in excess of 60 GW, making globally installed wind power capacity 651 GW, representing cumulative market growth of more than 19% with gross additions 10% below the high record in 2018, as illustrated in Figure 2.21. China, the largest overall market for wind power since 2009, retained the top spot in 2016. Installations in Asia-led global markets, with Europe ranked second, and North America closing the gap with Europe, in third place, as illustrated in Figure 2.22.

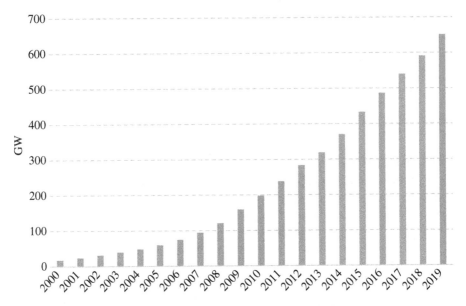

Figure 2.21 Total installed global wind power capacity, 2000–2019.

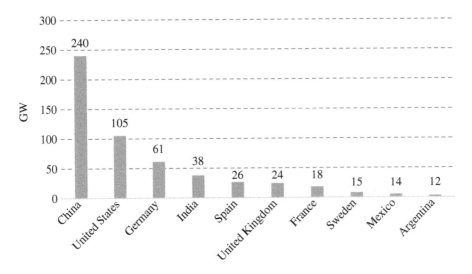

Figure 2.22 Installed wind power capacity, top 10 countries in 2019.

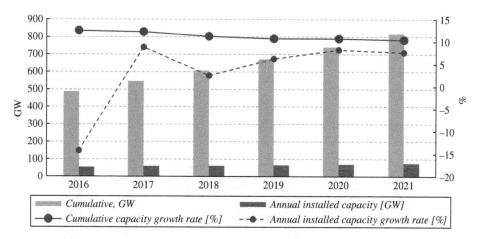

Figure 2.23 Wind power market forecast for 2017–2021. Ref [55]. Reproduced with permission from Global Wind Energy Council.

The Global Wind Energy Council (GWEC) believes that the wind power market will be going forward, as the technology is persistently becoming more advanced, prices continue to decrease and the demand for clean power and new industries becomes stronger with each passing year. The five-year 2017–2021 market forecasting is illustrated in Figure 2.23 [55].

2.3 Renewable Energy: Growth, Investment, Benefits and Deployment

RES integration is picking up ground. For the most part of renewable energy innovations, capacity and yield proceed to develop at a higher rate and their share in control division will be more than other routine advances, as outlined in Figure 2.23. Since 1990, RES has developed at a normal yearly rate of 2.2%. The development has been particularly tall for sun-based photovoltaic and wind control, which developed at normal yearly rates of 46.2 and 24.3% separately. Sun-oriented warmth developed with 11.7%, fluid biofuels developed with 10.4% and hydropower developed with 2.5% per year. Renewables worldwide add up to instated capacity and will proceed in its increment as shown by the 2040 projections in Figure 2.24. Renewables account for a rising share of the world's add up to power supply, the entire era from renewable assets increments by 2.9%/year, as the renewable share of world power era develops from 22% in 2012 to 29% in 2040, as outlined in Figure 2.25. Era from non-hydropower renewables is the overwhelming source of the increment, rising by a norm of 5.7%/year, [56] Renewable energy can give a parcel of benefits to society, as outlined in Figure 2.26. In expansion to the lessening of carbon dioxide (CO_2) outflows, governments have ordered renewable vitality (RE) arrangements to meet a number of goals including creation of neighborhood natural and wellbeing benefits; assistance of vitality get to, especially for country regions; headway of vitality security objectives by broadening the portfolio of vitality advances and assets; and progressing social and financial improvement through potential work openings. The combined financial, natural, wellbeing and

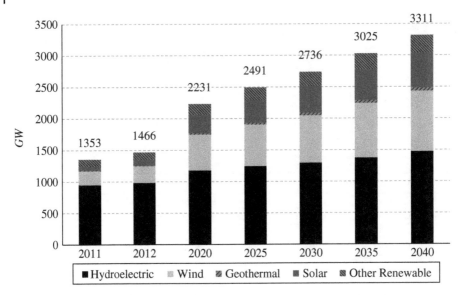

Figure 2.24 Predicted world total installed renewable generating capacity, 2011–2040, Adapted from [56].

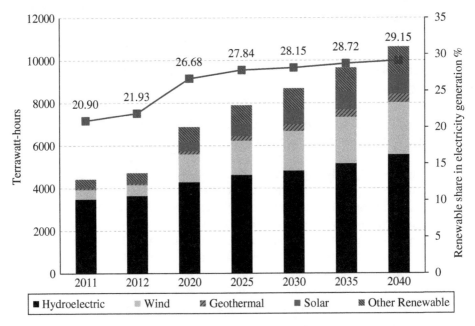

Figure 2.25 Predicted world renewable electricity generation and their global share, 2011–2040, Adapted from [56].

climate benefits give a solid case for approach intercession to quicken speculation in RE around the world. Expanding the share of renewables can influence the world's economy through speculation, exchange and power costs, The RES industry remains one of the foremost dynamics, fast-changing, and transformative divisions of the worldwide economy.

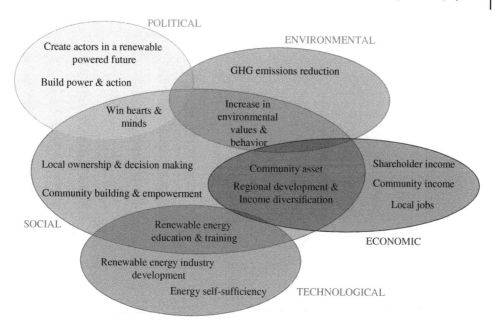

Figure 2.26 Potential features of renewable energy sources integration.

Technology advancements, cost drops, and the catalytic effect of new financing structures, have transformed the sector into a driver of economic growth globally.

World-wide investment in renewables exclusive of hydro decreased by 23% to $241.6 billion, the lowest total since 2013, yet, a record integration of renewable energy capacity worldwide in 2016. Wind, solar, biomass and waste-to-energy, geothermal, small hydro and marine sources among them contributed to an increase amounting to 138.5 GW, up from 127.5 GW in the previous year. This 2016-Gigawatt figure was almost 55% of all the producing capacity included globally, the greatest number to date. Installations increased in spite of the fact that the money invested declined because of the drastic drop in capital costs for solar PV, onshore and offshore wind. Figure 2.27 shows the trend of global new investment since 2004 in renewable energy. Moreover, the R&D investment levels for renewables in recent years are rising from private and governmental sectors as illustrated in Figure 2.28, [57, 58].

RES is attracting greater investments than fossil fuels. This suggests that green power is getting a large share of world capacity and generation, and could also be because the price of the project to produce energy from wind, solar, geothermal and small hydro is upfront. Fossil fuel plants cost less money to construct but have higher running costs, as the fuel has to be bought on a regular basis. Trade in renewable energy equipment and other investment goods and services will rise due to the scaled-up employment in power and end-use sectors. At the same time, this will decrease trade in other energy sources, particularly fossil fuels, which then can last longer for the earth. Therefore, due to the continuous employment in power and end-use industries, trade in clean energy equipment and other investment products and service would rise. In comparison, the trade in alternative energy sources, particularly fossil oil, would decrease, which will be continued for a long time.

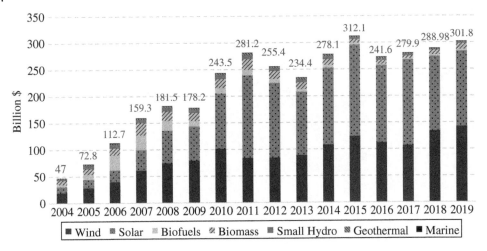

Figure 2.27 Financial investments in renewable energy by technology, 2004–2019, Adapted from [57].

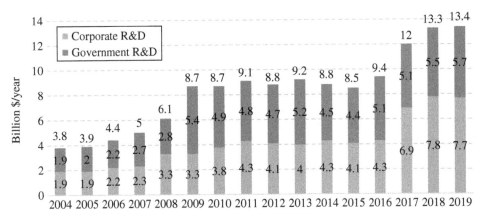

Figure 2.28 Research and development costs on renewable energy, 2004–2019, Adapted from [57].

The benefits of integrating RES are enormous especially in social and environmental levels. The environmental advantages include Greenhouse gas emission reduction and pollution decrease. The number of jobs created from RE integration significantly rose over the last few years. The PV sector has the lion's share by 3.6 created jobs as shown in Figure 2.29 followed by wind power by 1.16 million jobs. China is the largest RES manufacturer in the world with ever-growing twin of energy supply and demand which made it a leader in this alternative transition. The employment rate will be extended from 9.2 to 24.4 million in 2030 [59].

Figure 2.30 shows a diagram of a renewable energy market development. Technological influence plays a vital role in the dynamics of renewable energy advancement; hence, efficiency and effectiveness of investments are determined, mainly the cost and maturity of a certain technology. Moreover, through time and the utilization of a technology rises, costs will be decreased. During this development process, the government is helpful in many ways,

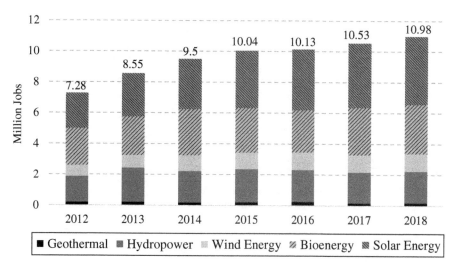

Figure 2.29 Global employment in renewable energy 2011–2018, Adapted from [59].

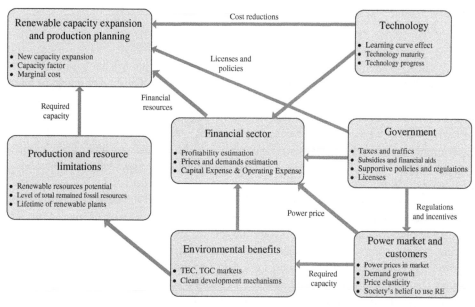

Figure 2.30 Renewable energy market development process, [9]. Reproduced with permission from ELSEVIER.

for instance, in supporting capacity expansions, setting regulations, and adopting worldwide implementation of renewable energy. Furthermore, market and consumers' satisfaction influence the producers of renewable energy in competing with conventional energy producers. Yet, financially, requirements are met with respect to profitability levels, estimated prices, demand, and costs. The market development process must also consider capacity expansion and short- to long-term production planning. Lastly, the equipment utilized in renewable resources and its lifetime must be taken into account during this process [9].

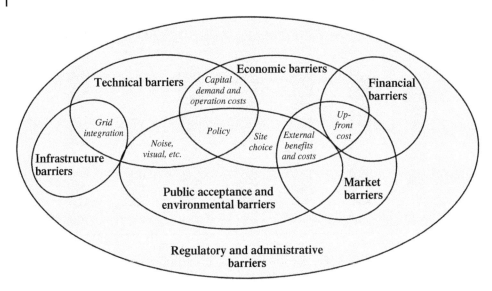

Figure 2.31 Barriers to renewable energy technology deployment.

The issues associated with implementing renewable energies can be summarized as shown in Figure 2.31. An economic barrier arises if the cost of a given technology is more than the cost of competing alternatives, even under optimal market conditions, with a direct connection between technological maturity and economic barriers. All other types of barriers are categorized as non-economic, although these barriers play a vital role in determining the cost of renewables. The importance of the barriers depends on the technology and market, and the priority varies as a technology matures along the direction to commercialization, [9].

A number of renewable energy technologies are present at different stages of maturity, which affects its odds of achieving commercialization, its cost of implementation, or – in the case of many mature technologies – its market cost and the level of subsidies needed to obtain market competitiveness with incumbent energy technologies as indicated in Figure 2.32. Hydropower and bioenergy are major sources of energy globally. Additional options, yet technically proven and available on commercial terms, still hold only a small percentage of their potential markets. Consequently, there are numerous chances to enhance performance and decrease costs [60–62].

2.4 Smart Grid Enable Renewables

Over the past decade, renewable energy technologies have moved into the mainstream of every technology. Annual capacity of RES has been added to 100 GW and exceeds all traditional energies since 2011. Figure 2.33 illustrates a general taxonomy of possible RES challenges. These challenges could be classified into: technical, financial and societal.

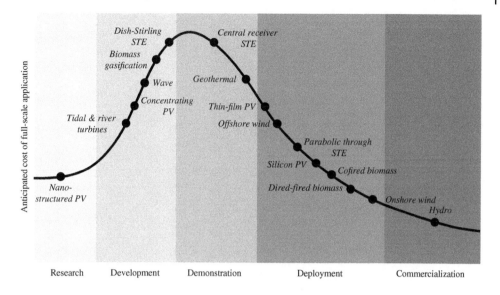

Figure 2.32 Maturity of selected renewable energy technologies, Adapted from [9].

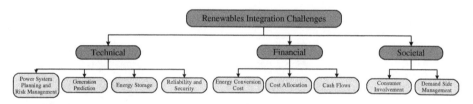

Figure 2.33 Common challenges in bulk implementation of RES into the SG [9]. Reproduced with permission from ELSEVIER.

From figure 2.33 two technical problems are reported:

1) Governing the energy mix and dealing with uncertainty during operational conditions.
2) Managing supply and demand during outages and abundances of energy demand.

Currently, existing power systems are not suitable for the bulk integration of the RES. Thus, it is vital to implement advanced technologies to support the stability and reliability of the power system. SG operations take advantage of information and communication technologies to add more flexibility in the energy mix deployment. The flexibility measures are classified into six categories as shown in Figure 2.34. To date, a variety of intelligent tools are proposed to support the integration of RES into the SG. Figure 2.35 proposes different options according to the location where they would be implemented [63, 64].

2.5 Conclusion

After the explanation of the different RES, it should be considered that each source of renewable energy has its ups and downs as summarized by Table 2.3. Moreover, Table 2.4 illustrates some undesirable environmental effects of RES [9].

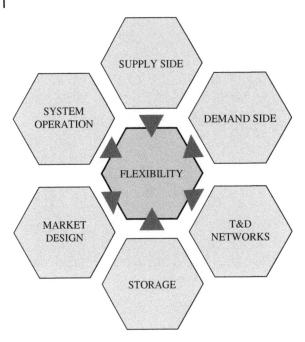

Figure 2.34 Modern power system flexibility measures.

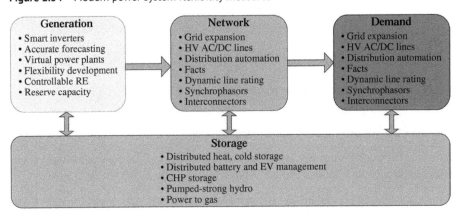

Figure 2.35 Overview of technical solutions for renewables integration into SG [63].

As a result of a shortage of inexhaustible resources, and environmental issues due to emissions, conventional energy production predicated upon fossil fuels are considered to be unsustainable in the long term. Therefore, numerous works are invested globally for integration of more renewable energies. RES are advanced options for electrical power generation and their chance to meet the world's energy need. This chapter presents an up-to-date and detailed current status and future projection of major renewable energy resources, and their benefits, growth, investment and implementation. To sum up, the deployment of renewable energy is one of the essential elements for a better reliability and resilience of the actual power grid. Moreover, the implementation of RES into SG system will aid in meeting the electrical power demands effectively.

Table 2.3 Advantages and disadvantages of different renewable energy resources. Ref Num [9]. Reproduced with permission from ELSEVIER

Energy source	Advantages	Disadvantages
Bioenergy Energy	• Abundant and renewable • Can be used to burn waste products	• Burning biomass can result in air pollution • May not be cost effective
Geothermal Energy	• Provides an unlimited supply of energy • Produces no air or water pollution	• Start-up/development costs can be expensive • Maintenance costs, due to corrosion, can be a problem
Hydropower Energy	• Abundant, clean, and safe • Easily stored in reservoirs • Relatively inexpensive way to produce electricity • Offers recreational benefits like boating, fishing, etc.	• Can cause the flooding of surrounding communities and landscapes • Dams have major ecological impacts on local hydrology. Can have a significant environmental impact • Can be used only where there is a water supply • Best sites for dams have already been developed
Marine Energy	• Ideal for an island country • Captures energy that would otherwise not be collected	• Construction can be costly • Opposed by some environmental groups as having a negative impact on wildlife • Takes up lots of space and difficult for shipping to move around
Solar Energy	• Potentially infinite energy supply • Causes no air or water pollution	• May not be cost effective • Storage and backup are necessary • Reliability depends on availability of sunlight
Wind Energy	• Is a free source of energy • Produces no water or air pollution • Wind farms are relatively inexpensive to build • Land around wind farms can have other uses	• Requires constant and significant amounts of wind • Wind farms require significant amounts of land • Can have a significant visual impact on landscapes • Need better ways to store energy

Table 2.4 Some negative environmental impacts of different renewable energy resources. Ref Num [9]. Reproduced with permission from ELSEVIER

Energy source	Potential negative impacts on the environment
Bioenergy Energy	May not be CO_2 natural, may release global warming gases like methane during the production of biofuels, landscape change, deterioration of soil productivity, hazardous waste.
Geothermal Energy	Subsidence, landscape change, polluting waterways, air emissions
Hydropower	Change in local eco-systems, change in weather conditions, social and cultural impacts
Marine Energy	Landscape change, reduction in water motion or circulation, killing of fish by blades, changes in sea eco-system
Solar Energy	Soil erosion, landscape change, hazardous waste
Wind Energy	Noises in the area, landscape change, soil erosion, killing of birds by blades

References

1 "BP Statistical Review of World Energy", June 2017, https://www.connaissancedesenergies.org/sites/default/files/pdf-actualites/bp-statistical-review-of-world-energy-2017-full-report.pdf (accessed on 30 June 2017).

2 Exxon Mobil (2017). 2017 Outlook for Energy: A View to 2040. http://oilproduction.net/files/2017_Outlook_for_Energy_highlights.pdf (accessed 30 June 2017).

3 BP (2017). BP Energy outlook, 2017 edition. https://safety4sea.com/wp-content/uploads/2017/01/BP-Energy-Outlook-2017_01.pdf (accessed 1 February 2021).

4 EIA (2017). Annual Energy Outlook 2017 with projections to 2050. http://www.eia.gov/aeo (accessed 1 July 2017).

5 Tromly, K. (2001). *Renewable Energy: An Overview*. Golden, CO (US): National Renewable Energy Laboratory.

6 Brown, A., Müller, S., and Dobrotková, Z. (2011). Renewable Energy: Markets and Prospects by Technology. International Energy Agency Information Paper.

7 Bull, S.R. (August 2001). Renewable energy today and Tomorrow. *Proceedings of the IEEE* 89 (8): 1216–1226.

8 European Renewable Energy Council (2010). RE-thinking 2050. http://citeseerx.ist.psu.edu/viewdoc/download?doi=10.1.1.638.5765&rep=rep1&type=pdf (accessed 10 February 2021).

9 Ellabban, O., Abu-Rub, H., and Blaabjerg, F. (2014). Renewable energy resources: current status, future prospects and their enabling technology. *Renewable and Sustainable Energy Reviews* 39: 748–764.

10 "Deploying Renewables 2011", https://webstore.iea.org/deploying-renewables-2011-best-and-future-policy-practice (accessed 1 July 2017).

11 IRENA (2017). RE thinking Energy. www.irena.org (accessed 1 July 2017).

12 REN21 (2020). Renewables 2020 Global Status Report. https://www.ren21.net/wp-content/uploads/2019/05/gsr_2020_full_report_en.pdf (accessed 1 July 2020).

13 Srirangan, K., Akawi, L., Moo-Young, M., and Chou, C.P. (December 2012). Towards sustainable production of clean energy carriers from biomass resources. *Applied Energy* 100: 172–186.

14 Sriram, N. and Shahidehpour, M. (12-16 June 2005). Renewable biomass energy. *IEEE Power Engineering Society General Meeting* 1: 612–617.

15 Reddy, B.Y. and Srinivas, T. (2013). Biomass based Energy Systems to Meet the Growing Energy Demand with Reduced Global Warming: Role of Energy and Exergy Analyses. *International Conference on Energy Efficient Technologies for Sustainability (ICEETS)*, Nagercoil, India (10–12 April 2013). IEEE.

16 Hall, D.O. and Scrase, J.I. (1998). Will biomass be the environmentally friendly fuel of the future? *Biomass and Bioenergy* 15: 357–367.

17 Klass, D.L. (2004). Biomass for renewable energy and fuels. In: *Encyclopedia of Energy*, vol. 1 (ed. C. Cleveland), 193–212. Elsevier.

18 Ruiz, J.A., Juárez, M.C., Morales, M.P. et al. (2013). Biomass gasification for electricity generation: review of current technology barriers. *Renewable and Sustainable Energy Reviews* 18: 174–183.

19 IRENA (2015). Renewable Power Generation Costs in 2014. https://www.irena.org/publications/2015/Jan/Renewable-Power-Generation-Costs-in-2014 (accessed 10 February 2021).

20 IRENA (2012). Renewable Energy Cost Analysis - Biomass for Power Generation. https://www.irena.org/publications/2012/Jun/Renewable-Energy-Cost-Analysis---Biomass-for-Power-Generation (accessed 10 February 2021).

21 Hammons, T.J. (2003). Geothermal power generation worldwide. *IEEE Bologna Power Tech Conference Proceedings*, Bologna, Italiy (23–26 June 2003). IEEE.

22 Fridleifsson, I.B. (2001). Geothermal energy for the benefit of the people. *Renewable and Sustainable Energy Reviews* 5: 299–312.

23 Yan, Q., Wang, A., Wang, G. et al. (2010). Resource Evaluation of Global Geothermal Energy and the Development Obstacles. *2010 International Conference on Advances in Energy Engineering*, Beijing, China (19–20 June 2010). IEEE.

24 Sheth, S. and Shahidehpour, M. (June 2004). Geothermal energy in power systems. *IEEE Power Engineering Society General Meeting* 2: 1972–1977.

25 Geothermal: International Market Overview Report, May 2012, https://repit.files.wordpress.com/2012/12/2012-gea_international_overview.pdf.

26 Matek, B. (March 2016). *Annual U.S. & Global Geothermal Power Production Report*. Geothermal Energy Association.

27 Bertani, R. (2015). Geothermal Power Generation in the World 2010–2014 Update Report. *Proceedings World Geothermal Congress 2015*, Melbourne, Australia (19–24 April 2015). IGA.

28 Năstase, G., Șerban, A., Năstase, A.F. et al. (December 2017). Hydropower development in Romania. A review from its beginnings to the present. *Renewable and Sustainable Energy Reviews* 80: 297–312.

29 Cambridge University Press (2012). Renewable Energy Sources and Climate Change Mitigation. https://www.cambridge.org/gb/academic/subjects/earth-and-environmental-science/climatology-and-climate-change/renewable-energy-sources-and-climate-change-mitigation-special-report-intergovernmental-panel-climate-change?format=PB (accessed 10 February 2021).

30 "Hydropower and the Environment: Present Context and Guidelines for Future Action", May 2000, IEA Hydropower Agreement, http://www.ieahydro.org/reports/HyA3S5V2.pdf.

31 Berga, L. (2016). The role of hydropower in climate change mitigation and adaptation: a review. *Engineering* 2 (3): 313–318.

32 World energy council (2016). World Energy Resources 2016. https://www.worldenergy.org/publications/entry/world-energy-resources-2016 (accessed 10 February 2021).

33 "Ocean Energy Technology: Overview", July 2009, Federal Energy Management Program. http://large.stanford.edu/courses/2013/ph240/lim2/docs/44200.pdf (accessed 3 July 2017).

34 Magagna, D. and Uihlein, A. (2015). Ocean energy development in Europe: current status and future perspectives. *International Journal of Marine Energy* 11: 84–104.

35 Uihlein, A. and Magagna, D. (2016). Wave and tidal current energy – a review of the current state of research beyond technology. *Renewable and Sustainable Energy Reviews* 58: 1070–1108.

36 Khan, N., Kalair, A., Abas, N., and Haider, A. (May 2017). Review of ocean tidal, wave and thermal energy technologies. *Renewable and Sustainable Energy Reviews* 72: 590–604.

37 Segura, E., Morales, R., Somolinos, J.A., and López, A. (September 2017). Techno-economic challenges of tidal energy conversion systems: current status and trends. *Renewable and Sustainable Energy Reviews* 77: 536–550.

38 ARENA (2016). Ocean energies, moving towards competitiveness: a market overview. https://arena.gov.au/knowledge-bank/ocean-energy-report-by-seanergy-and-ernst-and-young/ (accessed 10 February 2021).

39 Ocean Energy Forum (2016). Ocean Energy Strategic Roadmap 2016: building ocean energy for Europe. https://www.wavehub.co.uk/downloads/OceanEnergyForum_Roadmap_Online_Version_08Nov2016.pdf (accessed 10 February 2021).

40 IRENA (2013). Solar Photovoltaics: Technology Brief. https://www.irena.org/-/media/Files/IRENA/Agency/Publication/2015/IRENA-ETSAP-Tech-Brief-E11-Solar-PV.ashx?la=en&hash=229A7B44B1FD2456A671462327F089C5AA847689 (accessed 10 February 2021).

41 Pandey, A.K., Tyagi, V.V., Selvaraj, J.A./.L. et al. (January 2016). Recent advances in solar photovoltaic systems for emerging trends and advanced applications. *Renewable and Sustainable Energy Reviews* 53: 859–884.

42 International Energy Agency (2014). Technology Roadmap: Solar Photovoltaic Energy. https://www.iea.org/reports/technology-roadmap-solar-photovoltaic-energy-2014 (accessed 10 February 2021).

43 Brankera, K., Pathaka, M.J.M., and Pearcea, J.M. (2011). A review of solar photovoltaic levelized cost of electricity. *Renewable and Sustainable Energy Reviews* 15: 4470–4482.

44 Sheikh, N. and Kocaoglu, D.F. (2011). A Comprehensive Assessment of Solar Photovoltaic Technologies: Literature Review. *Technology Management in the Energy Smart World, PICMET*, Portland, USA (31 July–4 August 2011). IEEE.

45 EPIA (2012). Global Market Outlook for Photovoltaics Until 2016. http://large.stanford.edu/courses/2012/ph240/vidaurre1/docs/masson.pdf (accessed 10 February 2021).

46 Machinda, G.T., Chowdhury, S.P., Chowdhury, S. et al. (Dec. 2011). Concentrating solar thermal power technologies: a review. *Annual IEEE India Conference (INDICON)* 16-18: 1–6.

47 IEA-ETSAP and IRENA (2013). Concentrating Solar Power Technology Brief. https://www.irena.org/publications/2013/Jan/Concentrated-Solar-Power (accessed 10 February 2021).

48 IRENA (2012). Renewable Energy Technologies: Cost Analysis Series; Concentrating Solar Power. https://irena.org/publications/2012/Jun/Renewable-Energy-Cost-Analysis---Concentrating-Solar-Power (accessed 10 February 2021).

49 Sioshansi, R. and Denholm, P. (October 2010). The value of concentrating solar power and thermal energy storage. *IEEE Transactions on Sustainable Energy* 1: 173–183.

50 IEA-ETSAP and IRENA (2015). Solar Heating and Cooling for Residential Applications: Technology Brief. https://www.irena.org/publications/2015/Jan/Solar-Heating-and-Cooling-for-Residential-Applications (accessed 10 February 2021).

51 Kaygusuz, K. (2009). Wind power for a clean and sustainable energy future. *Energy Sources, Part B: Economics, Planning, and Policy* 4 (1): 122–132.

52 IEA-ETSAP and IRENA (2016). Wind Energy: Technology Review. https://www.irena.org/publications/2016/Mar/Wind-Power (accessed 10 February 2021).

53 Islam, M.R., Mekhilef, S., and Saidur, R. (2013). Progress and recent trends of wind energy technology. *Renewable and Sustainable Energy Reviews* 21: 456–468.

54 Blaabjerg, F. and Ionel, D.M. (2015). Renewable energy devices and systems – state-of-the-art Technology, Research and Development, challenges and future trends. *Electric Power Components & Systems* 43 (12): 1319–1328.

55 Global Wind Energy Council (2016). Global Wind 2016 Report. https://gwec.net/?s=global+wind+report+2016 (accessed 10 February 2021).

56 EIA (2016). International Energy Outlook 2016 With Projections to 2040. https://www.osti.gov/biblio/1296780-international-energy-outlook-projections (accessed 10 February 2021).

57 Frankfurt School-UNEP Centre/BNEF (2020). Global Trends in Renewable Energy Investment 2020. https://wedocs.unep.org/bitstream/handle/20.500.11822/32700/GTR20.pdf?sequence=1&isAllowed=y (accessed 10 Februaru 2021).

58 IRENA (2016). Roadmap for A Renewable Energy Future (2016 edition). https://irena.org/publications/2016/Mar/REmap-Roadmap-for-A-Renewable-Energy-Future-2016-Edition (accessed 10 February 2021).

59 IRENA (2019). Renewable Energy and Jobs – Annual Review 2019. https://www.irena.org/publications/2019/Jun/Renewable-Energy-and-Jobs-Annual-Review-2019 (accessed 10 February 2021).

60 IEA (2011). Deploying Renewables: Best and Future Policy Practice. https://webstore.iea.org/deploying-renewables-2011-best-and-future-policy-practice (accessed 10 February 2021).

61 Doner, J. (2007). Barriers to Adoption of Renewable Energy Technology. Institute for Regulatory Policy Studies. Illinois State University.

62 Wiuff, A., Sandholt, K., and Marcus-Møller, C. (2006). Renewable Energy Technology Deployment: Barriers, Challenges and Opportunities. http://iea-retd.org/archives/publications/barriers-challenges-and-opportunities (accessed 10 February 2021).

63 IRENA (2015). The age of renewable power: designing national roadmaps for a successful transformation. https://www.irena.org/publications/2015/Oct/The-Age-of-Renewable-Power-Designing-national-roadmaps-for-a-successful-transformation (accessed 10 February 2021).

64 Verzijlbergh, R.A., De Vries, L.J., Dijkema, G.P.J., and Herder, P.M. (August 2017). Institutional challenges caused by the integration of renewable energy sources in the European electricity sector. *Renewable and Sustainable Energy Reviews* 75: 660–667.

3

Power Electronics Converters for Distributed Generation

Qianwen Xu, Pengfeng Lin and Frede Blaabjerg

In this chapter, power electronics technology for distributed generation integrated into smart grid (SG) is reviewed. First, an introduction to typical distributed generation systems with the power electronics is presented. Next, power electronic converters in grid-connected AC systems and their control technologies are introduced. Then power electronics enabled autonomous AC power systems are discussed with the coordination and power management schemes. Then, autonomous DC power systems are illustrated. Finally, conclusions are drawn with future works.

3.1 An Overview of Distributed Generation Systems with Power Electronics

The world electricity demand has increased continuously in recent years. Figure 3.1 shows the world net electricity generation by different energy sources from 2012 to 2040 [1]. As can be observed, the electricity generation will increase by an estimated value of 69% from 2012 to 2040 with an average increase of 1.9% and renewable energy sources (RESs) are the fastest-growing sources with an average annual growth of 2.9%. However, the world electricity is dominantly generated by fossil fuels, resulting in CO_2 emissions and global warming. Due to environmental concerns and sustainable requirements, there is great interest in the utilization of the RESs (such as solar PV, wind, fuel cell, etc.) and energy storage systems (ESSs) (e.g. batteries, supercapacitors, etc.) around the world [2]. Power electronics, being the technology of efficiently converting electrical power, is an enabling technology for large utilization of RESs [3, 4]. It has changed rapidly over the past 30 years due to the development of semiconductor devices and microprocessor technology [5]. This section presents an overview of distributed generation systems including PV, wind energy and different ESSs using power electronics, which is the key technology for grid integration.

3.1.1 Photovoltaic Technology

Photovoltaics (PV) is the conversion of light into electricity with semiconducting materials using PV effects. Solar PV is a sustainable and RES as it is omnipresent, freely available, environmentally friendly and has low operational costs once installed. The improvement of PV conversion efficiency, advancement in manufacturing technology and reduced cost of PV

Smart Grid and Enabling Technologies, First Edition. Shady S. Refaat, Omar Ellabban, Sertac Bayhan, Haitham Abu-Rub, Frede Blaabjerg, and Miroslav M. Begovic.
© 2021 John Wiley & Sons Ltd. Published 2021 by John Wiley & Sons Ltd.
Companion website: www.wiley.com/go/ellabban/smartgrid

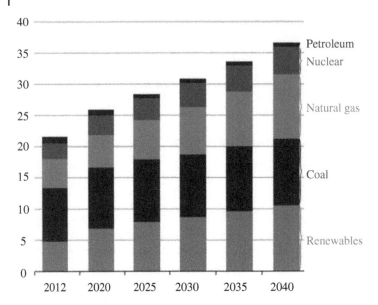

Figure 3.1 The world net electricity generation from 2012 to 2040 (trillion kilowatt-hours). Adapted from Ref num [1].

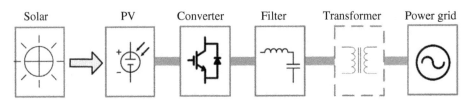

Figure 3.2 A general grid connected PV power system.

modules, are the main driving forces for the intensive utilization of PV power. By the end of 2018, world-installed PV capacity increased to 510 GW, with a growth of 27% from 2017. This means PV power production is sufficient to supply 3% of electricity demand worldwide.

PV can either be grid-connected or isolated. Figure 3.2 shows a general grid-connected PV power system [6]. The power electronic converter is a key element for the efficient and reliable utilization of PV energy. It can be a single-stage converter (e.g. DC-AC converter) [7] or a two-stage converter (e.g. DC-DC and DC-AC converter) [5] with or without galvanic isolation. Output filters are utilized in the PV system to reduce the injection of harmonics.

In general, there are three configurations for PV power systems, i.e. module, string and central inverters [6]. For the module topology, each PV module has a dedicated inverter to interface with a common AC bus (typically the grid). This is flexible for system expansion with a plug-and-play (PnP) feature. For string topology, PV panels are connected in series to form a string, and the string is connected to the grid through an inverter. The multi-string inverter topology is a further development of string topology with several strings interfaced with DC/DC converter to a common DC converter. The string topology is widely used for residential applications with lower power ratings (several kWs up to several tens of kWs). For central inverter topology, PV strings are connected in parallel to a common DC bus, and a central inverter is connected as the interface. The central topology is generally adopted for the residential or utility-scale applications of tens of kWs to several hundred kWs due to its simple and cheap construction. Furthermore, multilevel power converters can also be applied for a high power and high voltage PV power system [8].

The output characteristics of the PV module are nonlinear and change with solar radiation and the temperature. Due to the variability and non-dispatchability of PV power, the control of power electronic interface converter is significant. The control objectives of a PV power system can be divided into three levels [4]. The basic requirement is the current/voltage control and grid synchronization to ensure stable and reliable operation. Considering specifics of the PV system, more functions are required, including maximum power point tracking (MPPT), anti-islanding protection and active power limiting. Extensive MPPT algorithms are explored in the literature, as reviewed and compared in [9]: hill climbing and perturb and observe, curve fitting, lookup table, ripple correction control, incremental conductance, sliding mode control, neuro network based, particle swarm optimization based, etc. For further utilization of PV, many advanced control objectives are explored with extensive research work in this area, for example, weather forecasting, power oscillation damping, protection, grid supporting, PV panel monitoring, power quality, etc. [10].

3.1.2 Wind Power Technology

Wind power uses the kinetic energy of air to drive wind turbines to generate electricity. With the advancements in power electronics, aerodynamics, and mechanical drive train designs, wind-turbine technology has experienced a dramatic transformation, which makes wind power become a generally accepted utility generation technology. Started with a few tens of kilowatt power in the 1980s, now more and more large wind turbines with multimegawatt power level are installed widely [11]. The global wind power capacity reached 591 GW in 2018, and the annual wind energy production accounts for 4.4% of global electricity power usage in 2017.

The main components of a wind turbine system include a turbine rotor, a gearbox, a generator, power electronic system and a transformer for grid connection, as shown in Figure 3.3 [12]. For the generator, the two most common types are induction and synchronous generators. Power electronics play essential roles in wind turbine power systems, which include controlling the wind turbine generator and decoupling the generator from the electrical grid, maximizing the power extraction and balancing the power flow, improving dynamics and steady-state performance, etc. With the rapid growth in the capacity of individual wind turbines, the roles of power electronics become more critical.

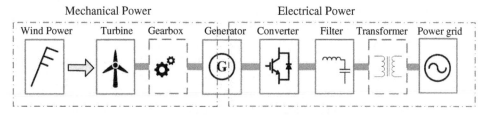

Figure 3.3   A general wind power system. Adapted from Ref Num [12].

In early wind turbine systems, fixed-speed wind turbines are used, consisting of a squirrel-cage induction generator connected to the grid via a transformer. In this case, a thyristor soft starter is connected to limit the inrush current. The advantages of fixed-speed wind turbines include the simple and cheap construction and reliability. But they have to operate in a constant speed and a stiff power grid is required for stable operation meaning that MPPT is difficult.

Variable-speed wind turbines can increase energy production and reduce power fluctuation into the grid; thus, they are widely used nowadays. In the variable-speed system, a power electronic system is necessary as the interface for the integration of the wind turbine generator. There are different topologies with different generators in the wind turbine systems. For a wounded rotor induction generator, the stator is connected to the grid directly and the rotor is connected to a power electronics-controlled resistor [13]. For a doubly fed induction generator (DFIG), the stator is connected to the grid and the rotor is connected to the grid through slip rings with a power electronic converter in between [14]. For the above two topologies, the partial-scale power electronics are adopted and the DFIG have dominated the wind power market over the past decades. For squirrel-cage induction generators and synchronous generators, they are connected to the grid through a full rated power electronic system where the total power generation is fed through the power electronic system [15]. The topology with full-scale power electronics dominates the current wind power market due to the high controllability and power processing flexibility, but the DFIG has also a large market share.

For control of a wind turbine system, both the mechanical and electrical parts should be involved. Generally, the control includes three levels. The basic control requirements are to manage the power flow with controlled current/voltage and grid synchronization. Moreover, considering specifics of wind power systems, more functions should be achieved, like MPPT, power limiting, fault ride through and grip support [4]. There are several methods for MPPT control, including tip speed ratio (TSR) control, power signal feedback (PSF) control, hill climbing searching (HCS) control. With the great efforts of numerous researchers in this field, other advanced control features have been explored, including inertia emulation, power quality, protection, black start, power oscillation damping, etc. [16].

3.1.3 Energy Storage Systems

The intermittent renewable resources (e.g. PV and wind turbine) bring the issues of system stability, reliability and power quality. To address these issues, ESSs are widely applied for renewable energy systems like PV, wind turbine, transportation systems [17] and other power system applications.

According to different energy storage technologies, ESSs can be divided into five types: electrochemical (e.g. batteries), electromagnetic (e.g. supercapacitor), mechanical (e.g. pumped hydro, compressed air, flywheel), chemical (e.g. hydrogen) and thermal (e.g. water tank) [9]. Different ESSs have various features in terms of power density, energy density, response rate, life cycle, cost, etc. [17]. These features should be taken into account for different application requirements. Table 3.1 presents some typical ESSs, which can also be practically divided into two categories: capacity-oriented and access-oriented ESSs [18]. Capacity-oriented ESSs have high energy density but low power density. Such ESSs are

Table 3.1 Characteristics of various energy storage systems.

Type	Energy density (Wh/Kg)	Power density (W/Kg)	Life(cycles or years)	Response time(s)
Li-on battery	100–200	360	500–10000	< 1/4cycles
CAES	10–30	—	20–40 years	360
Pumped hydro	0.3	—	20–60 years	10
SC	2–5	800–2000	10 years	0.01
Flywheel	5–30	>2000	10 years	0.1
SMES	10–75	—	>100000	0.01

usually applied for long-term electric energy balancing which helps compensate the low-frequency fluctuations caused by RESs. Examples of capacity oriented ESSs are batteries, compressed air energy storage (CAES) and pumped hydroelectric power systems. Access-oriented ESSs have high power density, fast dynamic response but low energy density. Therefore, they are responsible for high frequency components of the imbalanced power during transients. Typical access oriented ESSs include supercapacitors (SC), flywheels and superconducting magnetic energy storage (SMES), etc.

As can be observed, capacity-oriented ESSs and access-oriented ESSs have complementary characteristics and thus hybridization of the two types can provide fast dynamic performance, high power supply and large energy supply at the same time. There are growing interests in the utilization of hybrid energy storage systems (HESSs) and extensive research works have been conducted in HESSs, including the topology, sizing and power management strategies [19, 20].

Power converters are the key technology for the utilization of ESSs, for power flow control, voltage regulation and ensuring that the grid codes and standards are met when supplying grid services [3]. The primary requirement is to manage power flow in a bidirectional way to control the charging and discharging with high efficiency. To interface ESSs like batteries, SCs and regenerative FCs between two different DC voltage levels, the most widely used topology is the bidirectional boost converter. For high voltage ratio and high-power requirement, the interleaved boost converter is an attractive solution [21]. If isolation is required, isolated boost converters can be considered similar to the dual active bridge converter and its variants [22]. To interface ESSs like batteries, SCs and regenerative FCs directly to the grid or standalone AC load, voltage source inverters are widely used. For high power applications, multilevel topologies can be utilized [23].

3.2 Power Electronics for Grid-Connected AC Smart Grid

For the grid integration of DG units, voltage-source converters (VSC) are widely used as the interface converters, to guarantee a stable output, high efficiency, grid synchronization, high power quality and low cost [24]. In the medium- and high-voltage applications, the multilevel converters can be adopted and among them, the modular multilevel converter

(MMC) attracts much attention. In this section, control strategies of VSC will be presented, followed by an overview of multilevel converters and MMCs.

3.2.1 Voltage-Source Converters

DG interface converters can operate in current control mode to inject required power/current into the grid (i.e., grid feeding), or they can operate in voltage control mode to regulate the desired frequency and voltage amplitude (i.e. grid forming). To control the VSCs, two classical techniques are widely used, one is the synchronous reference frame control with PI controllers and the other is the stationary reference frame control with PR controllers.

3.2.1.1 Synchronous Reference Frame

Because three phase variables are time varying variables, a common approach is to use Park transformation to transform them into dq reference frame which rotates synchronously with the grid voltage. Then, the three phase AC variables will be DC and the simple proportional integral (PI) controller can be applied to regulate the dq components.

Figure 3.4a shows the grid feeding VSC under the synchronous reference frame. Current PI controllers are implemented to regulate the dq currents at their nominal values in dq reference frame. Due to the coupling of the circuit after transformation, decoupling terms are added back to get the single input single output of the PI control. The current PI controller is expressed as

$$G_{c1}(s) = k_{pc} + \frac{k_{ic}}{s} \tag{3.1}$$

where k_{pc} and k_{ic} are the proportional and integral gains of the current PI controller.

Figure 3.4b shows the grid forming VSC under the synchronous reference frame, including the outer voltage PI controllers and inner current PI controllers. The voltage PI controller is given by

$$G_{v1}(s) = k_{pv} + \frac{k_{iv}}{s} \tag{3.2}$$

where k_{pv} and k_{iv} are the proportional and integral gains of the voltage PI controller. The control bandwidth of the voltage loop is limited to less than 1/10 of the current loop to minimize the interactions between the current loop and voltage loop. The control bandwidth of the current loop should be designed as 1/20–1/10 of the switching frequency in order to get good dynamic response.

One disadvantage of synchronous frame PI control is that its compensation capability of low-order harmonics is very poor. The detailed design of the synchronous frame PI controllers can be referred to [25].

3.2.1.2 Stationary Reference Frame

The three-phase variables in the abc reference frame can be transformed to the stationary αβ reference frame through Clarke transformation. Then, the three-phase system can be modeled in two independent single-phase systems in αβ frame. As the PI controller cannot

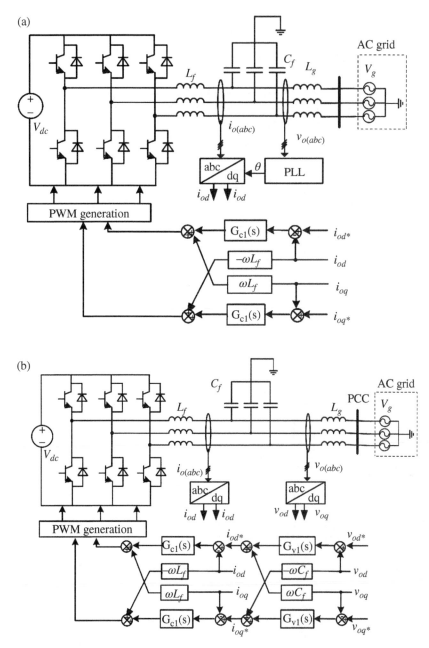

Figure 3.4 Voltage source converter with synchronous reference frame control (a) grid feeding; (b) grid forming.

remove the steady-state error when controlling sinusoidal waveforms, proportional resonant (PR) controllers have attracted much attention.

The grid feeding VSC under stationary reference frame is shown in Figure 3.5a. The current PR controller is implemented to regulate the real-time currents at their

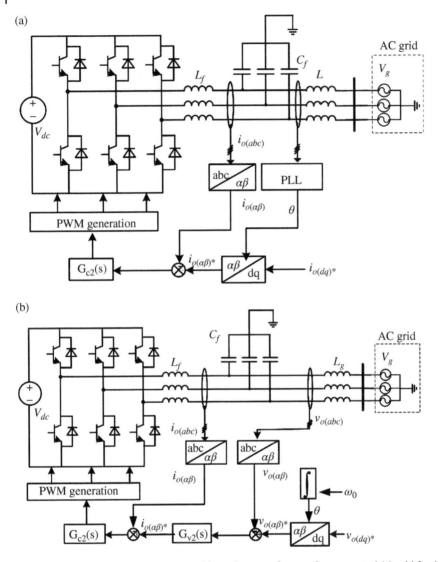

Figure 3.5 Voltage source converter with stationary reference frame control (a) grid feeding; (b) grid forming.

nominal values in αβ reference frame to inject the desired currents. Transfer function of the current PR controller is expressed as

$$G_{c2}(s) = k_{pc} + \frac{k_{rc}s}{s^2 + \omega_0^2} \tag{3.3}$$

where k_{pc} and k_{rc} are the proportional and resonant gains, ω_0 is the system fundamental frequency.

The grid forming VSC under stationary reference frame is shown in Figure 3.5b, including the outer voltage PR controllers and inner current PR (or P) controllers. The outer voltage controllers generate current reference values for the inner current loops. The current controllers are to improve control dynamics in this case, and they can be PR controllers or P controllers. The voltage PR controller is expressed as

$$G_{v2}(s) = k_{pv} + \frac{k_{rv}s}{s^2 + \omega_0^2}$$ (3.4)

Where k_{pv} and k_{rv} are the proportional and resonant gains, ω_0 is the system fundamental frequency.

As the stationary reference PR controller achieves a very high gain around the resonance frequency, it is able to eliminate the steady-state error. The detailed design of the PR controllers can be referred to [26, 27].

3.2.1.3 Grid Synchronization

It is widely recognized that the phase angle of the grid voltage is critical for the control performance of the grid-connected converters to achieve grid synchronization, as is also shown in Figures 3.4 and 3.5. Phase-locked loop (PLL) is the most popular grid synchronization technique and has been adopted for a few decades. PLL has wide applications, like the estimation of fundamental parameters (phase, frequency and amplitude), measurement of harmonics, control of machines, islanding detection, fault detection, measurement of power quality indices, designing robust controllers, etc. Figure 3.6 shows a standard PLL in three-phase applications [28], i.e. the synchronous reference frame PLL (SRF-PLL), which is also the building block of almost all advance PLLs. It consists of Phase Detector, Low Pass Filter and Voltage Controlled Oscillator. To enhance disturbance rejection capability and improve dynamic behavior, many advanced PLLs are also proposed, e.g. moving average filter-based PLLs, notch filter-based PLLs, multiple SRF filtering-based PLLs, complex-coefficient-filter-based PLLs, second-order generalized integrator-based PLLs, delayed signal cancelation-based PLLs, type N and quasi-type N PLLs, etc. as discussed and compared in [28].

In recent years, the adverse impacts of PLLs on the stability of a power electronics-based power system has been reported [29]. To avoid the instability issues caused by PLLs, various direct power control techniques have been proposed to control the real and reactive power of the grid-connected converter directly without using a PLL, like lookup-table DPC, sliding mode DPC, passivity based DPC [30]. However, to the current stage,

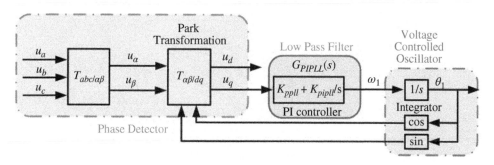

Figure 3.6 A standard PLL structure for grid synchronization.

they still have their respective limitations, and PLL based vector control techniques are still dominant in the market.

3.2.1.4 Virtual Synchronous Generator Operation

Virtual synchronous generator (VSG) technology has attracted much attention in recent years as a promising inertia support technique of the future electricity grid, to deal with the frequency stability issues caused by the high penetration of renewable generations in the smart grid. In the past, grid frequency is regulated by synchronous generators, which can provide inertia by absorbing or delivering the kinetic energy stored in their rotors and turbines [31]. However, with the high penetration of DG interface converters in the grid, which have very small or no inertia and damping properties, power system inertia is reduced and the frequency stability is becomes a considerable concern [32]. The VSG technology has been proposed by mimicking the essential characteristics of the synchronous generators so that the rotating inertia can be emulated in power electronics interface converters [32, 33]. There are a number of implementations of VSGs in literature, for example, virtual synchronous machine [34], VSG [35, 36], synchronverter [37], etc. The inertia characteristics emulated by VSGs can contribute to the total inertia of the grid and enhance the transient frequency stability.

There are several different implementations of the VSG. Figure 3.7 shows a relatively common topology. An energy storage unit is applied with the VSG control technique to mimic the swing equation of the SGs in order to support the grid frequency fluctuation and provide inertia. The inverter operates in the voltage source mode with the inner voltage and current regulation, where PR controllers are in the stationary reference frame or PI

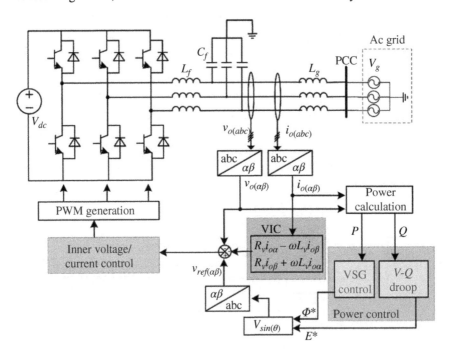

Figure 3.7 A grid connected virtual synchronous generator system.

controllers in the synchronous reference frame can be implemented. The virtual impedance control (VIC) is implemented to emulate an impedance connected in series with the grid line impedance between the inverter output and the PCC to make the equivalent line impedance inductive. The VSG control is implemented in the power control loop to provide virtual inertia, given by

$$P_{in} - P - D(\omega - \omega_g) = J\omega \frac{d(\omega - \omega_g)}{dt} \tag{3.5}$$

where P_{in} is the set-point value of active power, P is the measured output power of VSG, ω is the angular frequency of the virtual rotor, ω_0 is the angular frequency of the grid; J represents the moment of inertia of the virtual rotor.

Along with the increasing utilization of renewable sources and power electronics, the power system is gradually transitioning toward power electronics-dominated power systems. It is possible that VSGs will serve as inertia provision and frequency control in replacement of SGs in the future [32].

3.2.2 Multilevel Power Converters

The conventional two-level voltage source converters/inverters are the main building blocks for the integration of DG units into the smart grid. However, for medium- and high-voltage applications, the voltage ratings of power semiconductor devices are not sufficient to meet the requirement in the conventional two-level converters [38]. The multilevel converters, which use a series/parallel of power semiconductor switches with several low DC sources to perform power conversion by synthesizing a staircase voltage waveform, are good solutions for high power applications [39]. DGs such as PV, wind, FC and energy storage units can easily be interfaced by multilevel converters for high power applications.

Compared with two level converters, multilevel converters have advantages of improved output power quality, higher efficiency and reduced dv/dt rate that dramatically reduces device stress, common mode current and electromagnetic interference (EMI). The concept of multilevel converters has been introduced since 1970s. Then many multilevel converter topologies have been developed. Three classical multilevel converter structures are neutral-point clamped converter (NPC), flying capacitor converter (FC) and cascaded H-bridge converter (CHB), as shown in Figure 3.8 [39]. In 2003, a promising multilevel converter topology, namely the MMC, was introduced and has attract great attention [40]. The advantages of MMCs, such as modularity, scalability, ability to handle high power with relatively low voltage rating semiconductors, filter-less configuration, easy redundancy, capacitor energy distribution, make the MMC one of the most popular multilevel topologies in recent years [41]. The applications include the high-voltage direct current systems, medium voltage drives, integration of large-scale battery ESSs, PV and wind turbines, etc. [40]. However, the drawback of MMC cannot be ignored: due to the great number of semiconductor switches used, each switch requires a r gate drive circuit, making the overall system more expensive and complex.

Figure 3.9 shows a typical configuration of a three-phase MMC. There are three phase legs connected to a common DC bus. Each phase leg has two arms, upper arm and lower arm, consisting of series connected submodules and an arm inductor. The arm inductors

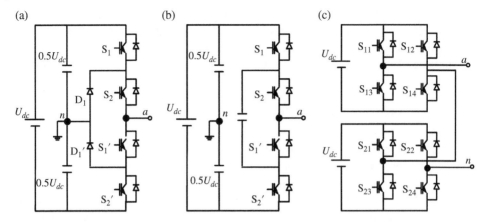

Figure 3.8 Classical multilevel converter topologies: (a) three-level neutral-point clamped converter (NPC); (b) three-level flying capacitor converter (FC); (c) five-level cascaded H-bridge converter (CHB). Adapted from Ref Num [39].

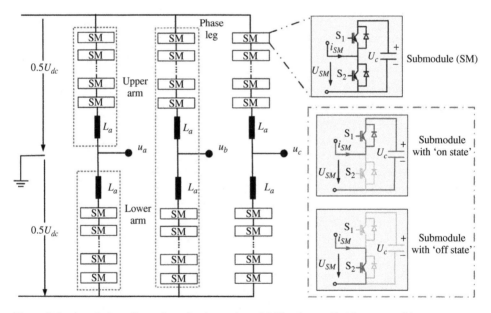

Figure 3.9 A typical configuration of a three-phase MMC to be applied in smart grid.

are to suppress circulating current and limit the fault current rising rate. The submodules (SMs) are identical and different topologies can be applied for SMs, e.g. half-bridge, full-bridge, neutral-point-clamped, flying capacitor, etc. [41]. Among them, half-bridge SM is the most commonly used SM topology and is depicted in Figure 3.9. A half-bridge SM consists of two IGBT switches (S_1 and S_2) and two anti-parallel diodes. DC capacitor voltage of the SM is defined U_c and the output voltage of a SM is defined as U_{sm}, by switching S_1 and S_2 in different patterns, the SM would have different states (on/off) and the output voltage will be different (U_c or 0). An MMC with N SMs in each arm could generate N + 1 level output voltage. The voltage value of a SM capacitor should be equal and DC bus voltage U_{dc} should be N times U_{sm}. To make the DC voltage constant, the total number of SM in operation in one phase (i.e. the sum of SMs with on state in the upper arm and lower arm) has to be equal to N.

The design of MMC, including the selection of the capacitor, inductor and the number of SMs, which is based on some performance indices like current ripple, short-circuit current, voltage ripple, reliability, voltage rating and power loss. The size of the inductor depends on the arm current ripple and short-circuit current limits. The SM capacitor is selected based on the voltage ripple requirement and cost/size. The number of SMs connected in each arm is determined by the common bus voltage and the voltage rating of semiconductor devices.

The control objectives of MMC include average capacitor voltage control, capacitor voltage balancing control, output current control and circulating current control, etc. Extensive research works have been conducted to achieve these objectives. Although many advanced control methods are developed for the control of MMC, including the model predictive control [42], feedback linearization [43], sliding mode control [44], etc. the classical cascaded control schemes are still mostly adopted [45]. Figure 3.10 shows the general block diagram of a classical control method for MMC [45]. The submodule voltage control includes the average capacitor voltage control and voltage balancing control. The average capacitor voltage control regulates the average capacitor voltage at a constant value of U_c by using a double-loop PI controller to generate a compensating signal to the upper and lower arm modulation signals. The voltage balancing can be achieved by a PI controller with the output added to the modulation signal of the corresponding SM [46] or after the modulation using the SM selection method [47]. To regulate the output current at a required AC output current, synchronous frame-based PI controller can be employed to generate a compensating signal. Circulating current is caused by the difference between capacitor voltages in the upper arm and lower arm of each phase leg, which increases the root mean square value of the arm current and results in an increased power loss and decreased system efficiency. To suppress the circulating current, various methods are applied, such as cascaded repetitive control+PI controller [48], resonant controller in stationary reference frame [49], etc. Then, the modulation signals can be generated by adding the above commands together. There are many modulation

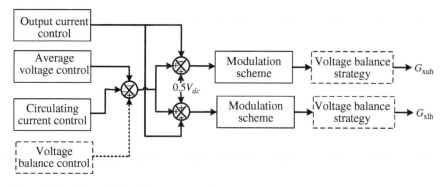

Figure 3.10 General block diagram of a classical control method for MMC. Adapted from Ref [45].

techniques that can be utilized for MMC, like phase-shifted carrier modulation, level-shifted carrier modulation, sampled average modulation, space vector modulation, staircase modulation, etc. as reviewed and compared in [41].

3.3 Power Electronics Enabled Autonomous AC Power Systems

Different from high-voltage AC systems which could be utilized in generation and transmissions networks, low voltage AC systems are more than likely suitable for distribution systems. With an increasing number of RESs penetrating into end-user applications and to accommodate existing AC power devices, AC microgrids can be locally formulated to balance generations and consumptions more flexibly. Moreover, by actively interacting with the utility grid, local AC microgrids would obtain power and inertia support from the utility so that the microgrid operating security can be improved. Figure 3.11 shows the system architecture of a typical AC microgrid. Both AC and DC power sources are interfaced with a common AC bus via power converters. A DC load is assigned with an AC/DC converter transforming the AC voltage to the DC that could meet the input voltage requirement by the DC load, whereas an AC load is directly coupled to the AC bus without any intermediary devices. The utility grid will dominate the AC bus voltage once the switch is on, otherwise the microgrid is working in islanding mode.

Notice that the correct operation of an AC microgrid necessitates fair controls in each source/load interfaced power converter and system level coordination between them. In what follows, converter level control methodologies and system level coordination will be carefully reviewed and explained.

3.3.1 Converter Level Controls in Microgrids

Section 3.2.2 presents the control of voltage source converters, in voltage control mode or current control mode and the voltage and current of the converter can be regulated by PI controllers in synchronous reference frame or PR controllers in stationary reference frame. They serve as the internal controllers to drive the converter system output tracking the given commands. Then the advanced control objectives, such as power sharing and impedance compensation, etc. can be reasonably accomplished.

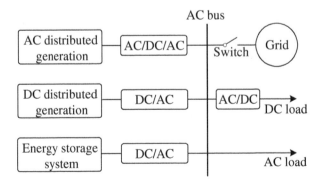

Figure 3.11 Typical structure of an AC microgrid having power electronic interfaces.

3.3.1.1 Master–slave Operation

In the case that the AC microgrid is connected to the utility grid, since AC bus voltage magnitude and frequency are dominated by the utility, all source-fed converters should be working in power control mode such that active and reactive power would either be injected into or absorbed from the main grid. Differently, in islanded mode, some of the source converters should be scheduled to regulate the AC bus, and these sources are playing master roles in the system. As for other converters, they operate as grid system followers which simply generate active and reactive powers according to setpoints.

It is critical to clarify that the multiple master converters are operated in droop mode to avoid possible control conflicts when regulating the AC bus voltage. For the masters, droop controllers construct voltage references to the inner controllers whose plants should be LC-type filters. The derivations of AC droop mechanisms will be provided in subsequent contexts. Regarding the slaves, power references are given to the controllers of L-type and LCL-type filters because the intention here is to force the currents over inductors to track specified values ensuring high-power qualities. Note that the power references can be either set as constants or determined by inverse-droop controllers so that the overall droop characteristics of the AC system is reserved and autonomous power sharing among slaves can also be attained.

3.3.1.2 *f-P* and *V-Q* Droops

Droop concepts are actually inherited from conventional power system wherein system frequency is naturally drooped in the case of synchronous generators are supplying more power. In power electronics-dominated systems, there are only power converters as well as their interfaced filters. Hence, it is hard to intuitively draw a parallel between synchronous generators and power converters to figure out the droop controllers tailored for microgrid systems. However, droop controllers in AC system can be derived by investigating the typical power flow distribution on an inductive transmission line. Illustrative description is shown in Figure 3.12 where $E{<}0$ stands for the amplitude and angle of AC bus voltage, $V{<}\delta$ denotes the voltage at the power converter terminal, and $Z{<}\theta$ represents the line impedance. Notice that the transmission line is purely inductive when θ equals 90°, whereas it is resistive when θ is zero.

In view of Figure 3.12, the active power P and reactive power Q can be calculated as follows,

$$P + jQ = V\angle\delta\left[\frac{V\angle\delta - E\angle 0}{Z\angle\theta}\right]^*$$

$$\Rightarrow \begin{cases} P = \dfrac{V^2}{Z}\cos\theta - \dfrac{EV}{Z}\cos(\delta + \theta) \\ Q = \dfrac{V^2}{Z}\sin\theta - \dfrac{EV}{Z}\sin(\delta + \theta) \end{cases} \tag{3.6}$$

Firstly, considering a pure inductive transmission line, the substation of $\theta = 90°$ into (4.6) gives

$$\begin{cases} P = \dfrac{EV}{Z}\sin\delta \\ Q = \dfrac{V^2}{Z} - \dfrac{EV}{Z}\cos\delta \end{cases} \tag{3.7}$$

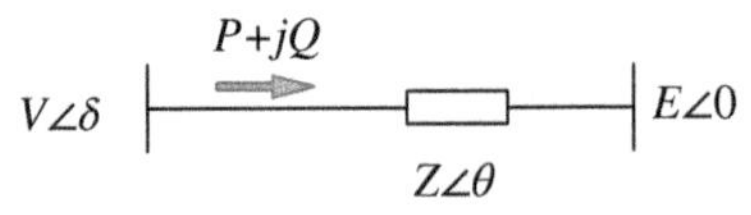

Figure 3.12 Power flow on a transmission line.

To show that active power and reactive power are respectively dominated by phase angle and voltage magnitude, performing partial derivatives of P and Q with respect to V and δ respectively results in,

$$
\begin{cases}
\dfrac{\partial P}{\partial \delta} = \dfrac{EV}{Z}\cos\delta, & \dfrac{\partial P}{\partial V} = \dfrac{EV}{Z}\sin\delta \\[2mm]
\dfrac{\partial Q}{\partial \delta} = \dfrac{EV}{Z}\sin\delta, & \dfrac{\partial Q}{\partial V} = \dfrac{2V}{Z} - \dfrac{E}{Z}\cos\delta
\end{cases}
\tag{3.8}
$$

By means of the assumption that δ is comparatively small and three approximations, i.e. $\cos\delta \approx 1$, $\sin\delta \approx 0$, $V \approx E$, the four partial derivatives in (3.8) can be simplified into

$$
\frac{\partial P}{\partial \delta} \approx \frac{EV}{Z}, \quad \frac{\partial P}{\partial V} \approx 0, \quad \frac{\partial Q}{\partial \delta} \approx 0, \quad \frac{\partial Q}{\partial V} \approx \frac{V}{Z}.
\tag{3.9}
$$

According to (3.9), it is apparent that $\dfrac{\partial P}{\partial \delta} \gg \dfrac{\partial P}{\partial V}$ and $\dfrac{\partial Q}{\partial V} \gg \dfrac{\partial Q}{\partial \delta}$, which means active power is much more sensitive to phase angle than voltage magnitude, and reactive power is much more sensitive to voltage magnitude than phase angle. Furthermore, noting that E, V, and Z are all positive values and the quantity EV/Z is hence positive. This fact indicates the partial derivative of P with respect to δ is positive. The increment in phase angle will linearly induce active power increment when the angle is near zero, which thereby founds the bases for controlling active power transfer by changing phase angle. Similar results can also be identified by investigating the last two expressions in (3.9) that the active power is linearly related to voltage magnitude and the former can be easily controlled by altering the latter.

However, by varying angle to regulate power flow is not always feasible if there is a lack of synchronous phase measuring system. Since an AC microgrid is a physical system, in steady state, there is only one prevailing frequency across the overall microgrid. Moreover, phase angle is actually the integration of system frequency f. Hence, its t is possible to relate f and P such that the phase angle δ sufficing the desired active power transmission can autonomously be formed during system transition. The f-P droop and V-Q droop for the ith converter can be written as follows,

$$
\begin{cases}
f_i = f_{rated} - m_i(P_i - P_{rated}) \\
V_i = V_{rated} - n_i(Q_i - Q_{rated})
\end{cases},
\tag{3.10}
$$

where f_{rated} and V_{rated} are the rated frequency and voltage amplitude that should be predefined in system planning stage. P_{rated} and Q_{rated} stand for the rated active and reactive powers. m_i and n_i are droop coefficients. The droop relations in (3.10) has been visualized in Figure 3.13. By consolidating f_i and V_i, reference voltage signals can be formulated and the signals will be sent

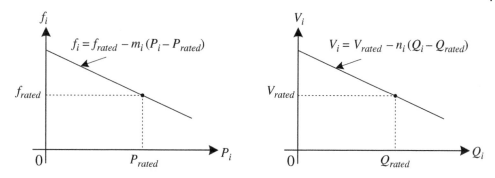

Figure 3.13 *f-P* and *V-Q* droops applied in power electronic based systems.

to internal controllers whose plant should be *LC*-type filters. When the AC microgrid is stably running, frequency is synchronized throughout the entire system. Proportional active power sharing among sources can thus be realized by properly designing m_i for each *f-P* droop, and as indicated by *V-Q* droop, reactive power sharing can also be accomplished if all source converters share the same terminal voltage.

3.3.1.3 *V-P* and *f-Q* Droops

The scenario would be different when it comes to resistive transmission lines (in low voltage distribution system). Revisiting (3.6), θ should now be updated as 0° and P, Q expressions are accordingly modified as

$$\begin{cases} P = \dfrac{V^2}{Z} - \dfrac{EV}{Z}\cos\delta \\ Q = \dfrac{EV}{Z}\sin\delta \end{cases}. \tag{3.11}$$

The power descriptions in (3.11) have the inverse format to that in (3.7). By referring to the reasoning in (3.8)–(3.9), it is easy to infer that, in the case of a resistive transmission line, active power is much more sensitive to voltage magnitude while reactive power more responsive to phase angle. Furthermore, by analogy with (4.9), active and reactive power deliveries are respectively in linear relation to voltage magnitude and phase angle. In this sense, droop controllers suitable for AC microgrids with resistive transmission lines can be recognized as

$$\begin{cases} f_i = f_{rated} - m_i(Q_i - Q_{rated}) \\ V_i = V_{rated} - n_i(P_i - P_{rated}) \end{cases}. \tag{3.12}$$

f-Q and *V-P* droops in (3.12) have been illustrated in Figure 3.14 wherein both f_i and V_i will flag with the increase of active and reactive power generations.

3.3.1.4 Virtual Impedance Enabled Control

In practical engineering, transmission lines may not be purely inductive or resistive, and the impedance angle θ in would reside in (0°, 90°). For this situation, it is not appropriate to

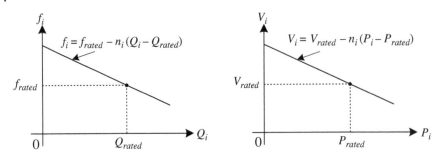

Figure 3.14 *f-Q* and *V-P* droops applied in power electronic based systems.

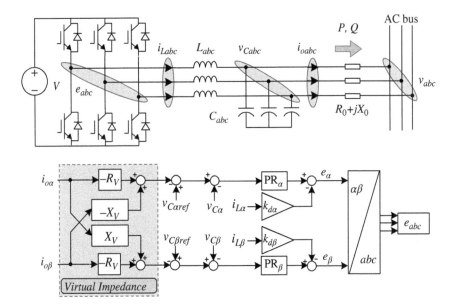

Figure 3.15 Schematic of virtual impedance in a voltage source inverter to be applied in smart grid.

utilize either droop controllers depicted in (3.10) or (3.12). To address this problem, virtual impedance techniques can be utilized to reshape the entire impedances of transmission lines at the designers' discretion. The illustrative schematic of a voltage source inverter has been plotted in Figure 3.15. $R_0 + jX_0$ is real-line impedance between the inverter and AC bus. Under $\alpha\beta$ coordinate frame, i_{Labc}, v_{Cabc}, i_{oabc}, and v_{abc} are represented as $i_{L\alpha\beta}$, $v_{C\alpha\beta}$, $i_{o\alpha\beta}$, and $v\alpha\beta$, respectively. $k_d\alpha$ and $k_d\beta$ are damping factors. According to Figure 3.15, the relation between $v_{C\alpha\beta}$ and $v\alpha\beta$ can be mathematically depicted as below,

$$v_\alpha + jv_\beta = (v_{C\alpha} + jv_{C\beta}) - (i_{o\alpha} + ji_{o\beta})(R_0 + jX_0) - (i_{o\alpha} + ji_{o\beta})(-R_V + jX_V). \tag{3.13}$$

where $-R_V + jX_V$ represents the virtual impedance. Obviously, in (3.13), the original line resistance can be offset when R_V is tuned equaling R_0. Then $v\alpha + jv\beta$ can be simplified into

$$v_\alpha + jv_\beta = (v_{C\alpha} + jv_{C\beta}) - (i_{o\alpha} + ji_{o\beta})j(X_0 + X_V), \tag{3.14}$$

which means that the line impedance between the inverter and bus has now been perfectly inductive no matter whether the rudimentary output impedance is resistive or complex. Then the conventional droops derived in (3.10) can still be used without any revision. On the other hand, if intended, it is also possible to employ X_V to neutralize X_0 such that the transmission line impedance is shaped to be purely resistive, and the droop controllers shown in (3.12) can thereby be implemented.

Notice that the virtual impedance technique shown in Figure 3.15 is presented under $\alpha\beta$ stationary frame. In fact, the virtual impedance in (3.13) has the same format as in the dq rotating frame. Relevant explanations are provided in the following. Expressing the complex expression in (3.13) into the matrix format gives,

$$\begin{bmatrix} v_\alpha \\ v_\beta \end{bmatrix} = \begin{bmatrix} v_{C\alpha} \\ v_{C\beta} \end{bmatrix} - B_0 \begin{bmatrix} i_{o\alpha} \\ i_{o\beta} \end{bmatrix} - B_V \begin{bmatrix} i_{o\alpha} \\ i_{o\beta} \end{bmatrix}, \tag{3.15}$$

where $B_0 = \begin{bmatrix} R_0 & -X_0 \\ X_0 & R_0 \end{bmatrix}$, $B_V = \begin{bmatrix} -R_V & -X_V \\ X_V & -R_V \end{bmatrix}$.

Applying rotating matrix, A^{-1} that converts signals under dq frame into the counterparts under $\alpha\beta$ frame to (3.15), then the following equation holds,

$$A^{-1} \begin{bmatrix} v_d \\ v_q \end{bmatrix} = A^{-1} \begin{bmatrix} v_{Cd} \\ v_{Cq} \end{bmatrix} - B_0 A^{-1} \begin{bmatrix} i_{od} \\ i_{oq} \end{bmatrix} - B_V A^{-1} \begin{bmatrix} i_{od} \\ i_{oq} \end{bmatrix}, \tag{3.16}$$

where $A^{-1} = \begin{bmatrix} \cos\phi & -\sin\phi \\ \sin\phi & \cos\phi \end{bmatrix}$. The above formulation can simply be manipulated into

$$\begin{bmatrix} v_d \\ v_q \end{bmatrix} = \begin{bmatrix} v_{Cd} \\ v_{Cq} \end{bmatrix} - AB_0 A^{-1} \begin{bmatrix} i_{od} \\ i_{oq} \end{bmatrix} - AB_V A^{-1} \begin{bmatrix} i_{od} \\ i_{oq} \end{bmatrix}. \tag{3.17}$$

For (3.17), with proper mathematical operations, it can be verified that $AB_0 A^{-1} = B_0$ and $AB_V A^{-1} = B_V$. As such, (3.17) evolves into

$$\begin{bmatrix} v_d \\ v_q \end{bmatrix} = \begin{bmatrix} v_{Cd} \\ v_{Cq} \end{bmatrix} - B_0 \begin{bmatrix} i_{od} \\ i_{oq} \end{bmatrix} - B_V \begin{bmatrix} i_{od} \\ i_{oq} \end{bmatrix}. \tag{3.18}$$

Comparing (3.15) and (3.18) suggests that the virtual impedance configuration is the same in both dq rotating frame and $\alpha\beta$ rotating frame, which largely facilitates its application in real engineering.

3.3.2 System Level Coordination Control

Upon having the building block of renewable source-based converter systems and loads in AC smart grids, various devices should be correctly cooperated to further achieve advanced operations, i.e. power quality enhancement, frequency/voltage restorations, economic

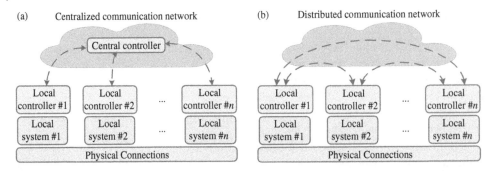

Figure 3.16 Grid structure control. (a) Centralized control scheme. (b) Distributed control scheme.

operations, etc. For these purposes, centralized and distributed control configurations are developed according to whether there exists a central controller in the system.

The two different control mechanisms can be seen in Figure 3.16. For centralized approaches, a central controller is used to collect data across the whole microgrid and based on the data, the controller will issue proper commands for each device to realize the desired system performances. However, as widely criticized, the central control method relies on intensive communications and it may suffer from single-point failures, which will stop all system services in the case of a part failure. Alternatively, the functions endowed by the central controller can also be achieved by distributed control scheme in which no central controller exists. The idea is to regard each local system as an autonomous agent and multiple agents are exchanging information with one another to reach consensus in steady state. In what follows, system frequency and voltage restorations will be shown, obtained by both centralized and distributed control schemes.

3.3.2.1 Centralized Control Scheme

In Figure 3.11, frequency and voltage amplitude at the AC bus will deviate from rated values since droop controllers are utilized. To restore frequency and improve voltage quality, the central controller can transmit control signals to compensate the original droops and these signals are represented below,

$$\begin{cases} \Delta f = k_{pf}(f_{rated} - f) + k_{if} \int (f_{rated} - f) \\ \Delta V = k_{pv}(V_{rated} - V) + k_{iv} \int (V_{rated} - V) \end{cases} \tag{3.19}$$

where k_{pf}, k_{if}, k_{pv} and k_{iv} are PI control parameters. Δf and ΔV are the compensation signals that should be superposed to (3.10) if an AC microgrid with inductive transmission lines is considered. In (3.19), due to the strong regulation of PI controllers, it can be inferred that system frequency f and bus voltage magnitude V in the AC main bus will be forced to be f_{rated} and V_{rated} respectively in steady state, which accomplishes f and V restorations.

3.3.2.2 Distributed Control Scheme

Different from the centralized control where the compensation signals are uniformly generated in the central controller, in the distributed control scheme, every source inverter is treated as an individual entity and acts as an agent node. A node can communicate with its nearby neighbors and finally, the compensation signals (Δf_i and ΔV_i) are produced locally. For the ith inverter, Δf_i and ΔV_i are produced as follows [50],

$$
\begin{cases}
\Delta f_i = \alpha_f \int \left(\sum_j (\Delta f_j - \Delta f_i) + (f_{rated} - f) \right) \\
\Delta V_i = \alpha_v \int \left(\sum_j (\Delta V_j - \Delta V_i) + (V_{rated} - V) \right)
\end{cases}
\tag{3.20}
$$

where Δf_j and ΔV_j are the compensation signals received from the jth node. α_f and α_v are control gains. f and V are the frequency and amplitude of AC bus voltage. In the first equation of (3.20), two types of errors are summed up and processed by an integrator. The errors will be regulated to zero in steady state which indicates that f is driven to track f_{rated}. Similar reasoning can also be applied to the second equation. After system transition, V will be fixed at V_{rated} and all voltage compensation signals are equalized.

3.4 Power Electronics Enabled Autonomous DC Power Systems

In recent decades, DC microgrids have been progressively developed with the increasing penetration of solar PV power and wide deployment of battery system which are naturally in DC. Together with the fact that most of the modern household appliances are also DC-compatible, DC systems need careful investigations from the perspective of both converter level control and system level coordination. Figure 3.17 shows a representative configuration of DC microgrids. There is one DC bus, which is linked to a utility grid via a DC/AC inverter. In the case of grid faults, the inverter will be disabled to evade unnecessary fault intrusions. DC loads are directly linked to the bus and DC loads can absorb power from the bus by a DC/AC inverter. DC sources and energy storages are equipped with DC/DC converters to improve their controllability and flexibility. AC sources supply the power to the DC through an AC/DC rectifier.

In general, DC microgrids have the following advantages: (i) There are no power quality problems induced by reactive power and frequency regulations; (ii) No harmonics exist in

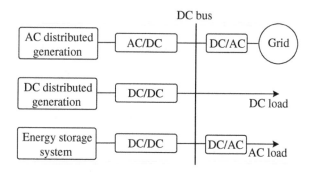

Figure 3.17 Typical structure of a DC microgrid connected to a utility grid and with power electronic interfaces.

the system; (iii) No inrush currents caused by transformers are found; (iv) Fewer power conversion stages will help save the operating costs of the entire DC system.

3.4.1 Converter Level Controls

Commonly used DC/DC converters in Figure 3.17 include a boost converter, buck converter, and buck-boost converter. According to diversified applications, different types of DC/DC converters can be adopted. For example, boost converters are capable of accommodating the sources with low voltage to the high voltage DC bus. Inversely, buck converters act as step-down transformers to transform high voltages to low ones. Notably, buck-boost converters merge the functionalities of both boost and buck converters. The circuitry topologies of these three types of DC/DC converters are depicted in Figure 3.18, where double-loop PI control scheme can be uniformly applied to the three topologies.

In Figure 3.18a–c, V_{dc} stands for the terminal voltage of power sources. v_C and i_L are the capacitor voltage and inductor current respectively. d is the duty cycle that should be modulated to generate PWM signals to control power switches. In spite of the differences in three circuitries, however, their PI control structures are identical, as shown in Figure 3.18d. As long as the voltage and current measurements follow the predefined directions in Figure 3.18a–c, the voltage tracking objective could be stably realized, i.e. the converter output voltage v_C will converge to V in steady state. Moreover, observing Figure 3.18d, i_L will also equal its reference value i_{ref} after system transition since the PI controller is employed in the inner current loop.

Different from the DC/DC converters, the AC/DC rectifier allows AC sources to be integrated into the DC microgrid. The control schematic of an AC/DC converter has been displayed in Figure 3.19. In this figure, the inductor current i_{abc} is expressed as i_d and i_q which are under the rotating frame of AC source voltage. Then, to prevent any reactive power from intruding the DC side, current reference in q axis is set as 0, whereas the reference for i_d is produced by a PI controller which addresses the difference between voltage reference

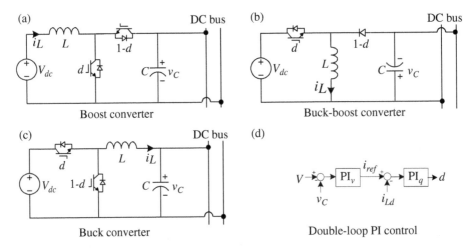

Figure 3.18 DC-DC power converters and the double loop PI control scheme.

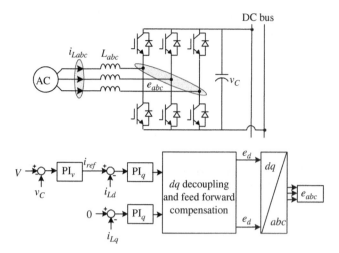

Figure 3.19 Control of the AC/DC rectifier shown in Figure 3.17.

V and real capacitor voltage v_C. By the strong regulation of PI control, voltage error can be suppressed to zero in steady state, which means v_C will stably track *V*.

3.4.1.1 *V-P* and *V-I* Droop Control
In Figures 3.18 and 3.19, the terminal voltage could be driven to track the assigned reference *V* by properly setting up the PI controllers in voltage source converters. For *V*, it is usually computed by droop controllers which have two forms, i.e. *V-P* droop and *V-I* droop,

$$V = V_{rated} - mP, \tag{3.21}$$

$$V = V_{rated} - RI, \tag{3.22}$$

where V_{rated} is the rated voltage. *P* and *I* are the converter output power and terminal current. *m* and *R* are droop coefficient respectively.

The *V-P* droop form in (3.21) is de facto deducted by the analogy of frequency-power relationship in conventional power systems, and simply replace frequency with voltage. *m* is designed as

$$m = \frac{\Delta V_{max}}{P_{max}}, \tag{3.23}$$

where ΔV_{max} is the maximum voltage deviation from the rated value. P_{max} is the maximum output power of the source. Eq. (3.23) means that as long as the real output power of the source is less than P_{max}, then the actual voltage fluctuation will always be within the allowable range. Assuming there are x *V-P* droop controllers are in the system, the droop in (3.21) can be extended to

$$V_1 = V_{rated} - m_1 P_1, \; V_2 = V_{rated} - m_2 P_2, \; \cdots, V_x = V_{rated} - m_x P_x. \tag{3.24}$$

By equalizing $V_1, V_2, \ldots, V_x$, the power-sharing pattern can be identified as

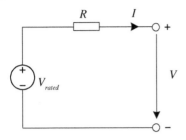

Figure 3.20 Equivalent circuit for V-I droop where V is the voltage reference in Figures 3.18 and 3.19.

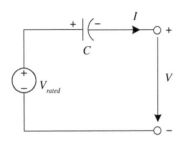

Figure 3.21 Equivalent circuit of the extended droop for HESS applications.

$$P_1 : P_2 \cdots P_x = \frac{1}{m_1} : \frac{1}{m_2} \cdots \frac{1}{m_x}. \tag{3.25}$$

By contrast, *V-I* droop control in (3.22) is more intuitive and can be understood from the aspect of physical circuit structure. The explanatory circuit for (3.22) has been displayed in Figure 3.20 wherein R is an output impedance. Similar to (3.23), to avoid undesirable overcurrent, R can be designed as

$$R = \frac{\Delta V_{max}}{I_{max}}, \tag{3.26}$$

where I_{max} is the maximum possible output current. For totally x *V-I* droops, the current sharing outcome can be given as

$$I_1 : I_2 \cdots I_x = \frac{1}{R_1} : \frac{1}{R_2} \cdots \frac{1}{R_x}. \tag{3.27}$$

3.4.1.2 Virtual Impedance Enabled Control

Ideally, in Figure 3.17, all source converters are directly coupled to the DC bus without any intermediary devices. However, different sources would be placed differently and far away from the DC bus. For this situation, there do exist line impedances between converters and the bus. Based on (3.21), the bus voltage V_b can be expressed in (3.28) considering the impacts of real line impedance r

$$V_b = V - rI = V_{rated} - mP - rI, \tag{3.28}$$

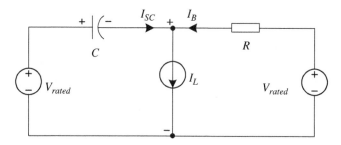

Figure 3.22 Coordination between V-I droop and the extended droop for the HESS in a DC microgrid.

where I denotes the output current. To retrieve the original droop characteristics defined by (4321), the rudimentary droop controller should be amended by involving virtual impedance r_V,

$$V = V_{rated} - mP + r_V I. \tag{3.29}$$

By tuning r_V to be r, then substituting (3.29) into (3.28) results in

$$V_b = V_{rated} - mP. \tag{3.30}$$

It can be seen that (3.30) is now the same as (3.21) which means the impacts of line impedance on droop controller have fully offset by virtual impedance r_V. Similar results can also be found in V-I droop, and the pertinent results are not planned here to avoid duplications.

3.4.1.3 Extended Droop Control

Eq. (3.27) has shown the current sharing among V-I droop controllers. This means the power source with a V-I droop will consistently generate power to build up DC bus voltage. However, this mechanism would not be suitable for HESSs where batteries and supercapacitors (SCs) coexist. It is widely accepted that batteries have comparatively high-energy density but limited power density because they store energy by internal electrochemical reactions. By contrast, SCs are believed to possess fast dynamics while they would be easily depleted, as interpreted in [51]. For batteries, they are capable of continuously supplying power to loads, whereas SCs can supply a large amount of power in a short time but would not convey power for the long run.

Different electrical features of batteries and SCs provided motivation to design a new droop for SCs. In the existing literature, many studies assign V-I droops, which contain embedded resistances in the mathematical expression to batteries for the reason that batteries naturally have output resistors. Inspired by this argument, for properly controlling SCs, V-I droop can be revised to be an extended droop controller by replacing the embedded resistance with a capacitance C, which is described as follows [52, 53],

$$V = V_{rated} - \frac{I}{sC}, \tag{3.31}$$

where s is Laplace operator. The equivalent circuitry of (3.31) is plotted in Figure 3.21. It is expected that, by utilizing the extended droop, SCs would only deliver power at the instances of load changes, while remaining in idle mode in steady state.

The schematic of a HESS with the *V-I* droop and the extended droop cooperatively feeding a lumped load is displayed in Figure 3.22. For clear explanations, the current of the extended droop is denoted as I_{SC} which means the current is from a SC. I_B is the current of battery regulated by *V-I* droop. Then by combining (3.22) and (3.31), the current sharing pattern between the battery and the SC can be derived as given below,

$$
\begin{cases}
V = V_{rated} - (sC)^{-1} I_{SC} \\
V = V_{rated} - RI_B \\
I_{SC} + I_B = I_L
\end{cases}
\Rightarrow
\begin{cases}
I_{SC} = \dfrac{s}{s+\omega} I_L \\
I_B = \dfrac{\omega}{s+\omega} I_L
\end{cases}
, \ \omega = \dfrac{1}{RC}.
\tag{3.32}
$$

It can be seen from (3.32) that a high pass filter (HPF) and a low pass filter (LPF) have been inherently formulated by the combination of *V-I* droop and the extended droop. ω is the cut-off angular frequency of HPF and LPF. The load current I_L is naturally decomposed into high and low frequency components, which are respectively responded to by SC and battery. The current sharing pattern in the HESS is visualized in Figure 3.23. In the case of load current step-up, I_{SC} bursts its current and gradually declines to zero, whereas I_B slowly increases its power in the transient state and consistently supply power to the load, thus achieving dynamic power sharing in the HESS.

Applying concepts underlying (3.31) to revise *V-P* droop, an integral droop controller is proposed in [54] which imposes an integrator to the power value, i.e.

$$
V = V_{rated} - n\frac{P}{s},
\tag{3.33}
$$

where n is the integral droop coefficient. Similar to the extended droop, in the HESS, batteries should be controlled by *V-P* droops as in (3.21) and SCs are with the integral droops. Then, by coordination of the two types of droops, dynamic power sharing between batteries and SCs can also be achieved.

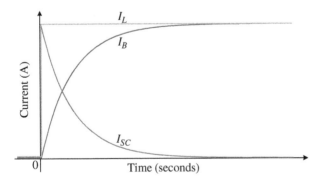

Figure 3.23 The current sharing pattern in the HESS by using the extended droop given in (4.31).

3.4.1.4 Adaptative Droop Control in DC Microgrids

For a practical autonomous DC microgrid, due to the uncertain generation of uncertain renewable power sources (PV and wind turbine), there would be power surplus or deficiency in the system all the time. To handle this scenario, a mode adaptive droop control is examined in [55]. For DC bus voltage (V) flowing within the allowable range, droop control strategies are respectively implemented in the utility gird interface converter, energy storages and the renewables. These droops have been minutely shown in Figure 3.24. The DC bus voltage is partitioned into three levels: V_1, V_2 and V_3 respectively.

For Mode 1, the DC bus voltage resides in $[V_2, V_3]$ which means the system is heavily loaded. In this mode, the utility is scheduled to main the DC bus voltage with droop characteristic. The energy storage is supplying the power at its maximum value. The RESs are working in MPPT mode. For Mode 2, the energy storage dominates the bus and the excessive power produced by the renewables are feedback to the utility. For Mode 3, this is an ideal case since the clean energy is conveying too many powers to be fully stored and injected to the storage and the utility even loads are fairly compensated. In this case, some renewable powers are purposely wasted as they cannot be completely absorbed by the system.

3.4.2 System Level Coordination Control

As a counterpart to the AC microgrid, system level coordination approaches in DC microgrids can also be classified into the centralized scheme and the distributed scheme. Explanations regarding these two schemes have been shown in Figure 3.16.

3.4.2.1 Centralized Control Scheme

Only having droop controllers may inevitably induce DC bus voltage deviating from the rated value. Furthermore, the existence of uncertain line impedances will also cause fluctuations to the voltage of critical buses. To improve the DC bus stability and power quality, in the centralized control scheme, a central controller will collect the voltage of the bus of interest and compare it with the rated value to find the voltage error. A PI controller is then used to process the error and transmit a compensation signal ΔV to all droop controllers. The signal is generated as follows,

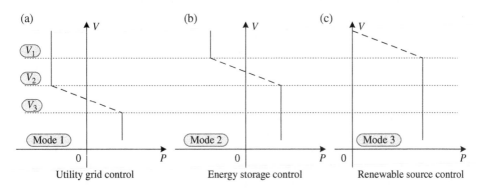

Figure 3.24 Mode adaptive droop control [55].

$$\Delta V = k_{pv}(V_{rated} - V) + k_{iv} \int (V_{rated} - V). \tag{3.34}$$

where k_{pv} and k_{iv} are PI control parameters. V stands for the voltage of the interested DC bus.

3.4.2.2 Distributed Control Scheme

For the distributed method, sparse communication links are established. Each droop controller portrays as an intelligent agent, which only exchange data with its nearby agents. Moreover, the distributed control enabled voltage restoration mechanism is more resilient against possible communication link failures. The consensus across the overall multi-agent system can always be achieved as long as the communication topology remains as a panning tree [50]. The distributed voltage recovery controller can be designed,

$$\Delta V_i = \alpha_v \int \left(\sum_j (\Delta V_j - \Delta V_i) + (V_{rated} - V) \right), \tag{3.35}$$

where V is the voltage of the bus that has high voltage quality requirement. ΔV_i is the locally generated compensation signal for the ith droop. It can be seen from (3.35) that the summed error is addressed by an integrator and the error will be driven to zero in steady state. All the compensation signals in droops will thus be equalized at the end of system transit and the voltage of the interested bus will also be stabilized at a rated value.

3.5 Conclusion

In this chapter, typical RESs (PV and wind) systems and ESSs are reviewed with their interface power electronics technologies. Then, the power electronics-based grid-connected AC power systems are illustrated, including the VSC with their control technologies and multilevel converters for high-power applications with MMCs specifically and their control technologies. Compared with grid-connected operation, standalone distributed generation systems are more challenging for the power balance issues, thus the power electronics enabled autonomous AC power systems are presented with the power management and their coordinated control schemes. Currently, autonomous DC power systems are gaining increased attention, and power electronics DC/DC converter technologies are illustrated with their control, coordination and power management.

For future works, there are several areas that can be explored. First is the resilience-related control strategies. RESs are variable and highly impacted by weather and locations. With the high penetration of RESs into the grid, the grid reliability is impacted due to the uncertainties of RESs especially with extreme weather conditions. To address this issue, resilience-related control strategies should be developed for power electronics interface converters of RESs with the coordination and management strategies to enhance the grid resilience. Next is the protection issue. The integration of RESs makes the power flow in distribution power systems bidirectional, which is different from the conventional unidirectional power low. Thus, the conventional protection scheme is not suitable and new protection schemes should be developed. Third is about the distributed control and distributed optimization in smart grid. The conventional centralized control and energy management suffer from single point of failure issues and are not flexible for system expansion. Considering the distributed nature of RESs and the development of information and communication technologies, the distributed control and optimization strategies can be implemented for the management of smart grid, with the advantages of lower communication/computation burden, higher reliability, higher scalability and the information privacy of individual customers.

References

1 EIA (2018). International Energy Outlook 2018. https://www.eia.gov/outlooks/ieo/executive_summary.php (accessed 11 February 2021).

2 Bose, B.K. (2017). Power electronics, smart grid, and renewable energy systems. *Proc. IEEE* 105 (11): 2011–2018.

3 Carrasco, J.M., Franquelo, L.G., Bialasiewicz, J.T. et al. (2006). Power-electronic systems for the grid integration of renewable energy sources: a survey. *IEEE Trans. Ind. Electron.* 53 (4): 1002–1016.

4 Blaabjerg, F., Yang, Y., Yang, D., and Wang, X. (2017). Distributed power-generation systems and protection. *Proc. IEEE* 105 (7): 1311–1331.

5 Blaabjerg, F., Chen, Z., and Kjaer, S.B. (2004). Power electronics as efficient interface in dispersed power generation systems. *IEEE Transactions on Power Electronics* 19 (5): 1184–1194.

6 Yang, Y., Kim, K.A., Blaabjerg, F., and Sangwongwanich, A. (2018). *Advances in Grid-Connected Photovoltaic Power Conversion Systems*. Woodhead Publishing.

7 Mastromauro, R.A., Liserre, M., and Dell'Aquila, A. (2012). Control issues in single-stage photovoltaic systems: MPPT, current and voltage control. *IEEE Trans. Ind. Informatics* 8 (2): 241–254.

8 Strasser, T., Andrén, F., Kathan, J. et al. (2014). A review of architectures and concepts for intelligence in future electric energy systems. *IEEE Transactions on Industrial Electronics* 62 (4): 2424–2438.

9 Chang, L. (2017). Review on distributed energy storage systems for utility applications. *CPSS Trans. Power Electron. Appl.* 2 (4): 267–276.

10 Yang, Y., Enjeti, P., Blaabjerg, F., and Wang, H. (2015). Wide-scale adoption of photovoltaic energy: grid code modifications are explored in the distribution grid. *IEEE Ind. Appl. Mag.* 21 (5): 21–31.

11 Yaramasu, V., Wu, B., Sen, P.C. et al. (2015). High-power wind energy conversion systems: state-of-the-art and emerging technologies. *Proc. IEEE* 103 (5): 740–788.

12 Chen, Z., Guerrero, J.M., and Blaabjerg, F. (2009). A review of the state of the art of power electronics for wind turbines. *IEEE Trans. Power Electron.* 24 (8): 1859–1875.

13 Chen, Z. (2005). Characteristics of induction generators and power system stability. *Eighth International Conference on Electrical Machines and Systems*, Nanjing, China (27–29 September 2005). IEEE.

14 Yao, J., Li, H., Liao, Y., and Chen, Z. (2008). An improved control strategy of limiting the DC-link voltage fluctuation for a doubly fed induction wind generator. *IEEE Trans. Power Electron.* 23 (3): 1205–1213.

15 Chen, Z. and Spooner, E. (2001). Grid power quality with variable speed wind turbines. *IEEE Trans. Energy Convers.* 16 (2): 148–154.

16 Blaabjerg, F. and Ma, K. (2013). Future on power electronics for wind turbine systems. *IEEE J. Emerg. Sel. Top. Power Electron.* 1 (3): 139–152.

17 Vazquez, S., Lukic, S.M., Galvan, E. et al. (2010). Energy storage systems for transport and grid applications. *IEEE Trans. Ind. Electron.* 57 (12): 3881–3895.

18 Nehrir, M.H., Wang, C., Strunz, K. et al. (2011). A review of hybrid renewable/alternative energy systems for electric power generation: configurations, control, and applications. *IEEE Trans. Sustain. Energy* 2 (4): 392–403.

19 Shen, J. and Khaligh, A. (2015). A supervisory energy management control strategy in a battery/ultracapacitor hybrid energy storage system. *IEEE Trans. Transp. Electrif.* 1 (3): 223–231.

20 Chen, Z., Mi, C.C., Xu, J. et al. (2014). Energy management for a power-split plug-in hybrid electric vehicle based on dynamic programming and neural networks. *IEEE Trans. Veh. Technol.* 63 (4): 1567–1580.

21 Muhammad, M., Armstrong, M., and Elgendy, M.A. (2016). A nonisolated interleaved boost converter for high-voltage gain applications. *IEEE J. Emerg. Sel. Top. Power Electron.* 4 (2): 352–362.

22 Lu, J., Wang, Y., Li, X., and Du, C. (2019). High-conversion-ratio isolated bidirectional DC-DC converter for distributed energy storage systems. *IEEE Trans. Power Electron.* 34 (8): 7256–7277.

23 Wang, G., Konstantinou, G., Townsend, C.D. et al. (2016). A review of power electronics for grid connection of utility-scale battery energy storage systems. *IEEE Trans. Sustain. Energy* 7 (4): 1778–1790.

24 Blaabjerg, F., Teodorescu, R., Liserre, M., and Timbus, A.V. (2006). Overview of control and grid synchronization for distributed power generation systems. *IEEE Trans. Ind. Electron.* 53 (5): 1398–1409.

25 Pogaku, N., Prodanovic, M., and Green, T.C. (2007). Modeling, analysis and testing of autonomous operation of an inverter-based microgrid. *IEEE Trans. power Electron.* 22 (2): 613–625.

26 Buso, S. and Mattavelli, P. (2006). Digital control in power electronics. *Lect. power Electron.* 1 (1): 1–158.

27 Vasquez, J.C., Guerrero, J.M., Savaghebi, M. et al. (2013). Modeling, analysis, and design of stationary-reference-frame droop-controlled parallel three-phase voltage source inverters. *IEEE Trans. Ind. Electron.* 60 (4): 1271–1280.

28 Golestan, S., Guerrero, J.M., and Vasquez, J.C. (2017). Three-phase PLLs: a review of recent advances. *IEEE Trans. Power Electron.* 32 (3): 1894–1907.

29 Wang, X., Blaabjerg, F., and Wu, W. (2014). Modeling and analysis of harmonic stability in an AC power-electronics- based power system. *IEEE Trans. Power Electron.* 29 (12): 6421–6432.

30 Gui, Y., Wang, X., Blaabjerg, F., and Pan, D. (2019). Control of grid-connected voltage-source converters: the relationship between direct-power control and vector-current control. *IEEE Ind. Electron. Mag.* 13 (2): 31–40.

31 Kundur, P., Balu, N.J., and Lauby, M.G. (1994). *Power System Stability and Control*, vol. 7. New York: McGraw-hill.

32 Fang, J., Li, H., Tang, Y., and Blaabjerg, F. (2018). On the inertia of future more-electronics power systems. *IEEE J. Emerg. Sel. Top. Power Electron.* 7 (4): 2130–2146.

33 Liu, J., Miura, Y., and Ise, T. (2015). Comparison of dynamic characteristics between virtual synchronous generator and droop control in inverter-based distributed generators. *IEEE Trans. Power Electron.* 31 (5): 3600–3611.

34 Chen, Y., Hesse, R., Turschner, D. and Beck, H.-P. (2011). Improving the grid power quality using virtual synchronous machines. *2011 International Conference on Power Engineering, Energy and Electrical Drives*, Malaga, Spain (10–12 May 2011). PSMA.

35 Driesen, J. and Visscher, K. (2008). Virtual synchronous generators. *IEEE Power and Energy Society General Meeting-Conversion and Delivery of Electrical Energy in the 21st Century*, Pittsburch, USA (20–24 July 2008). IEEE.

36 Shintai, T., Miura, Y., and Ise, T. (2014). Oscillation damping of a distributed generator using a virtual synchronous generator. *IEEE Trans. power Deliv.* 29 (2): 668–676.

37 Zhong, Q.-C. and Weiss, G. (2010). Synchronverters: inverters that mimic synchronous generators. *IEEE Trans. Ind. Electron.* 58 (4): 1259–1267.

38 Rodríguez, J., Bernet, S., Wu, B. et al. (2007). Multilevel voltage-source-converter topologies for industrial medium-voltage drives. *IEEE Trans. Ind. Electron.* 54 (6): 2930–2945.

39 Rodríguez, J., Leon, J.I., Kouro, S. et al. (2008). The Age of mli arrives. *IEEE Industrial Electronics Magazine* (June), p. 28–39.

40 Perez, M.A., Bernet, S., Rodriguez, J. et al. (2015). Circuit topologies, modeling, control schemes, and applications of modular multilevel converters. *IEEE Trans. Power Electron.* 30 (1): 4–17.

41 Debnath, S., Qin, J., Bahrani, B. et al. (2015). Operation, control, and applications of the modular multilevel converter: a review. *IEEE Trans. Power Electron.* 30 (1): 37–53.

42 Dekka, A., Wu, B., Yaramasu, V., and Zargari, N.R. (2016). Model predictive control with common-mode voltage injection for modular multilevel converter. *IEEE Trans. Power Electron.* 32 (3): 1767–1778.

43 Yang, S., Wang, P., and Tang, Y. (2018). Feedback linearization-based current control strategy for modular multilevel converters. *IEEE Trans. Power Electron.* 33 (1): 161–174.

44 Yang, Q., Saeedifard, M., and Perez, M.A. (2019). Sliding mode control of the modular multilevel converter. *IEEE Trans. Ind. Electron.* 66 (2): 887–897.

45 Dekka, A., Wu, B., Fuentes, R.L. et al. (2017). Evolution of topologies, Modeling, control schemes, and applications of modular multilevel converters. *IEEE J. Emerg. Sel. Top. Power Electron.* 5 (4): 1631–1656.

46 Hagiwara, M. and Akagi, H. (2009). Control and experiment of pulsewidth-modulated modular multilevel converters. *IEEE Trans. power Electron.* 24 (7): 1737–1746.

47 Fan, S., Zhang, K., Xiong, J., and Xue, Y. (2014). An improved control system for modular multilevel converters with new modulation strategy and voltage balancing control. *IEEE Trans. Power Electron.* 30 (1): 358–371.

48 Zhang, M., Huang, L., Yao, W., and Lu, Z. (2013). Circulating harmonic current elimination of a CPS-PWM-based modular multilevel converter with a plug-in repetitive controller. *IEEE Trans. Power Electron.* 29 (4): 2083–2097.

49 Li, Z., Wang, P., Chu, Z. et al. (2013). An inner current suppressing method for modular multilevel converters. *IEEE Trans. Power Electron.* 28 (11): 4873–4879.

50 Lin, P., Jin, C., Xiao, J. et al. (2018). A distributed control architecture for global system economic operation in autonomous hybrid AC/DC microgrids. *IEEE Trans. Smart Grid* (3): 1–1.

51 Jia, J., Wang, G., Cham, Y.T. et al. (2010). Electrical characteristic study of a hybrid PEMFC and ultracapacitor system. *IEEE Trans. Ind. Electron.* 57 (6): 1945–1953.

52 Xu, Q., Hu, X., Wang, P. et al. (Jul. 2017). A decentralized dynamic power sharing strategy for hybrid energy storage system in autonomous DC microgrid. *IEEE Trans. Ind. Electron.* 64 (7): 5930–5941.

53 Zhang, Y. and Li, Y.W. (2017). Energy management strategy for Supercapacitor in droop-controlled DC microgrid using virtual impedance. *IEEE Trans. Power Electron.* 32 (4): 2704–2716.

54 Lin, P., Wang, P., Xiao, J. et al. (2017). An integral droop for transient power allocation and output impedance shaping of hybrid energy storage system in DC microgrid. *IEEE Trans. Power Electron.* 8993 (c): 1–14.

55 Gu, Y., Xiang, X., Li, W., and He, X. (2014). Mode-adaptive decentralized control for renewable DC microgrid with enhanced reliability and flexibility. *IEEE Trans. Power Electron.* 29 (9): 5072–5080.

4

Energy Storage Systems as an Enabling Technology for the Smart Grid

The impact of the Energy Storage Systems (ESSs) on the future grid is gaining more attention than before from power system designers, grid operators and regulators. These technologies deliver the chance of shifting energy use farther away at peak periods and rising prices, enhance the reliability and resilience of power grids and aid the large amalgamation of renewable energy sources (RES). The goal of this chapter is to present a detailed review on different energy storage technologies, their current and future status, their share in different smart grid SG applications, and their technical and financial benefits as enabling technology for the deployment of the future SG.

4.1   Introduction

The traditional electric power system operates with a simple one-way power transmission from remote and large power generation plants to end consumers. The implication of that is electric power should constantly be utilized accurately when generated. Yet, the need for power fluctuates significantly on a daily basis and depending on the season, while the maximum demand could, in some instances, take place for a few hours per year. This results in inefficient, overdesigned, expensive and environmentally unfriendly power generation plants [1]. By having large-scale capacity ESSs, the organizers are required to produce just enough generating capacity to meet the average electrical demand instead of meeting the peak demands for ramping up and down. This would help avoid the need for construction of oversized plants with extra power capacity. Moreover, transmission and distribution systems are not sufficiently effective at peak times, they should be huge to report the infrequent peak periods demand. ESS are capable of relieving network contingency and hindering the susceptibility of ramifications of overloaded T&D network [2].

Furthermore, a precise power supply–demand balance at all times is a core principle for power grids, otherwise, the system frequency will not be stabilized. Frequency drops when demand surpasses supply and, on the contrary, escalates when supply surpasses demand. As a result, power quality will decline and has the potential to result in service interruption. Generally, frequency control is calibrated by fine-tuning the output of some

Smart Grid and Enabling Technologies, First Edition. Shady S. Refaat, Omar Ellabban, Sertac Bayhan, Haitham Abu-Rub, Frede Blaabjerg, and Miroslav M. Begovic.
© 2021 John Wiley & Sons Ltd. Published 2021 by John Wiley & Sons Ltd.
Companion website: www.wiley.com/go/ellabban/smartgrid

generating facilities to follow instantaneous fluctuations in demand and this indicates inefficient operation. ESSs can provide frequency control functions by locating ESSs at the end of a heavily loaded line, it could decrease the voltages applied through the discharging of electricity and hinder the voltage increases by the charging of electricity [3].

Energy technologies associated with RES have experienced fast changes over the previous years, particularly in the level of application and the residential and commercial attitude to their utilization. RES delivered approximately 11% of the final global energy consumption in 2018, and renewables encompassed approximately 26% of the global power generating capacity [4]. Nonetheless, a relatively large share of variable RES increases the variability and intermittency of the power supply, disturbing the ideal performance of the conventional power systems and grid reliability. In addition, it presents more influence in power quality and power system dynamics. The ESS could be implemented in several methods in order to improve the operation of RES (intermittency and power quality challenges). By storing energy from variable RES that entails wind and solar, energy storage has the potential to deliver robust generation from these units, permitting the energy generated to be utilized in an efficient manner, and offer ancillary transmission benefits [5].

SG coordinates the requirements and capabilities of all power generators, grid operators, end-users, and electricity market stakeholders to make all aspects of the system work in an efficient manner, which means bringing the costs and environmental impacts down to a minimum and maintaining system stability, reliability, and resiliency. Energy storage is one important aspect of the SG revolution that is taking place in all areas of electric power systems. SG may have numerous energy storage centers, large and small, that can be used as a buffer to tolerate the influence of sudden load variations in RES generation and to shift energy consumed away at peak periods by implementing energy balancing, load following, and dynamic compensation of reactive and real power. Therefore, energy storage combined with the SG are projected to experience a huge leap forward in the next decade [6].

Recently, numerous storage technologies have been improved, depend on mechanical, electrochemical, thermal, electrical or chemical energy. However, a large percentage of these storage technologies are clustered at the demonstration or early commercial deployment phases. These ESS technologies vary in the context of investment costs per power capacity and per energy capacity, lifetime, storage losses, efficiency, ramping rates and reaction times. ESS technologies can provide multiple services at each stage of the electricity supply chain (generation-transmission-distribution-consumption), in addition to, mitigating the intermittency of integrating renewables and their power quality problems and enabling different SG applications. The goal of the chapter is to present a detailed review on different energy storage technologies, their current and future states, their share in different SG applications, and their technical and financial benefits as enabling technology for the SG infrastructure. The battery parameters will not be explained in this chapter. The parameters are: Capacity, State of Charge (SoC), Depth of Discharge (DoD), State of health (SoH), Age and history, Energy Density, Charging Efficiency, Float Voltage, and Cut-Off Voltage. Definitions of those parameters will be provided in Chapter six of the book (Section 6.4.1).

4.2 Structure of Energy Storage System

Energy storage includes a system used for storing energy in a certain form at a specific time and transferring it at another specific time. However, Electrical Energy Storage (EES) denotes the conversion of electrical energy imported from a certain power source into

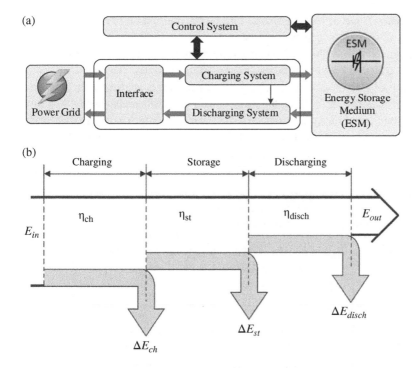

Figure 4.1 An electrical energy storage (EES) system structure (a) and energy losses (b).

another form that includes, chemical, mechanical, thermal or magnetic energy and storing it at off-peak demand for converting back to an electrical form at peak demand or when needed. An electrical energy storage system (EESS), as shown in Figure 4.1a, is composed of four components: (i) Energy Storage Medium (ESM), which is a system that is in charge of storing energy; (ii) charging system, which is a unit that allows energy to be transmitted to a power source at the ESM; (iii) discharging system, which guarantees the transfer of the energy stored at the time of need; and (iv) control system, which manages the whole EES [7]. Since energy storage is not an ideal energy source, losses at every stage of the storage process take place, as illustrated in Figure 4.1b. The round-trip efficiency of the EES is decreased by energy conversion processes inherent efficiencies and storage losses, therefore, the overall efficiency of ESS is defined by $\eta_s = E_{out}/E_{in}$, where E_{out} and E_{in} are output and input electrical energy, respectively [8].

4.3 Energy Storage Systems Classification and Description

There are many storage technologies for SG applications, but none of them are able to implement all the required applications. An assessment of each storage technology makes sense merely with respect to a certain application. Hence, it is better to utilize classification systems. Usually, the classification could be done depending on the way energy is stored, e.g. mechanical, electrical, or chemical. Yet, from an application standpoint, it would be more beneficial to classify the storage technologies in terms of the services they are able to deliver to the markets. One of the most commonly used methods for representing EES systems is predicated upon the form of energy stored as shown in Figure 4.2, which can be characterized as mechanical: Pumped Hydroelectric Storage (PHS), Compressed Air Energy Storage (CAES) and Flywheel Energy Storage (FES); chemical: Hydrogen, Methane and Synthetic Natural Gas (SNG); electrochemical: conventional rechargeable batteries, high temperature batteries and flow batteries; electrical: supercapacitors (SC) and Superconducting Magnetic Energy Storage (SMES); Thermal Energy Storage (TES): Molten Salt (MS) and thermochemical: Solar Fuels (SF) [9]. In terms of the application, ESS can be associated with power applications or energy applications, as shown in Figure 4.3. Some

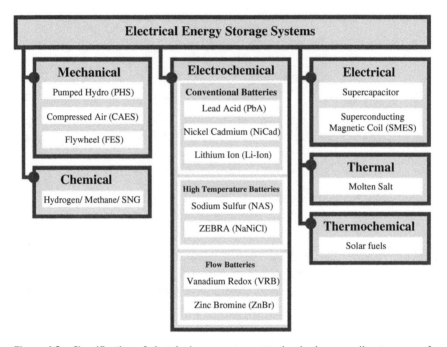

Figure 4.2 Classification of electrical energy storage technologies according to energy form.

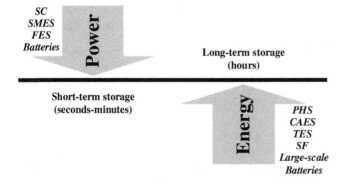

Figure 4.3 Electrical energy storage technologies classification according to application.

technologies such as: PHS, ACES, TES, SF and large-scale batteries are characterized by their ability to store energy over time (several hours) making those devices more suitable for energy management and stand-alone or mobile applications. Others, e.g. SC, SMES, FES and batteries are characterized by their ability to deliver power very quickly making them more suitable to power quality applications [10].

A brief portrayal of the principles and potential capability of energy storage technologies are presented in Table 4.1 [11–13].

Table 4.1 Commonly used energy storage technologies. Adapted from Ref Num [11–13].

Technology	Description
Pumped Hydroelectric Storage (PHS)	During periods of low demand, excess generation capacity is utilized to pump water from a low to a high reservoir. At times of higher demand, water flows into a lower reservoir by a turbine, producing electricity. It is widely known that pumped-storage hydroelectricity is the main capacity form of grid energy storage. The PHS has the highest nominal power and energy value, long lifetime, high efficiency and minute discharge losses. Additionally, PHS is the only commercially recognized, large-scale EES with no extra fuel required. Yet, the implementation of the PHS is reliant on the geographical conditions and one that influences the environment. Consequently, the flexibility of its application is unsatisfactory.
Compressed Air Energy Storage (CAES)	This technology stores a low-cost value off-peak energy as compressed air in an underground reservoir (salt cavern, abandon mines, rock structures) or a system consisting of vessels or pipes. After that the air flows during peak load periods and subjected to high temperatures via an exhaust heat from a combustion turbine. Then this air is transformed to energy by expansion turbines to produce electrical energy. The storage time can culminate up to 12 months due to tiny self-charge losses. Yet, the CAES installation is likewise constrained due to topographical situations moreover, it is a highly priced technology to obtain. CAES is the second commercially recognized, large-scale EES, after PHS.
Flywheel Energy Storage (FES)	This system operates by accelerating a rotor (flywheel) to great speeds and preserving the energy in the system as rotational energy and bringing frictional losses down to a minimum. In the event that energy is taken from the system, the flywheel's rotational speed is decreased; accumulating more energy to the system which in turn intensifies the speed of the flywheel. The key benefits of flywheels are the robust cycle stability, sufficiently high lifetime for delivering full charge–discharge cycles, low maintenance cost, great power density and efficiency. However, the disadvantages of FES are the limited operating time and increased self-discharge losses.
Superconducting Magnetic Energy Storage (SMES)	A SMES system is has the ability to store energy in a magnetic field in a manner that would allow an instantaneous discharge back, contributing to an electrical storage in a pure electrical form. SMES systems are known to possess a high energy storage efficiency, fast response, and long-life cycles. The SMES is very promising as a power storage system for load leveling or a power stabilizer. Had SMES became an application for utilities it had the potential to become a diurnal storage device, charged from baseload power at night and reaching peak loads by day. The problems that arise from the use of SMES is the dependency on a high capital cost and environmental reasons associated with high magnetic fields.

(Continued)

Table 4.1 (Continued)

Technology	Description
Rechargeable batteries	A conventional Rechargeable Battery entails of a set of low voltage/ power battery cells connected in parallel and series to obtain a certain electrical behavior. An individual cell consists of a liquid, paste or solid electrolyte along with the anode and the cathode. Presently, several types have been made for commercial use, and additional materials and advanced technologies being manufactured to enhance performance and decrease costs.
Flow batteries	A flow battery is also a rechargeable battery whereby the recharge capability is made possible by two chemicals dissolved in liquids confined within the system and detached by a membrane. Ion exchange arises through the membrane whereas as both of the liquids circulate in their vicinity. Therefore, the nominal values of power and energy has the potential to be considered independently: energy capacity is obtained by the amount of electrolyte found in external tanks and the power value is considered with the predication upon the active vicinity of the cell section. This implies that flow batteries are desired for both energy and power storage.
Supercapacitor (SC)	Supercapacitors store electricity as electrostatic energy; a SC uses an electrolyte solution between Its two solid conductors. It offers fast response including life cycles regularly of thousands and a relatively high efficiency, hence, they are primarily active in power quality services. The disadvantage of SC is its dependency on short storage time, low energy density, and high self-discharge loss.
Hydrogen	Hydrogen is not considered as a main energy source, yet, it is a flexible energy storage technique, as it is produced by other energy sources first to be utilized. Even though it has a small range of roundtrip efficiency, its employment is rising as a result of greater storage capacity as opposed to other energy storage technologies, consequently, substantially smaller storage reservoirs are essential relative to PHS and CAES. Presently, the rather small overall efficiency and high capital costs are significant barriers in commercial application of hydrogen-based storage in grid-scale applications.
Thermal Energy Storage (TES)	TES systems permits surplus electricity to be transformed into thermal energy and stored (short-term or seasonal) in a storage compartment at a certain temperature for future needs. Thermal energy storage systems are relevant in several industrial and residential purposes, for instance, space heating or cooling, process heating and cooling, hot water production, or electricity production. TES system could be categorized into three parts; sensible heat, latent heat, and adsorption system.

4.4 Current State of Energy Storage Technologies

Different energy storage technologies are at various maturity levels and are deployed in changing amounts. These differences should be treated when assessing certain technology and applications as the vulnerability of risk increases as the level of maturity declines or the level of deployment rises. In Figure 4.4, vital technologies are shown

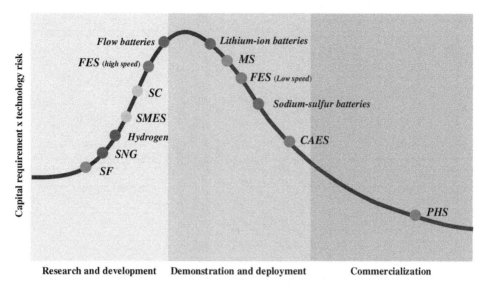

Figure 4.4 Maturity of electrical energy storage technologies.

according to their corresponding primary capital investment requirements and technology risk with respect to their current phase of development. Pumped hydro and compressed air energy storage are cutting-edge electricity storage technologies; others bring a cost and risk premium as result of small levels of commercial maturity. As technologies shift from the demonstration and deployment stage to commercialization, the price of the particular technology decreases and the technical features are usually improved. The period that technologies mature is determined by several reasons that entail market incentives, installation volumes, technical constraints and geographical restrictions [14–16].

Global grid-connected and stationary energy storage capacity in 2018 totaled approximately 172 928 GW, which accounts for approximately 5.5% of the installed generation capacity, with pumped storage hydropower responsible for the vast majority. Around 0.8 GW of novel developed energy storage capacity was employed in 2018, culminating in year-end capacity totaling up to 6.4 GW. This increase was mainly in the battery storage, where it rose by 0.6 GW for an added amount of 1.7 GW. The remaining additions were mostly stored as thermal energy form, higher by 0.2 GW, amounting to a total of 1.258 GW. A tiny amount of electro-mechanical storage was added in 2018, with a value reaching up to 1.003 GW, as illustrated in Figure 4.5. Yet, a large amount of surplus EES is anticipated to be employed considering that the developing market requires this for additional applications [4, 17].

Although a high percentage of the advanced storage capacity installed in 2018 were batteries (electro-chemical), thermal storage also plays a vital role alongside Concentrated Solar Power (CSP) plants. IRENA estimates that pumped storage hydropower will rise from 169.557 to 325 GW in 2030. Over the same time period, the available battery electrical storage will rise up to 1.6 GW, [18]. Energy storage is an essential means for allowing an effective implementation of RES and revealing the advantages of SGs. This technology repeatedly

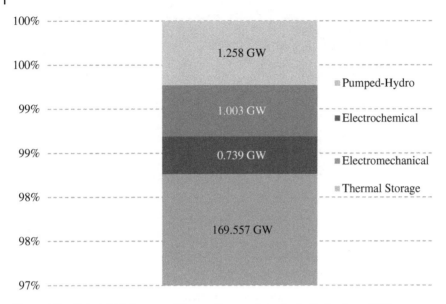

Figure 4.5 Global Grid-Connected Energy Storage Capacity, by Technology, 2018.

proves its worth to grid operators worldwide, moreover, the fast-decreasing costs and improving capabilities of ESSs technologies, along with growing industry expertise, will swiftly introduce new markets and cost-effective applications for energy storage [19].

4.5 Techno-Economic Characteristics of Energy Storage Systems

Different energy storage technologies have different applications in the energy system. Capacity, cost, energy density, efficiency level, and technical and economic life are factors that determine in which context the technologies are the most suitable. A number of energy storage technologies are now available in the market and each one of them offers a variety of performance parameters. The key performance parameters for evaluating energy storage technologies are discussed in this section [20–28].

Figure 4.6 demonstrates an overview of energy storage technologies with respect to their relative discharge times, power scale in the order of MW, and their respective efficiencies. Systems located at the right side of Figure 4.6 (shows a relatively high discharge time and energy storage capacity) demonstrate similarities to PHS and CAES, in terms of how perfect they are for product storage, arbitrage, rapid reserve, and area control-frequency responsive reserve. Systems located on the left side of Figure 4.6 (high power to energy ratio and low discharge time requirements), for instance, flywheels, SCs, and SMES, are applicable for power quality-reliability and transmission system stability implementations. Batteries are adequate enough to deliver short to medium term storage under a certain output capacity range. Because of battery flexibility batteries, delivery of a wide-range of

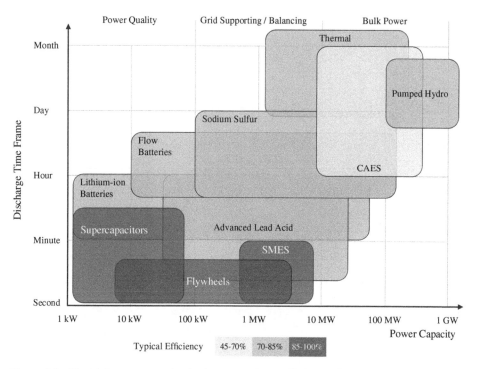

Figure 4.6 Electricity storage technologies comparison – discharge time vs power capacity.

power capacity values from the scale is possible along with the enhancement of technologies (e.g. Li-ion), that are competent in providing both slow and fast discharge rates. Batteries encompass applications ranging from power quality to the initial development of energy management, making flow batteries increasingly suitable for transmission and distribution deferral purposes.

Self-discharge conveys the losses of a storage system at the time of off-duty terms and obtains the suggested optimized storage time period. The self-discharge significance could be categorized into four sections; insignificant and low, benign and negligible self-discharge (approximately 5% each month), neutral, 5–30% each month, and high, if self-discharge losses surpass 30% each month. The relationship between self-discharge significance and suggested storage time span is shown in Figure 4.7. Na—S, FC, and bulk energy storage considering PHS, CAES, and flow batteries encounter negligible losses as opposed to SCs and flywheels which are restricted by the inherent self-discharge as flywheels has the potential to discharge fully at a time span of one day or even less). Restrictions in the storage time span does not include these systems from particular aspects, for instance, spinning reserve, which addresses the recurrence of cycling with relatively short and long-term intervals between two successive cycles. However, depending on other conditions, these systems could be relevant in power quality applications, such as the periodicity/recurrence of cycling, which is quite high. Conversely, bulk energy storage systems are vital for energy management uses, including rapid reserve and arbitrage.

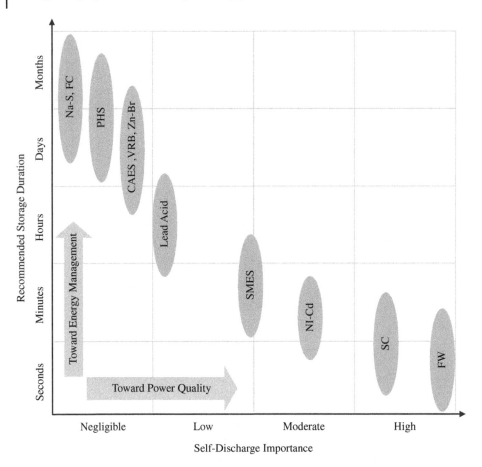

Figure 4.7 Self-discharge and suggested storage period of energy storage systems.

Considering a known value of energy, the greatest power and energy densities will be considered as the lesser volume of the needed energy storage system. Figure 4.8 shows the power density (per unit volume) of several EES technologies with respect to the energy density. Relatively compact EES technologies fit for mobile applications could be observed at the top right corner of Figure 4.8. A big area and volume utilizing storage systems are observed at the bottom left side of Figure 4.8. At this side PHS, CAES, and flow batteries possess a small energy density as opposed to other storage technologies. SMES, SC, and flywheel which bear quite large power densities, yet hold small energy densities. Li-ion has a large energy and power density, justifying the numerous applications that Li-ion is currently deployed in. NaS holds greater energy densities relative to the mature battery types that include lead-acid and NiCd, however, their power density is smaller relative to that of Li-ion. Flow batteries possess a relatively high potential for larger battery systems (MW/MWh) and holds neutral energy densities. The key benefit of SNG is the large energy density, exceeding each storage system mentioned.

Cost is perhaps the most essential and core challenge of energy storage, which could be explained by the unit cost of power ($/kW) and energy capacity ($/kWh) or the per-cycle cost

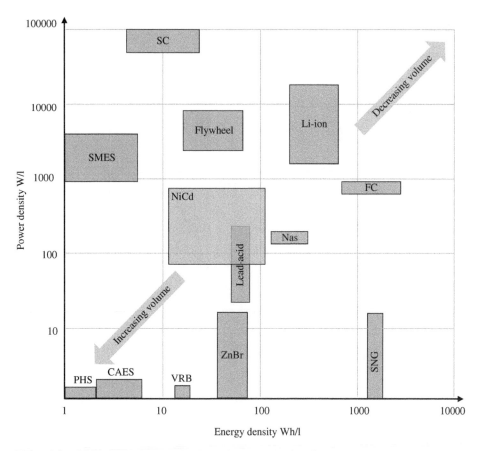

Figure 4.8 Comparison of power density and energy density of energy storage systems.

(¢/kWh). Unit cost of energy capacity is an improved way of determining the capital cost for power management, whereas $/kW or ¢/kWh is mainly applicable in energy arbitrage. Practically, two common units/metrics are utilized to link and evaluate the costs of energy storage systems, specific investment costs (SIC) and levelized cost of storage (LCOS). The metric of SIC represents the installation price for both power and energy storage capacity. Figure 4.9 conveys the storage technologies with their respective capital costs in $/kWh. The extent of the bars in Figure 4.9 shows the uncertainties, impact of location, and project magnitude including additional variables. The level of cost declines with regard to a particular technology determined by maturity and synergy influence, originating from cross-industry applications. For this reason, lithium-ion batteries, for example, are more likely to demonstrate a substantial cost decrease as time passes. The units of LCOS utilized the units of levelized costs of electricity (LCOE), which is normally utilized to evaluate the price of generated electricity from power plants. The LCOS is responsible for the net internal costs of EES systems excluding the impact of cost of charging electricity, Figure 4.10, presenting the LCOS for a number of EES systems [29]. Moreover, Figure 4.11, illustrates the cycle efficiencies of ESSs. Flywheels and electrical storage systems, including Na—S and Li-ion batteries are

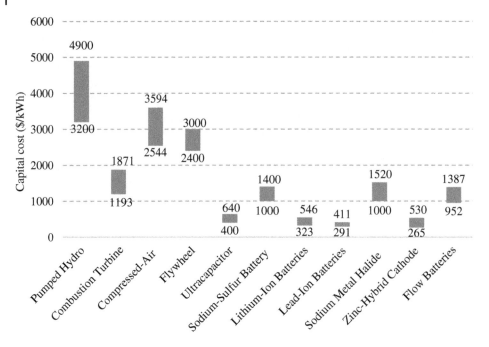

Figure 4.9 Energy storage technologies capital cost, 2018.

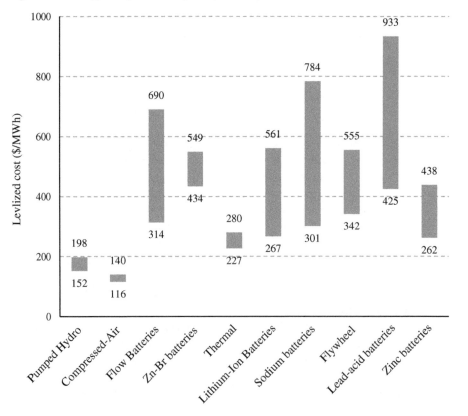

Figure 4.10 Levelized cost of storage (LCOS) for different technologies, 2016.

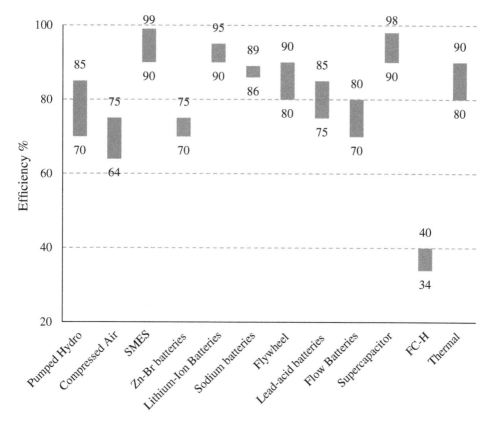

Figure 4.11 Typical cycle efficiency (max. and min.) of energy storage systems.

higher than 80%, but FC-HS was found to be under 50% [30]. Most EES technologies have high material substitutability (PHS, CAES, flywheels, capacitors, and SMES), therefore, are not at risk of a shortage since the choice in suitable materials is more flexible. Ordered from least to most limited by material availability, the technologies taken into account are: CAES, NaS, ZnBr, PbA, PHS, Li-ion, and VRB. Lastly, energy storage technologies relative to their respective energetic performance, utilizing the energy stored on energy invested (ESOI) ratio represents how much energetic benefit a certain society obtains for each unit of energy used in constructing an energy storage system. It also compares the energy output of an EES to the energy inputs needed to build and activate it. Figure 4.12 demonstrates the ESOI for several storage technologies. Ordered from least to most restricted by energy requirements, the technologies mentioned are: CAES, PHS, Li-ion, NaS, VRB, ZnBr, PbA, [31, 32]. Table 4.2 qualitatively lists the technical and economic behavior of each storage technology [33–38].

4.6 Selection of Energy Storage Technology for Certain Application

To choose the most economic storage technology for a certain application, a meticulous analysis of the working profile and a complete life-cycle cost calculation should be obtained. When associating a storage technology parameter set with a set of parameters representing

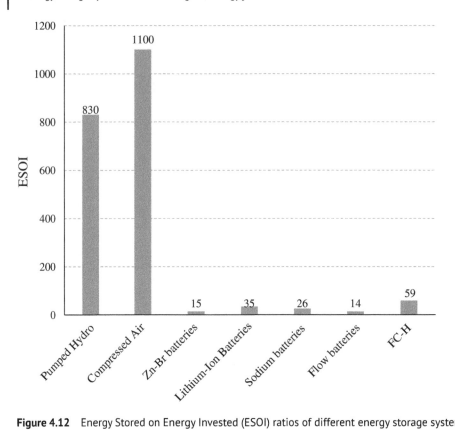

Figure 4.12 Energy Stored on Energy Invested (ESOI) ratios of different energy storage systems.

the application, one is then able to determine the full price of ownership for generating energy from the storage system along with the prices per installed kW power. Parameters should be selected prudently when considering the conditions for a certain application, because several storage technologies could be enhanced while conforming to their particular demands. The procedure observed in Figure 4.13 is used for comparing storage technologies under investigation [39].

As the ESS technology type is acquired, the ideal ESS allocation (sizing and siting) challenge could be unraveled by balancing the benefits and price. Ideal allocation and economic evaluation of ESS usually presents a complicated issue due to its intertemporal nature, i.e. the operation of the ESS in one-time step will influence its operation in the others. The sizing challenge entails the acquisition of the power and energy rating. Over-provisioning ESS size included expensive and underutilized assets, but under-provisioning decreases its working time span. Immediate acquisition of ESS location and size is a non-deterministic polynomial-time complicated question to solve and it has not been answered for large-scale problems up to this day. Numerous approaches have been made in literature for obtaining the size and siting of ESS. Procedures for the purpose of sizing and siting are categorized into four parts as follows: analytical methods (AM), mathematical programming (MP), exhaustive search, and heuristic methods. Siting and sizing of ESS could not be represented distinctly from their operation or the role/

Table 4.2 Technical and economic characteristics of energy storage technologies.

Technology		Maturity	Power rating (MW)	Discharge time	Energy density (Wh/kg)	Response time	Efficiency (%)	Self-discharge per day (%)	Lifetime years	cycles	Cost ($/kW)	Cost($/kWh)	Space requirements (m²/kWh)
PHS		Mature	100–5000	1–24h+	0.5–1.5	mins	65–80	Negligible	30–60	10000–30000	600–2000	5–100	0.02
CAES		Developed	5–300	1–24h+	30–60	s–min	40–50	Small	20–40	8000–12000	400–8000	2–50	0.01
FES		Demonstration	0–0.25	15 s – 15 min	10–30	s	90–95	100	15–20	20000	250–350	1000–5000	0.03–0.06
SMES		Demonstration	0.1–10	ms–8 s	0.5–5	ms	95–98	10–15	20+	20000+	200–300	1000–10000	6–26
Rechargeable batteries	Pb-Acid	Mature	0–20	s–3 h	30–50	<s	70–90	0.1–0.3	5–15	250–1500	300–600	200–400	0.058
	NiCd	Commercial	0–40	s–2 h	50–75	<s	60–80	0.2–0.6	15–20	1000–3000	500–1500	800–1500	0.03
Advanced batteries	Li-ion	Demonstration	0–0.1	min–2 h	75–200	<s	85–98	0.1–0.3	5–15	500–2000	1200–4000	600–2500	0.03
	NaS	Commercial	0.05–8	s–h	150–24	<s	75–90	20	10–15	2500–4500	1000–3000	300–500	0.019
Flow batteries	VBR	Developing	0.03–3	s–10 h	10–50	ms	75–85	Small	5–10	12000+	600–1500	150–1000	0.04
	ZnBr	Developing	0.05–2	s–10 h	60–80	ms	75–80	Small	5–10	2000	700–2500	150–1000	6–26
SC		Developed	0–0.3	ms–1 h	2.5–15	ms	85–98	2–40	8–20+	100000+	100–300	300–2000	0.04
H2 (FC)		Developing	0–50	s–24 h+	800–10000	ms	20–50	Negligible	5–15	1000+	500–1500	10–20	0.003–0.006

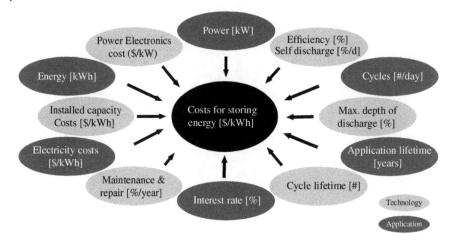

Figure 4.13 Cost calculation for energy storage system. Ref Num [39]. Reproduced with permission from SEFEP.

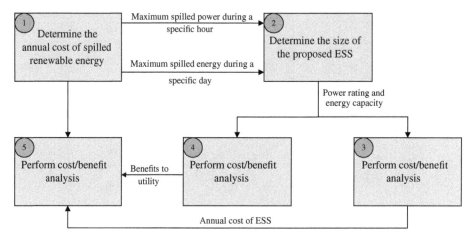

Figure 4.14 An example for optimal allocation procedure of ESS in distribution systems. Adapted from Ref [42].

services these units can deliver to stakeholders in the system, [40–42]. Figure 4.14 demonstrates a case of ESS optimal allocation method, [43].

4.7 Energy Storage Applications

Utility-scale energy storage systems are envisaged to form a crucial part of the SG. Relying upon the application, they can be employed in different segments of the electric system: generation, transmission, distribution and end-user. Each SG segment gives a number of potential opportunities to energy storage applications. Accurate location of the storage systems is essential to maximize the advantages offered. Large-scale, multi-MW, centralized storage may enhance generation and transmission load factors and system stability.

Smaller-scale, localized, or distributed storage may offer energy management and peak shaving services, and ultimately enhance the power quality and reliability. Distributed storage can be a perfect complement to distributed generation, more specifically on account of the rising levels of RES generation. ESS implementations are plentiful and diverse, ranging from larger scale generation and transmission systems to smaller scale implementations at the distribution network and the end-user's location. Energy storage can provide the following grid-services [44-47]:

- **Power quality services**, support utilization of electrical power while avoiding interferences or disturbances. Usually, power quality describes the maintenance of voltage and harmonics levels at certain limits.
- **Transient stability services**, supports synchronous operation of the grid at periods when the system faces sudden disturbances.
- **Regulation services**, correct short-term power imbalances that could potentially influence system stability.
- **Spinning reserve services**, offers on-line reserve capacity that can cover the electric demand in approximately 10 minutes.
- **Voltage control services**, generates or absorbs reactive power, and can maintain a certain voltage level.
- Arbitrage services, refer to utilizing generated power at off-peak hours to cover peak loads to accomplish load leveling/load shifting.
- **Load following/balancing services**, refer to modifying the power output as demand varies to ensure power balances in the system.
- **Firm capacity services**, offers energy capacity to cover peak power demand.
- **Congestion relief services**, decreases network flows in transmission constrained systems by raising the capacity of the lines or by offering different paths for the electrical energy.
- **Upgrade deferral services**, refer to deferring generation or transmission asset modifications by utilizing energy storage to decrease loading on the system for instance.

Figure 4.15 shows the allocation of these services' requirements in the power-discharge duration diagram, furthermore, these services can be used in different applications as illustrated in Figure 4.16 and summarized in Table 4.3 and the suitability of each ESS technology with different applications is summarized in Table 4.5. The application of ESS in SG can provide many benefits, Table 4.4 provides a sample list of energy storage technology benefits gained by different end users [48–56].

4.8 Barriers to the Deployment of Energy Storage

EES entails substantial investment and energy losses, that should be weighed against the advantages and compared to other non-storage solutions. There are three major barriers to ESS deployment: the first is associated with the economic feasibility of ESS business models, the second is the technological maturity and the third is the need for a regulatory system. The economic equation is complex to solve as ESS technologies require substantial initial investment expenditure and there is demand for combining many applications to achieve enough revenue. Therefore, the cost of each mature technology has to be reduced to be able

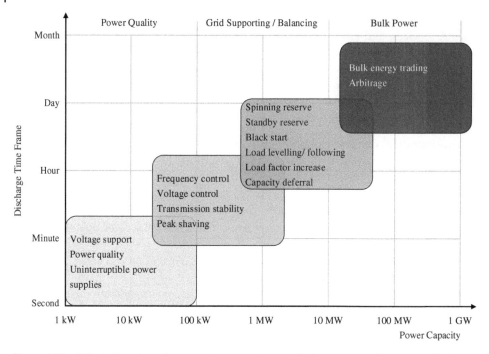

Figure 4.15 Selected services of energy storage systems with the corresponding power-discharge time requirements.

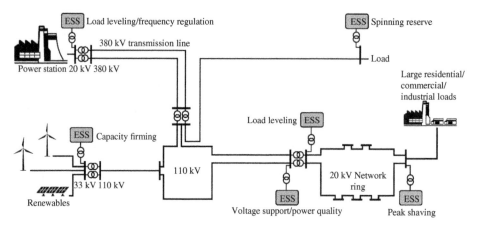

Figure 4.16 Typical grid energy storage applications at different voltage levels.

to compete with other alternatives. Furthermore, the ESS market has to be aligned with the SG needs. More large-scale deployment projects are required to push up the maturity level of different technologies. In addition, regulatory changes are required to quantify and monetize many applications of the ESS. There are a number of key barriers to increasing the use of storage systems as illustrated in Figure 4.17. Intensive research is needed to study the influence of these barriers on ESS implementation and methods to counteract them.

Table 4.3 Major energy storage applications.

Application	Description	Characteristics
Energy Arbitrage	Storing energy at low cost, discharged and sold in the event that costs are relatively high.	Discharge time of hours.
Generation Capacity Deferral	Discharge energy at the times of peak-load hours, decreasing the dependency on peaking generators.	Discharge times culminating up to a few hours.
Ancillary Services		
Regulation	The rise or decline of the total output of the storage to maintain the real-time balance between the system energy supply and demand.	Response time – seconds/minutes. Charge/discharge times are usually in minutes. Service is approximately zero net energy along a prolonged time span.
Contingency Reserves	A rise in the net output of storage to counteract contingencies, such as a generator or transmission outage.	Response time of minutes. Discharge time of up to a few hours.
Ramping	Follow hourly changes in electricity demand.	Response time – minutes/hours. Discharge time - hours.
T&D Capacity Deferral	Storing energy at low load of T&D and discharging at peak times.	Response time – minutes/hours. Discharge time - hours. Small-scale application is required, depending on a particular location.
End-User Applications		
Managing Energy Costs	Storing energy when retail price is small and discharging when cost is larger. Functionally comparable to Energy Arbitrage.	Response time of minutes to hours.
Power Quality and Reliability	Utilizing storage to enhance power quality (e.g. voltage, frequency, and harmonics). Discharging storage at a time-of-service outage.	Response and discharge time of seconds to minutes.
Curtailment service	Decrease the curtailment level of renewable systems for better economic benefit.	Response and discharge time – minutes/hours.

4.9 Energy Storage Roadmap

The traditional electric system – from central generation to end-user entailing the transmission and distribution grid in between – delivers several opportunities for energy storage's implementation. Storage allows for renewable integration by improving equilibrium between supply and demand. Once distributed generation is employed, storage is then decoupling the traditional value chain of the power sector. The prospect for energy storage

Table 4.4 Benefits of energy storage systems by users.

User	Benefit
Utilities	• Enhanced responsiveness of the supply and grid robustness. • Hinder the utilizations of peaking power plants. • Enhanced functions of transmission and distribution systems. • Grid upgrades can be avoided.
End-users	• Decreased electricity costs. • Decreased financial losses as a result of outages.
Independent System Operators	• Load balancing among regions. • Stabilization of transmission systems. • System control problems due to RES intermittent which could be alleviated. • More resiliency.

is massive; however, understanding the potential it possesses will need a technology guide to enable the development and integration of energy storage technology into SGs with thorough implementations from the respective individuals:

- **Technology Developers** are to turn their attention to authorizing and showing performance, decreasing costs, recognizing the most suitable applications for their respective technology, and improving the commercial relationships which will ultimately push them to start penetrating the market.
- **Adopters**, such as utilities, need to examine the potential for stacking storage applications, distribute knowledge obtained, and create new business models that optimize investment risks, and consumers on the other hand are required to comprehend the intention of storage, and the advancements that storage can present to the effectiveness of RES and smart appliances.
- **Regulators** are required to examine the importance of electricity storage in fulfilling the necessities of the systems and customers under their authority, and the extent at which regulatory
- changes could be wanted to allow for these new possibilities to be fit for that position efficiently and effectively, in the new world of opportunities stemmed from distributed generation and electricity storage. Long-term, stable policies are crucial in obtaining more investments as the investment community is reluctant to fund projects that may not deliver investment security.
- **Government Agencies** should provide funding and support for demonstration projects in high impact areas for energy storage industry. In addition, Government Agencies should promote storage technology and its potential to consumers, policy makers and regulators.
- **Research community** should do a complete study of interaction and optimization of storage with integrated grid elements and renewable sources.

This will decrease uncertainties in investigating the feasibility of energy storage technologies in the medium to long term and hence spearhead investment, advancement, and experience of utilizing energy storage technologies. Moreover, it is crucial that the

Application sector	Applications	Electrochemical				Mechanical				Electrical			Thermal
						CAES		PHS					
		Lead-acid	Lithium-ion	Nas	Flow battery	Underground	Above ground	Small	large	FES	SMES	SC	
Bulk energy	Energy arbitrage	●	●	●	●	▨	▨	○	○	●	●	●	○
	Peak shaving	●	●	▨	▨	▨	▨	○	○	●	●	●	○
Ancillary service	Load following	○	▨	○	○	▨	▨	▨	○	▨	●	●	○
	Spinning reserve	○	●	○	●	○	○	▨	●	▨	●	●	▨
	Voltage support	○	○	○	○	●	▨	●	●	●	●	●	●
	Black start	○	○	○	▨	▨	▨	▨	▨	▨	●	●	▨
	Frequency regulation Primary	○	○	○	○	▨	○	▨	●	●	●	●	●
	Secondary	○	○	○	○	○	○	▨	●	●	●	●	●
	Tertiary	○	○	○	○	○	○	○	▨	●	●	●	●
Customer energy management	Power quality	○	▨	▨	▨	●	●	●	○	●	○	○	●
	Power reliability	○	○	○	○	▨	▨	●	●	●	▨	●	●

(Continued)

Table 4.5 (Continued)

Application sector	Technology Applications	Electrochemical				Mechanical				Electrical			Thermal
						CAES	PHS						
		Lead-acid	Lithium-ion	Nas	Flow battery	Underground	Above ground	Small	Large	FES	SMES	SC	
Renewable energy integration	Time shift	▨	▨	▨	▨	⬤(gray)	⬤(gray)	⬤(gray)	⬤(gray)	⬤	⬤	⬤	⬤(gray)
	Capacity farming	▨	▨	▨	▨	⬤(gray)	⬤(gray)	⬤(gray)	⬤(gray)	▨	⬤	⬤	⬤

⬤ Suitable application   ⬤(gray) Possible application   ▨ Unsuitable application

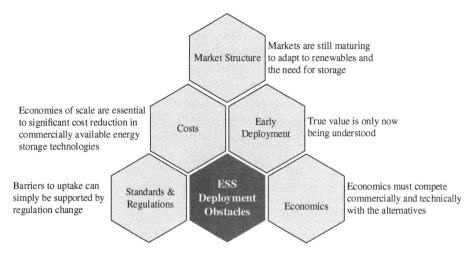

Figure 4.17 Energy Storage main deployment barriers.

importance of energy storage be distributed to renewable energy suppliers, and other grid's players and users. Teaching these stakeholders about transitions taking place in energy storage and its positive influence on people's lives is an essential endeavor that should be achieved.

Additionally, the future outlook for each energy storage technology is risky and not foreseeable, due to the fact that storage technologies are challenging one another in some applications and other solutions, for example, demand response (DR), improving the connection between power systems, and other simpler forms of power generation. When the storage market develops, it is projected that the market will concentrate on a decreased quantity of "winning solutions," profiting from mass production and the supply-chain scale.

4.10 Conclusion

The power grid should be improved to become more reliable, secure, and an efficient and clean network able to handle huge changes in the next 20 years. Energy storage will substantially alter the way the power industry functions. ESS will implement the SG concept which is one of the most advanced technologies in the future. Energy storage could alleviate grid congestion, decrease greenhouse gas emissions, enable the implementation of renewable energy systems, improve efficiency and deliver a more advanced approach to offer required capacity on the grid. The advantages of energy storage in power grids are complex to evaluate as they exceed the known boundaries of generation, transmission, distribution, and end-users. Hence, it is not yet clear how to distribute the costs and the profits of energy storage. Although, there are many barriers to overcome before ESS was an active part of the electric grid, many of these are associated with the present old-fashioned standard of power grid governance, regulatory, and market frameworks, which are unquestionably hindering investment in ESS implementation. Additional obstacles are of a technical nature, associated with the lack of comprehension of the technical advantages of employing ESS into the grid at numerous sections and with distinctive amalgamations of applications and control. Moreover, the price of the energy storing procedure is still relatively high, declining prices will lead to a rise in profitability of current applications of ESSs and offer new opportunities for more services and applications. Various ESSs with many technical features are found in the market. By knowing the pros and the cons of each technology, the most benefits can be achieved for many grid applications. Finally, integrating ESS into the electric grid is only one possibility of adding flexibility to the modern grid; many alternatives are there and will be investigated in the following chapters.

References

1 Miñambres, V.M.M., Guerrero-Martínez, M.Á., Barrero-González, F., and Milanés-Montero, M.I. (2017). A grid connected photovoltaic inverter with battery-supercapacitor hybrid energy storage. *Sensors* 17 (8): 1856.

2 Wen, Y., Guo, C., Kirschen, D.S., and Dong, S. (2014). Enhanced security-constrained OPF with distributed battery energy storage. *IEEE Transactions on Power Systems* 30 (1): 98–108.

3 Hajebrahimi, H., Kaviri, S.M., Eren, S., and Bakhshai, A. (2020). A new energy management control method for energy storage systems in microgrids. *IEEE Transactions on Power Electronics* 35 (11): 11612–11624.

4 IEA (2019). Renewables 2019- Market analysis and forecast from 2019 to 2024. https://www.iea.org/reports/renewables-2019 (accessed 11 February 2021).

5 Showers, S.O. and Raji, A.K. (2019). Benefits and Challenges of Energy Storage Technologies in High Penetration Renewable Energy Power Systems. *IEEE PES/IAS PowerAfrica Conference*, Abuja, Nigeria (20–23 August 2019). IEEE.

6 Mbungu, N.T., Naidoo, R.M., Bansal, R.C., and Vahidinasab, V. (2019). Overview of the optimal smart energy coordination for microgrid applications. *IEEE Access* 7: 163063–163084.

7 Akinyele, D.O. and Rayudu, R.K. (2014). Review of energy storage technologies for sustainable power networks. *Sustainable Energy Technologies and Assessments* 8: 74–91.

8 Zakeri, E. and Syri, S. (2015). Electrical energy storage systems: a comparative life cycle cost analysis. *Renewable and Sustainable Energy Reviews* 42: 569–596.

9 Luo, X., Wang, J., Dooner, M., and Clarke, J. (2015). Overview of current development in electrical energy storage technologies and the application potential in power system operation. *Applied Energy* 137: 511–536.

10 Chen, H., Cong, T.N., Yang, W. et al. (2009). Progress in electrical energy storage system: a critical review. *Progress in Natural Science* 19 (3): 291–312.

11 Parfomak, P.W. (2012). Energy Storage for Power Grids and Electric Transportation: A Technology Assessment. https://fas.org/sgp/crs/misc/R42455.pdf (accessed 11 February 2021).

12 Suberu, M.Y., Mustafa, M.W., and Bashir, N. (2014). Energy storage systems for renewable energy power sector integration and mitigation of intermittency. *Renewable and Sustainable Energy Reviews* 35: 499–514.

13 Guney, M.S. and Tepe, Y. (2017). Classification and assessment of energy storage systems. *Renewable and Sustainable Energy Reviews* 75: 1187–1197.

14 IRENA (2015). Renewables and electricity storage: A technology roadmap for REmap 2030. https://www.irena.org/-/media/Files/IRENA/Agency/Publication/2015/IRENA_REmap_Electricity_Storage_2015.pdf (accessed 11 February 2021).

15 IEA (2014). Technology Roadmap: Energy storage. https://www.iea.org/reports/technology-roadmap-energy-storage (accessed 11 February 2021).

16 IEA-ETSAP and IRENA (2015). Electricity Storage: Technology Brief. https://www.irena.org/publications/2012/Apr/Electricity-storage (accessed 11 February 2021).

17 World Energy Council (2016). World Energy Resources E-Storage 2016. https://www.worldenergy.org/assets/downloads/Resources-E-storage-report-2016.02.04.pdf (accessed 11 February 2021).

18 Mongird, K., Viswanathan, V.V., Balducci, P.J. et al. (2019). Energy storage technology and cost characterization report. No. PNNL-28866. Pacific Northwest National Lab, Richland, USA.

19 IFC, ESMAP (2017). Energy Storage Trends and Opportunities in Emerging Markets. https://www.esmap.org/node/57868 (accessed 11 February 2021).

20 San Martín, J.I., Zamora, I., San Martín, J.J. et al. (2011). Energy Storage Technologies for Electric Applications. *International Conference on Renewable Energies and Power Quality*, Las Palmas de Gran Caneria, Spain (13–15 April 2011). ICREPQ.

21 Sam, K.-K., Tyagi, V.V., Rahim, N.A. et al. (2013). Emergence of energy storage technologies as the solution for reliable operation of smart power systems: a review. *Renewable and Sustainable Energy Reviews* 25: 135–165.

22 Ibrahim, H. and Ilinca, A. (2013). Techno-Economic Analysis of Different Energy Storage Technologies. In: *Energy Storage - Technologies and Applications* (ed. A.F. Zobaa). IntechOpen.

23 Md, M.B., Md, S.A., Saha, T.K. et al. (2013). Towards implementation of smart grid: an updated review on electrical energy storage systems. *Smart Grid and Renewable Energy* 4 (1): 122–132.

24 Oberhofer, A. (2012). Energy storage technologies & their role in renewable integration. *Global Energy Network Institute*: 1–38.

25 Guney, M.S. and Tepe, Y. (2017). Classification and assessment of energy storage systems. *Renewable and Sustainable Energy Reviews* 75: 1187–1197.

26 Ould Amrouche, S., Rekioua, D., Rekioua, T., and Bacha, S. (2016). Overview of energy storage in renewable energy systems. *International Journal of Hydrogen Energy* 41 (45): 20914–20927.

27 Castillo, A. and Gayme, D.F. (2014). Grid-scale energy storage applications in renewable energy integration: a survey. *Energy Conversion and Management* 87: 885–894.

28 Ferreira, H.L., Garde, R., Fulli, G. et al. (2013). Characterisation of electrical energy storage technologies. *Energy* 53: 288–298.

29 Lazard (2016). Lazard's Levelized Cost of Storage-Version 2.0. https://www.lazard.com/media/438042/lazard-levelized-cost-of-storage-v20.pdf (accessed 11 February 2021).

30 Sabihuddin, S., Kiprakis, A.E., and Mueller, M. (2015). A numerical and graphical review of energy storage technologies. *Energies* 8: 172–216.

31 Barnhart, C.J. and Benson, S.M. (2013). On the importance of reducing the energetic and material demands of electrical energy storage. *Energy &Environmental Science* 6: 1083–1092.

32 Pellow, M.A., Emmott, C.J.M., Barnhart, C.J., and Benson, S.M. (2015). Hydrogen or batteries for grid storage? A net energy analysis. *Energy &Environmental Science* 8: 1938–11952.

33 Chen, B., Xiang, K., Yang, L. et al. (2019). Economic Analysis of Energy Storage System Based on LCC. *IEEE 3rd Conference on Energy Internet and Energy System Integration (EI2)*, Changsha, China (8–10 November 2019). IEEE.

34 Mahlia, T.M.I., Saktisahdan, T.J., Jannifar, A. et al. (2014). A review of available methods and development on energy storage; technology update. *Renewable and Sustainable Energy Reviews* 33: 532–545.

35 Kousksou, T., Bruel, P., Jamil, A. et al. (2014). Energy storage: applications and challenges. *Solar Energy Materials and Solar Cells* 120 (Part A): 59–80.

36 Zhao, H., Wu, Q., Hu, S. et al. (2015). Review of energy storage system for wind power integration support. *Applied Energy* 137 (1): 545–553.

37 ETSAP, IRENA (2012). Electricity Storage: Technology Brief. https://iea-etsap.org/E-TechDS/HIGHLIGHTS%20PDF/E18_HP_electr_stor_GS_April2012_rev4_GSOK%20EDIT_OK.pdf (accessed 11 February 2021).

38 "Thematic Research Summary: Energy Storage", Energy Research Knowledge Centre (ERKC), http://setis.ec.europa.eu/energy-research, September 2014, accessed on 12/07/2017.

39 Fuchs, G., Lunz, B., Leuthold, M., and Sauer, D.U. (2012). Technology Overview on Electricity Storage. *Smart Energy for Europe Platform GmbH*, Aachen, Germany (June 2012). SEFEP.

40 Carpinelli, G., Celli, G., Mocci, S. et al. (2013). Optimal integration of distributed energy storage devices in smart grids. *IEEE Transactions on Smart Grid* 4 (2): 985–995.

41 Khorrami, A., Mazidi, S., Nouri, A., and Safa, H. (2015). Optimal sizing of energy storage in order to improve microgrids reliability using contingency planning. *International Journal of Applied Engineering and Technology* 5 (4): 7–15.

42 Zidar, M., Georgilakis, P.S., Hatziargyriou, N.D. et al. (2016). Review of energy storage allocation in power distribution networks: applications, methods and future research. *IET Generation, Transmission & Distribution* 10 (3): 645–652.

43 Yun-feng, D. (2011). Optimal allocation of energy storage system in distribution systems. *Procedia Engineering* 15: 346–351.

44 Palizban, O. and Kauhaniemi, K. (2016). Energy storage systems in modern grids – matrix of technologies and applications. *Journal of Energy Storage* 6: 248–259.

45 Rodrigues, E.M.G., Godina, R., Santos, S.F. et al. (2014). Energy storage systems supporting increased penetration of renewables in islanded systems. *Energy* 75: 265–280.

46 Lucas, A. and Chondrogiannis, S. (2016). Smart grid energy storage controller for frequency regulation and peak shaving, using a vanadium redox flow battery. *International Journal of Electrical Power & Energy Systems* 80: 26–36.

47 Kyriakopoulos, G.L. and Arabatzis, G. (2016). Electrical energy storage systems in electricity generation: energy policies, innovative technologies, and regulatory regimes. *Renewable and Sustainable Energy Reviews* 56: 1044–1067.

48 Rastler, D. (2010). Electricity Energy StorageTechnology Options: A White Paper Primer on Applications, Costs, and Benefits. https://www.epri.com/research/products/000000000001022261 (accessed 12 February 2021).

49 Enescu, D., Chicco, G., Porumb, R., and Seritan, G. (2020). Thermal energy storage for grid applications: current status and emerging trends. *Energies* 13 (2): 340.

50 Vazquez, S., Lukic, S.M., Galvan, E. et al. (2010). Energy storage systems for transport and grid applications. *IEEE Transactions on Industrial Electronics* 57 (12): 3881–3895.

51 Safak Bayram, I., Tajer, A., Abdallah, M., and Qaraqe, K. (2015). Energy Storage Sizing for Peak Hour Utility Applications. *IEEE International Conference on Communications*, London, UK (8–12 June 2015). IEEE.

52 Eyer, J. and Corey, G. (2010). Energy Storage for the Electricity Grid: Benefits and Market Potential Assessment Guide. Sandia National Laboratories. https://www.sandia.gov/ess-ssl/publications/SAND2010-0815.pdf (accessed 12 February 2021).

53 Del Rosso, A.D. and Eckroad, S.W. (2014). Energy storage for relief of transmission congestion. *IEEE Transactions on Smart Grid* 5 (2): 1138–1146.

54 Battke, B., Schmidt, T.S., Grosspietsch, D., and Hoffmann, V.H. (2013). A review and probabilistic model of lifecycle costs of stationary batteries in multiple applications. *Renewable and Sustainable Energy Reviews* 25: 240–250.

55 Chambers, C. and Rozali, A. (2015). An overview of technology status and drivers for Energy Storage in Australia. Australian Renewable Energy Agency. www.aecom.com.

56 ABB (2012). Energy Storage: Keeping smart grids in balance. https://library.e.abb.com/public/59a2be960fdb777a48257a680045c04a/ABB%20Energy%20Storage_Nov2012.pdf (accessed 12 February 2012).

5

Microgrids

State-of-the-Art and Future Challenges

The microgrids have increased their penetration level in the existing power systems. Although their deployment is rapidly growing, there are still many challenges related to the design, control, and operation either in grid-connected or islanded modes. For these reasons, extensive research activities have been conducted to tackle these issues. This chapter presents a comprehensive review of microgrids including their control, operation, reliability, economic, protection, and communications issues.

5.1 Introduction

The microgrid, as defined by the U.S. Department of Energy, is "a group of interconnected loads and distributed energy resources (DERs) within clearly defined electrical boundaries that acts as a single controllable entity with respect to the grid and can connect and disconnect from the grid to enable it to operate in both grid-connected or island modes" [1]. According to this definition, the installations of DERs can be considered as a microgrid if they meet the following characteristics: they must have clearly defined electrical boundaries, to have a master controller in order to control and manage the DERs and loads as a single entity, and the installed generation capacity must exceed the peak critical load thus it could be disconnected from the utility grid, i.e. the islanded mode, and seamlessly supply local critical loads [2]. The microgrid can also be considered as a small-scale power grid that consists of DERs, loads, and controllers. One of the major advantages of microgrid is that it can operate in grid-connected or islanded modes that can generate, distribute, and regulate the power flow to local consumers. Furthermore, microgrids are completely different from uninterruptible power supplies and backup diesel generators which are known as backup generation. This kind of system can provide power to the local loads when the main power supply from the utility grid is interrupted. On the other hand, microgrids provide a wider range of benefits and are significantly more flexible than backup generation [2–4].

Although the microgrid concept has been gaining attention over the last decades, the first microgrid concept was built in 1882 by Thomas Edison. At that time, there were no centrally controlled and operated utility grids and his company installed around 50 DC

Smart Grid and Enabling Technologies, First Edition. Shady S. Refaat, Omar Ellabban, Sertac Bayhan, Haitham Abu-Rub, Frede Blaabjerg, and Miroslav M. Begovic.
© 2021 John Wiley & Sons Ltd. Published 2021 by John Wiley & Sons Ltd.
Companion website: www.wiley.com/go/ellabban/smartgrid

microgrids to supply local loads. However, these microgrids became part of the traditional grid by being connected due to the increased reliability requirements and economic benefits. On the other hand, nowadays, microgrid has become a popular topic due to its advantages that include increased reliability, higher efficiency, and a high penetration level of the renewable energy sources [5].

The simplified structure and components of the microgrid are depicted in Figure 5.1. It can be seen that the microgrid consists of various distributed generation units (DGs), energy storage systems (ESS), loads, control systems, intelligent switches, protective devices, and communication layers [4, 5]. The loads in the microgrid can be divided into two categories: critical and non-critical. The critical loads should be supplied constantly to ensure energy security for such loads while the non-critical loads can be curtailed or deferred according to the available power in the microgrid. DGs are the primary power source in the microgrid to supply loads [4]. The selection of the DGs (e.g. photovoltaic [PV], wind, diesel generator, microturbine) depends on the geographic location of the microgrid. ESSs also play a crucial role in the microgrid to guarantee microgrid generation adequacy especially when the microgrid operates in islanded mode. Furthermore, ESSs can be employed for energy arbitrage; for example, energy can be stored when energy is at a low price in the market and can be sold when the market price is high. In addition to these components, smart switches and protective devices are important for managing the connection among the DGs, ESSs, loads, and the main power grid. These devices are also important in preventing a fault in the microgrid. For example, in case of a fault in part of the microgrid, switches and protective devices only disconnect the faulty area. Thus, the microgrid can operate and loads can be supplied. Furthermore, the switch at the point of common coupling (PCC) performs microgrid islanding by disconnecting the microgrid from the utility grid. Control and communication systems are used to manage all microgrid components in harmony to guarantee system stability [5–8].

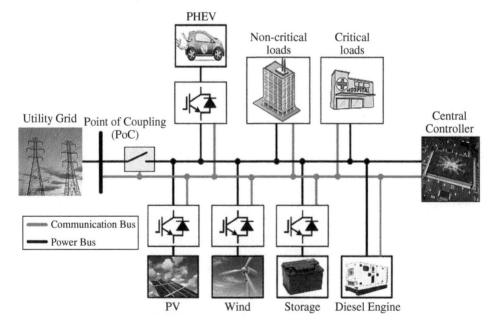

Figure 5.1 The simplified single-line diagram of a microgrid.

5.2 DC Versus AC Microgrid

Over the last few decades, electrical energy systems have faced stress conditions more often because of increased electricity demand. The traditional grid is not able to respond to the huge increase in electricity demand and to the integration of a large amount of stochastic in natural renewable energy resources, in addition to newly introduced advanced technologies such as electric vehicles (EVs). Undoubtedly, at some point within the next decade, intensive electricity demand particularly for clean energy, will exceed total electricity centralized generation and transmission capacity, which will force decision makers to improve the traditional power grid and convert it into an SG [9]. Fortunately, modern solutions – demand side management, flexible AC transmission system, ESS, renewable energy technologies, energy management systems (EMSs), and advanced control and communication systems – have made it possible for energy engineers and researchers to redesign the conventional power systems and convert them into SGs. On the other hand, many of these promised solutions are still in the developmental stages and more research is needed to make such advanced systems a reality [6].

The DC microgrid systems (DC distribution system) are mainly used in marine, automotive, and avionics for power distribution. Furthermore, development in power electronics has increased the use of DC microgrids in spacecraft, data centers, manufacturing industries, and modern aircraft since such systems need DC power supplies. Besides, most of the renewable energy sources generate DC power or AC power with variable frequency and voltage characteristics. Although, initial research has been focused on the AC microgrid, DC microgrid studies have been accelerated over recent years due to the development in power electronics technology [10, 11].

In this subsection, low-voltage alternating current (LVAC) and low-voltage direct current (LVDC) networks will be described. Then, AC and DC microgrids will be presented with comprehensive comparisons.

5.2.1 LVAC and LVDC Networks

Some of the DG units such as wind turbines, diesel generators, and low head hydros generate AC output power and these units can be connected to the AC bus either directly or through the power electronic converters to ensure the frequency and voltage level requirements. Furthermore, the DG units that generate DC output power such as solar PV modules, energy storage devices, and fuel cells can be connected to the AC bus of the LVAC networks using power converters. Similar to DG units, AC loads are directly connected while the DC loads need the AC/DC power converters in order to be connected to the LVAC networks. Figure 5.2a illustrates the typical configuration of the DG units with the AC power output (e.g. wind turbines) and that with the DC power output (e.g. PV systems and fuel cells) connected to a LVAC network. In this configuration, the LVAC network connects to the main system through the transformer [6].

As consumer equipment and power electronics impact is increasing in daily life, the DC distribution system may become an alternative way to supply all electrical equipment in the future's SG. In contrast to an LVAC network, AC DG units in the LVDC network must be connected to a DC bus through the power electronic converters (AC\DC) to ensure the voltage level and smooth power flow between DGs and loads.

Figure 5.2b shows a simplified configuration of the LVDC network with DG units. It can be seen that similar to the LVAC network, the LVDC network configuration consists of AC- and DC-based DG units. Although DC-based DG units connect to the DC bus directly, the AC-based DG units require power electronic converters to connect to the DC bus [12].

5.2.2 AC Microgrid

A typical AC microgrid system is shown in Figure 5.3. Both LVAC networks are connected at the PCC while the loads are supplied from the local DG units or the main utility network. The AC microgrid can operate either in grid-connected mode or

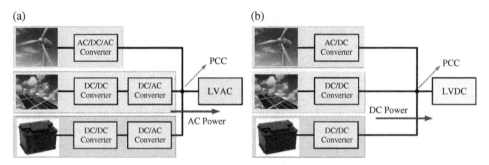

Figure 5.2 Typical configuration of the DG units with (a) LVAC network; (b) LVDC network.

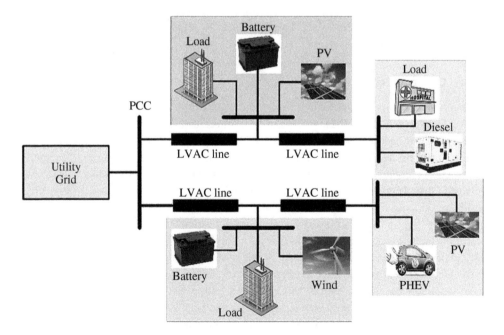

Figure 5.3 AC microgrid structure with DG units and mixed types of loads.

Table 5.1 Some examples of AC microgrid systems (http://microgridprojects.com). Adapted from

Location	Components and Capacity	References
Marcus Garvey Village, NY, US	400 kW Fuel Cell 400 kW Rooftop Solar PV Chem Batteries (300/1200 kW/kWh)	[15]
Ta'u, Manu'a, American Samoa	1.4 MW Solar PV 6 MWh battery storage	[16]
San Lorenzo, Panama	355 kW Solar PV 200 kW Diesel Generators	[17]
Kisii, Kenya	47 KW Solar PV	[18]
San Diego, CA, United States	8000 kW Solar PV 16 400 kW Gas/Diesel Generator	[19]
Milford, CT, United States	120 kW Solar 292 kW Gas/Diesel Generator 30 kW Storage	[20]
Philadelphia County, PA, United States	1000 kW Solar PV 9000 kW Gas/Diesel Generator 600 kW Fuel Cell	[21]
Azpeitia, Spain	3.6 kW Solar 6 kW Wind 160 kW Gas/Diesel Generator 1 kW Fuel Cell	[22]
Mannheim, Germany	33.8 kW Solar 14.5 kW Gas/Diesel Generator 4.7 kW Fuel Cell 1.2 kW Flywheel Storage	[23]
Zhejiang, China	120 kW Solar 120 kW Gas/Diesel 50 kW Fuel Cell 150 kW Capacitor and Batteries Storage	[24]

islanded mode. The operating mode of the AC microgrid depends on the power produced from DGs and the load demand in the microgrid. In islanded mode, the power generated by the DGs is sufficient to meet load demand. On the other hand, in grid-connected mode, load demand is either higher than the generated power or lower than the generated power [13]. If the generated power is higher than the local load demands, the excess power is injected into the main grid. The opposite of this situation is also true. If the load demand is higher than the power generated by the DGs, the difference between demand and generation is met by the utility. There are many installed AC microgrid systems all over the world [14]. Some examples of AC microgrid systems are given in Table 5.1.

5.2.3 DC Microgrid

Computers, LED lights, variable speed drives, house appliances, and industrial equipment are some of the examples of modern-day life devices that need DC power for their operation. On the other hand, these devices require a conversion stage to convert AC power into DC for operation. Furthermore, most of the DG units generate DC power. However, to be connected to the existing AC network, DG units need a DC-to-AC conversion stage. These DC–AC–DC power conversion stages result in substantial energy losses. To overcome these drawbacks, the LVDC distribution network is a new concept and it will play an important role in tackling the current power distribution problems and realizing the future power system [6, 25].

Recently, there has been a growing interest in the DC power system, which is gaining more importance in telecommunication systems, data centers, point-to-point transmissions over long distances through sea cables and for interconnecting AC grids with different frequencies [25, 26]. Figure 5.4 illustrates the typical DC microgrid system connected to the main utility system at PCC. The utility system can either be a medium voltage AC (MVAC) or high voltage DC (HVDC) transmission line connecting a long-distance DG unit such as an offshore wind farm. Some examples of the DC microgrid systems are given in Table 5.2.

Although the AC distribution network has dominated the traditional power system, the DC distribution system is becoming a more important concept for the future power grid. The main reason for this is the advantages of a DC power system over the AC power system. For example, DC power systems require less cable for transferring energy from one point to another. Furthermore, due to the lack of frequency, there is no reactance in the line which reduces line losses. Table 5.3 shows a comprehensive comparison between AC and DC distribution lines [6].

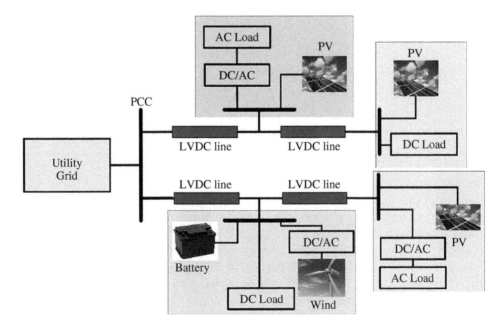

Figure 5.4 Concept of a DC microgrid system with the DG units and mixed types.

Table 5.2 Typical examples of DC microgrid systems.

Location	Voltage Range and Capacity	References
Karnataka 575 011, India	14 kW Solar PV 72 kWh Storage	[27]
Sweden UPN AB for Data center IBM	24–350/380 LVDC (bipolar DC-link) ≥5 MW	[28]
Japan NTT Group for data centers	380/400 LVDC (bipolar DC-link) ≥5 MW	[28]
Intelligent DC Microgrid Living Lab, Denmark	380/42/24/12 LVDC	[29]
Algeria Two Steam Turbines-Testing prototype	800 V – 1200 LVDC, 4.8–18 kW	[30]
China PV arrays, BESS & AC utility system	360 V – 420 LVDC, 150–945 W	[31]

Table 5.3 Comparison between AC distribution lines and DC distribution lines. Ref Num [6]. Reproduced with permission from ELSEVIER.

AC distribution line	DC distribution line
Less efficiency due to high line losses, hence less power could be transmitted	More efficiency and power could be transmitted for the same size
Requires more conductors for the three lines and neutral	Requires fewer conductors
Easily affected by external disturbances so it is less stable	More stable
Has reactance in the line due to the AC nature	No reactance in the line which allows for less voltage drop and more power transmitted
Frequency monitoring is mandatory	Frequency is zero, so monitoring is not required
Transient stability is an issue due line clearance	No transient stability problems
Electromagnetic interference is an issue which must be taken into consideration	No electromagnetic interference
Higher line resistance due to skin effect and hence high losses	Lower line resistance; hence, lower line losses
Involves complex numbers; hence difficult to analyze	Involves only real numbers, i.e. simpler

5.3 Microgrid Design

The requirement for the new power system structure due to economic and environmental concerns has inspired the development of microgrids. The microgrid concept will bring considerable benefits to the transmission and distribution networks because of its advantages that include a high degree of efficiency and reliability, high-level renewable energy sources (RES) penetration, and low power losses. Although the microgrid system has significant benefits, its needs to be tackled carefully to achieve optimal efficiency and reliability. This subsection presents microgrid design steps from different perspectives.

5.3.1 Methodology for the Microgrid Design

Although the microgrid structure takes various shape and procedures which depend on its aim, geographic location, and techno-economic analysis, the technical structure for almost all microgrid systems frequently remains constant. In general, microgrid systems consist of five main components that include a power source, ESS, loads, EMS, and utility connection.

i) *Power Source:* The fundamental component for any microgrid system is the power source. The selection of energy generation systems depends on geographic and climatic conditions, availability of conventional energy sources at the site, as well as the overall financial characteristics of the project. For example, a solar PV-based power source has been popular in Middle East countries whereas the wind-based power source is the main energy source in Denmark. For that reason, location plays a critical role for the selection of power sources in the microgrid. Furthermore, the installation and maintenance cost of the power sources in the microgrids are other arguments for selection of power sources. Recent price drops in PV solar-power modules made solar-based microgrids more favorable, and the accessibility of solar energy over the world further enhances the favorability of using solar power in many regions [32]. Figure 5.5 shows the price learning curve for c-Si and thin-film PV technologies between 2006 and 2017 [33]. It is clear that the price of PV panels regularly reduces. For more information about the different RESs are presented in Chapter 2.

On the other hand, the maintenance of the power source is critical in maintaining the optimum efficiency from the microgrid system. For example, high moisture levels can degrade the performance of the PV panel, so a regular maintenance should

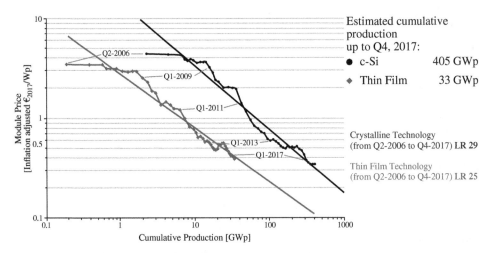

Figure 5.5 PV solar power modules price learning curve for different technologies [33]. Reproduced with permission from Fraunhofer Institute for Solar Energy Systems ISE.

be scheduled. Furthermore, intermittent and stochastic nature of RESs leads to instability and create power quality problems in microgrids. Therefore, further power sources that include microturbine, diesel generator, ESS should be employed to ensure high frequency and low frequency power stability in the microgrid. Each power source added in the microgrid improves system reliability whereas the overall microgrid cost is increased. For that reason, the selection of power sources in the microgrid is challenging so the power sources must be carefully analyzed to obtain effective microgrid system.

ii) *ESS:* ESS plays a vital role in the microgrid operation to maintain stability and robustness as well as to improve the power quality. For these reasons, an effective ESS must characterize high power density as well as high energy density. In recent years, various types of battery technologies are used for ESS. A review of common energy storage technologies for various capacities can be found here [9]. In spite of their maturity and variety, batteries still have a limited lifecycle and poor power and energy density, which are important elements for balancing the system when connected to renewable generation. Thus, to support and improve the battery performance, lifetime and system cost, hybrid energy storage systems (HESS) have been suggested while comprising supercapacitors and batteries [34–37]. There are several advantages of supercapacitors such as: (i) high efficiency (95%), (ii) high power density (up to 10 000 W/kg), (iii) tolerance for deep discharges, and (iv) long lifecycle (500 000 cycles at 100% depth-of-discharge) [37]. The disadvantage of the super-capacitor (SC) is the very low energy density. The combination of SCs and batteries permit the advantages of both solutions by obtaining high energy density, high power density, high lifecycle, high efficiency HESS and ensures better power stability when interfaced to the grid [38]. More information about ESS systems can be found in Chapter 5.

iii) *EMS:* The EMS coordinates the available energy in the microgrid. The main role of the EMS is to collect data from DGs, evaluate this data and send commands to the DGs to ensure demand-response balance among the power sources, ESSs, and loads. Modern microgrid systems often integrate software and control systems that can manage grid operation in an efficient and reliable manner [39].

iv) *Loads:* One of the main components of the microgrids are the loads that consume the energy. Loads in the microgrid can be divided into two categories; critical loads and non-critical loads. In islanded mode, the generated power from DGs is limited, and often it is not enough to meet load demand. Furthermore, the intermittent nature of RESs results in uncertainty on the generated power. For that reason, the control in islanded mode is commonly accomplished by load-shedding techniques. A non-critical load is shed gradually while the critical load is supplied continuously to prevent a complete outage [4].

v) *Utility Connection:* One of the advantages of microgrid is that it can operate in either grid-connected or islanded mode. A grid-connected operation indicates a situation where a microgrid is connected to the main grid, whereas an islanded operation indicates a situation where a microgrid is disconnected from the main grid in case of a fault condition in the main grid [4]. Therefore, it is very important to ensure utility connection. This connection enables the microgrid to exchange power with the utility grid when necessary.

5.3.2 Design Considerations

The electricity network is transforming, moving away from the traditional centralized grid and employing more distributed generations which are mainly based on RES. Thus, in the near future, the microgrids will be more active in the main utility network. Furthermore, the electricity network of many islands and rural areas around the world has been operated in complete isolation from the interconnected network, so these are also referred to as microgrids. As the microgrid begins to become so widespread, we need to consider the factors that affect the microgrid design.

    i) *Ownership:* Microgrids can be divided into five ownership categories: commercial (industrial); community (utility); campus; military; and off-grid. For example, the Borrego Springs' microgrid serves that community and is owned and operated by the local San Diego utility, SDG&E. The U.S. military has been implementing microgrids, both at stationary bases in the U.S. and at forward-operating bases overseas [7].

    ii) *Operating modes:* Most microgrids are connected to the utility grid with an option to disconnect and operate in islanded mode. On the other hand, some microgrids operate in islanded mode without interacting with the main utility grid such as in islands and remote villages and communities [40].

    iii) *Microgrid size:* Another key factor of microgrid is the size. Microgrid size can be expressed in terms of installed generation capacity, peak load, number of people served, or amount of land covered. The microgrid size depends on the system considered, it can be single home with a couple of kilowatts or islands that have several megawatts of peak load [7].

    iv) *Main purpose:* As another factor, microgrids may be built to serve different purposes. In some cases, microgrids are designed as an emergency backup for the critical loads that include hospitals, data centers, etc. In contrast, some microgrids are designed to operate at all times such as microgrids in islands and rural areas. It is quite possible that some microgrids fall somewhere in between. For example, the microgrid generates power in parallel with the grid to meet load demand during normal operation but can also meet the critical-load demand during a power outage in the main grid.

    v) *Covering range:* Along with its general purpose somewhere on the spectrum between emergency backup and continuous operation, another key question is how much load the microgrid is designed to cover. Some microgrids are designed to cover only the critical loads while others are designed to cover for all loads. Please note that microgrids in islands and rural areas are fairly designed to meet all kind of loads (critical and non-critical).

    vi) *Durability:* In some cases, microgrids consist of distribution generation and storage units. Therefore, such microgrids are designed to supply power for a certain amount of time. In some applications such as rural communities, microgrids are designed to provide continuous power delivery using local generation (and possibly some storage). Other types of microgrids are designed only for emergency backup, such as those that power critical infrastructures such as substations and hospitals, so they may heavily rely on storage availability or may have a fixed amount of fuel for traditional generation resources. Those microgrids are designed to provide backup power for a fixed period of time (e.g. 48 hours) during an emergency and until the main grid is restored [7].

5.4 Microgrid Control

In the control stage, centralized or decentralized control of the microgrid aims to optimize the production and consumption of electricity to improve overall efficiency. Local controls are the basic category of microgrid control. The challenge is that several sources must be able to connect or disconnect from the distribution grid whenever and wherever needed. Moreover, controlling a large number of different sources with different characteristics will be challenging because of possible conflicting requirements and limited communication. Furthermore, intermittent generation of certain microgrids, coupled with the unpredictable nature of the consumer demand, implies that these microgrids may, at certain points in time, come to exchange extra energy with each other instead of requesting it from the main grid. So, it has become an issue of great urgency that grid operators attempt to devise an excellent supply-and-demand management framework and reduce the amount of power wasted during transmission over the distribution lines. Such a requirement, together with recent developments in electric power systems and the interconnections between neighboring electric power systems to achieve a more economic and secure system, has led to tremendous difficulties in monitoring, controlling, and managing electric power systems [10, 40–46].

In the presence of microgrids, hierarchical distributed control is employed to provide a coordination strategy for the DERs, together with the storage devices and flexible loads. In hierarchical distributed control, the microgrid is divided into a number of areas where each area can include distributed energy sources, loads, and power lines. The primary motivation behind the implementation of a distributed control structure is the inability of the different areas' operators to share information and data [47]. A hierarchical control architecture can be divided into three control levels based on the required time frame: primary, secondary, and tertiary as shown in Figure 5.6.

5.4.1 Primary Control Level

The primary control level is the first control known as "internal control." This control level works on system variables such as voltage and frequency components to ensure that these variables track their set points. Primary control techniques in microgrids can be divided into two main categories: communication-based and droop characteristic based [46, 50]. The main advantage of communication-based control techniques is that the amplitude and frequency of the inverter output voltage are close to their reference level without using additional control levels. Concentrated control [46] master/slave control [51], and distributed control [53] techniques are good examples of communication-based control. Although communication-based control techniques provide proper power sharing among the DG inverters, these control strategies bring some drawbacks such as the requirement for long-distance communication lines, which increases the overall system cost and reduces system reliability. On the other hand, droop characteristic-based control techniques ensure voltage regulation and power sharing without communication medium [46, 49, 51, 52]. These control techniques are suitable for long-distance DG inverters to ensure proper power sharing while achieving good voltage regulation. These techniques also eliminate the complexity and high cost of communication-based control techniques. The droop-based control can be

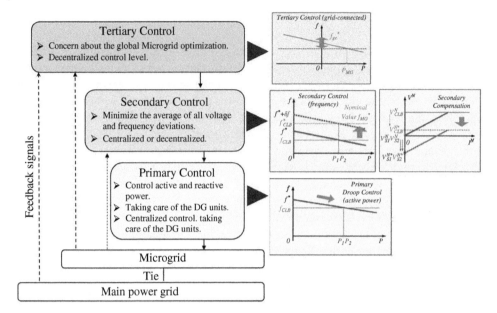

Figure 5.6 The hierarchical control structure of the microgrid. Adapted from Ref num [48].

divided into three main categories: (i) traditional droop control [54, 55]; (ii) virtual framework structure-based method [55–58]; (iii) the hybrid droop/signal injection method [58, 59].

5.4.1.1 Droop-Based Control

The basic idea of the droop control technique is to imitate the behavior of a synchronous generator whose frequency is reduced as active power increases [9]. Figure 5.7 depicts the simplified single-line diagram of the microgrid with two DG units. The active and reactive power of nth converter connected to the AC microgrid can be defined as:

$$\left. \begin{array}{l} P_n = \dfrac{VE_n}{X_n}\sin\varphi_n \\[2mm] Q_n = \dfrac{VE_n\cos\varphi_n - V^2}{X_n} \end{array} \right\} \tag{5.1}$$

where E_n is the inverter output voltage, V is the PCC voltage, X_n is the inverter output reactance, φ_n is the phase angle between the inverter output voltage and PCC voltage. Please note that these equations apply when the inverter output impedance is relatively inductive. The reactive power depends on the E_n, whereas the active power is dependent on the φ_n. Using this information, the P/ω and Q/E droop characteristics can be drawn, as shown in Figure 5.8 [8]. These characteristics can be explicated as follows: when the frequency decreases from ω_0 to ω, the DG is allowed to increase its active power from P_0 to P. A falling frequency is evidence of an increase in the active power demand of the system. In other words, parallel connected units with the same droop characteristic increase their active power outputs to handle the fall in frequency. Increasing active power of the parallel units will prevent the fall in frequency. Thus, the units will settle at active power outputs and

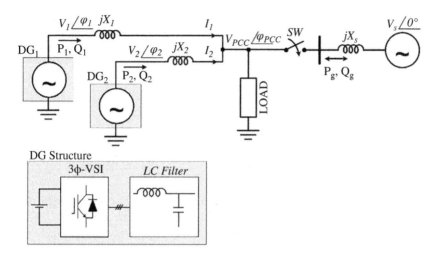

Figure 5.7 The simplified single-line diagram and DG structure of microgrid [45]. Reproduced with permission from IEEE.

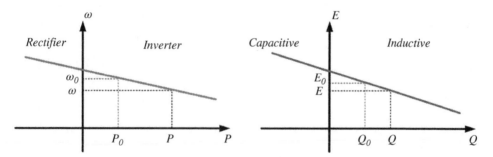

Figure 5.8 P/ω and Q/E droop characteristics [8]. Reproduced with permission from IEEE.

frequency at a steady-state point on the droop characteristic. To sum up, the droop characteristic based control technique allows parallel connected units to share load without units affecting each other.

Although this idea is common in synchronous generator control, it can be extended to control the voltage source inverters (VSIs). The P/Q droop technique is expressed as:

$$\left. \begin{array}{l} \omega_n = \omega_0 - k_P \left(P_n - P_0 \right) \\ E_n = E_0 - k_Q \left(Q_n - Q_0 \right) \end{array} \right\} \tag{5.2}$$

where n is the index representing each VSIs, ω_0 and E_0 are the base frequency and voltage, respectively, P_n and Q_n are the active and reactive power of the units, respectively. P_0 denotes the base active power while Q_0 denotes the reactive power. k_P and k_Q are the droop coefficients that directly affect the performance of the system.

The selection of k_P and k_Q affects the stability of the system, so these parameters should be designed appropriately. Each DG unit generates power proportional to its capacity with appropriately selected parameters. The active and reactive proportional droop coefficients (k_P and k_Q) can be expressed by:

$$\left.\begin{aligned} k_P &= \frac{\Delta\omega_{max}}{P_{max}} \\ k_Q &= \frac{\Delta E_{max}}{Q_{max}} \end{aligned}\right\} \tag{5.3}$$

where, $\Delta\omega_{max}$ is the maximum voltage frequency droop allowed; ΔE_{max} is the maximum voltage amplitude droop allowed; P_{max} is the maximum active power allowed; and Q_{max} is the maximum reactive power allowed.

The traditional droop control technique is depicted in Figure 5.9. The power stage consists of an input power source such as a PV panel, DC to AC converter, and LC filter with line inductor. The control structure consists of three control loops; (i) power-sharing loop; (ii) voltage-control loop; and (iii) current-control loop.

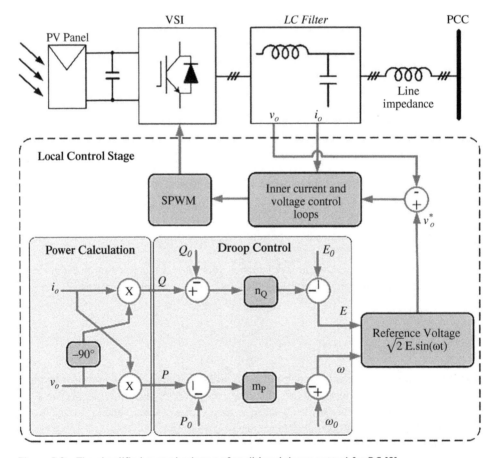

Figure 5.9 The simplified control scheme of traditional droop control for DG [8].

- To obtain the amplitude and frequency of the reference voltage according to the droop characteristic, a power-sharing loop is needed to be used.
- To control output voltage of the LC filter, a voltage-controller needs to be used.
- To control output current, a current controller is required.

Although the traditional droop control is easy to implement in the microgrid applications, it has several disadvantages. Due to the use of average active and reactive powers, there is a slow transient response and instability during fast- changing loads. This leads to a slow and oscillating dynamic response and steady state deviations. In choosing the droop coefficients, there is a tradeoff between the amplitude of the droop and the system stability. Large droops speed up the load sharing but can cause instability. Smaller droops slow down the control but are more stable. In addition, because the proportional controllers are without integral terms, the frequency and voltage in the microgrid are not constant but load dependent. It also has the inability to provide accurate power sharing among the DG units due to output impedance uncertainties as it does not account for current harmonics in the case of non-linear loads. Figure 5.10 summarizes the effect of the type of distribution network on this base relationship between the voltage, frequency, active power, and reactive power. When the impedance is highly inductive or resistive, the power angle δ is assumed to be small and therefore the following approximations are assumed, $\sin\delta \approx \delta$ and $\cos\delta \approx 1$.

It can be seen from Figure 5.10 that the ω/P characteristic changes to ω/Q droop, and the V/Q droop shifts to one with V/P as the line impedance changes from high to lower voltage lines. Therefore, impedance has to be tracked carefully to ensure droop control is working in the required mechanism. To resolve this issue, the implementation of virtual impedance

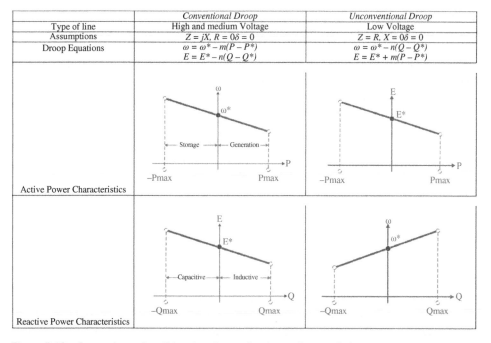

	Conventional Droop	Unconventional Droop
Type of line	High and medium Voltage	Low Voltage
Assumptions	$Z = jX, R = 0 \delta = 0$	$Z = R, X = 0 \delta = 0$
Droop Equations	$\omega = \omega^* - m(P - P^*)$ $E = E^* - n(Q - Q^*)$	$\omega = \omega^* - n(Q - Q^*)$ $E = E^* + m(P - P^*)$
Active Power Characteristics		
Reactive Power Characteristics		

Figure 5.10 Comparison of traditional and opposite droop characteristics.

has been explored in [56–58]. The above droop method, while producing reliable control over voltage and frequency, does not protect the network system from current harmonics and therefore proves to be inefficient in the presence of non-linear loads. In [59], a quadratic droop control is discussed for voltage regulation. While its steady state is the same as a traditional droop, its quadratic form allows the circuit theory technique to be applied for easier analysis purposes in islanded microgrids. However, this approach does not account for the R/X ratio of non-linear loads in a system.

5.4.1.2 Communication-Based Control

Control schemes that include master–slave control or multi-master control make use of high communication systems to regulate voltage and current between VSI [60]. They provide good power sharing in steady state and transients in contrast to non-communication-based methods mentions previously. However, while they provide sustainable frequency and voltage control, the use of cables and supervisory control make them less efficient and resistant to expansion. They require high bandwidth communication and degrade the modularity of the microgrid system [61].

In [62], a droop-free mechanism is proposed for the control of distributed VSI. The dependency of frequency and voltage on load is eliminated and it can also be used with non-linear loads. The controller uses three separate regulators for voltage, reactive power, and active power and, in addition, a sparse communication network. The control layer consists of a cyber physical network that enables sources to send information to their immediate neighbors, allowing synchronization of normalized power and frequencies. Note that not all sources need to communicate, therefore, is sparser and less complex. The voltage regulator and reactive power regulator each produce correction terms which are added to the rated voltage to produce voltage set points. Each of the voltage regulators allows for different bus voltages and account for individual impedance levels. Therefore, by working together, these two regulators allow load sharing for a variety of distribution networks. The active power regulator compares the local normalized active power of each inverter with its neighbors and uses the difference to update the frequency and phase angle of that inverter. Thus, it manages frequency correction without taking any frequency measurements.

In [62], a reduced communication technique is proposed for the coordination of intermittent DGs in a fully distributed multi-agent system. The load-sharing algorithm makes use of bus agents to communicate locally with its neighboring agents as well as a distributed information processing law for global information on the total power generation and demand available. The algorithm works in an aperiodic manner to reduce communication. Each bus agent updates its control protocol only when an agent or neighboring agent is triggered. An agent is triggered only when there is a significant deviation in the value of present time state from the last updated state.

5.4.2 Secondary Control Level

Secondary control is used to cope with the voltage and frequency deviations in the microgrid. It operates at a slower time frame than the primary control to reduce the communication bandwidth by using sampled measurements of the microgrid variables, and to allow enough time to perform complex calculations [9]. This control level is also considered as

EMS for the microgrid. The secondary control structure can be classified into two main categories: centralized and decentralized control. The centralized control approach requires extensive communication medium between DG units and a central controller. On the other hand, in the decentralized control approach, each DG unit is by the local controller. The general structure of the centralized and decentralized control approaches is depicted in Figure 5.11. In both approaches, some basic functions are centrally available, such as local production, demand forecasting and security monitoring [62].

Selection of the suitable control approach for microgrid control depends on the objectives and characteristics of the considered microgrid. Furthermore, the available resources (personnel and equipment) are an important parameter to decide control approach between the centralized and decentralized. The comparison between centralized and decentralized control approaches is summarized in Table 5.4 in terms of the various parameters.

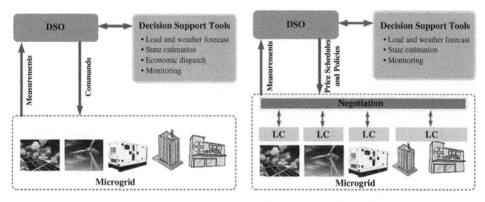

Figure 5.11 General structure of centralized and decentralized control approaches. Adapted from Ref [62].

Table 5.4 Comparison of centralized and decentralized control approaches. Adapted from Ref Num [62].

	Centralized Control	Decentralized Control
DG ownership	Single owner	Multiple owners
Goals	A clear, single task, e.g. • minimization of energy costs	Uncertainty over what each • owner wants at any particular moment
Availability of operating personnel (monitoring, low level management, special switching operations, etc.)	Available	Not available
Market participation	Implementation of complicated algorithms	Owners unlikely to use complex algorithms
Installation of new equipment	Requirements of specialized personnel	Should be plug-and-play (PnP)
Communication requirements	High	Modest
Market participation	All units collaborate	Some units may be competitive

5.4.3 Tertiary Control Level

The highest level of control in the microgrid control hierarchy is the tertiary control. The main role of tertiary control is to manage the power flow among the units to improve performance of the overall system. This control level also ensures optimal operation of the microgrid by considering technical and economic aspects. The main structure of this control can be either centralized or decentralized. However, cost is the key element in selecting a suitable control approach for the tertiary control level. The centralized control approach is more expensive than the decentralized control approach as its performance highly relies on a fast communication system and requires a special and strong coordination between protection and control systems. For that reason, the multi agent-based control system (MAS) is offered in the tertiary control level [62, 64].

5.5 Microgrid Economics

There are some key questions that microgrid investors need to know which are: Are microgrids economical? What is the amount of investment needed for what purpose? What are the factors influencing profitability within and outside the microgrid community? If the microgrids are not profitable currently then when and how they could be? Unfortunately, answering all these questions is not easy and requires complicated modeling and mathematics. On the other hand, understanding microgrid costs involves no less than two complex optimizations followed by cash flow analysis [62–66]. To the best of the authors' knowledge, this has not been done.

Investors expect to generate energy from microgrids at the lowest cost possible; with the lowest emissions; and continuous operation with high quality electricity. In fact, to design a cost-effective microgrid system, we can use existing ship design procedures. We can assume that a container ship is a floating microgrid. Its propulsion needs tens of MW engine power which is usually generated by diesel engines. Furthermore, the ship has tens of MW power system for a cargo refrigerator, appliances for the crew's living area, a communication infrastructure, and various loads similar to any community. In general, the main power source of the ship is its diesel engines. On the other hand, similar to a microgrid, modern ships also have batteries and waste heat recovery systems for cost saving and reliability. Therefore, similar design steps may be used for ships to design optimized microgrids.

5.5.1 Capacity Planning

The major challenge of microgrid design is capacity planning. What kind of DGs (solar, wind, diesel generator, etc.) should be used to meet the load demand? What is the power capacity of DG units required to obtain a cost-effective microgrid system? This is quite challenging. On the other hand, there is some commercial software programs to address important issues, thereby permitting budgeting to be done for capital costs. HOMER software is the global standard for optimizing microgrid design in all sectors, from village power and island utilities to grid-connected campuses and military bases. For example, by using this software, the designer can identify the number of solar panels needed in the proposed location, the wind turbine size for a given topography, and the size of diesel generator to meet a given load. When the capital required for the various generation sources is represented as payments spread over a number of years, it sets the first constraint on aggregate economics. The revenue estimated must cover that, plus the operating costs [65].

5.5.2 Operations Modeling

The second challenge is to ensure efficient operations. The deployed power sources should be utilized in the most efficient way to reduce overall operating cost. Furthermore, to reduce diesel engine use and maximize renewable energy sources, effective energy management techniques should be used. It is clear that this is not easy to fulfill all of these requirements as it is an integer-programming problem. Effective optimization requires proper control network that interfaces with the available resources which requires prototypes with successful analytics.

5.5.3 Financial Modeling

In addition to capacity planning and operations modeling, a financial model of the microgrid must be obtained. To do that, some assumptions about the microgrid are needed, for example, the differing life of the solar panels and the batteries; the financial cost of the business undertaking the project etc. Furthermore, the operating costs in any business plan – for people, offices, sales, maintenance, fuel consumption – should all be anticipated [65].

5.5.4 Barriers to Realizing Microgrids

The above-mentioned issues are the main technical challenges of establishing a business case. On the other hand, we must also consider the public and regulatory policies. For example, microgrid operators must check whether the microgrid service territory crosses public rights of way. Furthermore, if the microgrid operates in grid-connected mode, operators have to pay additional equipment costs as well as the advanced control technique for islanding detection. For that reason, all public policies and regulatory should be considered when the microgrid is designed [65].

5.6 Operation of Multi-Microgrids

The use of multi-microgrids is a novel concept consisting of several microgrids connected to medium voltage (MV) feeders. In this type of power system, a demand-side management technique is used to manage the power flow among these microgrids. This control technique also allows load management, which ensures load curtailment strategy in the multi-microgrids to optimize the power flow. The main challenge of this concept is to control many DGs and loads in harmony. Therefore, the system to be managed greatly increases in complexity and dimension, which requires a completely new control and management architecture.

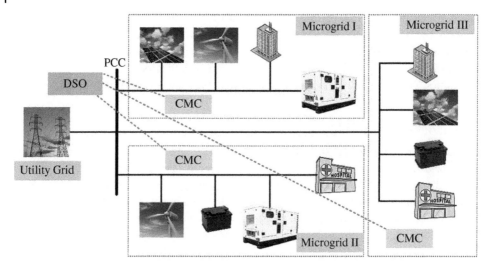

Figure 5.12 Control and management architecture of a multi-microgrid system.

The simplified single-line diagram of a multi-microgrid is depicted in Figure 5.12. It can be seen that the central management controller (CMC), which is an intermediate controller, is at the heart of the hierarchical control architecture. The distribution system operator (DSO) manages the CMCs to optimize power flow among the microgrids and main utility grid. By using this structure, the complexity of the system can be shared among several smaller individual controllers. Furthermore, the communication infrastructure plays a key role in obtaining full observability of the distribution network. To achieve this, a smart-meter infrastructure should be included in the system that allows a coordinated and integrated management of individual elements in the microgrids [62].

The major challenge of any control scheme for multi-microgrid systems is the use of individual controllers, which should be able to communicate with each other to perform explicit control actions. A decentralized scheme is a good candidate for the multi-microgrid system, as it offers a more flexible control and management architecture. On the other hand, the decision making should still adhere to a hierarchical structure.

5.7 Microgrid Benefits

A microgrid brings considerable economic, technical, environmental, and social benefits to both internal and external stakeholders. An overview of microgrid benefits in terms of economic, technical, and environmental aspects is illustrated in Figure 5.13. Each benefit item is mapped to the relevant stakeholder with dotted lines.

5.7.1 Economic Benefits

The economic benefits created by a microgrid can be classified into two sub-categories; locality and selectivity benefit. Locality benefit is mainly attributed to the ability to sell at prices higher than those at the wholesale level while end consumers buy at prices lower

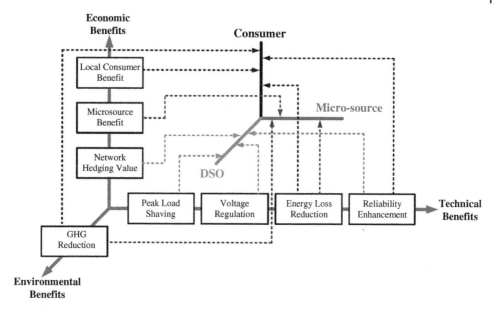

Figure 5.13 An overview of micro grid benefits Ref [67]. Reproduced with permission from John Wiley & Sons.

than the retail level. Selectivity benefit is associated with the optimization of real-time dispatch decisions which can minimize the opportunity cost of the microgrid system considering both technical and environmental constraints [67]. Furthermore, both benefits can be associated with the consumer side or microsource side. To secure a rational split of both benefits between consumer and microsource owner, a suitable market design/interest allocation strategy is required.

5.7.2 Technical Benefits

No doubt, the technical benefits are at the forefront of the benefits provided by the microgrid. Specifically, the microgrid enhances the performance of the local network from the following aspects:

- Demand energy is generated by local power supplies; this helps reduce transmission losses.
- The reactive power control and active power dispatch results in an improved voltage quality.
- Scheduling of microsource outputs relieve overloaded networks especially during the peak hours.
- The power security is maintained during the fault in the main grid by islanding operation capability.

The technical benefits of microgrid highly depend on the optimal source allocation and the degree of coordination among different players [67].

5.7.3 Environmental Benefits

As a result of increasing environmental awareness, and as a consequence of the exhaustible nature of fossil fuels, RESs are playing an important role in microgrids. The deployment of RESs in the microgrid results in lower greenhouse gas emission. Furthermore, the use of more energy efficient supply solutions (e.g. combined heat and power [CHP] applications) including demand side integration brings considerable environmental benefits to the microgrid. To accelerate RES deployment in the microgrids, decision makers from different countries have already prepared support policies for distributed RES. Furthermore, CHP applications and district heating and/or cooling concepts vary significantly from region to region and are expected to find considerably different levels of acceptance across Europe [67].

5.8 Challenges

Although the microgrid concept brings considerable benefits to the main utility grid, the implementation of microgrids is challenging. The first challenge is to obtain the proper operation of the microgrid since mismatch between generation and loads results in frequency and voltage control issues. Furthermore, the microgrids can operate either in grid-connected mode or islanded mode. The transition between these operating modes leads to a severe current and voltage transient. To overcome this issue, a special control technique, which is called seamless control, should be used to minimize the effect of this transition.

Microgrid components and their compatibility are another challenge for the implementation. It can be seen from Figure 5.1 that the typical microgrid consists of many components such as the diesel generator, microturbine, fuel cell, energy storage devices, power electronic converters, communication system, control structure and so on. These components have different characteristics in terms of capacity, operation cost, charge, and discharge rate. This variety leads to control and communication limits.

The intermittent and stochastic nature of RESs negatively affects the stability, reliability and power quality of microgrids. For instance, the solar PV system cannot produce energy at night or during cloudy conditions, and wind systems generate energy that depend on the wind condition. Therefore, the power output of these resources can vary abruptly and frequently and imposes challenges on maintaining microgrid stability. In addition, the high penetration level of the RESs will adversely affect the stability of the main grid.

So far, microgrid protection is the most significant and persistent challenge since it is not easy to design an appropriate protection system that must respond to both main grid and microgrid faults. Although the traditional power system is constructed with unidirectional fault current flow for radial distribution systems, this structure is not possible in the microgrid system. There are many DGs in the microgrid and this results in bidirectional fault current flow. Therefore, the connection between the main grid and a microgrid should be ensured by a fast-static switch to protect the microgrid in both modes of operation against all types of faults [68].

Another crucial challenge of microgrid implementation are the regulations that present guidance about RES integration, frequency and voltage requirements, and main utility connection. On the other hand, the implemented regulations are limited and still under development. Interconnection requirements especially between microgrids and the main utility grid should be designed properly to minimize the disturbance effect of RESs integration and disconnect the microgrid in case of any faults, blackouts, etc. However, high connectivity costs due to high connection fee policies are the most complained about challenge when interconnecting microgrids with the main grid [68].

Communication systems also play a key role in the operation of microgrids by providing a bidirectional connection between different components and the management unit. The communication technologies in microgrid can be classified into three categories: (i) home area networks (HANs), (ii) field area networks (FANs), and (iii) wide area networks (WANs). More details of these can be found in Chapter 8. There are still some issues with communication technologies which call for further research and analysis. First, economic analysis of high-data-rate and coverage technologies should be completed. Furthermore, the use of hybrid communication technologies such as optical wireless should be studied to provide reliable communication medium for the microgrids.

5.9 Conclusion

The microgrid has attracted increasing attention in both academic and industrial fields as an effective solution for the increasing energy and environmental problems. To secure future energy demands, the microgrid plays an important role in the existing power grid. Also, it acts as a facilitating technology for speeding up deployment of the SG concept. Through this chapter, the definition of the microgrid with different boundaries and metrics has been presented. The differences between DC and AC microgrids are comprehensively reviewed. Microgrid design steps with different perspectives have been discussed. The control schemes of microgrids have also been reviewed. Furthermore, the economy of microgrids has been reviewed in terms of capacity planning, operations modeling and financial modeling. The economical, technical, and environmental benefits have been discussed. Finally, implementation challenges of the microgrid are reviewed.

References

1 Parhizi, S., Lotfi, H., Khodaei, A., and Bahramirad, S. (2015). State of the art in research on microgrids: a review. *IEEE Access* 3: 890–925.
2 Percis, E.S., Manivannan, S., Nalini, A., and Sahoo, S.K. (2016). A self-sustained microgrid realized using coordinated control mechanism. *IEEE Region 10 Conference*, Singapore (22–25 November 2016). IEEE.
3 Katiraei, F., Zamani, A., and Masiello, R. (2017). Microgrid control systems: a practical framework [in my view]. *IEEE Power and Energy Magazine* 15 (4): 116–112.
4 Bayhan, S. (2018). Predictive load shedding method for islanded AC microgrid with limited generation sources. *IEEE 12th International Conference on Compatibility, Power Electronics and Power Engineering*, Doha, Qatar (10–12 April 2018). IEEE.

5 Ton, D. and Reilly, J. (2017). Microgrid controller initiatives: an overview of R&D by the U.S. Department of Energy. *IEEE Power and Energy Magazine* 15 (4): 24–31.

6 Justo, J.J., Mwasilu, F., Lee, J., and Jung, J. (2013). AC-microgrids versus DC-microgrids with distributed energy resources: a review. *Renewable and Sustainable Energy Reviews* 24: 387–405, ISSN:1364-0321.

7 Bunker, K. (2015). 10 questions to ask before you build a microgrid. https://www.greenbiz.com/article/10-questions-ask-you-build-microgrid (accessed 15 February 2021).

8 Bayhan, S. and Abu-Rub, H. (2017). Model predictive droop control of distributed generation inverters in islanded AC microgrid. *11th IEEE International Conference on Compatibility, Power Electronics and Power Engineering*, Cadiz, Spain (4–6 April 2017). IEEE.

9 Bayhan, S., Abu-Rub, H., Leon, J.I. et al. (2017). Power Electronic Converters and Control Techniques in AC Microgrid. *43rd Annual Conference of the IEEE Industrial Electronics Society*, Beijing, China (29 October–1 November 2017) IEEE.

10 Olivares, D.E., Mehrizi-Sani, A., Etemadi, A.H. et al. (2014). Trends in microgrid control. *IEEE Transactions on Smart Grid* 5 (4): 1905–1919.

11 Matos, M.A., Seca, L., Madureira, A.G. et al. (2016). *Control and Management Architectures, Smart Grid Handbook*, 1–24. John Wiley & Sons.

12 Colak, I., Fulli, G., Bayhan, S. et al. (2015). Critical aspects of wind energy systems in smart grid applications. *Renewable and Sustainable Energy Reviews* 52: 155–171.

13 Kish, G.J. and Lehn, P.W. (2012). Microgrid design considerations for next generation grid codes. *IEEE Power and Energy Society General Meeting*, San Diego, USA (22–26 July 2012). IEEE.

14 Microgrid Projects (2016). http://microgridprojects.com (accessed 15 February 2021).

15 Microgrid Projects (2016). New York Affordable Housing Microgrid. http://microgridprojects.com/microgrid/new-york-affordable-housing-microgrid (accessed 15 February 2021).

16 Microgrid Project (2016). Ta'u SolarCity Tesla Microgrid. http://microgridprojects.com/microgrid/tau-solarcity-tesla-microgrid-american-samoa (accessed 15 February 2021).

17 Microgrid Project (2016). Islas Secas Island Microgrid, http://microgridprojects.com/microgrid/islas-secas-island-microgrid (accessed 15 February 2021).

18 Microgrid Project (2016). Kissi County, Kenya Microgrids. http://microgridprojects.com/microgrid/kissi-county-kenya-microgrid (accessed 15 February 2021).

19 Microgrid Project (2016). Miramar Microgrid. Marine Base Military Microgrid. http://microgridprojects.com/microgrid/miramar-microgrid (accessed 15 February 2021).

20 Microgrid Project (2016). Milford Microgrid. http://microgridprojects.com/microgrid/milford-microgrid (accessed 15 February 2021).

21 Microgrid Project (2016). Philadelphia Navy Yard Microgrid. http://microgridprojects.com/microgrid/philadelphia-navy-yard-microgrid (accessed 15 February 2021).

22 Microgrid Project (2015). Tecnalia Microgrid. http://microgridprojects.com/microgrid/tecnalia-microgrid (accessed 15 February 2021).

23 Microgrid Project (2016). Mannheim-Wallstadt Microgrid. http://microgridprojects.com/microgrid/mannheim-wallstadt-microgrid accessed 15 February 2021).

24 Microgrid Projects (2016). Hangzhou University Microgrid. http://microgridprojects.com/microgrid/hangzhou (accessed 15 February 2021).

25 Gupta, A., Doolla, S., and Chatterjee, K. (2018). Hybrid AC/DC microgrid: systematic evaluation of control strategies. *IEEE Transactions on Smart Grid* 9 (4): 3830–3843.

26 Silva, P. and Marques de Sa Medeiros, C. (2017). A promising future to DC power system: a review. *IEEE Latin America Transactions* 15 (9): 1639–1642.

27 Microgrid Projects (2016). Kalkeri Sangeet Vidyalaya DC Microgrid. http://microgridprojects. com/microgrid/kalkeri-sangeet-vidyalaya-dc-microgrid (accessed 15 February 2021).

28 Symanski, D.P. 380 VDC Data Center Duke Energy Charlotte, NC. Electric Power Research Institute, Available at https://docplayer.net/12159313-380vdc-data-center-at-duke-energy.html.

29 Microgrid Projects (2016). Intelligent DC Microgrid Living Lab. http://microgridprojects. com/microgrid/intelligent-dc-microgrid-living-lab (accessed 15 February 2021).

30 Hatti, M., Meharrar, A., and Tioursi, M. (2011). Power management strategy in the alternative energy photovoltaic/PEM fuel-cell hybrid system. *Renewable and Sustainable Energy Reviews* 15 (9): 5104–5110.

31 Sun, K., Zhang, L., Xing, Y., and Guarrero, J.M.A. (2011). Distributed control strategy based on DC bus signaling for modular photovoltaic generation system with battery energy storage. *IEEE Transactions on Power Electronics* 26 (10): 3032–3045.

32 Santiago Miret, Berkeley Energy&Resources Collaborative. How to Build a Microgrid, 2015. https://blogs.berkeley.edu/2015/02/25/how-to-build-a-microgrid/

33 Froundhofer ISE (2020). Photovoltaics Report. https://www.ise.fraunhofer.de/content/ dam/ise/de/documents/publications/studies/Photovoltaics-Report.pdf (accessed 15 February 2021).

34 Ongaro, F., Saggini, S., and Mattavelli, P. (2012). Li-ion battery-supercapacitor hybrid storage system for a long lifetime, photovoltaic-based wireless sensor network. *IEEE Transactions on Power Electronics* 27 (9): 3944–3952.

35 Javed, K., Ashfaq, H., Singh, R. et al. (2019). Design and performance analysis of a stand-alone PV system with hybrid energy storage for rural India. *Electronics* 8 (9): 952.

36 Agarwal, A., Iyer, V.M., Anurag, A., and Bhattacharya, S. (2019). Adaptive Control of a Hybrid Energy Storage System for Wave Energy Conversion Application. *IEEE Energy Conversion Congress and Exposition*, Baltimore, USA (29 September–3 October 2019). IEEE.

37 Sufyan, M., Rahim, N.A., Aman, M.M. et al. (2019). Sizing and applications of battery energy storage technologies in smart grid system: a review. *Journal of Renewable and Sustainable Energy* 11 (1): 014105.

38 Tankari, M.A., Camara, M.B., Dakyo, B., and Lefebvre, G. (2013). Use of ultracapacitors and batteries for efficient energy management in wind–diesel hybrid system. *IEEE Transactions on Sustainable Energy* 4 (2): 414–425.

39 Wu, J., Xing, X., Liu, X. et al. (2018). Energy management strategy for grid-tied microgrids considering the energy storage efficiency. *IEEE Transactions on Industrial Electronics* 65 (12): 9539–9549.

40 Parida, A., Choudhury, S., and Chatterjee, D. (2018). Microgrid based hybrid energy co-operative for grid-isolated remote rural village power supply for East Coast zone of India. *IEEE Transactions on Sustainable Energy* 9 (3): 1375–1383.

41 Mohammed, A., Refaat, S.S., Bayhan, S., and Abu-Rub, H. (2019). Ac microgrid control and management strategies: evaluation and review. *IEEE Power Electronics Magazine* 6 (2): 18–31.

42 Mirsaeidi, S., Dong, X., Shi, S., and Tzelepis, D. (2017). Challenges, advances and future directions in protection of hybrid AC/DC microgrids. *IET Renewable Power Generation* 11 (12): 1495–1502.

43 Bhave, M.P. (2018). *Microgrid Economics: It Takes a Village, a University, and a Ship*, 25–28. Hamburg: Wind Energy.

44 Sahoo, S.K., Sinha, A.K., and Kishore, N.K. (2018). Control techniques in AC, DC, and hybrid AC–DC microgrid: a review. *IEEE Journal of Emerging and Selected Topics in Power Electronics* 6 (2): 738–759.

45 Bayhan, S. and Abu-Rub, H. (2018). A Simple Control Technique for Distributed Generations in Grid-Connected and Islanded Modes. *IEEE International on Industrial Electronics*, Cairns, Australia (12–15 June 2018). IEEE.

46 Han, H., Hou, X., Yang, J. et al. (2016). Review of power sharing control strategies for islanding operation of AC microgrids. *IEEE Transactions on Smart Grid* 7 (1): 200–215.

47 Han, Y., Li, H., Shen, P. et al. (2017). Review of active and reactive power sharing strategies in hierarchical controlled microgrids. *IEEE Transactions on Power Electronics* 32 (3): 2427–2451.

48 Ali, B. and Davoudi, A. (2012). Hierarchical structure of microgrids control system. *IEEE Transactions on Smart Grid* 3 (4): 1963–1976.

49 Rafique, S.F. and Jianhua, Z. (2018). Energy management system, generation and demand predictors: a review. *IET Generation, Transmission & Distribution* 12 (3): 519–530.

50 Solanki, A., Nasiri, A., Bhavaraju, V. et al. (2016). A new framework for microgrid management: virtual droop control. *IEEE Transactions on Smart Grid* 7 (2): 554–566.

51 Rodrigues, W.A., Oliveira, T.R., Morais, L.M.F., and Rosa, A.H.R. (2018). Voltage and power balance strategy without communication for a modular solid-state transformer based on adaptive droop control. *Energies* 11 (7): 1802.

52 Mortezaei, A., Simoes, M., Savaghebi, M. et al. (2016). Cooperative control of multi-master-slave islanded microgrid with power quality enhancement based on conservative power theory. *IEEE Transactions on Smart Grid* 99: 1–1.

53 Tayab, U.B., Roslan, M.A.B., Hwai, L.J., and Kashif, M. (2017). A review of droop control techniques for microgrid. *Renewable and Sustainable Energy Reviews* 76: 717–727.

54 Yao, W., Chen, M., Matas, J. et al. (2011). Design and analysis of the droop control method for parallel inverters considering the impact of the complex impedance on the power sharing. *IEEE Transactions on Industrial Electronics* 58 (2): 576–588.

55 Han, H., Liu, Y., Sun, Y. et al. (2015). An improved droop control strategy for reactive power sharing in islanded microgrid. *IEEE Transactions on Power Electronics* 30 (6): 3133–3141.

56 Sun, Y.H. and Hu, J. (2020). Optimized autonomous operation control to maintain the frequency, voltage and accurate power sharing for DGs in islanded systems. *IEEE Transactions on Smart Grid* 11: 3885–3895.

57 Khan, M.Z., Khan, M.M., Jiang, H. et al. (2018). An improved control strategy for three-phase power inverters in islanded AC microgrids. *Inventions* 3 (3): 47.

58 Chiang, H.C., Jen, K.K., and You, G.H. (2016). Improved droop control method with precise current sharing and voltage regulation. *IET Power Electronics* 9 (4): 789–800.

59 Zhang, Y. and Ma, H. (2012). Theoretical and experimental investigation of networked control for parallel operation of inverters. *IEEE Transactions on Industrial Electronics* 59 (4): 1961–1970.

60 Chen, T., Abdel-Rahim, O., Peng, F., and Wang, H. (2020). An improved finite control set-MPC-based power sharing control strategy for islanded AC microgrids. *IEEE Access* 8: 52676–52686.

61 Kantamneni, A., Brown, L.E., Parker, G., and Weaver, W.W. (2015). Survey of multi-agent systems for microgrid control. *Engineering Applications of Artificial Intelligence* 45: 192–203.

62 Shrivastava, S., Subudhi, B., and Das, S. (2018). Distributed voltage and frequency synchronisation control scheme for islanded inverter-based microgrid. *IET Smart Grid* 1 (2): 48–56.

63 Hatziargyriou, N. (2014). *Microgrids Architectures and Control.* Wiley.

64 Sanseverino, E.R. et al. (2015). Energy Management Systems and tertiary regulation in hierarchical control architectures for islanded microgrids. *IEEE 15th International Conference on Environment and Electrical Engineering*, Rome, Italy (10–13 June 2015). IEEE.

65 Bhave, M.P. (2018). Microgrid Economics: It Takes a Village, a University, and a Ship. https://www.renewableenergyworld.com/storage/microgrid-economics-it-takes-a-village-a-university-and-a-ship/#gref

66 Patel, H. and Chowdhury, S. (2015). Review of technical and economic challenges for implementing rural microgrids in south africa. *IEEE Eindhoven PowerTech*, Eindhoven, Netherlands (29 June–2 July 2015). IEEE.

67 Schwaegerl, C. and Tao, L. (2014). Quantification of technical, economic, environmental and social benefits of microgrid operation. In: *Microgrids Arctitectures and Control* (ed. N. Hatziargyriou), 275–313. Wiley https://onlinelibrary.wiley.com/doi/abs/10.1002/9781118720677.ch07.

68 Yoldaş, Y., Önen, A., Muyeen, S.M. et al. (2017). Enhancing smart grid with microgrids: challenges and opportunities. *Renewable and Sustainable Energy Reviews* 72: 205–214.

6

Smart Transportation

Recently, concerns about air pollution and exhaustible fossil fuels have led policy-makers and researchers to investigate alternatives to conventional internal-combustion-engine vehicles (CICEVs) in transportation. Electrifying transportation is a promising approach to alleviate the issues caused by CICEVs. On the other hand, electric vehicle (EV) deployment and the increasing trend of EV adoption in the market depend highly on the development of drivetrain, battery and charger technologies. This chapter presents an overview of EVs; their current status and also future opportunities, in addition to the challenges of integrating them into the smart grid (SG).

6.1 Introduction

Fossil fuel vehicles are of the highest greenhouse gases (GHG) sources worldwide although automobile companies are working intensively on new technologies to improve the efficiency of the CICEVs and to reduce the GHG. Transport accounts for about 19% of global energy use and 23% of energy-related carbon dioxide (CO_2) emissions and these shares are likely to rise in the future [1]. However, further preventive actions and climate policies are much needed to slow down the worsening environmental conditions and the impacts of climate change. The International Energy Agency (IEA) predicts that the amount of GHG caused by the transportation sector will be double by 2050 if no initiatives are taken by the decision makers. Only in China, the IEA predicts that the vehicle fleet could approach 600 million vehicles by 2050 [2]. The development of EVs technologies and government support have increased the sale of EVs dramatically. Figure 6.1 shows the global sales of EVs by year [2].

Although different efforts have been undertaken to reduce the effect of CICEVs, electrifying transportation is one of the most promising approaches to protect our environment and reduce the burning of more fossil fuels. Replacing fossil fuel with electric engines results in a significant reduction in GHG emission, diminution of fuel consumption and an increase in energy security and stability worldwide. To maximize the potential benefits from EVs; including social, health, economic, and environmental, it is necessary to take specific steps to integrate this technology into the SG and transportation market [3–6].

Smart Grid and Enabling Technologies, First Edition. Shady S. Refaat, Omar Ellabban, Sertac Bayhan, Haitham Abu-Rub, Frede Blaabjerg, and Miroslav M. Begovic.
© 2021 John Wiley & Sons Ltd. Published 2021 by John Wiley & Sons Ltd.
Companion website: www.wiley.com/go/ellabban/smartgrid

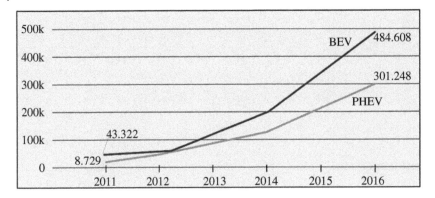

Figure 6.1 Global sales of electric cars (BEV and PHEV) by year.

EVs show better performance than CICEVs because they use a highly efficient drivetrain and electric motors instead of a low efficiency internal-combustion-engine. To accelerate EV usage, governments around the world are preparing supportive policies and initiatives for EVs. For example, developing charging infrastructure and offering incentive packages for EV purchase are some of the actions taken. Furthermore, the development of EV technologies play a crucial role in the improvement EVs performance and ensure that they can compete with traditional vehicles. Recently, most of the research has been focused on battery technologies, fast charging systems, and high efficiency powertrain systems as these technologies are the keys to the successful future of EVs'. For example, well-designed powertrain systems of EVs might enhance their driving range. Furthermore, implementation of fast charging infrastructure will help reduce the battery charge time so as to overcome the range anxiety issue among EV drivers [6]. The performance of EVs depend to a great extent on battery performance. Either a primary battery, such as a metal-air battery, or a secondary rechargeable battery can be used for propulsion as an EV battery or traction battery. Since the EVs batteries are needed to provide power over sustained periods of time, deep-cycle batteries can be used instead of other solutions such as starting, lighting, and ignition (SLI) batteries. A high ampere-hour capacity is necessary for the design of traction batteries. One of the features for EV batteries is that the power-to-weight ratio, energy-to-weight ratio, and energy density are relatively high. There are four main types of batteries, lead-acid, nickel metal hydride, zebra, and lithium-ion batteries that have been used in EVs and will be explained later in this chapter. The requirements and cost of on-board energy storage can be reduced by the availability of charging infrastructure. Many issues need to be resolved including vehicle charging time, distribution system extent, charging station demand policies, standardization, and regulatory procedures. Charging infrastructure and Electric Vehicle Supply Equipment (EVSE) are issues of great importance that must be addressed. The main components of EVSE include EV charge cords, residential or public charge stands, attachment plugs, power outlets, vehicle connectors, and protection.

EVs, as mobile energy storage, are capable of receiving power from the grid when the power demand is lower than power supply and provide power back to the grid when the grid supply is constrained which is of high importance for the future SG. With the growing penetration rate of EVs, there is a potential for EVs to participate in the vehicle-to-grid

(V2G) or grid-to-vehicle (V2G) modes in the utility's demand side management (DSM) program to help produce the desired load shape and relieve the electricity supply burden during high demand or contingencies, which includes peak clipping, valley filling, and load shifting. EV batteries could also help reduce the size of generating units for on-site power generation. This concept will be further discussed in the following subsections.

6.2 Electric Vehicle Topologies

EVs are classified into three main categories in terms of their powertrains: plug-in electric vehicles (PEVs), hybrid electric vehicles (HEVs), and fuel-cell electric vehicles (FCEVs). Furthermore, battery electric vehicles (BEVs) and plug-in hybrid electric vehicles (PHEVs) can be considered as PEVs since their batteries are charged by the electricity grid. In this section, each power train topology is explained with their advantages and disadvantages.

6.2.1 Battery EVs

The main characteristic of BEVs is that they are driven only by an electric motor. Although the power grid is the main power source for the battery charge, regenerative breaking technique is used to charge their batteries. Using only an electric motor in BEVs ensures that it is a zero-emission vehicle. On the other hand, BEVs require large battery packs to be able to drive in reasonable range. Table 6.1 presents the overview of commercially available BEVs in the market [7]. The size of battery packs varies from 16 to 60 kWh with all-electric-range (AER) of 62–260 miles.

6.2.2 Plug-in Hybrid EVs

Unlike the BEVs, PHEVs contain the electric motor and internal combustion engine that run with gasoline or diesel. As
a result of the use of the internal combustion engine the battery size is significantly decreased. The main advantage of PHEVs is that its driving range is higher than BEVs. For

Table 6.1 Overview of battery electric vehicles.

Model	AER (mi)	Engine (kW)	Battery (kWh)	5-year Cost Savings (USD)
BMW i3	81	125	21	6000
Chevrolet Spark	82	104	19	6000
Fiat 500e	87	83	24	5750
Kia Soul	93	50	16.4	5500
Nissan Leaf	84	80	24	5750
Smart Fortwo	68	55	17.6	5750
Tesla Model-S	265	300	60	5250
Volkswagen e-Golf	83	85	24.2	5750

example, the Chevrolet Volt can be driven for 38 miles with the electric motor and the driving range can be extended for additional 350 miles with the internal combustion engine [7]. The smart control unit manages the use of the electric motor and the internal combustion engine based on the requirement. For instance, the electric motor is employed in the drivetrain mostly in urban environments whereas the internal combustion engine is employed when rapid acceleration needed. Furthermore, the electric motors provide a good acceleration since they have better dynamics. The PHEVs offer a cleaner and more affordable transportation option for traveling long distances and at the same time ensuring reduced carbon emissions.

6.2.3 Hybrid EVs

Similar to PHEVs, HEVs contain the electric motor and internal combustion engine that run with gasoline or diesel. On the other hand, the internal combustion engine is used as primary traction unit in HEVs. These vehicles are equipped with battery packs, however, to charge these battery packs, regenerative braking systems is only employed. In other words, these vehicles cannot be charged from the power grid. Thus, the deployment of HEVs does not have major impact on the power grid. The main role of battery and electric motor is to improve the fuel efficiency of these vehicles. As of mid-2015 more than 10 million HEVs were deployed worldwide [7, 8].

6.2.4 Fuel-Cell EVs

The power plants of such vehicles convert the chemical energy of hydrogen to mechanical energy either by burning hydrogen in an internal combustion engine, or by reacting hydrogen with oxygen in a fuel-cell to run electric motors. The main advantage of FCEVs is that the fuel filling time is within 10 minutes and the driving range is around 300 miles even though their battery packs are small in size compared to other EVs. Furthermore, similar to HEVs, these vehicles do not need power grid to charge the battery pack. Although FCEVs offer considerable benefits over BEVs, PHEVs, and HEVs, the hydrogen filling station is limited, which reduces the deployment rate of these vehicles in the market. As of mid-2018, there are several hydrogen cars publicly available in select markets: the Honda FCX Clarity, the Mercedes-Benz F-Cell, The Hyundai Tuscon FACES, the Toyota Mirai, the Hyundai ix35 FCEV, and the Hyundai Nexo.

The comparison of different EVs is summarized in Table 6.2 in terms of drive components, energy source, advantages, and drawbacks [3].

6.3 Powertrain Architectures

PHEVs, HEVs, and FCEV are considered as HEV since these vehicles utilize at least two sources for propulsion system such as electric motor and internal combustion engine. The required electricity for the electric motor can be obtained by battery packs or fuel cell. The main aim of using HEVs is to reduce the fossil-based fuel consumption and the amount of GHG emission. To reduce the fuel consumption, some simple

Table 6.2 Comparison of different EV topologies. Ref Num [3].

EV Topology	Drive Component	Energy Source	Features	Drawbacks
BEV	Electric motor	Battery Ultracapacitor	No emission Range depends on the battery size and type	Battery price and capacity Range Charging time Availability of charging stations
PHEV	Electric motor internal combustion engine	Battery Ultracapacitor internal combustion engine	Reduced emission Long driving range	Management of the energy sources Battery and engine size optimization
FCEV	Electric motor	Fuel-cell Battery	No emission High efficiency	Cost of fuel cell • Feasible way to produce fuel • Availability of fueling facilities

control techniques have been developed by the car manufactures. For example, the internal combustion engine is switched off during idle times by the control unit; this is called stop–start control technique. Furthermore, to charge the battery packs kinetic energy is converted into the electric energy while breaking. This helps to increase the overall efficiency of the system. Different configurations of internal combustion engine and electric motor are possible in an HEV. They will be described in the following sections.

6.3.1 Series HEV Architecture

The typical powertrain configuration of series HEV is illustrated in Figure 6.2. In this configuration, the internal combustion engine is the main power source for the powertrain. The internal combustion engine drives the generator to charge the battery packs. The electric motor is supplied by the battery packs to drive the traction system. Please note that the traction system is driven directly by the electric motor even the internal combustion engine is the prime mover. The main advantage of this configuration is that the size of the internal combustion engine is smaller than a car that relies solely on the internal combustion engine [8, 9]. Furthermore, the internal combustion engine in this configuration operates at the most efficient operating point most of the time, resulting in an improvement of the efficiency of the entire system. In this configuration, the power required to drive the electric motor could be transferred directly from the generator, however, it is better to transfer power thru the battery packs so as to smooth the power transients or variable power demand from the electric motors. The ultracapacitors could be used with battery packs to minimize the stress on the battery packs. Furthermore, the use of ultracapacitors provide high transient currents. Series HEV configuration improves the efficiency at low speed and urban areas.

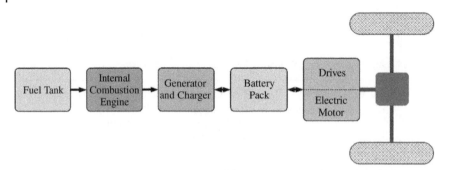

Figure 6.2 The typical powertrain configuration of series HEV.

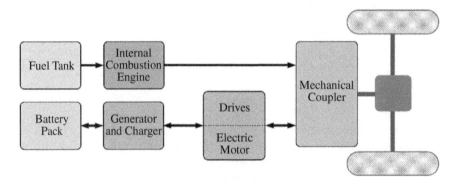

Figure 6.3 The typical powertrain configuration of parallel HEV.

6.3.2 Parallel HEV Architecture

The typical powertrain configuration of series HEV is illustrated in Figure 6.3. Unlike the series HEV architecture, the internal combustion engine drive directly the traction system similar to the electric motor. In other words, the traction system is driven by the electric motor or the internal combustion engine or both of them. This configuration offers high efficiency at high speeds such as driving on highways. In this configuration, the electric motor is of low power and usually less than 30 kW since the internal combustion engine is always in the loop. Furthermore, the battery packs are small-sized compared with other EVs [10, 11]. Similar to the series configuration regenerative braking can be supported in parallel HEVs.

6.3.3 Series–Parallel HEV Architecture

The typical powertrain configuration of series–parallel HEV is illustrated in Figure 6.4. In this configuration, a power split unit is used to divide the output mechanical power from the internal combustion engine to the drive shaft or electric generator. Thus, the system can operate either in series configuration or parallel configuration depending on the driving conditions and supervisory control strategy. It means that this configuration combines the features of both series and parallel HEV architectures. The series–parallel HEV ensures the highly efficient operation both in urban (low-speed) and on highways (high-speed). The comparison of series and parallel HEV configuration is given in Table 6.3 [3].

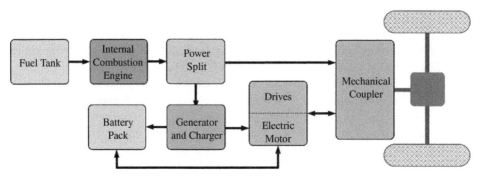

Figure 6.4 The typical powertrain configuration of series–parallel HEV.

Table 6.3 The comparison of series and parallel HEV configurations. Ref Num [3].

	Series HEV	Parallel HEV
Advantages	Efficient and optimized power-plant Possibilities for modular power-plant Optimized drive line Long lifetime Mature technology Fast response Capable of attaining zero emission	Capable of attaining zero emission Economic gain More flexibility
Limitations	Large traction drive system Requirement of proper algorithms Multiple energy conversion steps	Expensive Complex control Requirement of proper algorithms Need of high voltage to ensure efficiency

6.4 Battery Technology

Battery technology is the key element for the integration of EVs in the market since it has impacts on the vehicle performance, the driving range, and the reliability. The size and weight of battery packs in EVs should be optimized to satisfy certain physical requirements such as driving range, acceleration, and top speed. This section is devoted to discussing the present EV battery technologies.

6.4.1 Battery Parameters

The batteries parameters are: Capacity, State of Charge (SoC), Depth of Discharge (DoD), State of health (SoH), Age and history, Energy Density, Charging Efficiency, Float Voltage, and Cut-Off Voltage.

Capacity: This parameter is a measure of the charge stored by the battery. The battery capacity represents the maximum amount of energy that can be extracted from the battery under certain specified conditions. The battery capacity is measured in either watt-hours (Wh), kilowatt-hours (kWh), or ampere-hours (Ah). The most common measure of battery capacity is Ah, defined as the number of hours for which a battery can provide a current equal to the discharge rate at the nominal voltage of the battery. For example, a battery capacity of 10 Ah means that if a constant current of 10 A is drawn from the battery, it will become discharged completely after one hour [6]. However, this definition is acceptable for the theory. In practice, manufacturers provide some test results which have been performed under different constant current and constant power conditions. These test results provide much more information about battery capacity rather than standard definitions. Furthermore, the battery capacity varies from the nominal rated capacity since the capacity depends on the temperature, age and the past history of the battery.

SoC: This parameter can be assumed simply as the equivalent of a fuel gauge of the battery. In other words, the SoC represents the percentage of charge available from a battery relative to the entire capacity of the battery. To determine SoC, the rated battery capacity should be calculated regularly because the rated capacity of the battery reduces over time due to the aging.

DoD: The energy stored in the battery cannot be fully withdrawn which means that the battery cannot be fully discharged, otherwise irreparable damage can happen to the battery. The DoD of the battery describes the portion of power that can be taken from the battery. As an example, if the battery DoD is given by the manufacturer as 30%, then up to 30% of the battery capacity can be discharged.

SoH: is used to represent the ratio of the maximum charge capacity to its ideal conditions. The SoH indicates the performance degradations of the battery, which is also related to its remaining lifetime.

Age and history: The age and history work condition of the battery have significant impact on its actual capacity. Even when following the manufacturers recommendations on the DoD, the battery's capacity can maintain its rated capacity only for a limited charge and discharge cycles. The work history of the battery has also important impact on its capacity. In particular, if the battery was taken below its maximum DoD, then its capacity may experience an early-stage reduction, then its rated number of charge and discharge cycles will no more be the same.

Energy Density: In a simplest way, energy density is the capacity of the battery to store energy. i.e. a high energy density battery can store a large amount of energy. In a more specifically, energy density can be defined in two ways. The first definition is volumetric energy density that is defined as the amount of available energy from a fully charged battery per unit volume (Wh/l). The unit liter is used mainly for measuring the volume of liquids. The second definition is specific energy that is defined as the available energy from a fully charged battery per unit weight (Wh/kg). For EVs applications, the weight factor is usually more important than volume. Therefore, specific energy density is used in the EVs applications.

Charging Efficiency: There are chemical reactions inside the battery during the charge and discharge and these reactions leads to losses. It means that the energy used to charge the battery is not equal the energy while discharge. Some of this energy is wasted in other forms of energy dissipation such as heat energy dissipation. The charging efficiency can be defined as the ratio of available energy from the battery. The charge–discharge efficiencies of Li-ion, Pb-Acid, and NiMh batteries are 80–90, 50–95, and 66%, respectively. Li-ion efficiencies are extremely high, Pb-acid efficiencies have a huge range; NiMH efficiencies are low at 66%.

Float Voltage: is the steady state voltage level at which the battery is charged after reaching 100% SoC in order to compensate the self-discharge losses.

Cut-off Voltage: is defined as the minimum allowable voltage for the battery, which describes the empty state of the battery. It is determined by the manufacturer.

6.4.2 Common Battery Chemistries

Lead-Acid (Pb-acid): This battery is the oldest form of electrical battery, originally developed in the 1850s. These batteries have been used in different applications. Lead-acid batteries were suitable for the EVs since these were considered as maintenance-free batteries and have around 95–99% efficiency. On the other hand, these batteries have low specific energy (33–42 Wh/kg) which increase the weight of the battery packs compared with the other batteries [12]. In recent years, these batteries are not used in EVs.

Nickel-Cadmium (Ni—Cd): Although Ni—Cd batteries have higher energy density than Pb-acid battery, these batteries are not suitable for traction applications due to low specific energy (40–60 Wh/kg). The main applications for this type are portable devices, but in cases that demand high instantaneous currents, there use is desirable. On the other hand, the market share of Ni—Cd batteries is falling dramatically as Cadmium is carcinogenic.

Nickel-Metal Hydride (Ni-MH): These nickel-based batteries have negative electrodes made of a hydrogen-absorbing metal alloy. These batteries have higher specific energy (60–120 Wh/kg) than Ni—Cd batteries, however, their life cycle is lower than Ni—Cd. In general, for the batteries of the same size, Ni-MH batteries can have up to two or higher times the energy of Ni—Cd type. This type is widely used in EVs and PHEVs.

Lithium-Ion (Li-ion): These batteries are using lithium, not in a metallic form but ionically bound to other materials. There are several types of lithium-ion batteries distinguished by specific chemistries. The main advantage of Li-ion battery is that it has higher specific energy than other batteries. This feature makes these batteries good candidate for EVs applications. Their specific energy is in the range of 100–265 Wh/kg. Another important advantage of these batteries is that they can be charged and discharged at a faster rate than other batteries due to their nature. Furthermore, these batteries show wonderful potential for long life when they are used in proper conditions. On the other hand, the safety issues are weak point of these batteries. Especially, charge conditions of these batteries should be carefully managed in order to prevent them from explosion because of the overheating caused by overcharging [13].

Lithium-Polymer (Li—Po): Although these batteries have the same specific energy as Li-ion batteries, Li—Po batteries are cheaper. Furthermore, recent improvements on these batteries enhances the maximum discharge rates of these batteries from 1C to about 30C. In addition, these batteries can reach over 90% SOC within a couple of minutes, which can increase significantly their usage in EVs because of the noticeable reduction in charging

Table 6.4 Existing battery technology, energy density, specific density. Ref Num [14].

Battery Type	Specific Energy	Energy Density	Cycle Durability
Pb-acid	33–42 Wh/kg	60–110 Wh/l	500–800 cycles
Ni-Cd	40–60 Wh/kg	50–150 Wh/l	2000 cycles
Ni-MH	60–120 Wh/kg	140–300 Wh/l	500–1000 cycles
Li-ion	100–265 Wh/kg	250–620 Wh\l	400–1200 cycles
Li-Po	100–265 Wh/kg	250–730 Wh/l	2000 cycles

Table 6.5 Battery capacity and technologies by various EV manufacturers. Ref Num [15]. Reproduce with permission from ELSEVIER.

Car model/EV type	Company	Battery chemistry	Capacity (kWh)
Chevrolet Volt/PHEV	GM	Lithium manganese oxide spinel polymer	16.5
Prius Alpha/PHEV	Toyota	NiMH	1.3
Prius (ZVW35)/PHEV	Toyota	Lithium nickel cobalt aluminum oxide	4.4
Leaf/BEV	Nissan	Lithium manganese oxide	24
Tesla model S/BEV	Tesla	Lithium manganese oxide	85
Chevrolet Spark/BEV	GM	Nano Lithium iron phosphate	21.3
Fiat 500e/BEV	Chrysler	Lithium iron phosphate	24
Honda Accord/PHEV	Honda	Lithium manganese oxide	6.7

time. One of the other advantages of this type is that it can be produced in any size or shape, which offers great flexibility to vehicle manufacturers. These features nominate these batteries as good candidates for EVs applications.

The comparison of the discussed battery types is presented in Table 6.4 [14]. It can be seen that Li-ion and Li—Po batteries show similar characteristics, which are desirable for EVs applications. Li—Po batteries have significant advantages over the Li-ion batteries and these advantages make Li—Po batteries good candidate for the EVs manufacturers. The future expectation is to have the batteries with high energy and power capacities, small size and affordable purchasing cost. Table 6.5 presents the current battery technologies used by various automotive manufacturers [15].

6.5 Battery Charger Technology

Battery charger technology can be seen as an equivalent structure of petrol stations for EVs. These technologies play a crucial role for deploying of EVs since they directly influence charging durations of EVs batteries, which is the primary anxiety of using EVs. Battery chargers can be divided into two groups; onboard battery chargers and off-board battery chargers. As the name implies, onboard battery chargers are located in the EVs. These

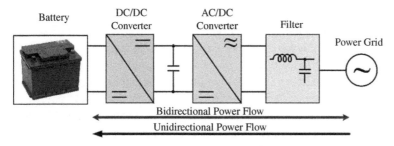

Figure 6.5 The simplified block diagram of battery charger.

chargers are used in low-power levels (up to 10 kW) and help to prevent batteries health since they do not cause battery heating. These kinds of chargers are suitable if there is not enough physical infrastructure for off-board chargers. Thus, the user can charge their EVs batteries in their home and public parking lots with standard electrical outlet. Off-board chargers are in the second category and designed for high-power charging levels typically between 10 and 80 kW. To obtain high-level power for off-board chargers, three-phase power electronic converters are required. These off-board chargers mainly are used in charging stations. Although the off-board chargers reduce the charging duration of EVs, they degrade battery lifetime more quickly than the onboard ones, due to involving high-voltage energy transfer [16].

The simplified block diagram of battery charger is illustrated in Figure 6.5. This structure can be used for both onboard and off-board chargers. However, the used converter and inverter topologies can be different depends on the power level. Furthermore, the power flow can be either unidirectional or bidirectional, which depends on the application. Usually, unidirectional chargers employ a filter, a diode bridge, and a DC-DC converter. On the other hand, bidirectional chargers employ a filter, an active rectifier, and a DC-DC converter. Hardware architecture and control structure of unidirectional chargers are simpler than bidirectional chargers. However, bidirectional chargers allow to inject power from vehicle to grid, which provides voltage control, frequency control, and emergency backup for the power system. These benefits increase the system cost since bidirectional chargers are more costly than unidirectional ones. Furthermore, to allow for bidirectional power flows, battery chargers should comply with standards, such as IEEE 1547, SAE-J2894, and IEC 1000-3-2, which regulate the allowable harmonics and DC injections into the grid [17–19]. The vehicle to grid operating mode will be discussed in the following section.

6.5.1 Charging Rates and Options

To maximize potential co-benefits; including social, health, economic, and environmental; from EVs, it is necessary to take specific steps to integrate this technology into social life. However, some challenges from customers' point of view have not all been taken into account until now. Dissemination of EVs depends not only on awareness of environment but also on a level of social acceptance of this new technology. Therefore, EVs could not become as prevalent s as fossil fuel vehicles without social acceptance. One of the key factors for social acceptance is anxiety, directly related to people's previous practice. Some factors result in anxiety can be counted in EVs dissemination context: (i) purchasing EV, (ii) usage and service of EV, (iii) reselling EV, and (iv) charging process. Among these factors, charging process especially holds an important place. Refueling is easy and petrol stations can be found anywhere for fossil fuel consumption vehicle; on the other hand, for the EVs it is almost opposite: the time of battery charging may be significantly longer, and charging stations are not wide spread like petrol stations [18, 19].

For that reason, to promote acceptance of EVs and to meet customer needs, availability of charging stations with various charging rates are critical. Especially, to overcome the driving range anxiety, public charging facilities should be built at various locations. For example, cars are parked at different locations – workplaces, schools, and shopping malls – during the weekdays. Therefore, building charging stations at these locations could offer power source for EVs to extend the driving range.

Furthermore, various charge methods should be adopted so as to reduce the charging duration, which helps to overcome the charging anxiety. Table 6.6 presents the charge methods of EVs. For example, the driving range of an EV charged one hour with AC level 2 is more than an EV charged one hour with AC level 1. Furthermore, DC fast charging provides high power charging, which results in more faster charging. To be more specific, AC level 1 charger is designed for individual users to charge their own car at home with standard home outlets. On the other hand, AC level 2 charging units are designed for public places such as shopping malls, schools, workplaces. AC level 3 and DC fast charging units are designed for charging stations to provide charging services in short time.

The two figures illustrate the setup of the facilities at the charging point and embedded EV kits for charging scenarios by considering the AC and DC charging levels as depicted in Table 6.6. With the AC Level 1 and 2 configurations in Figure 6.6a, the EVSE is provided at the charging point by supplying the AC power to an on-board charger. However, with the DC fast charging configurations in Figure 6.6b, the charging point supplies the DC current to the EV battery pack [15–19].

The US EV charging sites and stations are given in Table 6.7 [7]. As of December 31, 2017, there are an estimated 20 178 EV public and private charging locations (sites) in the US, with 17 526 (86.9%) being available to the public. With 48 472 total public stations, each location has an average of 2.75 stations/outlets.

Europe EV charging sites and stations are given in Table 6.8 [20]. It can be seen that the total EV charging stations in Europe is almost triple times higher than the US charging stations. Furthermore, the charging stations are more common in Netherlands, Germany, and France due to the support policies of government.

Table 6.6 Charge methods electrical ratings.

Method	Connection	Power	Max. current
AC Level 1	1-phase	3.7 kW	10–16 A
AC Level 2	1 or 3-phase	3.7–22 kW	16–32 A
AC Level 3	3-phase	>22 kW	>32 A
DC fast charging	DC	>22 kW	>3225 A

(a)

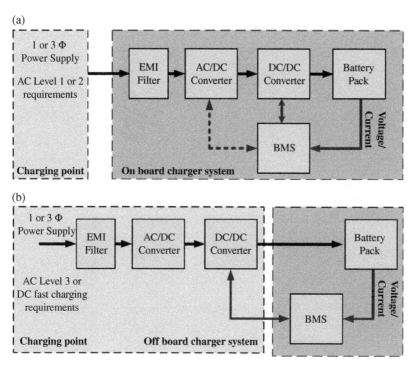

(b)

Figure 6.6 EV charging configuration at (a) AC Level 1 and 2 setups; (b) DC fast charging.

Table 6.7 US electric vehicle charging sites and stations. Adapted from Ref Num [7].

| Public/Private | Locations/Sites | EV Charging Stations | | | |
		AC Level 1	AC Level 2	DC Fast Charging	Total Stations
Public	17526	2619	38630	7223	48472
% by Type-Public	86.9%	5.4%	79.7%	14.9%	100.0%
Private	2652	639	5445	66	6150
% by Type-Private	13.1%	10.4%	88.5%	1.1%	100.0%
Public & Private	20178	3258	44075	7289	54622
% by Type – Both	100%	6.0%	80.7%	13.3%	100.0%

6.5.2 Wireless Charging

One of the most important challenges is limited charging infrastructures and anxiety from the use of these charging systems. Because electric safety is of concern, therefore some aspects should be carefully designed. An example is that electric shock is possible due to the weather conditions. Plug and cable can be also easily damaged and stolen by the thieves'. One way to get rid of these drawbacks is to use wireless power transfer (WPT)

Table 6.8 Europe electric vehicle charging sites and stations. Adapted from Ref Num [20].

Country	AC Level 1-2	AC Level 3 DC fast charging	Total
Austria	3429	699	4128
Belgium	1485	329	1814
Croatia	418	63	481
Denmark	2124	492	2616
Estonia	193	194	387
France	22308	2019	24327
Germany	22213	3218	25431
Italy	2582	542	3124
Netherlands	33616	920	34536
Norway	8739	2414	11153
Poland	410	172	582
Spain	4312	777	5089
Sweden	2718	2722	5440
Switzerland	3460	580	4040
Turkey	69	7	76
United Kingdom	13668	3708	17376
Others	5061	1460	6521
Total	**126805**	**20316**	**147121**

technology in the charging infrastructures. The WPT technology, which eliminates all the charging troublesome mentioned above, is more attractive solution for the EV owners. The charging process will be the easiest task by the WPT technology [21–25].

Although the fundamentals of this technology were reported by Nicola Tesla a century ago [26], it has been becoming a hot research area after the research team from Massachusetts Institute of Technology (MIT) published a research paper in Science [27]. In this paper, 60 W power was transferred at 2-m distance using strongly coupled magnetic resonance theory. Following this innovative study, several interesting studies have been succeeded with different types of methods and their control algorithms [28–30]. In some WPT applications, the resonant frequency is usually chosen at MHz level to transfer power more efficiently with midrange distance (up to several meters). However, in terms of operating frequency, human safety should be considered since the transmission power is much larger than that used in wireless communications. In particular, the human exposure limits have to comply with international safety guidelines (ICNIRP 1998 [31], IEEE C95.1-2005 [32]). For that reason, the resonance frequency at MHz level is hard to meet the power and efficiency criteria for high power WPT applications such as EVs charging system. It is quite difficult to transfer a few hundred kilowatts power at MHz frequency level with high efficiency. Furthermore, the transmission coils are too sensitive against ferromagnetic objects.

This drawback leads to significant variations in the coil parameters as well as increase the eddy current losses. In order to reduce the losses and parameter variations in the transmission coils, some design considerations should be taken into account. Ferrite and aluminum plate are usually adapted in the coil design stage to make it practical for EV charging systems [33]. The inductively coupled WPT method is a well-known method which has a ferrite-based core. This method operates at below MHz frequency (approximately several tens of kHz~hundreds of kHz). It shows higher efficiency with a several mm air gap when compared with the magnetic resonance method.

The operating principle of the WPT technology is similar to traditional transformers except weak inductive coupling coefficient that is approximately 0.95 for transformers; on the other hand, for the typical WPT system, this value is between 0.01 and 0.5 because of the large air gaps [34]. The simplified structure of the WPT system is shown in Figure 6.7. It has three main stage consisting of transmitting circuit, transmission coils, and receiver circuit. Each of these circuits include several sub-circuits to ensure proper operation. In the first stage, the diode rectifier is employed to convert utility AC power into DC power. After that, to drive the primary coil, a high-frequency DC to AC converter is used with a compensation network. Using this converter generates high frequency alternating magnetic field that induces an AC voltage on the secondary coil (pickup coil). Primary and secondary compensation circuits ensure the system works at resonance frequency. Hence the transferred power from primary coil to secondary coil is significantly increased. In the final stage of this system, AC power is converted to the DC power to supply battery charge unit, which is located in the EVs.

The WPT systems can be divided into two groups consisting of static WPT system, and dynamic WPT system. For static WPT system, the drivers just need to park their cars and leave; on the other hand, the dynamic WPT system allows the drivers to charge their EVs while driving. The dynamic WPT technology could help to reduce battery capacity of the EVs, which significantly decrease the EVs price. Figure 6.8 illustrates the static and dynamic WPT concepts.

Here, some of high-power WPT applications will be summarized. In the late 1970s, the dynamic WPT technology was carried out by the Partners for Advance Transit and Highways (PATH) program at the UC Berkeley [35]. In this study, 60 kW bus was employed with double powered units throughout 213 m pathway. The current of the transmitting circuit was 1200 A at 400 Hz. Furthermore, the distance between transmitting and receiving coils was 7.6 cm. The efficiency of the system under these conditions was

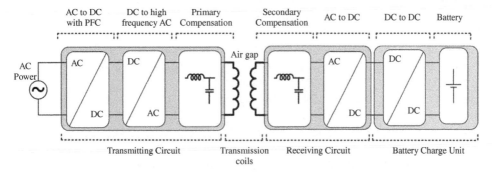

Figure 6.7 The general structure of wireless power transfer technology for EVs.

(a)

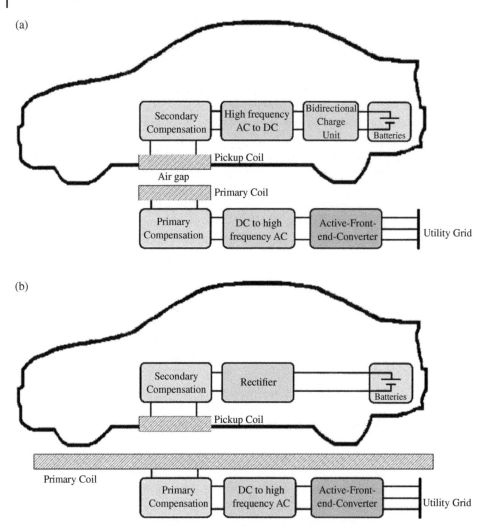

(b)

Figure 6.8 Concepts of (a) static WPT system; and (b) dynamic WPT system.

approximately 60%. One of the key factors of the low efficiency is limited semiconductor technology in 1970s. The high efficiency inductive power supply for EVs charging system has been presented by researchers from Auckland University [36]. In this study, proposed WPT system allowed 5 kW power transfer with over 90% efficiency under 200 mm distance. Furthermore, misalignment tolerance of this application is 250 mm lateral and 150 mm longitudinal with the 766 × 578 mm pad.

One of the important projects for WPT technologies is the on-line electric vehicle (OLEV) project, which is conducted by Korea Advanced Institute of Science and Technology (KAIST). This project also brings considerable benefits in terms of knowledge and WPT design. Three different size OLEV systems; a golf cart as the first generation; a bus for the second; and an SUV for the third; have been built from this project. The accomplishment of the second and the third is noteworthy: 60 kW power transfer for the buses and 20 kW for the SUVs with efficiency of 70 and 83%, respectively; allowable vertical distance and lateral misalignment up to 160 mm and up to 200 mm, respectively [35, 37].

In the United States, more and more public attention was drawn to the WPT since the publication of the 2007 Science paper [27]. The WiTricity Corporation with technology from MIT released their WiT3300 development kit, which achieves 90% efficiency over a 180 mm gap at 3.3 kW output. Recently, a wireless charging system prototype for EV was developed at Oak Ridge National Laboratory (ORNL) in the United States [38]. The tested efficiency is nearly 90% for 3 kW power delivery. The research at the University of Michigan–Dearborn achieved a 200 mm distance, 8 kW WPT system with dc-to-dc efficiency as high as 95.7% [39]. From the operational point of view, it is evident that the WPT for EV is available in both stationary and mobile applications. However, to make it possible for large-scale commercialization and application, there is still significant gap to be filled in term of performance optimization, standardization, cost reduction, etc.

6.6 Vehicle to Grid (V2G) Concept

EVs are good candidates for the future transportation sector due to the considerable benefits of EVs. On the other hand, the integration of EVs into the current power grid leads to several challenging problems that include demand management and source management. For example, EV charging process results in the extra load for the power system in case of large penetration level of EVs. Nevertheless, the energy stored in EVs could help power system operators to manage the demand at peak hours, blackouts, and other emergency situations. The technologies that depend on the energy stored in EV batteries are referred to as V2G technologies. In other words, V2G denotes to the control of EV batteries by the power system operators thru the communication between EVs and power system. Apart from the V2G concept, there are vehicle to home (V2H) and vehicle to vehicle (V2V) concepts, which can be considered under the same concept. In V2H concept, battery of EV works as backup energy source that can support home power system in case needed. Furthermore, V2V is a local EV community that can charge or discharge EV battery energy among EVs. The general structure of V2G, V2H, and V2V concepts are illustrated in Figure 6.9. The structure consists of traditional and renewable energy sources, aggregator, transmission and distribution system, loads, EVs, charge units [40–43].

In this concept, power flow can be unidirectional or bidirectional, which depends on the applications. It means that power can flow either from battery to grid or vice versa. The unidirectional and bidirectional power flow concepts are described in the following subsections. Table 6.9 shows the comparison between the unidirectional V2G and bidirectional V2G in various ways, such as the hardware infrastructure, power levels, costs, available services, benefits, and drawbacks [6].

6.6.1 Unidirectional V2G

As its name suggests, unidirectional V2G technology controls the charging rate of EV battery in a single power flow direction. To control the charge rate of battery in unidirectional V2G technology, a simple controller should be included in the management level.

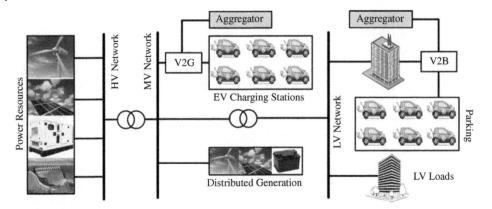

Figure 6.9 The general structure of V2G, V2H, and V2V concepts in power system.

Table 6.9 Comparison of V2G technologies.

V2G Technology	Services	Benefits	Limitations
Unidirectional	Load leveling	Maximized profit Minimized power loss Minimized emission Minimized operation cost	Limited service available
Bidirectional	Active and reactive power support Power factor correction Peak power shaving Voltage and frequency regulation	Maximized profit Minimized power loss Minimized operation cost Prevention of grid overloading Minimized emission Improved load profile Maintain voltage level Maximization of renewable energy generation	Battery degradation Complex hardware infrastructure High investment cost Social barriers

This technology gives the system operator to manage the power grid in more flexible way. However, the success rate of this technology mainly depends on the trading policy between the EV owners and the system operators. For that reason, the trading policy should offer reasonable revenues to the EV owners which depends on the charging time of their vehicles. For example, the revenues should be attractive when EV owners charge their vehicles during the off-peak hours and limit the charge during the peak hours. Although this technology improves the flexibility of the power system, its contribution is limited due to the nature of the unidirectional power flow. The power system operator often needs extra ancillary services such as reactive power support, voltage, and frequency regulation. To use full functionalities of the EVs, bidirectional V2G technology should be employed instead of unidirectional V2G technology [44, 45].

6.6.2 Bidirectional V2G

Bidirectional V2G technology allows bidirectional power flow between EVs and the power system. This bidirectional power flow allows for several grid ancillary services that give the system operator the possibility to manage the power system. To use bidirectional V2G technology, the power electronic converters in the battery charger should be designed to allow bidirectional power flow. The typical bidirectional power converter topology is depicted in Figure 6.10. It can be seen that it consists of bidirectional DC\DC converter and AC\DC converter. The bidirectional DC\DC converter operates as buck converter mode during charging and operates as boost converter mode during the discharging. The AC\DC converter operates as Pulse width modulation (PWM) rectifier mode during battery charging and it converters AC power into DC power with power factor correction functionality, which is recommended by international standards. On the other hand, the AC\DC converter operates as inverter mode during the discharging and it injects power from batteries to the power grid [46–50].

As mentioned above, the bidirectional Vehicle-to-Grid (V2G) technology offers larger flexibilities and possibilities to improve the power system operations. Some of the benefits of this technology are active and reactive power support, power factor correction. Similar to the unidirectional V2G technology, the trading policy should offer reasonable revenues to the EV owners to maximize the benefits of bidirectional V2G technology.

Although bidirectional V2G technology offers numerous benefits, the implementation of this technology encounters several challenges. The main issue with this technology is the battery degradation owing to frequent charging and discharging cycles. This issue increases the operational cost of the system. Furthermore, to enable this technology, bidirectional power converters that needs complex design and control structure. This also increases the system cost. Another significant barrier for the implementation of this technology is social acceptance. Most of the EV owners do not want to participate in the bidirectional V2G services due to the driving range anxiety. EV owners want to keep their battery state of charge (SoC) at high level for the unexpected traveling usage [51–53].

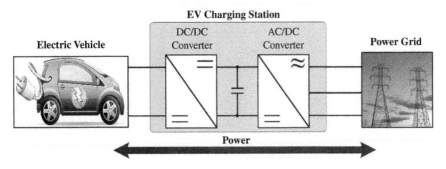

Figure 6.10 The typical bidirectional power converter topology for bidirectional V2G technology.

6.7 Barriers to EV Adoption

Although EVs offers considerable benefits, they are still not commonly adopted in the car market due to the technological, social, and economical problems. Figure 6.11 presents the barriers to EV adoption under three main categories; (i) technological; (ii) social; (iii) economical. In this section, barriers to EV adoption are described.

6.7.1 Technological Problems

The major barrier to EV adoption in the market is technical issues that include the driving range limit, long charging duration, and safety concerns. The batteries are the main source of these technical issues.

Limited Range: The driving range of the EVs mainly depends on the energy stored in the battery. Moreover, driving style, the terrain it is being driven on, and carrying cargo in the EV impact the driving range of the EVs. All these parameters result in range anxiety among the potential EV owners. Recently, the EV manufacturers have worked on the new techniques to increase the driving range of EVs. For example, Tesla Model S 100D offers around 500 km driving range, which is competitive with the internal combustion engine cars [7]. However, range anxiety remains a major barrier for EV deployment.

Long Charging Period: Another barrier to EVs adoption is long charging period for EVs. The charging duration takes from a few minutes to hours, which is depending on the battery chemistry and charge method. On the other hand, the filling petrol to the internal combustion engine only takes a couple of minutes. Although fast charging technologies have been developed and more are being studied, the long charging period is still a barrier to EV adoption.

Safety Concerns: The concerns about safety are rising mainly about the fuel cell vehicles (FCVs) nowadays. There are speculations that, if hydrogen escapes the tanks it is kept into, can cause serious harm, as it is highly flammable. It has no color either, making a leak hard to notice. There is also the chance of the tanks to explode in case of a collision [3].

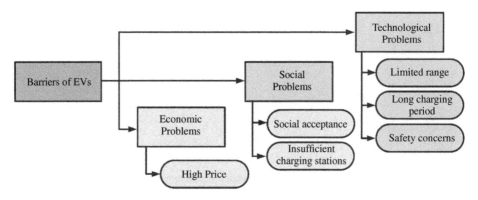

Figure 6.11 Technological, social, and economic problems of EVs.

6.7.2 Social Problems

The acceptance of a new technology takes long time in the society due to the high initial cost of this technology and the difficulty of changing habits. For example, someone who has been driving a conventional car for many years mostly will not accept the EV due to the requirement of different driving patterns, refueling habits. Another barrier to social acceptance is insufficient charging stations, which are very few compared with the conventional petrol stations. Furthermore, not all the public charging stations are compatible with every car as well; therefore, it also becomes a challenge to find proper charging stations. There is also the risk of getting a fully occupied charging station with no room for another car since charging duration is quite long. Taking account of the long charging duration, it is of utmost importance to eliminate potential EV owners' concerns.

6.7.3 Economic Problems

The price of EVs is quite higher than conventional cars. The main reason of high price is battery packs and drive system. To make EVs more attractive for people, governments in some countries including, Germany, Netherlands, and France, have prepared support policies and incentives. Furthermore, recent technological developments in the battery technology lead to a decrease in the prices of batteries.

6.8 Trends and Future Developments

The use of EVs has brought considerable benefits to mankind in terms of social, economic, and environmental. Although recently many developments have been announced in EVs market, the adoption of EVs has still opened doors for new possibilities especially for the power system. According to many market reports, EVs will have a significant share in future transportation industry. Furthermore, the use of EVs will have serious impact on the power system. To reduce their impact on the power system, the smart grid, which is being considered as the future electric energy paradigm, will play an important role. Vehicle to grid technologies are the link between the EVs and smart grid and as described above EVs and power systems may get several benefits from V2G technologies. However, to get maximum benefits from V2G technologies, the trading policy should offer reasonable revenues to the EV owners. For that reason, the decision makers should prepare support policies to encourage the EV owners to be participate in V2G technologies.

Another open area in the EVs adoption is the charging infrastructures since these technologies make EVs widely acceptable by the community. Especially important goal is to reduce the charging duration, which can make EVs competitive with the conventional cars. In addition, the chargers should have capability to communicate with the grid to schedule charging time to improve the benefits of V2G technologies.

The development of battery technologies also plays a crucial role to take the EV technology further. The batteries should have higher energy density and capacity with less weight and less charging time. The batteries should also be designed with non-toxic chemicals and the cost should be affordable since the main cost of the EVs is the battery systems. Therefore, the research on this area mainly have focused on the new chemicals to obtain the mentioned features.

The development of electric motors, which is the main part of the EVs, have a great impact on the EV performances. In recent years, induction motor, synchronous reluctance motor, and switched reluctance motor are being attractive choice for the EVs. For example, Tesla is using an induction motor for the traction system of Tesla Model S. However, further research and development should be carried out to obtain more efficient electric motors.

The wireless charging technology offers the EV owners a seamless and convenient alternative to charging conductively. In addition, it provides an inherent electrical isolation and reduces on-board charging cost, weight and volume as well. Over the last few years, the wireless charging technology has gained global acceptance and it is very popular for supplying power to various electronic devices without physical contacts. There are countless wireless charging technology research studies going on around the world to improve the existing technologies, however, this technology has many limitations yet to be resolved. These issues are mostly linked to the power electronic converters and transmission coil, which is the key part of this technology. Some of the issues are related to energy efficiency, power density, and magnetic coupling which have been studied extensively with many optimization techniques.

The future research topics will of course, revolve around making the EV technology more efficient, affordable, and convenient. A great deal of research has already been conducted on making EVs more affordable and capable of covering more distance: energy management, materials used for construction, different energy sources, etc. [3].

6.9 Conclusion

Developments in technology have increased the share of EVs in the future transportation industry. It is no doubt that EVs are a feasible alternative to conventional vehicles that depend on the fossil-based fuels. To foster the EV deployment and acceptance, further development is needed on drivetrain, battery storage, and charger technologies. The EV types, configurations, battery technologies, power conversion and charging technologies have been discussed in this chapter. The key technologies for EVs have been reviewed and some of characteristics have been summarized. Furthermore, the barriers of current EVs have been listed along with possible solutions to overcome these shortcomings. Finally, trends and ways of EVs future developments have been presented.

References

1 Tanaka, N. Transport, Energy, and CO_2. International Energy Agency. https://www.iea.org/reports/transport-energy-and-co2.

2 European Parliament (2018). Battery-powered electric vehicles: market development and lifecycle emissions. http://www.europarl.europa.eu/RegData/etudes/STUD/2018/617457/IPOL_STU(2018)617457_EN.pdf (accessed 15 February 2021).

3 Fuad, U., Sanjeevikumar, P., Lucian, M. et al. (2017). A comprehensive study of key electric vehicle (EV) components, technologies, challenges, impacts, and future direction of development. *Energies* 10: 8. https://doi.org/10.3390/en10081217.

4 Alshahrani, S., Khalid, M., and Almuhaini, M. (2019). Electric vehicles beyond energy storage and modern power networks: challenges and applications. *IEEE Access* 7: 99031–99064.

5 Pelletier, S., Jabali, O., and Laporte, G. (2014). *Battery Electric Vehicles for Goods Distribution: A Survey of Vehicle Technology, Market Penetration, Incentives and Practices*. Interuniversity Research Centre on Enterprise Networks, Logistics and Transportation.

6 Shafiei, A., Carli, G., and Williamson, S.S. (2014). *Electric and Plug-In Hybrid Electric Vehicles*. Wiley.

7 Bayram, I.S. and Tajer, A. (2017). *Plug-in Electric Vehicle Grid Integration*. Norwood, MA: Artech House.

8 Yong, J.Y., Ramachandaramurthy, V.K., Tan, K.M., and Mithulananthan, N. (2015). A review on the state-of-the-art technologies of electric vehicle, its impacts and prospects. *Renewable and Sustainable Energy Reviews* 49: 365–385.

9 Rind, S.J., Ren, Y., Hu, Y. et al. (2017). Configurations and control of traction motors for electric vehicles: a review. *Chinese Journal of Electrical Engineering* 3 (3): 1–17.

10 Andwari, A.M., Pesiridis, A., Rajoo, S. et al. (2017). A review of battery electric vehicle technology and readiness levels. *Renewable and Sustainable Energy Reviews* 78: 414–430.

11 Shaukat, N., Khan, B., Ali, S.M. et al. (2018). A survey on electric vehicle transportation within smart grid system. *Renewable and Sustainable Energy Reviews* 81 (Part 1): 1329–1349.

12 Cano, Z.P., Banham, D., Ye, S. et al. (2018). Batteries and fuel cells for emerging electric vehicle markets. *Nature Energy* 3 (4): 279–289.

13 Hannan, M.A., Hoque, M.M., Hussain, A. et al. (2018). State-of-the-art and energy management system of lithium-ion batteries in electric vehicle applications: issues and recommendations. *IEEE Access* 6: 19362–19378.

14 Herron, D. (2017). Energy density in battery packs or gasoline. https://greentransportation. info/energy-transportation/energy-density.html (accessed 15 February 2021).

15 Mwasilu, F., Justo, J.J., Kim, E. et al. (2014). Electric vehicles and smart grid interaction: a review on vehicle to grid and renewable energy sources integration. *Renewable and Sustainable Energy Reviews* 34: 501–516.

16 Tan, K.M., Ramachandaramurthy, V.K., and Yong, J.Y. (2016). Integration of electric vehicles in smart grid: a review on vehicle to grid technologies and optimization techniques. *Renewable and Sustainable Energy Reviews* 53: 720–732.

17 Williamson, S.S. (2013). *Energy Management Strategies for Electric and Plug-In Hybrid Electric Vehicles*. New York: Springer.

18 Kong, P. and Karagiannidis, G.K. (2016). Charging schemes for plug-in hybrid electric vehicles in smart grid: a survey. *IEEE Access* 4: 6846–6875.

19 Williamson, S.S., Rathore, A.K., and Musavi, F. (2015). Industrial electronics for electric transportation: current state-of-the-art and future challenges. *IEEE Transactions on Industrial Electronics* 62 (5): 3021–3032.

20 European Alternative Fuels Observatory (2020). Electric vehicle charging infrastructure. http://www.eafo.eu/electric-vehicle-charging-infrastructure (accessed 15 February 2021).

21 Moon, S., Kim, B., Cho, S., and Moon, G. (2013). Analysis and design of wireless power transfer system with an intermediate coil for high efficiency. *IEEE ECCE Asia Downunder*, Melbourne, Australia (3–6 June 2013). IEEE.

22 Ahmad, A., Alam, M.S., and Chabaan, R. (2018). A comprehensive review of wireless charging technologies for electric vehicles. *IEEE Transactions on Transportation Electrification* 4 (1): 38–63.

23 Moon, S., Kim, B., Cho, S. et al. (2014). Analysis and design of a wireless power transfer system with an intermediate coil for high efficiency. *IEEE Transactions on Industrial Electronics* 61 (11): 5861–5870.

24 Li, S. and Mi, C.C. (2015). Wireless power transfer for electric vehicle applications. *IEEE Journal of Emerging and Selected Topics in Power Electronics* 3 (1): 4–17.

25 Mi, C.C., Buja, G., Choi, S.Y., and Rim, C.T. (2016). Modern advances in wireless power transfer systems for roadway powered electric vehicles. *IEEE Transactions on Industrial Electronics* 63 (10): 6533–6545.

26 Tesla, N. (1907). Apparatus for transmitting electrical energy. US Patent 1119732, filed 4 May1907 and issued 1 December 1914.

27 Kurs, A.K.A., Moffatt, R., Joannopoulos, J.D. et al. (2007). Wireless power transfer via strongly coupled magnetic resonances. *Science* 317 (5834): 83–86.

28 Cannon, B.L., Hoburg, J.F., Stancil, D.D., and Goldstein, S.C. (2009). Magnetic resonant coupling as a potential means for wireless power transfer to multiple small receivers. *IEEE Transactions on Power Electronics* 24 (7): 1819–1825.

29 Kainan, C. and Zhengming, Z. (2013). Analysis of the double-layer printed spiral coil for wireless power transfer. *IEEE Journal of Emerging and Selected Topics in Power Electronics* 1 (2): 114–121.

30 Yiming, Z., Zhengming, Z., and Kainan, C. (2014). Frequency decrease analysis of resonant wireless power transfer. *IEEE Transactions on Power Electronics* 29 (3): 1058–1063.

31 (1998). Guidelines for limiting exposure to time-varying electric, magnetic and electromagnetic fields (up to 300 GHz). *Health Physics* 74 (4): 494–522.

32 IEEE SCC28 C95.1-2005 (2005). *IEEE Standard for Safety Levels with Respect to Human Exposure to Radio Frequency Electromagnetic Fields, 3 kHz to 300 GHz*. International Committee on Electromagnetic Safety.

33 Nagatsuka, Y., Ehara, N., Kaneko, Y. et al. (2010).Compact contactless power transfer system for electric vehicles. *International Power Electronics Conference*, Sapporo, Japan (21–24 June 2010). IEEE.

34 Budhia, M., Boys, J.T., Covic, G.A., and Huang, C.-Y. (2013). Development of a single-sided flux magnetic coupler for electric vehicle IPT charging systems. *IEEE Transactions on Industrial Electronics* 60 (1): 318–328.

35 Cirimele, V., Diana, M., Freschi, F., and Mitolo, M. (2018). Inductive power transfer for automotive applications: state-of-the-art and future trends. *IEEE Transactions on Industry Applications* 54 (5): 4069–4079.

36 Song, B., Shin, J., Lee, S. et al. (2012). Design of a high power transfer pickup for on-line electric vehicle (OLEV). *IEEE International Electric Vehicle Conference*, Greenville, USA (4–8 March 2012). IEEE.

37 Shin, J., Shin, S., Kim, Y. et al. (2014). Design and implementation of shaped magnetic-resonance-based wireless power transfer system for roadway-powered moving electric vehicles. *IEEE Transactions on Industrial Electronics* 61 (3): 1179–1192.

38 Puqi, N., Miller, J.M., Onar, O.C., and White, C.P. (2013). A compact wireless charging system development. *IEEE Energy Conversion Congress and Exposition*, Denver, USA (15–19 September 2013). IEEE.

39 Nguyen, T.-D., Li, S., Li, W., and Mi, C. (2014). Feasibility study on bipolar pads for efficient wireless power chargers. *Applied Power Electronics Conference and Exposition*, Fort Worth, USA (16–20 March 2014). IEEE.

40 Xiong, R., Cao, J., Yu, Q. et al. (2018). Critical review on the battery state of charge estimation methods for electric vehicles. *IEEE Access* 6: 1832–1843.

41 Sarlioglu, B., Morris, C.T., Han, D., and Li, S. (2017). Driving toward accessibility: a review of technological improvements for electric machines, power electronics, and batteries for electric and hybrid vehicles. *IEEE Industry Applications Magazine* 23 (1): 14–25.

42 Mukherjee, J.C. and Gupta, A. (2015). A review of charge scheduling of electric vehicles in smart grid. *IEEE Systems Journal* 9 (4): 1541–1553.

43 Zhang, F., Zhang, X., Zhang, M., and Edmonds, A.S.E. (2016). Literature review of electric vehicle technology and its applications. *5th International Conference on Computer Science and Network Technology*, Changchun, China (10–11 December 2016). IEEE.

44 Habib, S., Khan, M.M., Abbas, F. et al. (2018). A comprehensive study of implemented international standards, technical challenges, impacts and prospects for electric vehicles. *IEEE Access* 6: 13866–13890.

45 Zhou, Z., Sun, C., Shi, R. et al. (2017). Robust energy scheduling in vehicle-to-grid networks. *IEEE Network* 31 (2): 30–37.

46 Khosrojerdi, F., Taheri, S., Taheri, H., and Pouresmaeil, E. (2016). Integration of electric vehicles into a smart power grid: A technical review. *IEEE Electrical Power and Energy Conference*, Ottawa, Canada (12–14 October 2016). IEEE.

47 Vepsäläinen, J. (2017). Driving Style Comparison of City Buses: Electric vs. Diesel. *IEEE Vehicle Power and Propulsion Conference*, Belfort, France (11–14 December 2017). IEEE.

48 Ahmadi, M., Mithulananthan, N., and Sharma, R. (2016). A review on topologies for fast charging stations for electric vehicles. *IEEE International Conference on Power System Technology*, Wollongon, Australia (28 September–1 October 2016). IEEE.

49 Lehtola, T. and Zahedi, A. (2016). Electric vehicle to grid for power regulation: A review. *IEEE International Conference on Power System Technology*, Wollongong, Australia (28 September–1 October 2016).IEEE.

50 Silvas, E., Hofman, T., Murgovski, N. et al. (2017). Review of optimization strategies for system-level design in hybrid electric vehicles. *IEEE Transactions on Vehicular Technology* 66 (1): 57–70.

51 Deb, S., Kalita, K., and Mahanta, P. (2017). Review of impact of electric vehicle charging station on the power grid. *International Conference on Technological Advancements in Power and Energy*, Kollam, India (21–23 December 2017). IEEE.

52 Deng, L. and Liu, M. (2017). A review of research on electric vehicle charging facilities planning in China. *2nd International Conference Sustainable and Renewable Energy Engineering*, Hiroshima, Japan (10–12 May 2017). IEEE.

53 Afshari, E., Moradi, G.R., Ramyar, A. et al. (2017). Reactive power generation for single-phase transformerless Vehicle-to-Grid inverters: A review and new solutions. *IEEE Transportation Electrification Conference and Expo*, Chicago, USA (22–24 June 2017) IEEE.

7

Net Zero Energy Buildings

The topic of zero energy buildings (ZEBs) has gained increasing attention in recent years, due to its potential benefits for decreasing energy consumption in building sector, in addition to reducing green gas emission and improving environmental comfort inside the building. The goal of this chapter is to shed light on its definition, design, modeling, control, and optimization. Furthermore, generalizing its concept into the smart grid (SG) community will be discussed. Its benefits, barriers, current state and future trends will also be highlighted.

7.1 Introduction

The urge to decrease greenhouse gas emissions and mankind's dependency on depleting fossil energy resources depicts the crucial role of enhancing the effectiveness of buildings, as the building sector is responsible for 31% of final energy consumption and for 29% of CO_2 emissions into the atmosphere [1]. Consequently, buildings which consume the least amount of energy possible and produces low levels of carbon emissions, have been the main goal of the power industry. Many areas worldwide began to seek enhancement of their building codes and implementing energy-saving procedures for buildings. Moreover, buildings in the near future are strongly encouraged to generate the magnitude of energy they consume, i.e. become or approach ZEBs. ZEBs are buildings that operate in synergy and coordination with the grid, sidestepping extra stress applied on the power infrastructure. This can be achieved by decreasing required energy by implementing efficient actions and meeting the decreased energy needs by employing renewable energy sources, a sequence of optimized and balanced processes between consumption and production associated with effective SG employment. Smart grids may establish a revolution in the building sector, they can provide the chance to advance power quality and the reliability of energy sources because it can deal with decentralization of supply and improved balance between supply and demand. Therefore, implementation of SG is the key to accomplishing the previously outlined ZEB objectives. This chapter will review the net Zero Energy Buildings (nZEB)' definition, design, modeling, controlling, and optimization. The net Zero Energy Community (nZEC) will be presented as an extension for applying the nZEB concept. The benefits, barriers, current state, investment trends for the nZEB will be presented.

Smart Grid and Enabling Technologies, First Edition. Shady S. Refaat, Omar Ellabban, Sertac Bayhan, Haitham Abu-Rub, Frede Blaabjerg, and Miroslav M. Begovic.
© 2021 John Wiley & Sons Ltd. Published 2021 by John Wiley & Sons Ltd.
Companion website: www.wiley.com/go/ellabban/smartgrid

7.2 Net Zero Energy Building Definition

The net-zero method is known to be a technique applied for achieving equilibrium between the resource drawn from the natural environment (e.g. energy or water) and its consumption at homes, buildings, blocks or communities at a pre-defined time span, usually every year. The net Zero Energy Concept (nZEC) implements the idea of energy self-sufficiency that attempts to decrease energy demand (decrease final energy use, optimize energy efficiency, employ energy recovery and cogeneration opportunities) and expand the employment of on-site renewable energy resources to meet the remaining load. Furthermore, occupancies energy conservation behavior and their awareness and engagement are essentials for achieving the nZEC, as illustrated in Figure 7.1 [2].

In spite of enthusiasm over the term "zero energy," there is no or little common definition, or yet a common comprehension, of its definition. The zero-energy description influences the way buildings are built to accomplish the required objective. It can highlight energy efficiency, supply-side plans, bought energy sources, utility rate structures, or if fuel-switching and conversion accounting has the potential to achieve the objective. A ZEB normally utilizes common energy sources, for example, electric and natural gas utilities when on-site generation is not capable of sufficiently supplying the loads. When the on-site generation is higher than the building's loads, additional power is transmitted back to the utility grid. Figure 7.2 provides an outline of the nZEB and its corresponding terminology, these terminologies can be summarized as [3, 4]:

- *Building system boundary* defines the building boundary at which the energy balance is computed. It entails, the physical boundary (for a single building or a community) which identifies whether renewables are generated "on-site" or "off-site," and the Balance boundary which finds out which energy consumed (e.g. heating, cooling, ventilation, hot water, lighting, appliances) is considered in the balance.

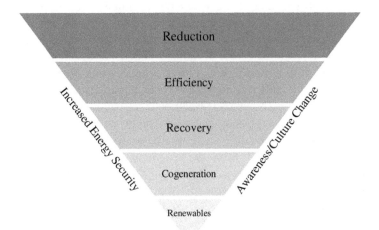

Figure 7.1 The net-Zero Energy Concept (nZEC). Adapted from Ref Num [2].

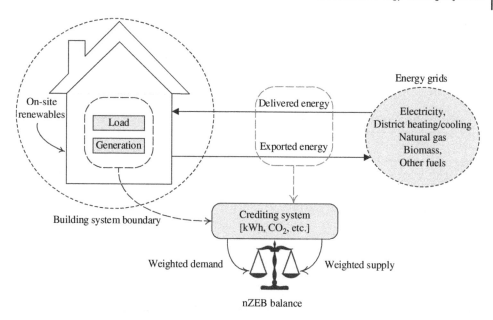

Figure 7.2 Net Zero Energy Building (nZEB) overview and relevant terminology. Adapted from Ref Num [4].

- *Energy grids* are attributed to the supply system of energy carriers including electricity, natural gas, thermal networks for district heating/cooling, biomass, and other fuels. A grid could also be implemented as a two-way grid, as power grids and as thermal networks.
- *Delivered energy* refers to the energy transferred from power grids to buildings, quantified per each energy carrier in (kWh/y) or (kWh/m^{2}y).
- *Exported energy* refers to the energy transferred from buildings to power grids, quantified per each energy carrier.
- *Load* refers to the building's energy demand, characterized per each energy carrier. The load might not concur with the transferred energy as a result of self-consumption of energy produced on-site.
- *Generation* refers to the building's energy production, quantified per each energy carrier. The production might not match the transferred energy as a result of self-consumption of energy produced onsite.
- *Crediting system* is a weighting system that transforms the physical units into a metric convenient to the user to obtain, produce, and transfer the energy. Weighting aspects could be more convenient with regard to political preferences instead of 100% scientific or engineering intentions.
- *Weighted demand* refers to the total produced energy from each energy carrier multiplied by its corresponding weighting factor.
- *Weighted supply* refers to the total transferred energy to each energy mover and each is multiplied by its corresponding weighting factor.
- *Net ZEB balance* refers to the requirement fulfilled as the weighted supply achieves or surpasses weighted demand at a certain time span, supposedly a year. The nZEB balance

can be represented graphically as shown in Figure 7.3 [4], the reference building could describe the effectiveness of a new building constructed while conforming to conditions of the national building code, in the majority of cases crucial energy efficiency procedures are required due to the fact that on-site energy production choices are inadequate, for example, by appropriate surface areas of solar systems, particularly in tall buildings.

Four usually known definitions for the ZEB are as follows: net zero site energy, net zero source energy, net zero energy costs, and net zero energy emissions [5].

- *Net Zero Site Energy*: A site ZEB generates, at a minimum, the same energy it utilized in 365 days, when considering it at the site.
- *Net Zero Source Energy*: A source ZEB generates at least the same energy it utilizes in a year, when considering it at the source, which denotes the main energy utilized to produce and transmit the energy to the site. To determine a building's whole source energy, applied and transferred energy is multiplied by the applicable site-to-source conversion multipliers.
- *Net Zero Energy Costs*: In a cost ZEB, the price the utility transfers to the building owner with regard to the energy the building transfers to the grid is at least the same as the money the owner gives to the utility for the energy amenities and energy utilized in the course of 365 days.
- *Net Zero Energy Emissions*: A net-zero emission building generates at least the same emissions-free renewable energy as it utilized from emissions-producing energy sources. To determine a building's emissions, applied and transferred energy are multiplied by the applicable emissions multiplier, predicated upon the utility's emission and on-site production emissions.

Table 7.1 highlights key characteristics of each definition.

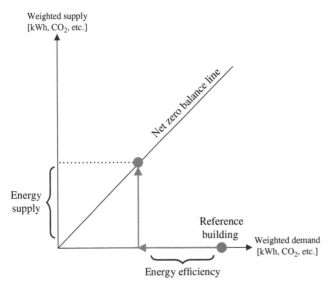

Figure 7.3 The nZEB balance concept. Ref Num [4]. Reproduced with permission from ELSEVIER.

Table 7.1 ZEB definitions summary.

Definition	Advantages	Drawbacks	Other concerns
Site ZEB	• Simple to employ; • Certifiable by on-site measurements; • Cautious method to accomplish nZEB; • No externalities influencing effectiveness, capable of following success through time; • Simple for the building community to comprehend and establish communication; • Promotes energy efficient building plans;	• Needs additional Photovoltaics (PV) generation to offset natural gas; so, won't take into account all utility costs (could hold a small load factor); • Incapable of associating fuel types; • Does not consider non-energy differences among fuel types (availability of supply, pollution).	
Source ZEB	• Capable of associating energy value of fuel types utilized at the site; • Improved model for influence on national energy system; • Simpler nZEB to achieve.	• No consideration for non-energy differences among fuel types (availability of supply, pollution); • Source calculations very comprehensive (no consideration for regional or daily fluctuations); • Source energy usage accounting and fuel switching could possess a greater influence than efficiency technologies; • Not all utility costs are taken into account (could possess a small load factor).	• Required to advance site-to-source conversion factors, which need substantial quantities of information to describe.
Cost ZEB	• Simple to employ and determine; • Market forces a reasonable balance among fuel types; • Permits demand-responsive control; Certifiable from utility bills.	• Incapable of sufficiently demonstrating influence on the national grid for demand, as additional PV production may be increasingly treasured for decreasing demand with on-site storage than transferring it to the grid; • Demands net-metering agreements in a manner that the delivered power may offset energy and non-energy charges; • Relatively volatile energy rates make it tough to follow as time passes by.	• Offsetting monthly service and infrastructure charges need to surpass ZEB. • Net metering is not sufficiently recognized, usually with capacity limits and at buyback rates smaller than retail rates.

(Continued)

Table 7.1 (Continued)

Definition	Advantages	Drawbacks	Other concerns
Emissions ZEB	• Improved model for green power; • Responsible for non-energy differences among fuel types (pollution, greenhouse gases); • A simpler ZEB to accomplish.		• Requires acceptable emissions factors

Different supply-side renewable energy technologies are present for ZEBs, for example: PV, solar hot water, wind, hydroelectric, and biofuels. On-site renewables such as Rooftop PV and solar water heating are the most suitable supply-side technologies for extensive application of ZEBs. Other off-site (community) renewable technologies for instance, parking lot-based wind, PV systems, biomass, hydroelectric, and biofuels may be available for limited applications and may not be available over the lifetime of the building. Furthermore, utility renewables can be purchased and used in-site; these renewables will be subject to the availability of the source and transmission losses [6]. The nZEB can be explained, according four categories, predicated upon the renewable energy sources listed in Table 7.2.

7.3 Net Zero Energy Building Design

The notion of nZEB has been extensively studied over the previous 10 years to attain energy efficiency in the building sector and motivate renewable energy incorporation on-site. However, the design strategies of the nZEB lack clarity because the buildings could entail complicated coordination or management of several energy systems, which includes renewable energy generations, energy appliances, energy storages, and could also have a certain connection with the SG. The optimized nZEB design begins with understanding of the building use, its interior comfort needs, and the investigation of the natural and environmental resources found on the building site. As illustrated in Figure 7.4, there are three main design elements for attaining the nZEB goal, which are passive design strategies, advanced energy efficiency technologies and renewables integration. Passive design strategies must be employed to decrease energy consumed, as their prices are quite low. Developed building technologies must then be implemented, as their prices are greater, but they can substantially decrease the building's energy use. Finally, renewable energy sources should be utilized after the inactive or passive design strategies and developed building technologies are utilized efficiently, due to high initial prices [7–12].

A decent passive design for the building, could entail improved orientation, a high performance thermal-isolation envelope, reasonable tightness and well-designed shade for windows, which could hinder thermal and electrical consumption of buildings. To meet the decreased loads, several HVAC (heating, ventilation, air-conditioning) systems, DHW

Table 7.2 Classifying nZEBs by renewable energy supply.

nZEB Classification		Description	Comments
On-site Renewables	A	Utilize renewable energy sources present in the vicinity of the building's footprint. Examples: PVs, solar hot water, building-integrated wind systems.	Applicable for: Site, Source, and Emissions nZEBs. Suitable as a cost nZEB could be complex according to certain net metering policies in the region.
	B	Utilize renewable energy sources present in the vicinity of the building's footprint (nZEB-A) and the building site. Examples: PVs, solar hot water, low-impact hydroelectric, and wind found on parking lots or adjacent open space, yet not physically installed on the building.	Applicable for: Site, Source and Emissions NZEBs. Suitable as a cost nZEB could be complex according to certain net-metering policies in the area.
Off-site Renewables	C	Utilize renewable energy sources as nZEB-A and nZEB-B and utilize renewable energy sources present off site to produce energy on site. Examples: biomass, wood, ethanol, or biofuel that can be introduced from off-site, or gathered from waste streams from on-site procedures that could be implemented on-site to produce electricity and heat.	Applicable for: Site nZEBs; Less applicable for: Source, Cost, and Emissions nZEBs; nZEB-C buildings usually do not arrive at a cost nZEB position due to the fact that renewable materials are brought to be introduced on-site, so it may be extremely complicated to recover these costs by any compensation acquired from the utility for renewable energy production.
	D	Utilize renewable energy sources for example, nZEB-A, nZEB-B and cZEB-C and buy off-site utility renewable green energy to preserve nZEB status. Examples: Utility-scale wind, photovoltaic, emissions credits, green energy purchasing choices.	Applicable for: Source NZEBs, Emissions NZEBs; Less applicable for: Site NZEBs, Cost NZEBs.

(domestic hot water) systems, intelligent lighting systems, smart plugs, smart thermostat, temperature, moisture and occupancy sensors and smart communication protocols, and algorithms, etc. are put forward. The practical intention of these systems is to effectively establish a calm and indoor environment for people living. Certainly, different energy sources, that include natural gas or electric power, are required to operate the building service system. Therefore, renewable energy production system should be implemented to offset the energy consumption. Consequently, a nZEB has the potential to be implemented with electricity and thermal production from renewable energy sources, given that it possible to employ the energy capacity. To optimize the performance of the nZEB operation and to facilitate the interaction between the in-building energy system and the energy grid

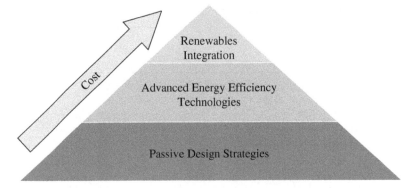

Figure 7.4 nZEB main design elements.

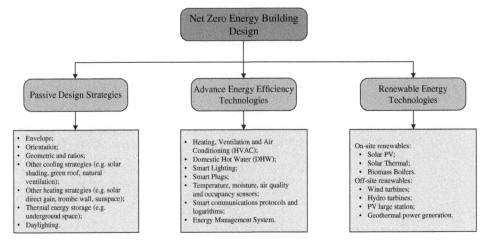

Figure 7.5 Main design elements for the nZEB.

an Energy Management System (EMS) will be an essential part in nZEB. So, EMS provides a more convenient method of decreasing energy requirements of buildings and preserving indoor environmental quality meeting the global standard of sustainability, which was certified as reliable for many applications including HVAC control, thermal comfort control, and renewable production, etc. An explanatory diagram of the main elements used in the design process of the nZEB is shown in Figure 7.5.

Three passive design plans denote the energy efficiency plan at nZEB:

1) Reduce heat transmittance: decreasing heat flow in the building during the summer helps in lessening the stress on the air conditioning system.
2) Daylighting: in numerous buildings, lights are on during the day even when sunlight is at its peak. At nZEB, sunlight enters deep into the space decreasing the need for artificial lighting.
3) Natural ventilation: more than 40% of energy used in a common building is due to air conditioning. Natural ventilation at nZEB helps decrease the demand for air conditioning.

Energy efficiency technologies entail effective technologies with automated controls, networked sensors and meters, developed building automation, data analytics software, energy management and information systems, and monitoring-based commissioning. Smart technologies deliver significant energy savings as illustrated in Table 7.3 [13]. EMS provides a more convenient method to decrease the energy needed for buildings and to preserve indoor environmental quality complying with the global standard of sustainability which, as certified, as reliable for different applications including HVAC control and thermal comfort control, plug load, smart appliance, on-site renewable generation, on-site storage, Plug-in electric vehicles, etc. in addition to the interaction with the utility through smart meters, as illustrated in Figure 7.6 [14].

Options range from decentralized (On-site) renewable energy generation, producing energy within buildings, to centralized (Off-site) renewable energy generation, where energy is produced elsewhere and then distributed to buildings by energy networks, as illustrated in Table 7.4. On-site choices entail solar thermal collectors, solar PV panels, biomass boilers, and modern cook stoves utilizing bioenergy. Off-site choices entail utilizing renewable energy applications to produce heat or cold generated to buildings via district energy networks, and renewable power, which could be utilized for cooking, lighting and appliances, and heating or cooling.

During 2010 and 2014, rooftop solar power capacity tripled worldwide, from 30 to 100 GW, sufficient enough to meet the power demand of approximately 30 million houses. By 2030, IRENA approximates the PV used may increase up to 580 GW. This paradigm shift offers houses the chance to generate their own electricity, and decreases the vulnerability associated with shortages, blackouts, and volatility in electricity costs [15].

Table 7.3 Energy savings potential using smart technology. Ref Num [13]. Reproduced with permission from ELSEVIER.

System	Technology	Energy savings
HVAC	Variable frequency motor drive	15–50% of pump or motor energy
HVAC	Smart thermostat-based HVAC system	5–10% HVAC
Plug type	Smart load plug	50–60%
Lighting	Advanced lighting controls	45%
Lighting	Remote-based lighting management system	20–30% above controls savings
Window shading	Automated shading system	21–38%
Window shading	Switchable film system	32–43%
Window shading	Smart glass system	20–30%
Building automation	Smart building Automation System (BAS)	10–25% whole building
Analytics	Cloud-based energy information system (EIS)	5–10% whole building

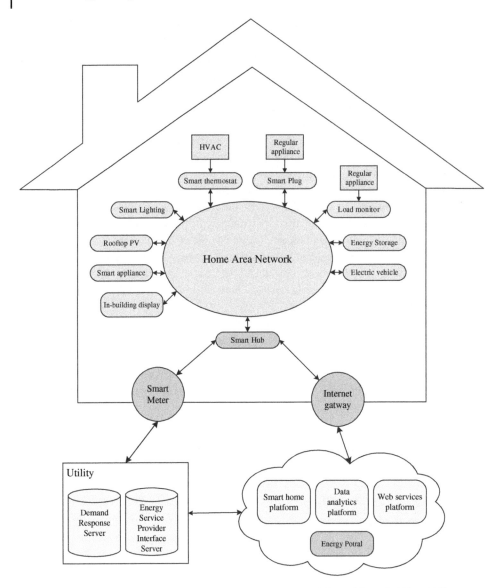

Figure 7.6 Fully integrated nZEB energy management system. Adapted from Ref Num [14].

7.4 Net Zero Energy Building: Modeling, Controlling and Optimization

Several procedures have been proposed to create energy consumption models that imitates a building for load forecasting or price saving approximations. To consider the thermo-visual comfort of the people living, the HVAC systems, lighting systems, electric motors are the main contributors for energy consumption in the building sector. The top four end uses are space heating, space cooling, water heating, and lighting – responsible for approximately

Table 7.4 Renewable energy technologies applications in nZEB.

Energy use	Technology	Renewables Type
Space and water heating	District heating	PV, Solar thermal, Residual heat, Waste, Biomass, Geothermal
	Decentralized boilers	
	Solar thermal heating	
	Heat pump	
Cooling	District cooling	Solar, Geothermal
	Solar cooling	
Electricity production	Rooftop solar PV	Solar
	Building integrated PV	
Cooking	Improved cook stoves	Solid biomass, ethanol, biogas

70% of site energy consumption. The most critical aspect in control strategies for building energy control and optimization is modeling. Building energy modeling and control combines a number of fields of study which considers concepts and studies of electrical and electronics engineering, mechanical engineering, civil engineering, and architecture. Building energy system modeling could be divided into three categories, as follows [16]:

- Detailed physical (white-box, physics-based) models, which might not be applicable with regard to on-line building operation, in spite of the fact that they can give a reasonable estimation precision, as they need hundreds of parameters and many iterations. Yet, these models are effective and useful for operation strategy assessment and building operation data acquisition;
- Statistics (black-box, data-driven) models, that could be sufficiently fast for on-line building operation; yet they require high quantities of training data, and the training data must consider the projecting building operation range, confined within the building operation strategies. Luckily, such operation data could be obtained by detailed physical models;
- Simplified physical (gray-box, combination of physics based and data-driven) models are more applicable for a practical building model predicated upon employment application. They possess lower parameters to be calculated and require less computation time, which is an advantage in building energy model for energy and cost saving.

Advancements were possible by employing the building energy model for building control and implementation to save energy or cost. Yet, lots of work must be invested to implement these methods and ensure the necessary performance, from a practical point of view. Reference [17] reviewed all the significant developed and adopted modeling methodologies used to model the energy systems of buildings, with the following gaps and needs:

- A systematically developed simplified building model with temperature, heat, and relative humidity is absent.
- Methods utilized in modeling do not possess the computational efficiency and most of the proposed models are projected models.

- Although models have been established for occupancy pattern behavior, thermal behavior of building components and of room zone, as well as for HVAC and lighting systems. Yet, an integrated modeling method to study the influence of the net energy consumption aspects has not been conveyed.
- Influence of building thermal mass and many other procedures (decreasing lighting levels, raising thermostat set points, regulating supply air temperature, changing chilled water temperature) on net energy consumption pattern of buildings requires extensive research.
- Improving the dynamic response and decreasing the computational price and memory demand for building energy modeling, while ensuring the accuracy level, are among the critical problems for on-line practical employment.

Advanced control approaches deliver a more convenient way of decreasing the energy requirements of buildings and preserving indoor environmental quality complying with the global standard of sustainability, which was certified as reliable for several applications, for example, HVAC control and thermal comfort control, etc. There are mainly four categories of control strategies in buildings, which are classified according to [18]:

- Controlled elements and advanced strategies in buildings entail HVAC control, smart shading control, artificial lighting control, heat recovery smart systems, cooling smart systems, and brightness control. The HVAC system is among the most essential controlled component in buildings.
- Controlled parameters, parameters in buildings include indoor temperature, relative humidity, CO_2%, occupancy count, air velocity, heating demand, and mass air flow rate. The indoor temperature control was examined by numerous literature in the previous decade considering control strategies in buildings.
- Control modes, there are: simple control, e.g. manual control and on/off control and mixed-mode control.
- Control algorithm, there are: bang–bang (on/off) control, iterative control, proportional–integral–derivative (PID), Model Predictive Control (MPC), nonlinear control, pole-placement control, optimal control, fuzzy logic control, artificial neural network (ANN), adaptive control, and also mixed control.

Furthermore, developed control strategies implemented with HVAC technologies for improved multi-purpose applications have recently been the new shift for building energy conservation and indoor environment quality research.

To obtain an energy efficient building design, optimization algorithms are crucial. The optimization approaches can be implemented to several building design difficulties for example, massing, orientation, façade design, thermal comfort, daylighting, life cycle analysis, structural design analysis, energy and of course cost.

Optimization is the method for determining the minimum or maximum value of a function by selecting a pre-defined quantity of variables associated with some constraints. The optimization function is known to be the cost or any other function and is normally determined by simulation tools. When one or more objective functions are used for optimization, the so-called multi-objective optimization issue takes place. This is the case with many building design issues and these particular functions usually entail anomalies.

Generally, two well-known methods for multi-objective optimization problems are utilized. The first method implements a weighted sum function where each objective is normalized and summed up with its corresponding weight factors to obtain one cost function. The second method is the Pareto Optimization, which includes the existence of conflicting (possibly infinite) number of Pareto optimal solutions.

The resources utilized for the optimization of building design can be divided into three categories:

- Custom programed algorithms, which require enhanced programing skills and flexibility.
- General optimization packages, such as: mode FRONTIER, Matlab and GenOp.
- Special optimization resources for building design.

Examination of the cost function by a whole building simulation program needs sufficient processing time to obtain the figures for the following iterations to be successful. Hence, the crucial features of an optimization tool are:

- Good performance.
- Delivery of the graphical user interface.
- Approximate multiple solutions with comparable performance.
- Parallel processing.

A building design optimization analysis consists of the following steps:

- Recognizing the design variables and their corresponding restraints.
- Choosing a building performance simulation tool and constructing the building model. An assessment among 20 widely used building simulation tools: EnergyPlus, TRNSYS, DOE-2, ESP-r, EQUEST, ECOTECT, DeST, Energy-10, IDE-ICE, Bsim, IES-VE, PowerDomus, HEED, Ener-Win, SUNREL and Energy Express is presented in [19] and their utilization share is presented in Figure 7.7.

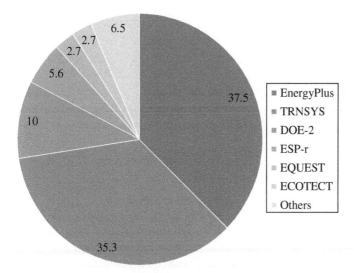

Figure 7.7 Utilization share of major simulation programs in building optimization research. Ref Num [19] Reproduced with permission from ELSEVIER.

- Choosing a suitable objective function.
- Choosing a suitable optimization method, a representation of the widely utilized optimization algorithms in building research is offered in Table 7.5 [19]. Furthermore, a set of performance indices, namely, stability, robustness, validity, speed, coverage, and locality

Table 7.5 Classification of algorithms for building performance optimization. Ref Num [19]. Reproduced with permission from ELSEVIER.

Family	Strength and weakness	Typical algorithms
Direct search family (including generalized pattern search [GPS] methods)	• Derivative-free methods • May be utilized even if the cost function consists of minor discontinuities • A number of algorithms may not deliver the precise minimum point • Could determine a local minimum point. • Managing search procedures usually come with issues regarding non-smooth functions	Mesh adaptive search, Generating set search, Exhaustive search, Hooke–Jeeves, Coordinate search.
Integer programming family	Solving problems with integers or mixed-integer variables	Branch and Bound methods, Simulated annealing, Tabu search, CONLIN method, Hill climbing method.
Gradient-based family	• Rapid convergence; a stationary point can be provided • Sensitive to discontinuities in the cost function • Sensitive to multi-modal function	Bounded BFGS, Levenberg–Marquardt, Discrete Armijo Gradient, CONLIN method.
Meta-heuristic method Stochastic population-based family	• Does not lean toward a local optimum • Numerous cost function evaluations • Global minimum might not be provided	• Evolutionary optimization family: Genetic Algorithms, Genetic programming, Evolutionary programming, Differential evolution, Cultural algorithm • Swarm intelligence: Particle swarm optimization (PSO), Ant colony algorithm, Bee colony algorithm, Intelligent water drop.
Trajectory search family		Simulated annealing, Tabu search, Hill climbing.
Others		Harmony search method, Invasive weed optimization.
Hybrid family	An amalgam of the strengths and less weaknesses of the aforementioned methods.	PSO-HJ, GA-GPS, CMA-ES/HDE, HS-BFGS algorithm

had been used to evaluate the most commonly used optimization algorithms, namely Hooke-Jeeves algorithm, Multi-Objective Genetic Algorithm II (MOGA-II), and Multi-Objective Particle Swarm Optimization (MOPSO) as illustrated in Table 7.6 [20]. The MOGA-II and the MOPSO algorithm is more superior than the Hooke-Jeeves algorithm except for the locality. Consequently, in the event of encountering an energy efficient design optimization problem, the algorithm must be prudently chosen depending on the problem and the performance indices.

Optimization techniques should be used in the early and late building design stages of advanced building design process, as illustrated in Table 7.7. As an example, Figure 7.8 illustrates the use of multi-objective MOGA-II optimization algorithm to obtain the

Table 7.6 Summary of the three optimization algorithms' performance under six indices. Ref Num [20]. Reproduced with permission from ELSEVIER.

	Stability	robustness	validity	speed	coverage	locality
Hooke-Jeeves	G	P	P	-	P	G
MOGA-II	G	G	A	G	A	A
MOPSO	G	A	G	A	G	P

G, Good; A, Average; P, Poor.

Table 7.7 A building design framework using the optimization methods as a decision aid. Adapted from Ref Num [21].

Design step	Description
1	Preliminary design using simple building models and simple calculation method; Expected simulation runtime: a few minutes; Results: several near optimal solutions.
2	Selected solutions from the previous step will be further analyzed with detailed building performance programs; Expected simulation runtime: a few hours; Results: extraction of the basic architectural form.
3	Using information from step 2, a more detailed model is designed; Expected simulation runtime: a few hours; Results: a more detailed building geometry model.
4	Optimization of the detailed model with building performance programs. Expected simulation runtime: a few hours; Results: performance improvement of the given form.
5	Building design process continues as usual; Initiating construction detailing planning.
6	Building model calibration with the help of optimization methods; Expected simulation runtime: a few hours; Results: a calibrated building model to be used for building diagnostics and control.

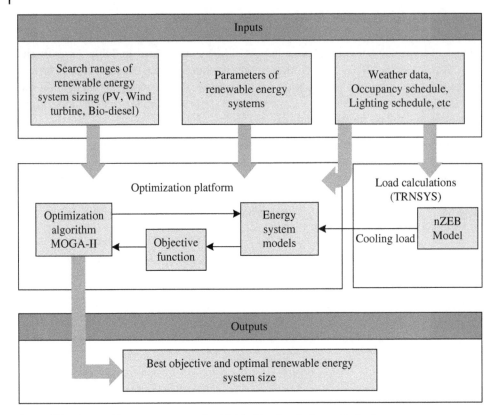

Figure 7.8 Renewable energy system size Multi-objective design optimization.

optimal size of the renewable energy system by optimizing a three objective cost function: full cost, CO_2 emissions and grid interaction index. Several trial values are researched and implemented to determine the ideal results [22, 23].

7.5 Net Zero Energy Community

Considering the advancement of nZEBs and the complications of community-level energy problems, the formation of Net-Zero Energy Communities (nZECs) is a developing goal for local energy efficiency and sustainability. By the difference between the nZE's effectiveness in building-scale and community-scale, including the vast number of renewable energy sources for instance, solar-thermal electricity generation, solar-assisted heating/cooling systems, and wind energy, it shows that community-based net-zero energy can achieve the same overall energy-performance objective with economic advantages. Commonly, a Zero Energy Community (ZNEC) decrease in energy use within certain boundaries by encouraging energy efficiency and balancing the energy used (heat sharing, load diversity and mismatch among the on-site renewable production and the nZEB demand) and the facilitation of nearby renewable energy sources,. [24, 25].

By adapting the definition of the nZEB, the nZEC "substantially decreases energy demand by efficiency rises in a manner that the balance of energy for vehicles, thermal, and electrical energy among the community is met by renewable energy sources. Like nZEB, the energy performance of an nZEC could be compensated by many methods entailing net-zero site energy, net-zero source energy, net-zero energy costs, and net-zero energy emission. Table 7.8 demonstrates a hierarchy of choices to transfer to zero-energy communities and this classification has been made to motivate community managers and developers to use all available cost-effective energy efficiency plans and then assess the choices to transfer to net zero [26].

Figure 7.9 presents a general overview of nZEC with N buildings (N≥2) connected through the local community electrical grid. Every building produces its renewable energy

Table 7.8 Community efficiency and renewable supply hierarchy.

Option Number	Option Name
0	Energy efficiency and energy demand decrease.
1	Employ renewable energy sources in the built environment & on unusable brownfield (established land not currently in use) sites
2a	Employ renewable energy sources on community greenfield sites (a greenfield site is a site that has not been developed or built on before, which may provide open space, habitat or agriculture).
2b	Employ renewable energy produced off-site and on-site.
3	Buy new off-site renewable energy certifications

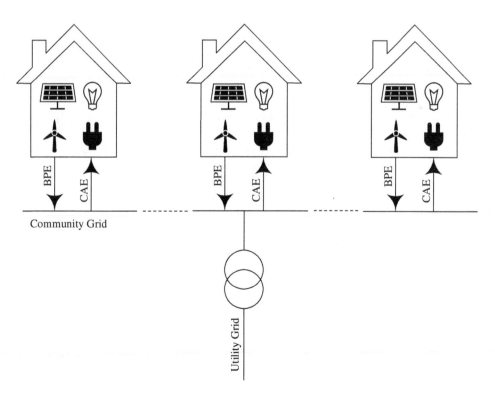

Figure 7.9 Net Zero Energy Community: general overview.

on-site, known as Building Produced Energy (BPE), and is in charge of compensating the energy consumption of the corresponding devices. The BPE of a particular nZEB is consequently offered to the whole community, contributing to the Community Available Energy (CAE). The CAE is then spread by the community power grid with a building producing or gaining energy from the CAE to account for the cumulative demand of all buildings with the goal of enhancing the full load matching the nZEC. Moreover, in the event that the CAE is not adequate enough to meet the community energy need, the outstanding required energy is delivered from the power grid. Hence, the surplus energy of CAE is transferred to the power grid. The nZEC accomplished the best load matching results with a yearly load and supply cover factors of 46 and 39%, respectively [27].

7.6 Net Zero Energy Building: Trends, Benefits, Barriers and Efficiency Investments

Based on the World Green Building Council, around 500 net zero energy commercial buildings and 2000 net zero energy housing units are present around the globe, which is much less than 1% of the total global building stock. Considering geography, the European Union (EU) has the most nZEB, because of government-sponsored retrofit programs, progressive policies, and market interest. North America is known to possess the largest concentration of net zero buildings. Its share in 2018 is 79.1% of the global net zero energy buildings market. Regarding building type, residential projects are considered to be the highest quantity and type of net zero projects constructed until now. A lot of these residential projects are single-family homes constructed as demonstration projects by renewable energy financing programs and, in some instances, with the aid of publicly sponsored home energy retrofit programs. It was found that commercial office projects are considered to be the second highest number of net zero projects globally. Furthermore, the previous few years have observed an increase of nZEB in developed sectors, such as the education sector (primary schools and university campus) [28, 29].

According to the Energy Performance of Building Directive (EPBD) at the EU, all new buildings occupied and owned by public authorities are nZEBs after 31 December 2018 and each new building is an nZEBs by 31 December 2020 [30].

The nZEB provide numerous benefits, including, [31, 32]:

- A reduced total net monthly cost of living for occupying building space, energy savings range from 57 to 74% annually compared to standard building. The total savings indicated at year 30 for a single-family home range from $31 000 for the single-family home to $16 000 for the quadplex [33].
- Insulation from future energy price rises.
- Improved comfort as a result of more uniform interior temperatures and enhanced air quality.
- By using passive resources for more efficient buildings, for example, natural daylight and ventilation, the building intrinsically transforms to a healthier and more comfortable place to live, work, and play.
- Reduced requirement for energy austerity.

- Greater resale value as potential owners require more ZEBs than available supply.
- Reduced total cost of ownership due to improved energy efficiency.
- Many tax breaks and incentives exist for nZEB owners.

The nZEB Key barriers entail:

- Initial cost, the most efficient technologies usually result in high initial cost to customers. The on-site renewable energy production required to decrease total consumption, reaching extremely low levels currently requires more money relative to the power bought from utility grids. The additional cost could be funded with longer term mortgages or mortgages that offer a special rate for more efficient houses.
- Longevity of the building stock, present energy-inefficient building stocks will probably continue to offer shelter for decades to come. Modifying the present building stock includes substantial cost and many problems, and hence NZE efforts must be to invest in new buildings. Older buildings will probably be withdrawn gradually or the cost of modifications will be implemented at the right time.
- Reliable information, recently, customers have been given incorrect information or are misinformed with regard to the energy-related implications of their purchase and operational decisions.

The global building efficiency revenue has maintained a steady rise with an average annual growth rate of 11%. Building efficiency revenue nearly doubled from 2011 to 2018 and the building efficiency observed a rise in each product category, as illustrated in Figure 7.10. In the global market, Lighting remained the largest subsegment in revenue in 2018 representing 46.7%, followed by Heating, Ventilation, and Air Conditioning (HVAC)

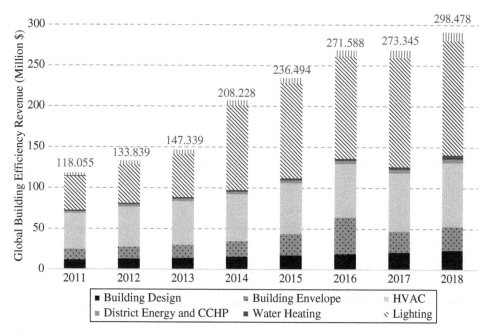

Figure 7.10 Global building efficiency revenue, 2011–2018.

representing 26.4%. Furthermore, the two fastest growing subsegments globally relative to last year were Appliances and Electronic Equipment and Demand Response & Enabling IT with growth rate 60 and 16%, respectively. The Building Envelope experienced double-digit growth of 13%, and Combined Cooling, Heating, and Power (CCHP) also grow with 13%. As forecasted by Navigant Research, this market will grow at a compound annual growth rate (CAGR) of 52%, to a total of $1.6 billion in 2025 [34].

7.7 Conclusion

The nZEB has attracted attention in academic and professional fields as an effective solution to the increasing energy and environmental problems. Also, it acts as a facilitating technology for speeding up the deployment of the SG concept. The deployment of the nZEB concept and its extension to the community will decrease the energy consumption of the building sector, which represents over 30% of the total final energy consumption, due to its optimized design through passive and active energy saving techniques, the integration of renewable energy resources and the application of smart energy management algorithms. Throughout this chapter, the definition of the NZEB with different boundaries and metrics has been presented. Different design approaches using passive and active energy saving and energy efficient algorithms have been reviewed. Modeling methods and simulation programs have also been presented and integrating optimization in all of the design steps of the nZEB have been discussed. The nZEC has been presented as a solution for the mismatch between on-site renewable energy generation and nZEB demand. The benefits, barriers, current investments and different targets for the nZEB have been discussed.

References

1 World Green Building Council (2016). Towards zero-emission efficient and resilient buildings: Global Status Report 2016. https://www.worldgbc.org/sites/default/files/GABC_Global_Status_Report_V09_november_FINAL.pdf (accessed 15 February 2021).

2 CRAVEzero (2020). D6. 4: Co–Benefits of nZEBs. https://www.cravezero.eu/wp-content/uploads/2020/05/CRAVEzero_D64_CoBenefits.pdf (accessed 15 February 2021).

3 Marszal, A.J., Heiselberg, P., Bourrelle, J.S. et al. (2011). Zero energy building – a review of definitions and calculation methodologies. *Energy and Buildings* 43 (4): 971–979.

4 Sartori, I., Napolitano, A., and Voss, K. (2012). Net zero energy buildings: a consistent definition framework. *Energy and Buildings* 48: 220–232.

5 Bourrelle, J.S., Andresen, I., and Gustavsen, A. (2013). Energy payback: an attributional and environmentally focused approach to energy balance in net zero energy buildings. *Energy and Buildings* 65: 84–92.

6 Kylili, A. and Fokaides, P.A. (2015). European smart cities: the role of zero energy buildings. *Sustainable Cities and Society* 15: 86–95.

7 Lu, Y., Wang, S., and Shan, K. (2015). Design optimization and optimal control of grid-connected and standalone nearly/net zero energy buildings. *Applied Energy* 155: 463–477.

8 Deng, S., Wang, R.Z., and Dai, Y.J. (2014). How to evaluate performance of net zero energy building – a literature research. *Energy* 71: 1–16.

9 Rodriguez-Ubinas, E., Montero, C., Porteros, M. et al. (2014). Passive design strategies and performance of net energy plus houses. *Energy and Buildings* 83: 10–22.

10 Rodriguez-Ubinas, E., Rodriguez, S., Voss, K., and Todorovic, M.S. (2014). Energy efficiency evaluation of zero energy houses. *Energy and Buildings* 83: 23–35.

11 Bejder, A.K., Knudstrup, M.-A., Jensen, R.L., and Katic, I. (2014). *Zero Energy Buildings – Design Principles and Built Examples: For Detached Houses*. Denmark: SBI forlag.

12 Aksamija, A. (2016). Regenerative design and adaptive reuse of existing commercial buildings for net-zero energy use. *Sustainable Cities and Society* 27: 185–195.

13 King, J. and Perry, C. (2017). Smart Buildings: Using Smart Technology to Save Energy in Existing Buildings. https://www.aceee.org/sites/default/files/publications/researchreports/a1701.pdf (accessed 15 February 2021).

14 Zhou, B., Li, W., Ka, W.C. et al. (2016). Smart home energy management systems: concept, configurations, and scheduling strategies. *Renewable and Sustainable Energy Reviews* 61: 30–40.

15 IRENA (2016). *Renewable Energy in Cities*. Abu Dhabi: International Renewable Energy Agency (IRENA) www.irena.org.

16 Li, X. and Wen, J. (2014). Review of building energy modeling for control and operation. *Renewable and Sustainable Energy Reviews* 37: 517–537.

17 Harish, V.S.K.V. and Kumar, A. (2016). A review on modeling and simulation of building energy systems. *Renewable and Sustainable Energy Reviews* 56: 1272–1292.

18 Wang, Y., Kuckelkorn, J., and Liu, Y. (2017). A state of art review on methodologies for control strategies in low energy buildings in the period from 2006 to 2016. *Energy and Buildings* 147: 27–40.

19 Nguyen, A.-T., Reiter, S., and Rigo, P. (2014). A review on simulation-based optimization methods applied to building performance analysis. *Applied Energy* 113: 1043–1058.

20 Si, B., Tian, Z., Jin, X. et al. (2016). Performance indices and evaluation of algorithms in building energy efficient design optimization. *Energy* 114: 100–112.

21 Machairas, V., Tsangrassoulis, A., and Axarli, K. (2014). Algorithms for optimization of building design: a review. *Renewable and Sustainable Energy Reviews* 31: 101–112.

22 Lu, Y., Wang, S., Zhao, Y., and Yan, C. (2015). Renewable energy system optimization of low/zero energy buildings using single-objective and multi-objective optimization methods. *Energy and Buildings* 89: 61–75.

23 Hamdy, M., Nguyen, A.-T., and Hensen, J.L.M. (2016). A performance comparison of multi-objective optimization algorithms for solving nearly-zero-energy-building design problems. *Energy and Buildings* 121: 57–71.

24 Javaid, N., Hussain, S.M., Ullah, I. et al. (2017). Demand side management in nearly zero energy buildings using heuristic optimizations. *Energies* 10 (8): 1131.

25 Zhao, J. and Du, Y. (2019). A study on energy-saving technologies optimization towards nearly zero energy educational buildings in four major climatic regions of China. *Energies* 12 (24): 4734.

26 Carlisle, N., van Geet, O., and Pless, S.D. (2009). *Definition of a "Zero Net Energy" Community*, 1–14. National Renewable Energy Laboratory.

27 Lopes, R.A., Martins, J., Aelenei, D., and Lima, C.P. (2016). A cooperative net zero energy community to improve load matching. *Renewable Energy* 93: 1–13.

28 World Green Building Council (2017). From Thousands to Billions: Coordinated Action towards 100% Net Zero Carbon Buildings by 2050. https://www.worldgbc.org/sites/default/files/From%20Thousands%20To%20Billions%20WorldGBC%20report_FINAL%20issue%2020310517.compressed.pdf (accessed 15 February 2021).

29 New building institute (2016). 2016 list of zero net energy buildings. https://newbuildings.org/wp-content/uploads/2016/10/GTZ_2016_List.pdf (accessed 15 February 2021).

30 D'Agostino, D., Zangheri, P., Cuniberti, B. et al. (2016). Synthesis Report on the National Plans for Nearly Zero Energy Buildings. https://ec.europa.eu/jrc/en/publication/eur-scientific-and-technical-research-reports/synthesis-report-national-plans-nearly-zero-energy-buildings-nzebs-progress-member-states (accessed 15 February 2021).

31 Yadav, M., Jamil, M., and Rizwan, M. (2018). Accomplishing approximately zero energy buildings with battery storage using FLANN optimization. *International Conference on Advances in Computing, Communication Control and Networking*, Greater Noida, India (12–13 October 2018). IEEE.

32 Babu, K.R. and Vyjayanthi, C. (2017). Implementation of net zero energy building (NZEB) prototype with renewable energy integration. *IEEE Region 10 Symposium*, Cochin, India (14–16 July 2017). IEEE.

33 Architects, M. (2015). Net Zero Energy Feasibility Study Full Report. https://www.efficiencyvermont.com/news-blog/whitepapers/net-zero-energy-feasibility-study (accessed 15 February 2021).

34 Navigant Research (2019). Advanced Energy Now 2019 Market Report. https://info.aee.net/hubfs/Market%20Report%202019/AEN%202019%20Market%20Report.pdf (accessed 15 February 2021).

8

Smart Grid Communication Infrastructures

8.1 Introduction

Smart grid (SG) offers multi-way communication among energy generation, transmission, distribution, and usage facilities. The reliable, efficient, and intelligent management of complex power system requires the employment of high-speed, reliable, and protected data information and communication technology into the SG to manage and regulate energy generation and consumption. Advanced metering infrastructure (AMI) with the employment of smart meter (SM) and communication techniques is crucial to manage the SG. In this context, this chapter explains the evolution and benefits of the AMI and the essential functionalities that SMs should deliver. Adding to that, this chapter will shed light on the communication challenges and requirements for SG and describes the most suitable communication architecture and technologies including wired and wireless technologies.

8.2 Advanced Metering Infrastructure

AMI is a fundamental infrastructure for the integration of SGs and is utilized for other SG applications to give the functional advantages within the utility. AMI is a two-way communication infrastructure among the utility, SMs, and customers. AMI gathers and distributes information for improving customer empowerment characteristics for energy consumption and regulation. AMI is the employment of computer software, hardware, data management, monitoring systems, and SMs. Old systems, which used one-way communications to gather meter data were transferred to Automated Meter Reading (AMR) Systems. The AMI is considered to be an advanced form of AMR as it entails many advanced technologies including SMs, home area networks (HANs), neighbored area networks (NANs) or neighbored networks [1, 2]. Figure 8.1 illustrates the improvements made in electrical metering, from electro-mechanical accumulation metering to advanced smart metering.

AMI is used for financial benefits, enhanced services, and opportunities for taking into account environmental concerns. AMI will improve SG operation in terms of time-based pricing; accessibility to consumption information for consumers, and providers; issue and

Smart Grid and Enabling Technologies, First Edition. Shady S. Refaat, Omar Ellabban, Sertac Bayhan, Haitham Abu-Rub, Frede Blaabjerg, and Miroslav M. Begovic.
© 2021 John Wiley & Sons Ltd. Published 2021 by John Wiley & Sons Ltd.
Companion website: www.wiley.com/go/ellabban/smartgrid

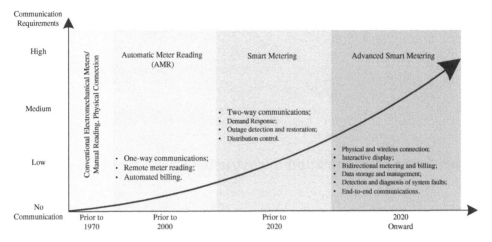

Figure 8.1 Evolution of electricity metering.

outage notifications; remote guidelines; load control for demand response (DR); energy quality checks; energy theft identification; coordination with additional smart devices; and reduced harmful emissions by efficient energy use [3]. The advantages of AMI are outlined in Table 8.1. Short-term advantages, mainly for the energy suppliers and metering operators, could be acquired from AMR. Longer-term advantages originate from other functions of smart metering that result in the implementation of SMs in SGs.

Typical SM consists of information exchange and measurement infrastructures, as illustrated in Figure 8.2. The former allows for bidirectional data flow giving data on consumers and providers for each end of the system, while the providers possess an AMR framework, which is a pricing control mechanism, and an infrastructure for handling data. Furthermore, the communication infrastructure consists of two components: a network connection and control infrastructure. They enable meters to accept commands and notify control centers. Moreover, SMs consist of a power supply, a control, an indicator, an encoding, and a timing module [3].

In practice, SMs can measure energy consumption in real time, including voltage, frequency, and phase angle. Thus, the information to control centers is securely communicated. By implementing bidirectional communication of data, SMs may gather data of the electricity consumption values at the consumer's vicinity. The gathered information by SMs entail a number of parameters; unique meter identifier, timestamp of the data, and electricity consumption values. Depending on the received data, SMs can monitor and implement control commands for all home devices remotely and locally. In addition, SMs can communicate with other meters in their vicinity by utilizing HAN to gather diagnostic data regarding appliances at the consumer and the distribution power grid. Furthermore, SMs could be programmed in a way that only the power consumed from the grid is billed while the power consumed from the Distributed Generation (DG) sources or storage devices possessed by the consumers is not billed. Therefore, they minimize the electricity consumed and can reconnect the power supply to any consumer remotely. Figure 8.3 illustrates an architectural model of an SM system [5]. SM gateway is the main communication element of SG infrastructure that links a Wide Area Network (WAN) with several devices of one or more SMs. They maintain communication among the prosumer and his consuming and producing devices, while being guarded from physical threats. SMs need a

Table 8.1 Benefits of advanced metering infrastructure. Adapted from Ref [4].

Time Horizon	Energy suppliers and network operator benefits	All benefit	Customer benefits
Short-term	• Decreased metering costs and increased frequent and precise values measured;	• Improved customer service; • Variable pricing plans;	• Energy savings due to improved information;
	• Hindering commercial losses as a result of simpler detection of fraud and theft;	• Enabling the employment of DG and flexible loads;	• Frequent and accurate billing;
Longer-term	• Decreasing peak demand via Demand-Side Integration (DSI) programs and hence, lowering cost of buying wholesale electricity at peak periods;	• Increased reliable energy supply and decreased consumer complaints;	• Easier payments for Distributed Generation (DG) Output;
	• Improved planning of generation, network, and maintenance;	• Utilizing Information and Communication Technologies (ICT) infrastructure to regulate DG, recompense customers and decrease costs for utility;	• More payments for wider system benefits;
	• Enhancing real-time system operation by staying at distribution levels; • Potential to sell more services (e.g. broadband and video communications);	• Enabling the integration of electric vehicles and heat pumps, while lowering peak demand;	• Enabling the implementation of home area automation for a better calm life while reducing energy cost;

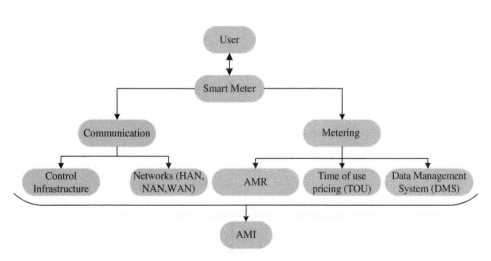

Figure 8.2 Illustration of typical smart meter systems.

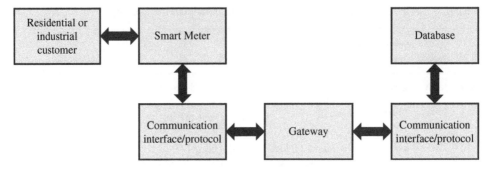

Figure 8.3 Architectural model of smart meter system.

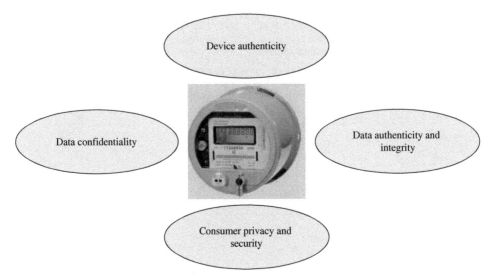

Figure 8.4 Smart meter security objectives. Ref Num [6]. Reproduced with permission from ELSEVIER.

comprehensive life-cycle security from beginning to end and must meet four security conditions as illustrated in Figure 8.4 [6]:

- Device authenticity: SM accessing services (i.e. manufacturing tests, software debugging) should be permissible with an approved procedure. False SM requiring a wireless network should not be granted.
- Data confidentiality: It is associated with generating, transmitting, handling, and storing consumer information, either dynamically generated information, for example, meter readings and power consumption profiles. This information should be hidden from eavesdroppers, and only authorized systems are to be granted access to certain customer information.
- Data authenticity and integrity: Amid AMI applications, making sure that the customer information and transactions are reliable. Moreover, it is crucial to confirm that the identity of both parties is as claimed by them for authenticity purposes.
- Consumer privacy and security: Consumer privacy protection is the main condition of current SG infrastructure. Hence, SM must possess the advanced storage components with high security abilities. It is not to be accessed by unlicensed individuals.

Furthermore, by implementing SMs in power grids, utility companies can monitor and determine electricity theft and unauthorized consumption for enhancing power quality and distribution efficiency. Therefore, SMs have the potential to manage the operation and the energy usage features of the load on the power distribution grid in the days to come.

8.3 Smart Grid Communications

8.3.1 Challenges of SG Communications

The challenges and characteristics of the communication needed for SG applications are summarized is this section [7].

- *Harsh wireless communication conditions:*
  The SG encloses numerous settings inside houses to outdoor areas, and on to power distribution stations. The main challenge is wireless communication in electrical power equipment, which usually consists of coils and power electronics switching devices. The high switching noise from these switching power electronics components can affect wireless communications due to electromagnetic interference. Furthermore, wireless communication from devices in houses to outdoor meters may experience high path losses.
- *Application demands for reliable and low-delay communication:*
  Many SG applications need strict communications quality of services (QoSs). Reliable bidirectional communication between houses and the utility is needed for more efficient operation of the DR control and direct load control, furthermore, managing and regulating the electric grid mainly depends on the low-delay transfer of real-time information.
- *Heterogeneous SG network structure and traffic:*
  The electric grid transfers power to regions with vastly heterogeneous population densities, mainly from sparsely populated rural regions to densely populated urban regions. The SG communication network should also consider these vastly heterogeneous service regions, mainly from sparsely populated and connected networks to densely populated and connected network structures. Simultaneously, the network traffic for SG applications considers a wide range of traffic rates; from tens or hundreds of bytes at periodic (few seconds or minutes) intervals to high bit rates for applications associated with multimedia.

8.3.2 Requirements of SG Communications

SG communications have certain conditions from technical and economic standpoints, such as [8–11]:

- *QoS:* The SG communications infrastructure should deliver a pre-defined level of QoS that meets the required applications, the level of QoS can be quantified using the following parameters:
  - Latency is the End-to-End (E2E) delay of the data. Some mission-critical applications might not accept latency, for instance, substations protection and control applications and distribution automation system. Yet, in other applications, which

include AMI applications for the residential sector or home energy management (HEMs), latency is not crucial.

- Bandwidth is associated with the communication infrastructure that should deliver an aggregated data rate sufficient enough to transfer the traffic related to the required application. Usually, this is predicated upon the number of devices and size of the exchanged packets and traffic pattern. A number of SG applications that transfer video and audio information need high data-rate values to deliver a reliable and accurate information transfer. Yet, the communication data rates for distribution automation and AMI can potentially be small. Hence, the channel bandwidth can become a serious concern to maintain small transmission delays and decrease packet losses on transfer buffers in certain SG applications.
- Reliability is a measure of the level of reliability a communication system may implement data transfers depending on certain conditions. The more critical the application is, the greater the reliability level will be. Moreover, data could possess different levels of criticality. Consequently, the communication network must run applications with the capability to choose between different priority levels for data transfers.

Table 8.2 summarizes typical bandwidths and delays needed by many representative traffic types in SG communications network [12].

- **Scalability:** The communication infrastructure should maintain scalability from both technical and economic standpoints. An SG can include millions of customers and many devices. Therefore, scalability among the most crucial conditions for the SG communication system is vital. The SG communications architecture should have the ability to implement advanced devices to employ additional services. Yet, the SG scalability is worsened because most of the power grid devices will be restricted in terms of storage, computing, and communication capabilities.
- **Flexibility:** Flexibility involves the support provided in implementing heterogeneous SG services that possess a number of reliability and timeliness conditions, and also delivers different communication models. Therefore, the SG communication infrastructure should meet the conditions of different applications with a high degree of flexibility and self-adaptability.
- **Interoperability:** Interoperability is when two or more networks, systems, devices, applications, or components externally exchange and readily utilize data securely and effectively. It is essential that the diverse devices, communication technologies, networking protocols, and applications in an SG achieve interoperation ability to ensure harmonization between them. The developed SG technologies used in the entire grid demonstrate substantial challenges to accomplishing interoperability.
- **Security:** An SG is a critical infrastructure that requires protection in the event of failures and threats. Therefore, security (both physical and cyber-security) and privacy possess the main features for the SG communication infrastructure: resilience to cyber-attacks and protects the consumer's information. Yet, as these two features normally increase with cost, a trade-off is needed to secure achievable solutions.

Table 8.2 Usual QoS conditions of few representative kinds of traffic in SG communications network. Ref Num [12]. Reproduced with permission from ELSEVIER.

Traffic Types	Descriptions	Bandwidth	Latency
Meter reads	Meters provides a statement about the energy used (for example the 15-min interval reads are typically transmitted every 4 h).	Up to 10 kbps	2–10 s
Demand response (DR)	Utilities to coordinate with consumer devices to permit consumers to decrease their power consumption at peak periods.	Low	500 ms to min
Connects and disconnects	To connect/disconnect customers to/from the grid	Low	A few 100 ms to a few minutes
Synchro-phasor	The main primary measurement technologies implemented for wide-area situational awareness (WASA)	A few 100 kbps	20–200 ms
Substation supervisory control and data acquisition (SCADA)	Periodical polling by the master to IEDs in substations	10–30 kbps	2–4 s
Inter-substation communications	Evolving applications like Distributed energy resources (DER) may require GOOSE communications outside the substation	–	12–20 ms
Substation surveillance	Video site surveillance	A few Mbps	A few seconds
Fault Location, Isolation, and Restoration for distribution grids	To regulate protection/restoration circuits	10–30 kbps	A few 100 ms
Optimization for distribution grids	Volt/VAR optimization and power quality optimization on distribution networks	2–5 Mbps	25–100 ms
Workforce access for distribution grids	To deliver video and voice access to field workers	250 kbps	150 ms
Protection for microgrids	To address faults, isolate them, and make sure that loads are not disturbed	–	100 ms–10 s
Optimization for microgrids	To direct and control the functions of the entire microgrid to enhance the energy exchanged between the microgrid and the primary electric grid.	–	100 ms to min

8.3.3 Architecture of SG Communication

The communication infrastructure in SG should support its expected functionalities and meet its performance conditions. As the communication infrastructure of the SG links a vast number of devices and regulates complex device communications, it should have a

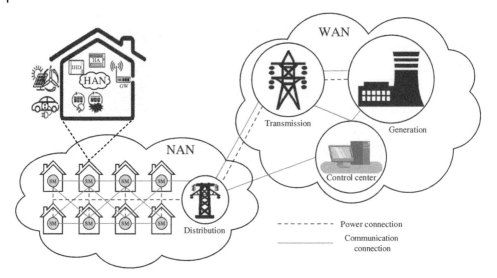

Figure 8.5 SG communication: a hierarchical structure with three major network types: HAN, NAN, and WAN. Adapted from Ref [13].

hierarchical architecture with interconnected individual subnetworks and each should be accountable for different areas. Usually, the communication networks may be divided into three classes: HAN, NAN or Field Area Network (FAN), and WAN, as illustrated in Figure 8.5, which are described briefly as [13]:

- *HANs* are private networks located in the customer premises or domain. The HAN comprises of the SM interconnecting a number of home appliances, sensors, In-Home Display (IHD), gas meter, water meter, renewable energy resources, energy storage devices, Plug-in Hybrid Electric Vehicle (PHEV) and HEM system and this will allow consumers to be conscious of electricity usage prices and control their consumption accordingly and manage smart appliances to enable DR application. Furthermore, HAN is an essential part of the Automatic Metering Infrastructures (AMI). HAN supports low-bandwidth (between 10 and 100 kilobits per second [kbps]) communication between home electrical appliances and SMs and covers a distance 1–10 m. In addition, there is no need for low latency. HANs are comparable to other private networks, for example, Industrial Area Networks (IANs) and Building/Business Area Network (BANs).
- *NANs/FAN,* NAN/FAN is one important component of communication network infrastructure, it imitates a bridge among consumer premises and substations with collectors, access points and data concentrators to transmit consumption measurements from SMs and to enable diagnostic notifications, firmware improvements and real-time or near real-time notifications for the power system support. This network can be formed according to wired and wireless communication technologies. It covers long distances from 10 m up to 100 and has a data rate around 10–1000 Kbps [14–16].
- WAN offers communication between electric utility and substations. It gathers information from many NANs and transfers it to the control center. It connects the highly distributed smaller area networks that serve the power systems at different locations. WAN must span all over the substations, distributed power generation, and storage facilities,

and distribution to be sufficiently operational and scalable. It entails two types of networks: backhaul and core network. In addition, it is a high bandwidth backbone communication network that regulates data transmissions for long distances with advanced managing and detecting applications. WAN offers a two-way communication network for communication, automation, and management reasons for SG applications. The WAN encompasses a distance ranging from 10 to 100 km, hence, data transmission rates could range from 10 to 100 Mbps.

These three segments are interconnected by gateways: an SM between HAN and NAN and a data aggregation points data aggregation points (DAP) between NAN and WAN. An SM gathers power usage information of a house or building by communicating with the home network gateway or functioning as the gateway itself. The data concentrator accumulates data from a collection of SMs and passes it on to the grid operator's control centers. Requirements for improving the electric grid and customer energy use may be transmitted from control centers to intelligent electronic devices (IEDs) and customer devices through WAN, NAN, and HAN in the opposite direction. These segments could implement a number of communications technologies and protocols to cover the conditions associated with data rates, communications latencies, deployment/maintenance costs.

8.3.4 SG Communication Technologies

Wired and wireless communication technologies are usually implemented in the SG. Choosing a communication technology is predicated upon many aspects, for instance, users or devices, network architecture, data type, environmental conditions, and price.

Wired communication technologies are usually associated with physical cabling in a network. The three types that may take one of the following shapes. Fiber optic cable, Digital Subscriber Line (DSL), and Power Line Communication (PLC) which is a highly recognized technology used in an SG scheme. Table 8.3 delivers a summary comparison of different wired communication technologies [17].

In wireless communication technologies, the data could be spread across the air without the need of a certain medium. These technologies are useful in an SG scheme. The main four wireless communication technologies are: IEEE 802.15.4 standard which is known as Wireless Personal Area Network (WPAN), Wi-Fi, Global System for Mobile Communications (GSM), and satellite. Table 8.4 compares the core types, benefits and drawbacks of wireless communication technologies [17].

8.4 Conclusion

This chapter focused on smart metering and SG communication techniques taking into account the associated technologies, applications, and challenges. This chapter showed a survey of smart metering for SG infrastructure, beginning from the potential advantages and past development steps to providing pathways of future SMs infrastructure. Furthermore, this chapter gives an insight into the SG communications network by illustrating its layered architecture, typical types of traffic that it could transmit and the corresponding QoS conditions, and candidate wireless communications technologies that could be integrated.

Table 8.3 Comparison between wired communication technologies. Ref [17]. Reproduced with permission from ELSEVIER.

Technology	PLC	Fiber optic	DSL
Speed	1~5 Mbps	0.1~1 Gbps	8 Mbps
Coverage	150 km	20 km	5 km
Architecture	• Large scale: NAN, FAN, WAN. • Small scale: HAN, BAN, IAN	WAN	AMI, NAN, FAN
Advantage	• Utilized the present infrastructure for communication reasons; • Long-distance communications; • Low operation & maintenance cost;	• Relatively high bandwidth; • Withstand and overcome radio interferences and electro-magnetic;	• Widely distributed broadband; • Utilized the existing infrastructure for communication purposes;
Disadvantage	• More signal losses and interferences; • Hard to transfer information with increased bit rates.	• Increased employment costs; • Not suitable for metering applications.	• Not suitable for long distances; • The utility pays additional price to communication operators.

Table 8.4 Comparison between wireless communication technologies. Ref [17]. Reproduced with permission from ELSEVIER.

	WPAN	Wi-Fi	Wi-Fi	Satellite
Speed	4~256 Mbps	50~500 Mbps	• 2G: 14.4 kbps • 3G: 14.4 Mbps • 4G: 326 Mbps • 5G:10 Gbps	0.1~1 Mbps
Coverage	0.1~1 km	0.3~1 km	1~5 km	100~6000 km
Architecture	HAN, BAN IAN, NAN, FAN, AMI	HAN, BAN, IAN, NAN, FAN, AMI	HAN, BAN, IAN, NAN, FAN, AMI	WAN, AMI
Advantage	• Low operation and maintenance price; • Little consumed power.	• Low operation and maintenance price; • High resiliency.	• Low energy consumed; • Great resiliency; • Aids numerous devices.	• Incredibly long distance; • Extremely reliable
Disadvantage	• Big network construction deals with numerous restraints; • Small bandwidth.	• Horrible interferences; • High consumed power	• Costly due to the utilization of service provider networks.	• Operation and maintenance are costly; • Large response time.

References

1 Shaukat, N., Ali, S.M., Mehmood, C.A. et al. (2018). A survey on consumer's empowerment, communication technologies, and renewable generation penetration within smart grid. *Renewable and Sustainable Energy Reviews* 81: 1453–1475.

2 Rashed Mohassel, R., Fung, A., Mohammadi, F., and Raahemifar, K. (2014). A survey on advanced metering infrastructure. *International Journal of Electrical Power & Energy Systems* 63: 473–484.

3 Avancini, D.B., Rodrigues, J.J.P.C., Martins, S.G.B. et al. (2019). Energy meters' evolution in smart grids: a review. *Journal of Cleaner Production* 217: 702–715.

4 Sun, Q., Li, H., Ma, Z. et al. (2016). A comprehensive review of smart energy meters in intelligent energy networks. *IEEE Internet of Things Journal* 3 (4): 464–479.

5 Le, T.N., Chin, W.-L., Truong, D.K. et al. (2016). Advanced metering infrastructure based on smart meters in smart grid. https://www.intechopen.com/books/smart-metering-technology-and-services-inspirations-for-energy-utilities/advanced-metering-infrastructure-based-on-smart-meters-in-smart-grid (accessed 15 February 2021).

6 Sharma, K. and Mohan Saini, L. (2015). Performance analysis of smart metering for smart grid: an overview. *Renewable and Sustainable Energy Reviews* 49: 720–735.

7 Khan, A., Rehmani, M.H., and Reisslein, M. (2016). Cognitive radio for smart grids: survey of architectures, spectrum sensing mechanisms, and networking protocols. *IEEE Communications Surveys & Tutorials* 18 (1): 860–898, First quarter.

8 López, G., Moreno, J.I., Amarís, H., and Salazar, F. (2015). Paving the road toward smart grids through large-scale advanced metering infrastructures. *Electric Power Systems Research* 120: 194–205.

9 Cagri Gungor, V., Sahin, D., Kocak, T. et al. (2013). A survey on smart grid potential applications and communication requirements. *IEEE Transactions on Industrial Informatics* 9 (1): 28–42.

10 Yan, Y., Qian, Y., Sharif, H., and Tipper, D. (2013). A survey on smart grid communication infrastructures: motivations, requirements and challenges. *IEEE Communications Surveys & Tutorials* 15 (1): 5–20, First Quarter.

11 Ancillotti, E., Bruno, R., and Conti, M. (2013). The role of communication systems in smart grids: architectures, technical solutions and research challenges. *Computer Communications* 36 (17): 1665–1697.

12 Ho, Q-.D. and Le-Ngoc, T. (2013). Smart Grid Communications Networks: Wireless Technologies, Protocols, Issues, and Standards. In: *Handbook of Green Information and Communication Systems* (ed. M.S. Obaidat, A. Anpalagan and I. Woungang), 115–146. Elsevier.

13 Bian, D., Kuzlu, M., Pipattanasomporn, M. et al. (2019). Performance evaluation of communication technologies and network structure for smart grid applications. *IET Communications* 13 (8): 1025–1033.

14 Yoldaş, Y., Önen, A., Muyeen, S.M. et al. (2017). Enhancing smart grid with microgrids: challenges and opportunities. *Renewable and Sustainable, Energy Reviews*.

15 Bajer, M. (2019). Building Advanced Metering Infrastructure using Elasticsearch database and IEC 62056-21 protocol. *7th International Conference on Future Internet of Things and Cloud*, Istanbul, Turkey (26–28 August 2019). IEEE.

16 Le, T.N., Chin, W-.L., Truong, D.K. et al. (2016). Advanced metering infrastructure based on smart meters in smart grid. https://www.intechopen.com/books/smart-metering-technology-and-services-inspirations-for-energy-utilities/advanced-metering-infrastructure-based-on-smart-meters-in-smart-grid (accessed 15 February 2021).

17 AbouBakr, M., Atallah, A.M., Abdel-Sattar, A., and El-Dessouki, A. (2019). Various Communication Technologies Used in Smart Grid. *21st International Middle East Power Systems Conference*, Cairo, Egypt (17–19 December 2019). MEPCON.

9

Smart Grid Information Security

The structure of the SG can be divided into two main parts, SG infrastructure and SG applications. SG infrastructure entails three main layers the power system layer, the information layer and the communication layer. The power system layer is in charge of the development of the energy generation, transmission, and consumption. The information layer is responsible for the development of information metering, management, and controlling of SGs. The communication layer is in charge of delivering the means to exchange data electronically and can be viewed as electronic languages. Although the cyber system has made the grid more energy efficient, it has introduced threats of cyber- attack such as operational failures, loss of synchronization, damage power components, and loss of system stability. For that, information security is an essential element for the information and communication infrastructure in SG to improve grid efficiency and reliability taking privacy into consideration.

9.1   Introduction

To achieve a high level of management and control of the SG, a communication network is needed for the power grid support [1]. However, due to the strong dependency on communication networks in SG, these networks induce new threats on the cyber security system. The attack on SG infrastructure does not impact customers alone, rather, the utility provider's business as well as the economic growth of the nation. Understanding the SG infrastructure and the different attack scenarios and studying their effect on SG networks will provide a platform to design adequate protection for the grid. This chapter therefore offers novel insights and comprehensive analysis on the cyber security of the SG to ensure that the SG is highly secure against the cyber-attack threat for both communication and physical networks, without impacting customer confidence in the utility provider, and without substantially inconveniencing customers. One of the important future directions would be deploying strong security control in SG work in both directions, communication network security (securing the data transferred) and physical security (securing the electric grid against physical attacks). This chapter presents the challenges of building resilience control in the power system and highlights the governmental efforts needed to secure the SG.

Smart Grid and Enabling Technologies, First Edition. Shady S. Refaat, Omar Ellabban, Sertac Bayhan, Haitham Abu-Rub, Frede Blaabjerg, and Miroslav M. Begovic.
© 2021 John Wiley & Sons Ltd. Published 2021 by John Wiley & Sons Ltd.
Companion website: www.wiley.com/go/ellabban/smartgrid

9.2 Smart Grid Layers

The structure of the SG from the viewpoint of security can be divided into three layers; the power system layer, the information layer, and the communication layer [2]. A general description of the SG structure is presented in this section to understand each layer and provide the security system needed for each layer.

9.2.1 The Power System Layer

The power system layer is in charge of advanced electricity generation, delivery, and consumption. The power system layer in the SG describes a two-way flow of the interconnected electric network transmit electricity and digital information from power generation systems to customers to efficiently and reliably manage various appliances at the consumers side [3]. It includes generating stations, renewable energy sources, generators, diesel engines, gas turbines, power transformers, substations, motors, compressors, pumps, switchgear, transmission lines, distribution lines, circuit breakers, etc.

9.2.2 The Information Layer

The information layer is in charge of advanced information metering, monitoring, and management in the context of the SG. The main components of electric utility information system are shown in Figure 9.1. They include; the Supervisory Control and Data Acquisition (SCADA) system that obtains data from the utility field, then uses it to manage the electrical grid infrastructure, Customer Information System (CIS), Geographic Information System (GIS), Advanced Metering Infrastructure (AMI) and Meter Data Management System (MDMS) handles the data transmitted from energy consumers and exchanges data between each other. Demand Response Management System (DRMS) and Outage Management System (OMS) are the main systems in the grid as they coordinate with every other system and together maintain the reliability of the grid and customer satisfaction [2, 4].

9.2.3 The Communication Layer

The communication layer is responsible for communication connection and information transmission among SG systems, sub-systems, individual devices, and various applications. The communication layer should realize the support of the quality of data service, and should be highly reliable, pervasively available and should have a high coverage with guaranteed security and privacy. The communication layer is built in a centralized and/or distributed data topology. The data produced in smart meters arrives at the data collector by other network nodes. The data collector transfers this data by the access point of the communication network, after that, the data is transferred to the centralized or decentralized control centers or to the service provider as shown in Figure 9.2 [5].

Such a sophisticated infrastructure has to be supported by system analysis, communication architectures and automation, control concepts, evaluation and testing in order to achieve balancing to overall grid operations. These communication and control architectures in SG are considered as an attack spot to the grid from any hacker who may cause catastrophic threats to the entire grid.

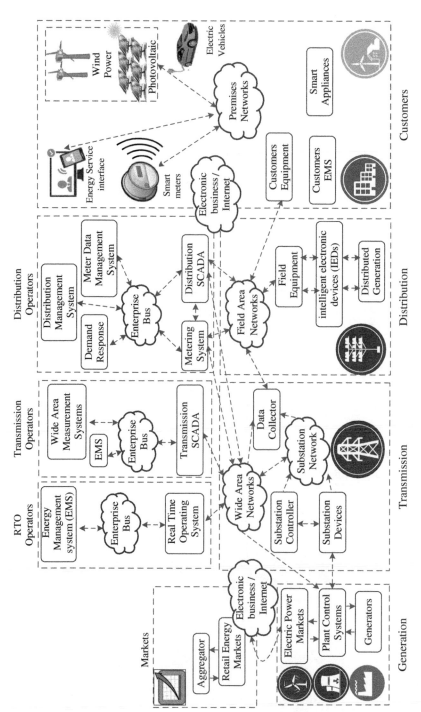

Figure 9.1 Basic information layer of SG.

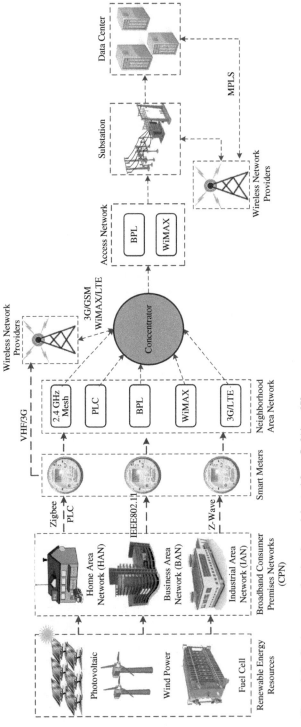

Figure 9.2 Basic communication layer. Adapted from Ref Num [5].

9.3 Attacking Smart Grid Network Communication

Smart grids are electric grids with digital information/communication capabilities. Attacks on an SG could be destructive to the power grid. If the attack and defense strategies are considered, the attack could be even more smart and complicated. Attacks on the SG network can be divided into three categories as follows:

1) Attacks on the physical layer
2) Data injection and data integrity attacks.
3) Network-based attacks.

9.3.1 Physical Layer Attacks

Physical layer attacks have different types such as attacking the sensitive information transmitted from a smart appliance. Eavesdroppers are readily available for such attacks which is known as Eavesdropping [6]. One way to protect the data against this form of attack is to use data encryption but some hackers may use the pattern to decode the transferred messages. Another type of attack works on disturbing the wireless medium by jamming it with noise signals to impede the smart meters communication with the utility; such an attack is called Jamming. When the jammer emits noise signals continuously, the wireless channels become completely blocked which give an incorrect result on the smart meter function. Injection requests is another type of SG attack which works on disturbing the regular operations of SG devices. The last type is the injection attack, which inserts false determined messages into the wireless network. To avoid this attack, suitable security techniques are used to obtain unauthorized access to a wireless network.

9.3.2 Data Injection and Replay Attacks

The targeted data in any malicious attack is the SG infrastructure such as measurements or monitoring data. In case of data injection attacks, falsified data is injected into the surrounding area observed by the network operator. Message replay attacks happen when the attacker injects fabricated control signals into the system after gaining elevated privilege to smart meters. Message replay attacks are mainly used by the attacker to manage the energy by guiding power to the wrong location and causing system damage [7].

9.3.3 Network-Based Attacks

A network-based attack is a case of topology attack on a SG. In this attack, the hacker captures the network data and measured data from terminal units, then changes part of the it into a different format and forwards the new version of data to the control center. Some techniques are used to identify this attack on SG such as a fusion-based defense technique [8], which is predicated upon the feedback received from individual nodes in the network. A game theoretic analysis is considered as a defense strategy to maintain local observation by individual critical nodes. It deals with issues where multiple players with differing objectives compete with each other, and it can give us an insight about the mathematical framework for analysis and for modeling network security issues.

9.4 Design of Cyber Secure and Resilient Industrial Control Systems

9.4.1 Resilient Industrial Control Systems

To monitor and control any industrial process, a control system must be provided. Resilient control is a new topic of control technologies that is based on achieving the balance between an acceptable level of operation and undesirable incidents. This description can be inferred as a definition of a resilient control system. A resilient control system is one that ensures state awareness and a reasonable level of operational normalcy in response to disturbances, entailing threats of a sudden and malicious nature [9].

Figure 9.3 illustrates the resilient control system framework considering the flow of a digital control system. The first stage contains the operational and controls considering measures of performance. The second stage combines both human and automation control responses in order to understand performance. Finally, the last stage works on informing the consumer about the required action for a desirable response. These technologies are applied to offer a protection for the infrastructure such as for water treatment systems, the oil and gas industry, and the three stages of power grid (generation, transmission, and distribution). Safe and secure operation of an industrial facility depends on the design of the cyber-physical control system taking into consideration reliable computing. To understand the design of a resilient control system, all areas of resilience must be discussed.

9.4.2 Areas of Resilience

9.4.2.1 Human Systems

It is essential to achieve the ability of humans to understand the problems on the system, to find solutions, then be able to adapt the solution to unexpected conditions. This ability can provide flexibility of system control. However human behavior can't be predicted and this is considered a challenge [10]. Bayesian Networks are robust tools that use algorithms for inference and associated learning. Considering the historic human preferences in the Bayesian inference to tailor the amount of automation necessary [11]. This allows optimal resilience for this mixed initiative response. The drawback of the Bayesian approach is its computational load, which makes optimal Bayesian strategies more applicable as guidelines to implement and gauge the effectiveness of more lightweight heuristic and suboptimal methods [12].

9.4.2.2 Cyber Security

Cyber security has become among the main concerns with regard to the integration of SGs. Any system must have a security framework to protect data from cyber attacks. Hackers have the ability to undermine the desired control system behavior which affects the control system resilience [13]. Designing a control system without taking cyber security into consideration will represent a weak point in the system for which the hackers look for to provide a pathway to undermine it. Therefore, when considering resilient control system architecture, cyber security must be involved, whether performed separately or within the control system architecture.

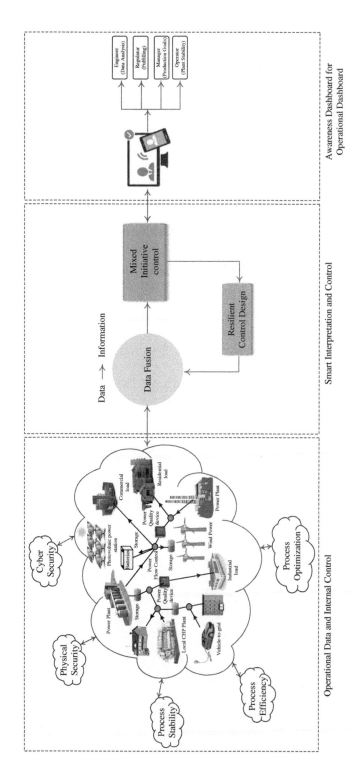

Figure 9.3 Resilient control system framework.

9.4.2.3 Complex Networks and Networked Control Systems

The traditional power grid infrastructure is controlled by a network control system, distributed control system, or data acquisition system. Moving toward SG means a more complicated control system due to the complex interconnected nature of the grid, assorted types of generated power, and storage process. Strengthening the security and resilience of the grid is a mandate in this case [14]. This creates a challenge to the control system used to ensure greater resilience to threats which requires mechanisms to holistically design to coordinate between the human side and cyber security toward achieving a system that is secure, interoperable, and effective. This could be realized through managing the assets with constrained controllers to maintain resilience for rapidly changing conditions.

9.5 Cyber Security Challenges in Smart Grid

Recently, the number of documented cyberattacks in the electricity sector has been increased worldwide. The Oklahoma-based Memorial Institute for the Prevention of Terrorism has presented statistics about the attacks on the three sectors of electrical power grid and has shown that the most common target for cyber-attack is the transmission system as shown in Figure 9.4 [15]. As SG relies more heavily on communication and control, the security of both physical and communication networks is fundamental to the reliable operation of the SG.

The impact of cyber-attack on the grid is dangerous. The United States Department of Energy (DoE) report in 2019 found that the cost of outages and disturbances from all events is at least $150 billion per year [16], which has a huge negative impact on economy. Hacking is a creative activity and hackers are trained to look for the weakest and most vulnerable link that engineers may have omitted in their design. Cyber security is a complex and complicated infrastructure that plays an essential role in delivering power securely and efficiently and in maintaining stable grid performance. Therefore, many technical challenges and problems related to effective and secure communication and information processing

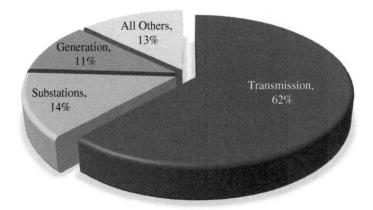

Figure 9.4 Percentage of attacks on grid components. Ref Num [15]. Reproduced with permission from ELSEVIER.

should be addressed to realize the vision of a smarter power grid. This section summarizes various factors that make cyber security a complex challenge to achieve a SG [17].

- Modern SCADA/EMS/DMS systems are complex and incorporate an increasingly large number of distributed elements and connectivity information that could be sources of attacks. SCADA systems are mainly vulnerable to attack because of issues in design, human interactions, and configuration. Furthermore, attacks may propagate from one system to another due to the high connectivity [18].
- The distributed nature of SCADA/EMS/DMS systems result in complex network architectures that are required to provide detailed enough data through different metering devices. Such metering devices have significant implications on the privacy of the customers [19].
- SCADA/EMS/DMS systems and networks employ variety of cyber security equipment and software.
- Efficient and secure communication protocols must be provided for power systems where the communication system is very complicated.
- The number of devices that need to be managed will increase which is associated with more maintenance, management, and monitoring for any cyber-intrusions which, in turn, need more time and processing capacity that may not be available [20].
- Secure Routing and Aggregation Protocols is needed. More specifically, a modern power grid depends on a complex communication system using sensors to control the power transfer through the grid. Compromising the operation of sensors, communication, and control for generators, substations, or power lines is a complex process that may lead to attacking, jamming, or sending improper commands that can lead to system blackout and damage system components. Choosing a defensive technology to achieve secure communication system such as firewalls, passwords, encryption, physical barriers, and authentication mechanisms is essential but not a big challenge for the attacker because breaches may still occur.

The emerging challenge is to find the most predictive methods that can distinguish the problems, decrease the impact of these problems on the system level, and manage cyber security operations. Some concerns have to be recognized to tackle these challenges [21].

1) The goal is protecting the whole grid in an economical way and doesn't need to protect every asset or all the data.
2) Balancing between privacy and protection from obtrusion must be calculated.
3) Develop a set of baseline requirements for security in SGs.

The challenge is great and requires the adoption of a special architecture for the security methodology in SG. Examples of the publications that deal with these issues are [22].

9.6 Adopting an Smart Grid Security Architecture Methodology

Malware and viruses have been identified as ways that hackers use to gain access to power grid [19]. To maintain secure and reliable operations, it is crucial to comprehend the objectives and conditions that must be provided in the security system in the context of energy delivery and management.

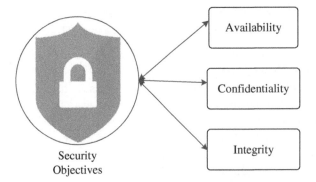

Figure 9.5 Main security objectives in SG. Adapted from Ref [24].

9.6.1 SG Security Objectives

The National Institute of Standards and Technology (NIST) has determined a smart comprehensive guideline for SG cyber security [23]. As shown in Figure 9.5, the three core principles of information security in SG security system are as follows [24]:

- **Availability**: means a guarantee of reliable access to the information by authorized persons only in suitable time; in other words, "timely and reliable access to the use of information" [25]. The time latency depends on the application of SG as shown in Table 9.1.
- **Integrity**: means commitment and trustworthiness from the utility side and authorized access to save the information and guarding against the improper and destructive information which would induce an incorrect decision.
- **Confidentiality:** a set of authorized restricted information to protect personal privacy and proprietary information where the collected data can reveal information about consumer daily activities.

9.6.2 Cyber Security Requirements

Security requirements for the SG entails cyber and physical security. Cyber security means thorough security problems and conditions associated with SG information and network

Table 9.1 Time latency for SG applications [26]. Reproduced with permission from Taylor & Francis Group.

Access time requirements	Availability of data for certain applications
≤4 ms	Protective relaying
Sub seconds	Transmission wide-area monitoring
Seconds	Substations and feeder SCADA system data
Minutes	Equipment monitoring and short-term market pricing information
Hours	Meter reading and longer-term market pricing information
Days/Weeks/Months	Power Quality information

systems while physical security means conditions related to physical equipment and environmental protection. These requirements are mentioned in this section in detail [27].

9.6.2.1 Attack Detection and Resilience Operations
The SG has a feature of open communication network which requires consistent profiling, testing, and comparison to monitor networks and identify any abnormal incidents which is called attack detection. Providing resilience operation in the security system is essential for sustaining grid availability.

9.6.2.2 Identification, and Access Control
The SG network infrastructure depends on the incorporation between different electronic devices which requires the system to identify the devices and to have strict access control to prevent unauthorized users from accessing network information and controlling critical infrastructures.

9.6.2.3 Secure and Efficient Communication Protocols
Transferring information contains two objectives, time-criticality and security in the SG which require the system design to have the ability to balance between communication efficiency and information security. Cyber security requirements for the SG can be summarized in Table 9.2 [22].

Taking these requirements, with the previous objectives, into consideration when designing a cyber-security system for communication networks and operation policies for SG will give the SG comprehensive security capabilities to fulfill the goal of a SG security system. Numerous efforts have been exerted in the power community to come up with protected architectures for SGs [28]. In the next section, the secure protocols and security architectures for the SG will be presented. Generally, two proposed secure network architectures for the SG have been used; Trust computing-based architecture, and Role-based network architecture [29]. Trust computing-based architecture was proposed for SCADA network in power communication and information systems to validate input data, information, determine security risks, and detect the improper data and respond with the attack situation through response actions. Trust computing architecture is used as a security infrastructure for smart networks. Role-based network architecture is responsible on network

Table 9.2 Main cyber security requirements for the SG. Adapted from Ref Num [22].

Security Function	Communication network in SG
Access Control	Strictly enforced for all communication flows throughout the system
Authentication	
Countermeasures	Essential and widely deployed in the network
Attack Detection	
Security for network protocols	From Application Layer to Media Access control layer security
Node	Basic cryptographic functions

structure based on dividing the power communication network into multiple control centers for several microgrids. Each control center for microgrid will follow a number of rules which can be executed by the authorized users only to enhanced security.

From the previous discussion, many security methods can be applicable to the SG and can achieve the permissible level of security in the Markets/Customer/Service provider domains, but still more demanding requirements are needed for security design in the generation/transmission/distribution domains. The practical solution for network security systems not only protects information exchange, but also meets the requirements for data communication and processing.

9.7 Validating Your Smart Grid

Before using smart devices and algorithms in SG, it is of paramount importance to test and validate their capabilities, functionalities, and accuracy. To validate the SG, security objectives and requirements must be the main points to be assessed. Testing and validation of SG must be based on studying how cyber-attacks in SG can affect the cyber security requirements which, in turn, decrease the grid protection against cyber-attack which can be done through modeling, simulation and testing [30], as described in Figure 9.6.

Smart grid should ideally have some functionalities [31]:

- Self-healing
- Motivating and including the consumer
- Resisting attacks
- Improving power quality
- Accommodating all generation and storage options
- Enabling electrical markets
- Optimizing assets and operates efficiently

These functionalities can be affected by some threats from cyber-attack. Validating SG means examining architectures and protocols to keep its functions which will help to protect the power grid and ensure national security. One of the methods used in validating specific architectures and protocols in SG is applying them on the so-called "smart grid test-bed". The following requirements must be accomplished when implementing a testbed [32].

1 *Modeling of smart appliances*
All smart appliances must have the ability to test their functionalities through modeling them to be a functional requirement for both operational and technical reasons. The test bed should implement appliance emulation for operational reasons to guarantee that smart appliances can respond to demand signals and, for technical reasons, to ensure security assessment purposes. That can be done through modeling the smart appliances to fulfill SG objectives and characteristics.

2 *Graphical User Interface*
To ensure an effective interaction with the test bed in the design, development, and fulfillment of the test, a graphical user interface must be used to enable participation and facilitate the implementation of the test bed.

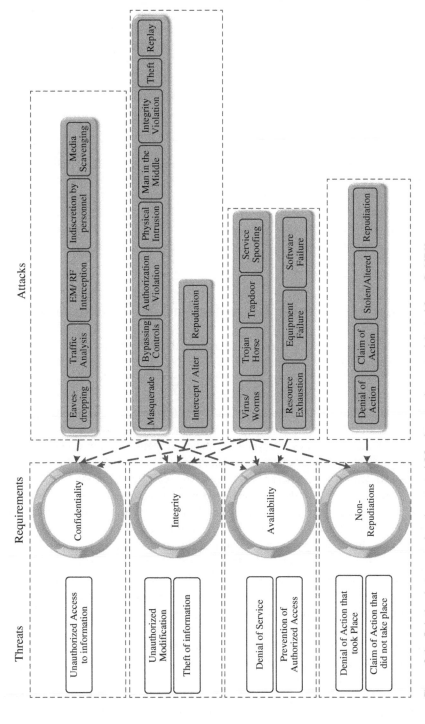

Figure 9.6 Relation between cyber threats and SG cyber requirements. Adapted from Ref [30].

3 *Hardware Integration*

Testing needs to enable the actual hardware to integrate or at least deliver an interface to integrate with the actual hardware. A realistic implementation of the test bed will allow for the evaluation and testing of real-time characteristics. In addition, the possibility of evaluation and testing of real-time characteristics is doable with test beds. Also, this requirement will also allow for the testing of potential hardware, i.e. the test bed could be beneficial for testing new smart appliances or conventional hardware in the vicinity of a controlled environment impacting external users.

4 *IP-based communication*

IP-based communication is an inveterate element in the test bed. IP-based communication between all major nodes is used, where most network communication paths in SG are based on IP networks, enabling distributed use and remote access which allows distributed emulation, i.e. implemented and shared elements of the test bed from geographically diverse networks for improved utilization of resources. In addition, SG networks will gain greater flexibility. The demand for IP-based communication is essential from a technical employment standpoint and to employ the functional and system management requirements [33].

9.8 Threats and Impacts: Consumers and Utility Companies

Cyber security is an essential element of the information infrastructure toward obtaining improved security, efficiency, and reliable operation of the SG. Different types of cyber threats including the electricity industry and its transmission and distribution systems are growing and affect both consumers and utility companies. Cyber threats are mainly divided into two types, cyber-attacks via communication networks and network attacks. For communication network attacks, the Information Systems Audit and Control Association (ISACA) identifies the type of malicious communication cyber-attacks and categorizes them into targeted and untargeted attacks. This includes organized criminal groups, hackers, information warfare and phishers as summarized in Figure 9.7 [34]. The most common source of cyber threats is **Hackers** who break into the network for different reasons such as monetary gain and political activism. Hackers can download attack protocols from the internet and use them to gain unauthorized access to networks. Hackers represent a high threat to any network because of the serious damage that they can cause. **Criminals groups** is the second source that is based on organized criminal groups who use spyware/malware or spam and online fraud. This type of hacking is more dangerous and creates a huge threat because of its ability to conduct industrial espionage. **Bot-network operators** use network or a bot-network to coordinate the attack. **Insiders** are another organization responsible of computer crime which represents a big threat for company utilities. **Spammers and phishers** represent a small group or individuals who work on stealing information or attacking organizations. **Nations** use cyber tools to gather information in any field of communication and economy which threatens the daily lives and interests of citizens across the country. Some nations work to develop information warfare programs for political concerns. **Spyware or malware authors** are individuals or organizations attacking users by producing spyware and malware which can

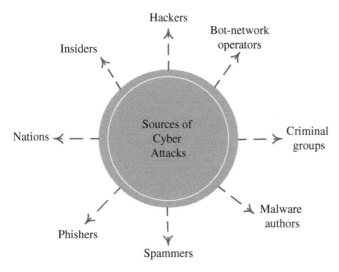

Figure 9.7 Common sources of cyber-attacks. Adapted from Ref [34].

destroy computer files and hard drives. These sources of cyber threats utilize different methods that could impact computers, software, networks, organization's operations, industries, or the internet itself.

Network attacks or physical attacks in power systems can be defined as power injection attacks that alter generation and loads in the network. The main risks from such physical attacks are the power line towers and transformers. In addition, the malicious codes have the ability to directly control electricity substation switches and circuit breakers and has the capability to turn off power distribution or to harm the equipment in the power distribution grid [35]. From the previous discussion, it has been observed that different threats can affect the SG from the domain that interacts with consumers (i.e. the Markets/Customer/Service Provider domains) and from the domain that interacts with utility companies. Cyber-attacks target consumers through the following:

- **Financial loss:** any attack aimed at obtaining financial information of customers may threaten their banking accounts through stealing from their online banking accounts.
- **Identity theft:** full information such as full name, addresses, telephone numbers and other information can be used by attackers for malicious purposes.
- **Blackmail:** Hackers can use the leaked information to extort money from customers.
- **Interrupted service:** Hacking on the power grid may cause interrupted power or cut off electricity which, in turn, affects consumer comfort level.

Attacks could impact not only the customers, but also targeted organizations. The most common effects of targeted attacks on companies are following:

- **Business Disruption:** is the most dangerous threat that can affect any organization and can render it incapable of completing its daily tasks and operations as the attack causes system downtime.

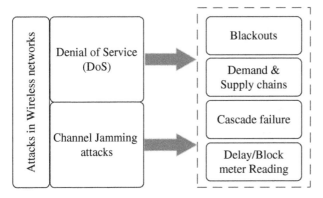

Figure 9.8 Impact of attacks on the power grid.

- **Customer Information Loss:** means a breakthrough to the company's information database which contains the personal identification information of customers which puts their customers at risk such as identity theft, blackmail, extortion.
- **Financial Losses:** either due to spending millions of dollars in fixing the damage and to prevent future attacks or as customers' loss due to the effects caused by the attack which may also cause reputation loss.
- **Intellectual Property Loss:** due to hacking the company's own intellectual property.

Figure 9.8 describes the impact of various attacks in the communication network on the SG and the damages that may affect the power grid.

9.9 Governmental Effort to Secure Smart Grids

Smart grid has all features that qualify it to be a prime target for cyber-attack. In addition, the development of SG has made more access points for penetrating into the grid computer systems which may cause destructive cyber-attacks affecting all areas of life. Figure 9.9 shows the cyber incidents in different sectors of life. It's clear that a higher percentage of cyber incidents take place in the energy sector. Cyber-attack on the power grid sector is significant and may lead to loss of grid synchronization, failure in critical power system devices, and interruption in power supply to critical infrastructure. In addition, there is an increase in the numbers of cyber-attack events every year. For example, the worldwide Marsh & McLennan (MMC) markets present a ranking of the industries, and it has been noted that the number of cyber-attack events worldwide on the energy sector increased in 2020 with 30% [28]. The global cyber security market is currently worth $173B in 2020 [36]. Over recent years, sophisticated security events targeted the power grid such as the one in California at Metcalf Substation which led to widespread outage and took many weeks to repair and to get the station operating again. On December 23, 2015, Ukraine experienced power outage in substations from cyber security problem [37], that caused loss of synchronization, equipment corrupted, and power disconnected for one to six hours affecting 225 000 customers. Indeed, as electric utilities and governments become increasingly

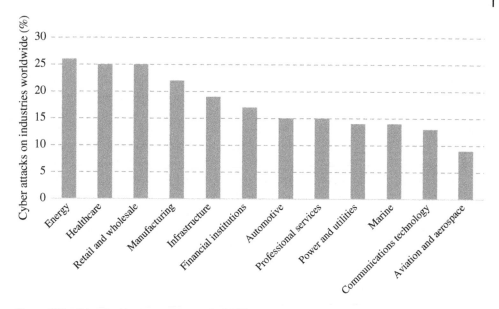

Figure 9.9 Cyber incidents percentages in 2017.

reliant on the SG which, in turn, increases the level of risk, security and protection must be of highest priority. As a result, protecting the electric grid is the responsibility of national security, and therefore, governments should play a vital role in directing the security efforts to SG sector. Governments should take serious steps to help secure the electricity grid and perform several activities that are intended to secure the energy sector. The National Academy of Sciences report outlined the rigidity of the power system and its inability to resist or quickly recover from attacks on multiple elements, to secure the grid from cyber-attack. Using collaboration between academic research and industry, under governmental supervision, will help promote the security and resilience of the grid. In August 2019, GAO (*U·S Government Accountability Office*), reported that in spite of cyber security occurrences, reports have mentioned that no outages took place domestically but cyber-attacks on industrial control systems have interfered with grid functions. The cyber-attacks can lead to widespread power outages, the scale of power outages that could be the outcome of a cyber-attack is ambiguous because of limitations in those assessments. Eight cyber security standards, along with additional documents, were relatively comparable to NIST (National Institute of Standards and Technology guidance) to guarantee the effective and efficient decrease of risks to the reliability and security of the power system. The advanced eight crucial infrastructure standards for protecting the power utility-critical and cyber-critical assets are as follows [38]:

1) Cyber security-related controls;
2) Critical cyber asset identification;
3) Security management controls;
4) Personnel and training;
5) Electronic "security perimeters;"

6) Physical security of critical cyber assets;
7) Systems security management;
8) Incident response plan for critical cyber assets.

There are a number of ways for industry and governments to continue to share information about cyber security events and identify the magnitude of the outages, the level of sophistication, and the kind of hacking to prevent such event. In addition, including the management level is necessary to prevent the spread of these threats across the entire network. A key role is that electric grid infrastructure must be developed to overcome cyber security threats. Governments offer incentives for developing cyber insurance. Presently, insurance plays a vital but limited role in the electricity sector for transmission and distribution. The government plays an essential role in maintaining the viability of private insurance by making suitable legislative and regulatory frameworks focus on violations leading to actual harm, by the loss of load or by some other ways, that force utilities to enhance security levels. Governmental roles to secure SG could be summarized as follows:

- Coordination with industry to unite efforts toward attacks that threatens grid security.
- Leveraging research to develop efficient infrastructure and protection tools to decrease the chance of attack incidents.
- Coordination between the energy sector and other infrastructure sectors such as transportation, water and communication to share the threat information as a way to help in planning the security system.
- Developing an evaluation strategy to judge the existing national security system and the security emergency exercises.
- Developing a security system for hazard preparedness with efforts from all executive departments and agencies.

9.10 Conclusion

Smart Grid technologies deliver numerous benefits to utility companies and consumers in the electricity field. However, these technologies are based on highly connected networks which have a great impact on system performance but it can be considered as a weak point in SG. This system allows hackers and attackers to determine and exploit vulnerabilities of the power grid which could result in widespread power outages, damages to property, repair costs, decreased system efficiency, and loss of life. For that, cyber security is extremely essential for the reliable and secured operation of a critical SG infrastructure. This chapter presents a brief description to the SG construction from the viewpoint of security and the areas that represent hacking spots, including power and communication/networking architecture. The cyber threat to the grid is real and of great importance, therefore, a number of observations for cyber threats on the grid and how they affect the whole areas of life are discussed. In addition, most challenges that face developing cyber-secure and resilient systems are discussed in this chapter. More should be performed to address challenges facing the industry in improving grid security. This chapter highlights the being made by governments and utility companies to employ new and effective security policies and methods for consumer authentication and data encryption to enhance security across the three main sectors of electric SG; generation, transmission, and distribution.

References

1 Bayram, I.S. and Papapanagiotou, I. (2014). A survey on communication technologies and requirements for internet of electric vehicles. *EURASIP Journal on Wireless Communications and Networking* 2014 (1): 223.

2 Refaat, S.S., Mohamed, A. and Abu-Rub, H. (2017). Big data impact on stability and reliability improvement of smart grid. *IEEE International Conference on Big Data*, Boston, USA (11–14 December 2017). IEEE.

3 Refaat, S.S. and Abu-Rub, H. (2016). Smart grid condition assessment: concepts, benefits, and developments. *Power Electronics and Drives* 1 (2): 147–163.

4 Daki, H., El Hannani, A., Aqqal, A. et al. (2017). Big data management in smart grid: concepts, requirements and implementation. *Journal of Big Data* 4 (1): 1–19.

5 Dragicevic, T., Dan, W., Shafiee, Q., and Meng, L. (2017). Distributed and decentralized control architectures for converter-interfaced microgrids. *Chinese Journal of Electrical Engineering* 3 (2): 41–52.

6 Xu, Q., Ren, P., Song, H., and Du, Q. (2016). Security enhancement for IoT communications exposed to eavesdroppers with uncertain locations. *IEEE Access* 4: 2840–2853.

7 Ghosh, S. and Sampalli, S. (2019). A survey of security in SCADA networks: current issues and future challenges. *IEEE Access* 7: 135812–135831.

8 Kordy, B., Piètre-Cambacédès, L., and Schweitzer, P. (2014). DAG-based attack and defense modeling: don't miss the forest for the attack trees. *Computer Science Review* 13: 1–38.

9 Rieger, C. and Manic, M. (2018). On Critical Infrastructures, Their Security and Resilience-Trends and Vision. https://arxiv.org/abs/1812.02710 (accessed 15 February 2021).

10 Lahcen, M., Ait, R., Caulkins, B. et al. (2020). Review and insight on the behavioral aspects of cybersecurity. *Cybersecurity* 3: 1–18.

11 Ghose, A. (2019). Grounding semantic cognition using computational modelling and network analysis. PhD dissertation. University of London.

12 Gallego, V., Naveiro, R., Redondo, A. et al. (2020). Protecting Classifiers From Attacks. A Bayesian Approach. https://arxiv.org/abs/2004.08705 (accessed 15 February 2021).

13 Le, D.-N., Kumar, R., Mishra, B.K. et al. (eds.) (2019). *Cyber Security in Parallel and Distributed Computing: Concepts, Techniques, Applications and Case Studies*. Wiley.

14 Abedi, A., Gaudard, L., and Romerio, F. (2019). Review of major approaches to analyze vulnerability in power system. *Reliability Engineering & System Safety* 183: 153–172.

15 Massoud Amin, S. and Giacomoni, A.M. (2012). Smart grid – safe, secure, self-healing. *IEEE Power and Energy Magazine* 10 (1): 33.

16 Casey Allen Shull, "Algorithm to Develop a model Providing Security and Sustainability for the us Infrastructure by Providing Incremental Electrical Restoration after Blackout." PhD diss., Purdue University Graduate School, 2019.

17 El Mrabet, Z., Kaabouch, N., El Ghazi, H., and El Ghazi, H. (2018). Cyber-security in smart grid: survey and challenges. *Computers and Electrical Engineering* 67: 469–482.

18 Jang-Jaccard, J. and Nepal, S. (2014). A survey of emerging threats in cybersecurity. *Journal of Computer and System Sciences* 80 (5): 973–993.

19 Kshetri, N. and Voas, J. (2017). Hacking power grids: a current problem. *Computer* 50 (12): 91–95.

20 Khurana, H., Hadley, M., Ning, L., and Frincke, D.A. (2010). Smart-grid security issues. *IEEE Security & Privacy* 2010: 81–85.

21 Line, M.B., Tøndel, I.A., and Jaatun, M.G. (2011). Cyber security challenges in Smart Grids. *2nd IEEE PES international conference and exhibition on innovative smart grid technologies*, Manchester, UK (5–7 December 2011). IEEE.

22 Ye, Y., Qian, Y., Sharif, H., and Tipper, D. (2012). A survey on cyber security for smart grid communications. *IEEE Communications Surveys & Tutorials* 14 (4): 998–1010.

23 Pillitteri, V.Y. and Brewer, T.L. (2014). Guidelines for smart grid cybersecurity. No. NIST Interagency/Internal Report (NISTIR)-7628 Rev 1.

24 Aloul, F., Al-Ali, A.R., Al-Dalky, R. et al. (2012). Smart grid security: threats, vulnerabilities and solutions. *International Journal of Smart Grid and Clean Energy* 1 (1): 1–6.

25 Pathan, A.-S.K. (ed.) (2014). *The State of the Art in Intrusion Prevention and Detection*. CRC Press.

26 Anwar, A. and Mahmood, A.N. (2014). Cyber security of smart grid infrastructure. https://arxiv.org/abs/1401.3936 (accessed 15 February 2021).

27 Pallotti, E. and Mangiatordi, F. (2011). Smart grid cyber security requirements. *10th International conference on environment and electrical engineering*, Rome, Italy (8–11 May 2011). IEEE.

28 Simonovich, L. (2020). Global Predictions for Energy Cyber Resilience in 2020. https://www.darkreading.com/attacks-breaches/global-predictions-for-energy-cyber-resilience-in-2020/a/d-id/1336746 (accessed 15 February 2021).

29 Shapsough, S., Qatan, F., Aburukba, R. et al. (2015). Smart grid cyber security: Challenges and solutions. *International conference on smart grid and clean energy technologies*, Offenburg, Germany (20–23 October 2015). IEEE.

30 Yahya, F., Walters, R.J., and Wills, G.B. (2015). Modelling threats with security requirements in cloud storage. *International Journal for Information Security Research (IJISR)* 5 (2): 551–558.

31 Camachi, B.E., Ichim, L., and Popescu, D. (2018). Cyber Security of Smart Grid Infrastructure. *IEEE 12th International Symposium on Applied Computational Intelligence and Informatics*, Timisoara, Romania (17–19 May 2018). IEEE.

32 Biswas, S.S., Shariatzadeh, F., Beckstrom, R. and Srivastava, A.K. (2013). Real time testing and validation of Smart Grid devices and algorithms. *Power and Energy Society General Meeting*, Vancouver, USA (21–25 July 2013). IEEE.

33 Kush, N., Clark, A.J., and Foo, E. (2010). Smart grid test bed design and implementation.

34 Doynikova, E., Novikova, E., and Kotenko, I. (2020). Attacker behaviour forecasting using methods of intelligent data analysis: a comparative review and prospects. *Information* 11 (3): 168.

35 Weise, E. (2017). Malware discovered that could threaten electrical grid. http://www.usatoday.com/story/tech/news/2017/06/12/malware-discovered-could-threaten-electrical-grid/102775998 (accessed 15 February 2021).

36 Australian Cyber Security Growth Network (2020). SCP - Chapter 1 - The Australian cyber security sector today. https://www.austcyber.com/resources/sector-competitiveness-plan/chapter1 (accessed 15 February 2021).

37 Whitehead, D.E., Owens, K., Gammel, D., and Smith, J. (2017). Ukraine cyber-induced power outage: Analysis and practical mitigation strategies. *70th Annual Conference for Protective Relay Engineers*, Texas, USA (3–6 April 2017). IEEE.

38 Cleveland, F., Small, F., and Brunetto, T. (2008). Smart grid: Interoperability and standards an introductory review. https://xanthus-consulting.com/Publications/documents/Smart_Grid_Interoperability_and_Standards_White_Paper.pdf (accessed 15 February 2021).

10

Data Management in Smart Grid

Today's energy utilities are confronted with an array of challenges such as uncertainties in energy demand, capital cost for building new stations, carbon gas emissions and the increase of the customers' expectations. As utilities strive to overcome these challenges and achieve high performance, smart grids (SGs) start to emerge as the way forward. The SG allows integration of renewable energy sources, distributed generation (DG) and storage systems. However, all these assets move the energy grid to be smart which is the main obstacle that makes it difficult to manage because of an unprecedented deluge of data. Unfortunately, utilities do not take advantage of this huge volume of data efficiently. To achieve high performance in SGs, a number of techniques and approaches must be used to manage all the data to drive the generating viable values from this big data which can improve utility's chances of reaping optimal long-term returns from its SG investment.

10.1   Introduction

A SG is unique relative to other grids as it is considered to be an intelligent electricity grid that organizes the power flow starting from generation to distribution to reach consumers through information and communication technologies. Due to the complexity of SG data by their nature: new types of data (electricity, information) are exchanged through the electricity grid. Such data relate to new consumers, for instance, electric vehicles (EVs) and smart households, new players including new renewable energies and new communication equipment such as sensors or smart meters. These massive and various amounts of data will assist utilities control energy consumption and avoid peak loads, understanding customer behavior and to achieve grid stability and better reliability. Big data process is categorized into two parts; data management and analytics. Data management contains Big Data storage, mining, and integration. In this part, the data is prepared for further analysis. Data analytics are responsible for managing and examining the data, making it a practical form for decision-making. The available huge mass of information is crucial to transform SGs into a more efficient, reliable, secure, independent, and supportive system at normal conditions and contingencies. However, this may also lead to problems for utilities

Smart Grid and Enabling Technologies, First Edition. Shady S. Refaat, Omar Ellabban, Sertac Bayhan, Haitham Abu-Rub, Frede Blaabjerg, and Miroslav M. Begovic.
© 2021 John Wiley & Sons Ltd. Published 2021 by John Wiley & Sons Ltd.
Companion website: www.wiley.com/go/ellabban/smartgrid

in managing these data and in providing cutting-edge analytics to convert data gathered into information that can be used in decision making or to set actionable plans. data from the SG is generated at a very high rate and volume and in real time. Extracting information from SG data which is required for specific applications calls for deep insight into its sources. In this chapter, we build on the concepts of data management and analytics in SG to build the foundation needed for data analytics to transform Big Data for high-value action. Significant content of this chapter is based on earlier authors' publication [1].

10.2 Sources of Data in Smart Grid

Smart grids are capable of producing a large amount of data from extensively installed measurement devices and many other sources. Big data sources in SG generally fall into two main categories; electric utility data sources and supplementary data sources as shown in Figure 10.1. Both sources are required for energy management and control decisions in SG DG systems. Electric utility data category includes data utility measurement such as phasor measurement unit (PMU) data, smart meters data, intelligent electronic devices (IEDs) data, asset management data, supervisory control and data acquisition (SCADA) data, digital protective relay (DPR) data, digital fault recorder (DFR) data, sequence of event recorder (SER) data, advanced metering infrastructure (AMI) data, control and maintenance data for equipment, and automated meter reading (AMR) operational data of utility [1]. Those are often in structured data forms [2–3].

The supplementary data source, which is widely used in decision making, includes data obtained through additional sources such as; Geographical Information System (GIS) data,

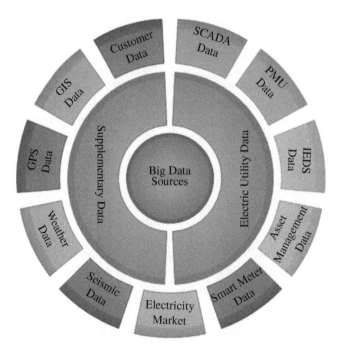

Figure 10.1 Data sources used by SG.

Global Positioning System (GPS) time-reference data, weather and lightning activity data, seismic reflection data, animal migration data, financial market data, social media data, and regulatory reporting data. Data originating from these different sources come in numerous types and formats, entailing rows and columns in usual databases, pictures, textual formats, video, PowerPoint and HTML files, electronic mail, messages, sensor data, web-based transactions, and IT systems logs. These data types are typically categorized into three comprehensive sections: structured, semi-structured, and unstructured data [4]:

- **Structured Data** is capable of being organized in rows and columns of commonly used database tables such as electric utility data, e.g. smart meters data, synchro-phasors data, sensors data, static offline consumer data, etc.
- **Semi-structured Data** refers to a partial structure or a standard format such as text documents, excel files or XML files, e.g. electricity distribution quality of service data, etc.
- **Unstructured Data** is not ordered in a recognized manner such as weather data, facility surveillance systems, distributed data storages system, and even social media and emails. Semi-structured data and unstructured data require preparation before being stored and further analyzed for making decisions on which data is useful. The data could be of varying quality in unstructured or semi-structured data formats. The quality and reliability of the diverse data sources are key factors for decision making in SG operation.

10.3 Big Data Era

Big data associated with SGs, generally refers to any massive volume, extremely fast, diverse, and complex of structured, semi-structured and unstructured information. It has the potential to be mined for information which can be utilized to achieve better decisions [5]. Big Data usually entails data sets with a variety of sizes exceeding the aptitude of frequently utilized tools for mining, capturing, selecting, managing, and analyzing data at an acceptable elapsed time period. It requires a number of methods and technologies with novel forms of integration to output visions originating from data sets. Big Data management for a power distribution system is intended to maintain an acceptable level of data quality and accessibility to make effective management and control decisions toward achieving higher reliability, efficiency, safety, and interoperability in SG operation. The SG is a power network composed of distributed nodes, which function under pervasive control of a smart subsystem, so-called smart microgrids. Microgrids can operate in a grid-connected mode, islanded mode, or both. With the advent of the SG, traditional distribution systems are facing a paradigm shift. The concept of DG is based on power generated from a variety of different scale sources, for instance, internal combustion (IC) engines, gas turbines, micro-turbines, photovoltaic (PV), fuel cells and wind power. The aim is to produce electricity for local distribution or inject electricity of sufficient quality into the main grid. The smart operation of distribution systems helps in improving voltage regulation, efficiency, reliability, availability, power quality, and possible operation under faulted conditions. DG in SG relies on complex communication, control, and management which are dependent on mass data and information between the utilities and customers. This mass of information is crucial to transform the grid into a more efficient, reliable, secure, independent, and supportive system at normal and contingencies operating conditions [6].

Table 10.1 shows the big data sources from these domains and their qualities from the application and analysis requirement viewpoint. There is a substantial rise in the amount of data that is incorporated into the processing and analytical units of current utilities. An example is shown in Figure 10.2, which shows the pattern of big data volume in electric utilities; they should manage a huge number of TB data, which continues to rise over time. Hence, electric utilities should give attention to increasing their ability to handle rising data to their common operations. Also, to devise appropriate plans to make value from the huge data to be managed.

Smart grids give a greater chance of energy delivery and power flows. It permits a two-way flow of electricity and digital information between utilities and customers. The grid is supported by smart meters, sensors, detectors, measurement units, etc. to gather data on a batch or real-time basis. In addition, the SG involves numerous applications which all create sources of Big Data such as; distribution system electricity marketplace and

Table 10.1 Typical big data sources [7]. Reproduced with permission from IEEE.

Data Source	Application	Data Scale	Type	Number of Users	Response Time
Walmart	Retail	PB	Structured	Large	Very fast
Amazon	E-commerce	PB	Semi-structured	Large	Very fast
Google Search	Internet	PB	Semi-structured	Very Large	fast
Facebook	Social network	PB	Structured, Un-structured	Very Large	fast
AT&T	Mobile network	PB	Structured	Very Large	fast
Health Care	Internet of things	TB	Structured, Un-structured	Large	fast
SDSS	Scientific	TB	Un-structured	Large	Slow

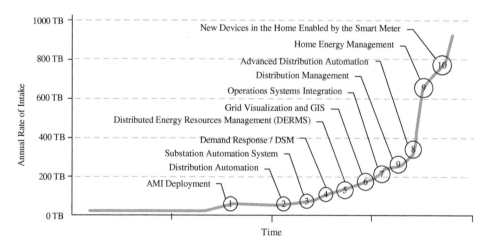

Figure 10.2 Pattern of big data volume in electric utilities. Adapted from Ref [8].

rates, managing power transmission and distribution through grid and substations, smart load switching, managing renewable energy intermittency, demand management of electric vehicles, electricity management in smart buildings, smart meters, effective demand response (DR) program, smart asset management, PMUs for disturbance recording, aggregation of energy consumption, utilities' data centers, loading and weather conditions forecasting, and social media programs to more effectively engage with customers. Such data needs effective management and analytics for better control decisions toward increasing grid reliability, reducing energy demand, improving operational efficiency, and increasing grid capability. Smart grid applications generate huge amounts of data, this Big Data helps to enhance SG applications. The Big Data system will store, process, and mine information in an efficient manner to enhance different SG services. Big Data is also used to reduce electric grid network losses, reduce reliance on the main electric grid, to aid in decreasing peak load, deliver required system care at time of contingencies, increase grid reliability, security, and efficiency through information exchange on a real-time basis.

The challenge, however, is the complexities involved in capturing the necessary data which leads to wrong control decisions [9–11]. Implementing Big Data system will play an important role for reliable energy management of SG that is combined of electrical and communication networks. Data plays a vital role in strategic decision making. With more data, more precise and effective control decisions can be achieved. Under the increase of dynamic and uncertain conditions of the distribution systems in SG, Big Data system is still evolving and not yet established. Most previous studies focus on data mining and big data analytics for SG [12–15]. A comprehensible understanding of the Big Data, its definition, and classification is needed.

Data in SG generally refers to the collection of facts, figures, statistics, or any other related material, which has the ability to serve as the information for basic, detailed analysis of operation, and for better performance of the SG during normal and emergency situations. They can be used to trigger, measure, and record of a wide range of monitoring, control, and management in the grid.

The importance of big data is not to revolve around the amount of data, but how to reduce complexity and online extract and use the useful information from the raw data. Data contains substantial information about grid status; such information would be useful in providing improved energy management and decision control. It enables informed decisions by presenting data in a way that can be interpreted by management tools for improved energy management and decision control for distribution generation systems in SGs.

The main driving benefits of Big Data systems in energy management and decision control for power distribution grids may be described as follows:

- Increasing flexibility for power consumption.
- Increased reliability, efficiency, and safety developments of power distribution networks.
- Big real-time information will allow power outage anticipation.
- Increased implementation of renewable energy sources into the power grid, and reduced reliance on conventional energy sources.
- Improved use of the current and developed distribution system assets to decrease system losses and elevate overall system efficiency.

- Large-scale optimization of the distributed energy system resources to decrease the whole energy cost, and the microgrids' operational costs.
- Maintained secure and reliable power supply for critical loads.
- Development of new network analysis, control and energy management topologies that can be implemented in both existing and future modern power systems.
- The smart operation of distribution system in SG reduces the power utility need for peak demand spinning reserve and for building new infrastructures. It also enables consumers to manage and reduce their energy bills. This, in turn, saves money for both the power utility and for consumers, thus increasing the economic benefits.
- Smart decision making.
- Determining root causes of failures.
- Support energy independence and the adaption to the changing energy scenery.
- Support the creation of private and public-sector partnerships in the energy sector.
- Allow a better exploring vision of the future grid.
- Raise the governments and local electric utilities' awareness, knowledge and ability to develop and establish the planning framework and the working plan for achieving an improved energy efficiency.
- Enable customers to reduce or eliminate their power consumption emissions, through the utilization of emission reduction credits (tariffs, incentives, or statutes that lead to remarkable cost savings).
- Improve grid energy efficiency through use of the proposed energy management strategies, thus benefiting the environment.

All the aforementioned benefits show that a Big Data system is a crucial step to improve SG elements reliability and stability across various sectors, end-users, and energy suppliers. The big data with high-powered analytics will lead to better management and control decision.

10.4 Tools to Manage Big Data

Big data includes a huge and complex information. A recent survey concludes that around 90% of data collected around the world are unstructured data which takes time and effort to make it structured and to figure out the required information to be analyzed and processed [16]. Structuring these data in a limited time is considered a big challenge. Another challenge is how to store these data to be used on time. As the data size can be in peta bytes or more, computationally high storage devices are required. Data is either stored on the servers/cloud or shared nodes or clusters based on the platform selected and the requirements of the application. Secure local area network (LAN) or wireless channels are used to migrate the data from the grid to the mining tools. Cyber risk is a major concern when it comes to data storage or transfer on the cloud. The following section presents a list of the most common big data tools utilized by successful analytics developers to store, manage, and analyze big data.

10.4.1 Apache Hadoop

Apache Hadoop is an open-source software utilized to effectively store and process huge amount of data in a cluster. This technique enables to process data across all nodes. In addition, it intelligently extracts data through analyzation and visualization and through Hadoop

Distributed File System (HDFS) and sorting it to be distributed across many nodes in a cluster which provides high availability, the processing component which is MapReduce programming model, and resource scheduler called Hadoop yet another resource negotiator (YARN) [17].

10.4.2 Not Only SQL (NoSQL)

Traditional Structured Query Language (SQL) is a database that offers an apparatus for storage and retrieval of structured data. However, for large, unstructured data a class of non-relational databases is needed which is called NoSQL. NoSQL databases are increasingly implemented in big data analysis. NoSQL Databases consists of two types [18].

- Column databases: A columnar database is a database management system that stores data in columns rather than rows. Consequently, it is effortless to add columns and these columns can also be added row by row. The databases offer great flexibility, performance, and efficiency. Also, the effectiveness of the column databases could be significantly enhanced by compression, late materialization and batch processing. Examples are BigTable, HBase in Amazon Dynamo, Google Bigtable, Apache HBase.
- Key-value stores: These are distributed data structures that provide key based access to data and are also called Distributed Hash Tables. An example is Apache Cassandra.

NoSQL Databases are very efficient when dealing with massive-scale data even with unstructured and semi-structured type data. However, the only disadvantage is that these do not offer SQL-like querying. To provide similar to querying SQL, many NoSQL databases have evolved with SQL-like interface (Cassandra query language [CQL] of Cassandra, Hive, Pig). There are developments in the form of SQL interfaces that can directly connect to the NoSQL databases (PrestoDB). The SQL-like interfaced NoSQL databases are termed as NewSQL and they possess the inherent capability of organizing large-scaled data and sorting which enables efficient offline analyses (H-Store, Google Spanner). Recently, a substantial rise in the availability of digital data and it has become in stream. Therefore, NoSQL databases have been evolved to cater to the stream-processing solution with fault-tolerant distributed data ingest systems such as Apache Kafka or Flume [19].

10.4.3 Microsoft HDInsight

The Microsoft Azure HDInsight powered by the Apache Hadoop technique can process unstructured or semi-structured data. This system stores the data in series files structured as a folder on Azure blob storage. The main advantage is the possible access of storage using the same technique which provides high availability with low cost. In addition, the option to create clusters to provide data processing and storage over multiple nodes is provided [20].

10.4.4 Hadoop MapReduce

Hadoop MapReduce technique is among the most used method for examining large data sets [21]. MapReduce method is utilized for static applications in SG and for all applications that need real-time response DR, real-time usage and pricing analysis, online grid control and monitoring, etc. Hadoop also has remarkable processing abilities, and possesses frequent updates and amendments within the system.

10.4.5 Cassandra

Cassandra is a tool that provides management of huge amounts of data effectively. The main advantage of this tool is using the database without compromising system performance or any change in the hardware, which offers high availability and scalability. This system has a number of different advantages such as performance, decentralization, professional support, durability, elasticity, scalability, and fault tolerance [22].

10.4.6 Storm

Apache storm is a stream processing technique which works by defining small, discrete operations on very large quantities of data. It composes them into a topology which means at least-once processing guarantees for any information which manages and secures data. This technique is suitable for real-time processing management, data monetization, cyber security analytics and threat detection with less latency.

10.4.7 Hive

Hive is a software used for managing large datasets. This can support SQL, while using HiveSQL (HSQL) to access big data is used for data mining purpose. Hive is not designed for online processing and doesn't offer real-time queries but is designed mainly for analyzing large data sets [23].

10.4.8 Plotly

This software creates great informative graphics by using online tools. In addition, it provides the ability to share the results through transforming them into different needed formats. There is availability to implement this software for data analytics in any company as it doesn't need enough time or competences to fulfill the necessities of big data.

10.4.9 Talend

Talend is an open-source environment used to load, extract, and process big data. It gives the opportunity to access to all components of NoSQL connectivity under an open-source Apache license which allows small businesses to benefit from customer data also helps managers and employees to improve the work plans through accessing the information after processing [24].

10.4.10 Bokeh

This software has great benefits for big data analytics experts as it has the ability to create easy and informative visualizations in an easy and quick way. It can be considered as the most advanced visual data representation tool.

10.4.11 Cloudera

This tool is mainly used for businesses to create data storage which is capable of being viewed by all users who need to use these data anytime for certain goals. This tool was created in 2008 and is, up to this day, the most common provider as it can be a supporter of Apache Hadoop which reduces business risk [25].

10.5 Big Data Integration, Frameworks, and Data Bases

A Big Data system compromises four primary stages after obtaining SG data from diverse sources. Data from these sources could be obtained in several formats from individual systems by utilizing certain protocols. The effectiveness of the individual systems is not related to the control of the Big Data system, and regularly these systems are external applications. Figure 10.3 gives the overall picture of the Big Data architecture and patterns. It is carried out in four main stages.

Data Acquisition stage, it contains many steps to concentrate the Big Data from diverse sources through large number of channels, then buffer this data long enough to apply an online trigger, and storing this Big Data. This is done through many steps; parsing and validation steps to check the statement for syntactic and semantic validity, cleansing and deduping steps to clean up data in a database that is incorrect, incomplete, or duplicated. Transformation step to transfer data to the required format; it is usually complicated, time-consuming, and the resource step in the data acquisition stage is usually consumed a great deal, once the transformed data is stored in persistent storage. Data storage stage includes data modeling; conceptual data model, logical data model, and physical data model. Next step is to determine data storage methodology according to the type and size of Big Data. The intention of the procedure chosen to store big data is to mirror the application and the corresponding usage patterns. Sharing and partitioning are other important functionalities that have a severe impact on the performance of the Big Data system. The stored data is now ready for processing and analysis. Data processing and analytical stages are considered to be the core of the big data system. Big data with no analysis has no value. This stage includes determining an appropriate data processing framework i.e. batch processing or real-time data processing. Determining the size of data transfer for data processing and analytics is important, data transfer function time depends on the data size. The

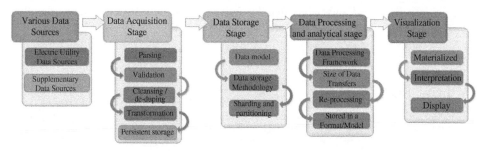

Figure 10.3 Big Data architecture and patterns.

re-processing step might need to occur on the same set of data due to some error/exception occurred in initial processing.

The outcome of processing performed is to be stored in a model. Finally, the Visualization stage, the final output summarized, forecasted, aggregated and the outcome in time-series intervals, presenting data in a sophisticated way means including infographics, dials and gauges, geographic maps, spark-lines, heat maps, and detailed bar, pie and fever charts for decision making. This stage is when the information is interpreted and then presented in a more meaningful context to make decisions based on it. This information is different from the data set entered to the big data system hence, data is the raw facts and statistics on SGs, whereas Information is data that possesses precision and appropriateness; explicit and ordered according to a pre-defined goal/reason; displayed in a certain context which delivers its meaning and applicability; and is capable of delivering an increase in comprehension and to reduce ambiguity with regard to energy management in SGs.

10.6 Building the Foundation for Big Data Processing

Big Data needs additional convincing methods to handle the massive amount of information in a limited time span. The challenge is to handle the SG by information management rather than focusing merely on the volumes of data. Regardless of the significance of volume in handling the entire data information, variety and velocity are still extremely essential to focus on. It is crucial to handle these data and acquire patterns from it to enable suitable decisions on time. Forming such a strategy needs a solid understanding of Big Data process as described in Figure 10.4. Big Data process is categorized into two central platforms; data management and data analytics. The data from the SG network equipment that includes meters, sensors, devices, substations, and mobile should be handled for acquisition of practical data before going into the analytics platform. Data management platform consists of Big Data storage, mining, integration, aggregation and representations to arrange and retrieve the data for analysis. Data analytics platform denotes the modeling and analysis of the managed data and display it in a clear form. The aim of Big Data analytics is to acquire practical values and information to support decision-making. In the next sections, each step of Big Data process will be explained. Then, discussion will focus on the crosscutting challenges, and end with a proposed system formed by utilizing the best methods that manage and analyze big data in SG. The aim is to reduce the challenges, enhance the efficiency of the grid, raise grid reliability and consumer satisfaction, and aid in online decision-making.

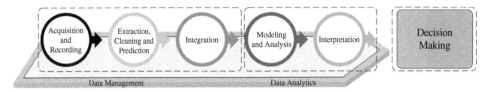

Figure 10.4 Big data process.

10.6.1   Big Data Management Platform

The Big Data management platform is an essential step in the process to arrange huge amounts of data for the data analysis platform. The data management platform includes two main characteristics; identify targeted plans while fulfilling the conditions, extract and proportionate practical information that accomplish the targeted plans to reduce cost and time consumed for analyzing and processing all the acquired data from installed measurement devices and other sources in SG.

10.6.1.1   Acquisition and Recording

The first step in the Big Data management process is defining, storing, and mining of huge amounts of data from diverse sources. SCADA systems gather data from sensors, perception devices, and substations. The core function of the SCADA system is serving the regulation of the power flow by offering remote monitoring and control of electric network devices. With the big data size raised, the complications regarding monitoring and controlling the power network will substantially rise. Cloud computing services and Fog computing are developing as crucial elements to host SCADA systems.

10.6.1.2   Extraction, Cleaning, and Prediction

Big Data mining is the ability to obtain practical information from huge datasets. Data mining process entails extraction, cleaning, annotation, and analyzing a huge array of information by following six steps [26]. Those steps are as follows: classification, estimation, prediction, association rules, clustering, and description. Classification step; the data is classified into groups. There are many algorithms utilized in data classification including the decision tree, Naive Bayes, AdaBoost, and Apriori [27]. Classification is predicated upon examining the properties of newly displayed data and transferring it to a given class. Classified data entails some unknown continuous variables that should be approximated by the Estimation step; obtain certain values that could be handled with the current state variables with their uncertainties. At the moment, the outcome of the classification step entails uncertain data, the approximation utilizing the displayed input measurements and the previously calculated state with uncertain data, which aids in arranging the data for prediction and to update the outcome. The Prediction step; denotes the expected outcome in the future depending on the knowledge and experience. The prediction step tries to create patterns that allows prediction of the next measurement given the available input data. Output data are comprised of various rules. Association Rules step; tries to form relationships between a set of data in the database. Complex rules are always produced by the same grouped data attributes recognized by variable clustering. Clustering step; tries to order a certain set of measured data into a partition or a set depending on their characteristics, aggregating with compliance to their similarities. The most popular Big Data's clustering methods that could be implemented in SGs are data mining clustering algorithms, dimension reduction, parallel classification, and the MapReduce framework [28]. The choice of the most applicable clustering methods for SG applications are reliant on the data set characteristics. The final step in the data mining process is the Description step; which analyzes the selected data in a structured form, patterns, or in row data appropriate to match with the analyzing methods to be employed to aid decision-making.

10.6.1.3 Big Data Integration

Smart Grid data needs an integration stage to acquire details of asset information and maintenance work. The main purpose of employing big data patterns is to support standards governing sources of Big Data that include; real-time pricing, distributed energy resources including DR, DG, energy storage, and consumer access to energy usage information. Architectural frameworks such as Zachman enterprise architecture framework, ISO Open-EDI (electronic data interchange) reference model, and the Common Information Model (CIM), are utilized to implement the big data in readiness for the analytic process. CIM architecture is among the most common methods fitting into SG applications because of flexible implementation and layered systems.

10.6.2 Big Data Analytics Platform

Present and future implementation of SG devices is generating a massive amount of data; these data depicted in certain patterns could be transformed into understandable information for operation decision-making utilizing analytics methods, which are categorized into advanced analytics and predictive analytics [29]. Advanced analytics are used to examine data that supports the enhancement of customer service. The time required in the analytics process can warn consumers and utilities during normal and contingency operating conditions. While, as fast the data could be analyzed and implement the action, fast recovery will be obtained. Predictive analytics is analyzing current data to form predictions about the future, expected outcomes and behaviors predicated upon the data and not on assumptions. Employing analytic methods on the big data in SG could enhance power distribution reliability and accomplish continuity to the power grid by decreasing power outages. The main qualities and advantages for big data analytics in SG are categorized into four parts. First, improving economic dispatch based on optimizing the asset life cycle, and decreasing the cost of condition-based maintenance. Second, improving reliability by enhanced regulation of operating parameters of SGs, and predicting and avoiding equipment failures. Third, improving efficiency by decreasing the number of maintenance visits by monitoring grid performance and offering the opportunity to predict any failure and take the appropriate timely decision. Fourth, improving customer satisfaction by offering field insight to consumer experience, and considering and resolving consumer complaints while increasing their comfort level. The Big data analytical process is categorized into two essential steps; modeling and analyzing; interpretation of which will be explained in the next section.

10.6.2.1 Modeling and Analysis

After the integration stage, data becomes ready to go by analytic methods. In this work, a brief description will be presented for the famous processing methods used to model and analyze different data sources in the SG. Moreover, it will select the most suitable technique that could fit big data analysis to accomplish enhanced performance and scalability in SGs. Three types of processing methods can be employed for big data analysis: batch, stream, and iterative processing. For batch processing, data is categorized into small sets to process by the Hadoop MapReduce method, which is among the most widely utilized methods for analyzing large data sets [30]. The MapReduce method is used for static applications in SGs and for all applications that need real-time response such as DR, real-time usage and pricing analysis, online grid control and monitoring, etc. Hadoop MapReduce is not efficient for most applications o SG. Stream processing or real-time processing deals with new data independently rather than waiting for the next stream of data to process. Processing data by the Apache Spark method requires more analytics; this method is considering batch, real-time, and iterative data processing requirements altogether [31]. This makes stream processing extremely useful in SG applications. Iterative Processing is a combination of batch and stream processing; the data is analyzed by the Apache Spark method. Yet, an iterative approach is processing all varieties of data but it is known to be a time-consuming method. The SG could be made more intelligent by selecting the optimal method that can analyze all diversity of SG data.

10.6.2.2 Interpretation

The output from the analytics process is collected into different forms that include binary codes, tables, or functions. Significantly, the output data contains information. Such information is accurate data that is timely; specific and arranged for a certain purpose; and such data are displayed in a manner that provides meaning and relevance; and could result in an improvement in comprehension and hinder uncertainty in SG. Transferring output data into clear and understandable information is the last step before decision-making. This can be accomplished by forming an effective infrastructure to deliver supplementary information, that is known through interpretation. There are several methods to display data that entail infographics, dials and gauges, geographic maps, spark-lines, heat maps, detailed bar, pie and fever charts. It's clear that a lot of effort has gone into figuring out the data that will serve the utility, but recently research has been directed to find out the procedure of displaying data according to a pre-defined context. Building the foundation for a data analytics program must consider operations, business functions, and customer service operations. In addition, seeking advanced analytics could be an evolutionary step for utilities in the era of the modernized grid with a complex and critical energy Internet. Smart grid data analytic architecture affects the future aptitude of the utility. It is important to reflect upon the requirements of each level for the utility taking into consideration the development and execution of each architecture, which can be mentioned in three requirements needed to build the big data analytic foundation.

- **Understanding the available data and the required data**: the real issue for the utility focusing on cost, reliable operation, strategic issues and customer service operations.
- **Organizing that data**: including internal and external data must be categorized and streamlined to make the transition of data easier.
- **Building trust in the data**: by making sure of the availability of the data with secure transfer by ensuring good data management, data quality and data cleansing processes are in place.

There are more approaches today that are used to manage and analyze Big Data. Modeling Big Data using an analytical oriented platform grants an intensive analytical framework applicable for performance prediction in scientific applications. Figure 10.5 shows a Big Data analytical oriented platform approach starting from determining the real issue for the utility by focusing on benefits that must be achieved to the grid, customer, economic, and business through a third party. Governance of the data through ensuring data availability

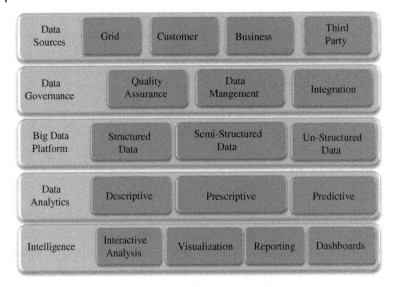

Data Sources	Grid	Customer	Business	Third Party
Data Governance	Quality Assurance	Data Mangement		Integration
Big Data Platform	Structured Data	Semi-Structured Data		Un-Structured Data
Data Analytics	Descriptive	Prescriptive		Predictive
Intelligence	Interactive Analysis	Visualization	Reporting	Dashboards

Figure 10.5 An approach to develop a platform-oriented analytical architecture.

with secure transfer through data managing and determining the quality of the presented data to restructure it into streamline to categorize the data into structured, semi-structured, or unstructured data to be analyzed and used in an intelligent way.

10.7 Transforming Big Data for High Value Action

Smart grid is rich with big data; these data are meaningless unless it can be employed through generating the value of the connecting data helps to take a decision quickly.

A SG can be viewed as an amalgamation of information and electrical power. The data sources in the SG fall under four categories, i.e. Historical (archived), Real-Time, multimedia, and timeseries. Data sources from SCADA, PMU, AMI, Smart Meter, DFR, DPR, IED, Asset Management, Operational and Utility and Weather or Lightning are real-time data sources. Massive amounts of high-Dimensional data produced out of the grid is bringing several new challenges and opportunities to the table. This leads to the research problem of Big Data Management in the SG. Managing the data generated off the grid, turning it into useful information and making decisions is one of the most important steps in managing SGs. The flow of information in the grid helps in the control and energy management of the grid. Big Data is treated as the foundation for the SG. Information from the Big Data being so valuable, many energy companies have invested in handling the data to perceive innovative and actionable insights.

All these massive amounts of data are gathered from the connected smart equipment and sensors at near real time in many types and formats. These data are always updated by rapid technology growth. To gain all advantages from these data, the ultimate value of data must be determined and useful attributes in the available data must be selected to predict outcomes. That leads to changing the concept of the four V's characterizing Big Data; Volume,

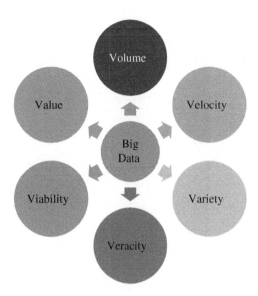

Figure 10.6 Big Data V's.

Veracity, Variety, and Velocity [32]. Yet, there are two missing V's may be more important, attributes to consider viability and value as shown in Figure 10.6.

Viability means selecting useful attributes in the data to predict outcomes that matter to utilities. Predictive analytics is analyzing present data to come up with predictions about the future, expected outcomes and behaviors predicated upon the data and not on assumptions. Predictive analytics have a huge influence on the electricity industry by enhancing power distribution reliability; achieving continuity to the power grid and by decreasing power outages. Value is the most important element of the big data to transfer the data from just information into valuable actions. These massive amounts of data have different effects on the system depending on data type.

Advanced data analytics tools, for example, data mining, pattern matching, and stream analytics are utilized to visualize, optimize, or automate grid operations. Furthermore, they form new values from the data. These values can be categorized as a path start from passive data passing through active data, proactive data, then smart data which can be used to decide, as depicted in Figure 10.7 [33]. The major attributes and benefits for producing practical values from big data by predictive and advanced analytics in SG are categorized into four parts. First, enhancing economic dispatch based on optimizing the asset life cycle, and decreasing the cost of condition-based maintenance. Second, enhancing reliability by improved management of operating parameters of SG, and allow predictions and measures regarding equipment failure. Third, working on having a better efficiency by decreasing maintenance visits by monitoring grid performance and offering the opportunity to predict any failure and take the appropriate timely decision. Fourth, working on customer satisfaction by offering field insight to customer experience, and considering customer complaints and their comfort level. To gain all these benefits, an integrated vision for analytics is needed to create value from the available data. Implementing this integrated analytics model is based on four steps as follows:

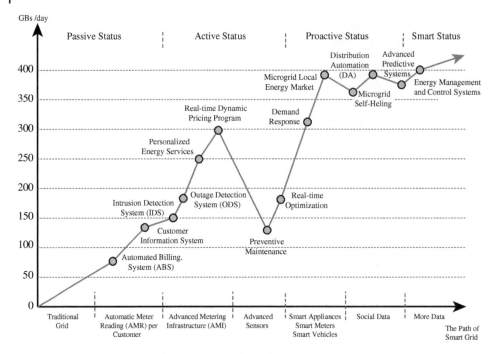

Figure 10.7 The path of SG. Adapted from Ref Num [33].

10.7.1 Decide What to Produce

The first step is determining the goal to be achieved. To determine the main goal, discrete and focused questions must be arranged based on the importance of the answer.

10.7.2 Source the Raw Materials

Row the available data starting by "small data" then the complex data to build a comprehensive "data warehouse" which is a great asset over the long term or "data mart" which is easier. After time, another layer of data can be added depending on the availability of additional data sets. These data can be social media data, location data, and financial information.

10.7.3 Produce Insights with Speed

All collected data used in the analytical process through suitable analytical techniques to produce quick decisions are built upon the prediction that can be concluded from the data value. Determine finite time limits to obtain structured output in less time, based on repeated analytical models. Create specific analytical and technical strategies to address each topic separately which is faster and relatively low in cost.

10.7.4 Deliver the Goods and Act

All the data can be used to inform specific actions through effective approaches to help utilities in prescribing actions, take the desired decisions by applying the right digital solution based on utilities insights which help with reducing risks, offer better service to

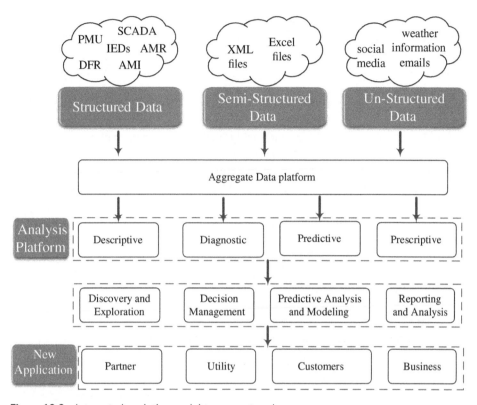

Figure 10.8 Integrated analytics model to generate value.

customers and increase revenue. Figure 10.8 illustrates how to build smart value action from the available data starting from collecting various types of data from different sources then implementing the integrated analytics vision to generate value.

10.8 Privacy Information Impacts on Smart Grid

Based on the nature of SG network, multi-directional communications and energy transfer networks allow consumers, utilities and third-party energy management assistance programs to open the data transferred. Organizations need to build internal control systems to facilitate consumer access and control their energy consumption. Furthermore, any data exchange in SG must be effectively protected through specific "Privacy Concerns" which have potential privacy impacts of SG and smart meter systems. These impacts can be listed as the following points [34]:

1) Identity Theft that could imitate a utility customer leading to serious consequences.
2) Determine Personal Behavior Patterns Access can be used by stockholders to perform "target" marketing or governments to tax certain activities, and hacking people.
3) Determine Specific Appliances used through smart meters which can be used by appliance manufacturers to impact appliance warranties or insurance companies to approve or decline claims.

4) Perform Real-Time Surveillance which gives almost real-time information about the energy used, the behavior of people in residence that is considered safety risk.
5) Reveal Activities Through Residual Data by keeping the unused data on the metering devices which may be used by activists or others to conduct activities with malicious intent.
6) Targeted Home Invasions through daily data transferred which can tell whether a home is totally unoccupied or vacant.
7) Provide Accidental Invasions through analysis of my meter data to discover the behavior of residents which may be detrimental to residents.
8) Decisions and Actions predicated on Inaccurate Data due to the massive amount of data accessed by different entities may affect SG power activities which may negatively affect grid operation.
9) Activity Censorship can express the real-time activities of residents so if these meters' data are shared due to any problem in the security system, residents may face embarrassment, harassment, and loss of vital appliances [35].
10) Profiling may occur through sis of energy consumption which may be used against the person which is against privacy legislation.
11) Undesirable Publicity and Embarrassment.
12) Behavior Tracking.
13) Public Aggregated Searches Revealing Individual Behaviors.

Recently, innovation in technology, such as cellphones, computers, instruments, smart appliances will improve the users' capacity to monitor, manage, and control the usage of electricity service. Smart grid systems using bi-directional communication among service providers will have a huge amount of data. In addition, utility companies are installing new meter technology to monitor customer power consumption which, in turn, offers a lot of information. This movement from an internet of people to the "internet of things" increases threats to security. A key issue is balancing security with ease of use through the study of how these devices are used to collect information, who can access that information, and reasons for using this data. Corporate management must make security a high priority to balance between privacy protection and beneficial use of data transfer.

10.9 Meter Data Management for Smart Grid

Meter Data Management (MDM) can be considered as a key strategic initiative for utilities aiming to make practical and accurate data to be utilized for revenue and system planning goals. In addition, consumers can use the data managed by MDM system to decrease their bills and use electricity more efficiently. MDM strategies depend on the software used to define data types, import the data, validate, process, analyze data sent by smart meters. MDM is integrated architecture that expands communication and management of data created by an AMI through the head end system which is a server application that sends the collected data to as MDMS software after categorized it by data collector layer based on some standardized communication protocol. Once the data delivered to MDMS is processed and estimated, then it is saved to the database to be used. Figure 10.9 show the MDM

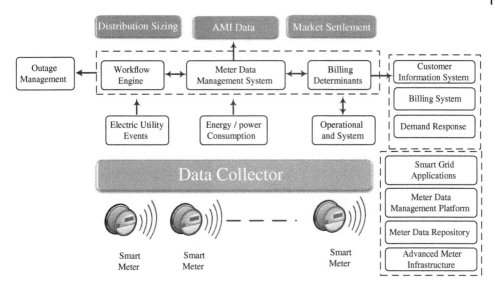

Figure 10.9 Meter data management.

architecture. However, meter data is used throughout the utility but it also serves different fields such as customer service; through helping customers to manage their power consumption which, in turn, decreases electricity bills, marketing; the knowledge of customer peak can help determine appropriate tariff, system planning; all the data needed for planning or remodifying is driven by analyzing consumption and DR in the grid. In addition, the data helps to support maintenance and purchasing decisions. To achieve all the previous advantages, the following requirements are crucial for MDMS:

- Provide security for the used data.
- Develop a backup and recovery system to prevent any data loss.
- Reliability to ensure system continuity as the used data is a business-critical service of utilities.
- Scalability and possibility to process a huge amount of data.

10.10 Summary

This chapter discussed how the real revolution is not in the volume of the collected data but in the value of such data and in how it is managed and used to gain all benefits that can serve the SG. In the beginning of this chapter, a focus on the types of data structures were made, showing 80–90% of future data rise proceeding from non-structured data types. These data containing information about grid status, such information that could be useful in providing better energy management and control decision through different management tools. Then, it outlines the most common tools used to manage these data and outlines the steps which are needed to build data process foundations. Then, the two main platforms are used to manage and analyze this massive data to optimize processes and derive more smart values which help to inform specific actions through effective approaches to help utilities in prescribing actions. This chapter covers the requirements for implementing MDM architecture for SG to serve utilities in different fields such as customer service, marketing, and system planning. Data management provides security for the used data, develops a backup and recovery system to prevent any loss in the data, improves the reliability to ensure system continuity and scalability to process a huge amount of data.

References

1 Mohamed, A., Refaat, S.S., and Abu-Rub, H. (2019). A review on big data management and decision-making in smart grid. *Power Electronics and Drives* 4 (1): 1–13.

2 Kezunovic, M., Pinson, P., Obradovic, Z. et al. (2020). Big data analytics for future electricity grids. *Electric Power Systems Research* 189: 106788.

3 Kezunovic, M. (2011). Translational knowledge: from collecting data to making decisions in a smart grid. *Proceedings of the IEEE* 99 (6): 977–997.

4 Azad, P., Navimipour, N.J., Rahmani, A.M., and Sharifi, A. (2019). The role of structured and unstructured data managing mechanisms in the internet of things. *Cluster Computing*: 1–14.

5 Jeble, S., Kumari, S., and Patil, Y. (2017). Role of big data in decision making. *Operations and Supply Chain Management: An International Journal* 11 (1): 36–44.

6 Michael I. Henderson, Damir Novosel, and Mariesa L. Crow. (2017). Electric power grid modernization trends, challenges, and opportunities. IEEE Power and Energy, pp.1-17. [online] Available: https://www.ieee.org/content/dam/ieee-org/ieee/web/org/about/corporate/ieee-industry-advisory-board/electric-power-grid-modernization.pdf.

7 Hu, H., Wen, Y., Chua, T.-S., and Li, X. (2014). Toward scalable systems for big data analytics: a technology tutorial. *IEEE Access* 2: 652–687.

8 Bhattarai, B.P., Paudyal, S., Luo, Y. et al. (2019). Big data analytics in smart grids: state-of-the-art, challenges, opportunities, and future directions. *IET Smart Grid* 2 (2): 141–154.

9 J. Fan, F. Han and H. Liu, "Challenges of Big Data analysis" in National Science Review, press, 2014.

10 H. S. Fhom, "Big data: Opportunities and privacy challenges," arXiv preprint arXiv:1502.00823, 2015.

11 Marinakis, V. (2020). Big data for energy management and energy-efficient buildings. *Energies* 13 (7): 1555.

12 Zhang, Y., Huang, T., and Bompard, E.F. (2018). Big data analytics in smart grids: a review. *Energy Informatics* 1 (1): 8.

13 Guerrero-Prado, J.S., Alfonso-Morales, W., Caicedo-Bravo, E. et al. (2020). The power of big data and data analytics for AMI data: a case study. *Sensors* 20 (11): 3289.

14 Ye, Feng, Yi Qian, and Rose Qingyang Hu. "Big data analytics and cloud computing in the smart grid." (2017).

15 Alejandro Sanchez, and Wilson Rivera. "Big data analysis and visualization for the smart grid." In 2017 IEEE International Congress on Big Data (BigData Congress), pp. 414–418. IEEE, 2017.

16 Sivarajah, U., Kamal, M.M., Irani, Z., and Weerakkody, V. (2017). Critical analysis of Big Data challenges and analytical methods. *Journal of Business Research* 70: 263–286.

17 Armoogum, S. and Li, X.M. (2019). Big data analytics and deep learning in bioinformatics with Hadoop. In: *Deep Learning and Parallel Computing Environment for Bioengineering Systems* (ed. A.K. Sangaiah), 17–36. Academic Press.

18 Oussous, A., Benjelloun, F.-Z., Lahcen, A.A., and Belfkih, S. (2013). Comparison and classification of NoSQL databases for big data. *International Journal of Database Theory and Application* 6 (4): 2013.

19 Liu, G., Zhu, W., Saunders, C. et al. (2015). Real-time complex event processing and analytics for smart grid. *Procedia Computer Science* 61: 113–119.

20 Taherkordi, A., Zahid, F., Verginadis, Y., and Horn, G. (2018). Future cloud systems design: challenges and research directions. *IEEE Access* 6: 74120–74150.

21 Cao, Z., Lin, J., Wan, C. et al. (2017). Hadoop-based framework for big data analysis of synchronised harmonics in active distribution network. *IET Generation, Transmission & Distribution* 11 (16): 3930–3937.

22 Malik, S.U.R., Khan, S.U., Ewen, S.J. et al. (2016). Performance analysis of data intensive cloud systems based on data management and replication: a survey. *Distributed and Parallel Databases* 34 (2): 179–215.

23 Liu, Xiufeng, and Per Sieverts Nielsen. "Streamlining smart meter data analytics." In Proc. of the 10th Conference on Sustainable Development of Energy, Water and Environment Systems, SDEWES2015, vol. 558, pp. 1–14. 2015.

24 Amalina, F., Hashem, I.A.T., Azizul, Z.H. et al. (2019). Blending big data analytics: review on challenges and a recent study. *IEEE Access* 8: 3629–3645.

25 Bhathal, G.S. and Singh, A. (2019). Big data: Hadoop framework vulnerabilities, security issues and attacks. *Array* 1: 100002.

26 Shady S. Refaat, Haitham Abu-Rub, and Amira Mohamed. "Big data, better energy management and control decisions for distribution systems in smart grid." In 2016 IEEE International Conference on Big Data (Big Data), pp. 3115–3120. IEEE, 2016.

27 Xindong, W., Kumar, V., Ross Quinlan, J. et al. (2008). Top 10 algorithms in data mining. *Knowledge and Information Systems* 14 (1): 1–37.

28 Btissam Zerhari, Ayoub Ait Lahcen, and Salma Mouline. "Big data clustering: Algorithms and challenges." In Proc. of Int. Conf. on Big Data, Cloud and Applications (BDCA'15). 2015.

29 Williams, S. (2016). *Business Intelligence Strategy and Big Data Analytics: A General Management Perspective.* Morgan Kaufmann.

30 Dittrich, J. and Quiané-Ruiz, J.-A. (2012). Efficient big data processing in Hadoop MapReduce. *Proceedings of the VLDB Endowment* 5 (12): 2014–2015.

31 Shyam, R., Bharathi Ganesh, H.B., Kumar, S. et al. (2015). Apache spark a big data analytics platform for smart grid. *Procedia Technology* 21: 171–178.

32 Dhamodharavadhani, S., Gowri, R., and Rathipriya, R. (2018). Unlock different V's of big data for analytics. *International Journal of Computational Science and Engineering* 6 (4).

33 M. Aiello and G. A. Pagani, "The smart grid's data generating potentials," in Computer Science and Information Systems (FedCSIS), Federated Conference on. IEEE, 2014, pp. 9–16.

34 Alvaro A. Cárdenas, and Reihaneh Safavi-Naini. "Security and privacy in the smart grid." Handbook on securing cyber-physical critical infrastructure (2012): 637–654.

35 Herold, R. and Hertzog, C. (2015). *Data Privacy for the Smart Grid.* CRC Press.

11

Demand-Management

In conventional power grids, the two sides of the electricity demand and supply are basically disconnected while the grid monitoring information is handled only at the operations side. Achieving stability in power grids requires that they meet electricity demand on a regular basis by the electricity supply, which requires planning and communication between both sides, generation and consumption. For that, an evolution in electric power generation and supply is needed to improve reliability, security, and the efficiency of power grids. The smart grid (SG) principle transfers the future generation electricity network to smarter and intelligent grid through enabling bi-directional information and active participation from all connected parties. Driven by concerns regarding electric sustainability, energy security, and economic growth, it is essential to have a coordination mechanism predicated upon heuristic rules to control energy demand and enhance the survivability of the system when failures take place or at peak periods. This is achieved through the principle of demand management systems.

11.1 Introduction

In recent years, emphasis has been put on obtaining suitable pricing schemes toward energy management in SGs. Pricing schemes can be categorized as time-of-use (TOC) pricing, day-ahead pricing, and real-time pricing (RTP) [1]. The TOC pricing means changing the price rate based on the time peak period, off-peak period, or shoulder period. Day-ahead pricing depends on offers presented by utilities to achieve equilibrium between supply and demand in hourly intervals. RTP is the tariff for actual delivery of the electric power, which changes hour-to-hour. To control and manage the pricing, demand management mechanisms have been used from wholesale market prices. Two important economic columns of demand management for smart distribution grid are Demand Response (DR) and Demand Side Management (DSM). DR is considered as a cost-effective solution for tackling uncertainty issues when operating the SG. Such a system allows the utilities to build automated, integrated, and flexible platforms for the regulation of efficient and smart

Smart Grid and Enabling Technologies, First Edition. Shady S. Refaat, Omar Ellabban, Sertac Bayhan, Haitham Abu-Rub, Frede Blaabjerg, and Miroslav M. Begovic.
© 2021 John Wiley & Sons Ltd. Published 2021 by John Wiley & Sons Ltd.
Companion website: www.wiley.com/go/ellabban/smartgrid

DR solutions [2]. To allow utilities to implement and deploy the DR solution on a large scale, and across a wide range of SG, the solution needs to be simple, cheap, and easily communicated with the rest of the DR system. DSM is the variations in electricity usage by end-use customers by changing their consumption behavior according to the variations in the price of electricity over time. DR and DSM bring great benefits for both the grid and customers, such as supporting grid interoperability which minimizes the integration cost when combining a vast number of end devices in SG. Such solutions promote policies that support capture and process huge amounts of data in real-time to achieve customer satisfaction through mitigating energy consumption and peak demand, providing consumers with better service and lower prices. Figure 11.1 divides the benefits of the demand management program into seven categories: economic, pricing, system reliability and risk management, market efficiency impacts, lower cost of electric system, customer service, and environmental benefits [3].

11.2 Demand Response

Federal Energy Regulatory Commission (FERC) defines the DR as: "Changes in electric usage by end-use customers from their normal consumption patterns in response to changes in the price of electricity over time, or to incentive payments designed to induce lower electricity use at times of high wholesale market prices or when system reliability is jeopardized" [4]. The concept of DR emerged with technology development, which increases the need for electricity and peak demand. To work under Smart Grid concept, the balance between generation side and demand side must be under consideration all the time. DR is founded as a solution to demand side control in SG. Figure 11.2 explains how DR can achieve the supply–demand balance through controlling the electric consumption which is monitored through smart meters. Controlling the loads through DR program depends on load size, load period and locations. A central unit (aggregator control) manages the electric consumptions after being measured and monitored through a smart meter. The aggregator makes the decision based on control signals from service provider after comparing the demand and generation in order to achieve market balance with supply–demand balance [5].

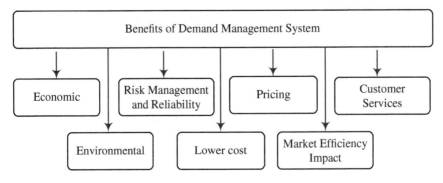

Figure 11.1 Benefits of demand side management. Adapted from Ref [3].

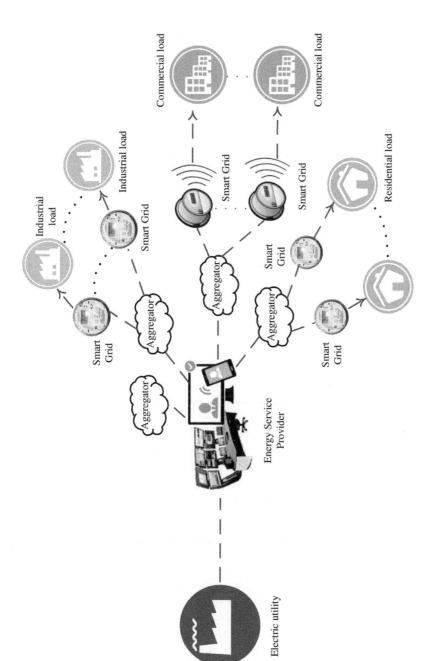

Figure 11.2 Demand response.

11.3 Demand Response Programs

Demand Response (DR) programs fall into the category of load-management programs to encourage customers to decrease their electric consumption in response to a change in the price of electricity over time which, in turn, addresses the economic use of the grid. DR programs help utilities lessen the power consumed, preserve energy, redistribute power consumed, enhance system reliability, reduce energy prices, and increase economic efficiency [6]. DR programs can be classified into two categories based on customer participation, load response and price response. The first category shows the criteria which is used by utilities to notify program participants to shed load. Utilities can call customers for reliability, economic, or emergency conditions which are named load response or time-based programs [7]. The second category shows the programs used by utilities to motivate customers to participate in managing their load, which are called price response or incentive-based programs.

11.3.1 Load-Response Programs

When the utility offers customers payments to reduce the demand in a determined period, it is called a load response program. Utilities use different ways to communicate with participants such as sending straight signals through cell phones or the internet as a notification to request a decrease in electricity demand. Load response programs include those where the customer responds to suggestions for suppressing the short-term peak load. Load response programs are also known as "reliability-driven" programs as opposed to "market-based pricing" programs [4]. This program notifies the customer of the reliability of their participation by using economic events or a merit load response in exchange for a utility payment or by using a pricing signal to identify the cost for providing power to the grid. Determining which program can be used depends on the program goals, customers interested, and the components used for the program design. Figure 11.3 summarizes the properties of each type of load-response program. Based on the information in Table 11.1, we can conclude that each of the following programs is suitable to be used in different types of load, based on the load size and consumption. For direct load control programs, the utility agrees with customers to include fixed monthly credited payments to the customer's bill and one additional participation payment when the load is reduced, which helps to reduce the loads when required. A direct load control program is relatively simple, reliable, and inexpensive to implement; only control switches or "smart" thermostats are needed to reduce the load. It is most effective when implemented for small loads. This program can achieve a 60% load reduction [8]. In contrast, a curtailable load program can be used for large commercial loads where the customers agree with utility to reduce their loads for a certain amount of time decided by the utility before a curtailable load event using different control technologies. Different incentives can be presented from the utility such as a monthly capacity credit or per event credit based on market pricing. In addition, the program imposes significant penalties for non-performance implementation. An interruptible load program needs particular capability from customers to participate especially for loads greater than 1 MW. These customers agree with the utility to switch off some or all loads for specified periods. Customers use backup generators during any interrupted period. They

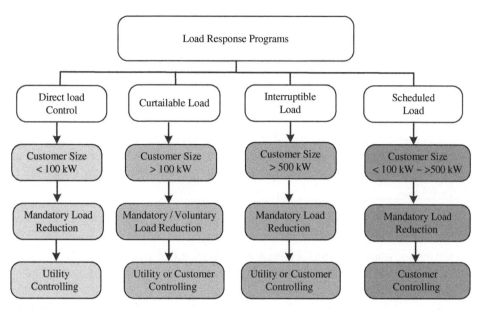

Figure 11.3 Major features load response programs.

Table 11.1 The properties of load-response programs [8]. Reproduced with permission from Rocky Mountain Institute.

	Direct load control	Curtailable load	Interruptible load	Scheduled load
Description	End-Use loads switched off for a certain amount of time	End-use loads decreased the	All or main parts of consumer total load turned off per periods of time	Load declines planned before between utility and consumer
Target Customer Size	<100 kW	>100 kW	>500 kW	All
Load Reductions	Mandatory	Mandatory or Voluntary	Mandatory	Mandatory
Party Controlling Reduction	Provider/Utility	Provider/Utility or Customer	Provider/Utility or Customer	Customer
Incentive (examples)	$0.014–0.40/ton of cooling	$7–45/kW $0.15–0.53/kWh	$7–45/kW $0.15–0.53/kWh	$7–45/kW
Advance notification	None	Minutes to hours	Minutes to hours	Months-contractual
Eligible enabling technology	Load-switches, two-way communications optional	Interval meters auto-mated controls, two- communications ways (optional)	Interval meters, backup generator (optional)	Not necessary
Billing system change?	No	No	No	No
Settlement	Fixed credit on monthly bill	Monthly bill amendments, penalties for non-performance	Monthly bill amendments, penalties for non-performance	Shown on monthly bills penalty for non-performance

receive different incentives from their utility such as lower electricity bills during normal operation or additional monthly capacity credit. Penalties are applied to participants to ensure performance or ensure binding contracts. Another program is the scheduled load program, which is considered the most efficient program compared to all DR programs. This is because customers have the opportunity to reduce their load in determined pre-planned period or regularly, such as during the summer season which presents energy efficiency strategy rather than DR. These periods may not coincide with unexpected emergencies and the utility can't call the customers to curtail loads which may be called inflexible system [9].

11.3.2 Price Response Programs

This program is when customers willingly decrease their demand according to economic signals through load reduction programs. These programs vary based on the pricing framework and the agreement with participating customers. Price response programs are categorized into the following main programs; Time of-Use, dynamic pricing, and demand bidding which depicted in Figure 11.4.

These programs depend mainly on the price in a mean time, to decide which program can be implemented a previous knowledge of each program should be available. Table 11.2. summarizes price response programs properties. Comparing the three

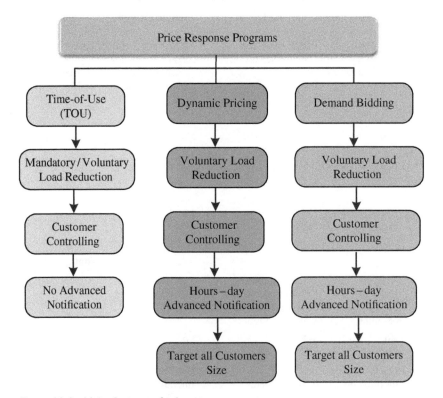

Figure 11.4 Major features of price response programs.

Table 11.2 Price response program overview [8]. Reproduced with permission from Rocky Mountain Institute.

	Time of-Use	Dynamic Price (Critical-Peak Pricing or Real-Time Pricing)	Demand Bidding
Description	Load management predicated upon a stepped rate structure entailing peak, off-peak, and occasionally shoulder peak rate	Load management dependent on a dynamic tariff	Customers bid load decline based on (i) provider proposed price, or (ii) customers Proposed Price
Target size	All	All	All
Mandatory or voluntary	Mandatory or Voluntary	Voluntary	Voluntary
Party controlling reduction	Customer	Customer	Customer
Incentives(example)	5% bill decrease premiere + payment per event	Locational marginal price.	Locational marginal price $0.15–0.50/kWh
Advance notification	None	Hours a day ahead	Hours a day ahead
Enabling system change?	TOU meters, real time or prior-day energy information by internet	Interval meters 2-way communication, energy information by internet	Interval meters, web-based market, spot price information by internet
Billing system change?	Yes, new TOU rate	Yes, new CPR or RTP rate	No
Settlement	Monthly bill adjustments	Monthly bill adjustments	Individual payments reliant on bid price, declines beyond bid paid, location marginal price

programs used for a price response program, it's concluded that: TOU pricing programs depend on a pre-determined period by the utility through dividing time into blocks based on load consumption, which entails the peak rate, an off-peak rate, a shoulder-peak rate, and price changes due to the consumption rate. This program allows consumers to observe the electricity prices, which is predicated on the actual price of power. All participants are obliged to decrease the power consumed in their premises amid peak periods. The utility determines the electricity prices based on the actual cost of the consumed power. TOU calculated for long periods based on expectation and marker conditions that can't be variable, only fixed, while the variation of TOU tariffs is the critical peak pricing (CPP) by providing the possibility of a customer call service from 24 hours to as little as a few-minutes earlier than required at critical peak events through the utility's observation toward the balance of supply and demand. Utilities use this program for emergency cases or cases of g high market prices being forecasted. The price of electricity increases during these periods. The CPP program has two designs which deal with different events, one deals with a predetermined increase for both prices and a specific period of time at critical events. The other strategy depends on changing the duration and the time of raising the price of electricity based on the electric grid's needs to achieve reduced loads. CPP reduces load by 20–60% compared by a TOU load reduction ranging from 4 to 17% [10]. The RTP program is a fully mutable program. Rates and prices changed 24 hours earlier than usual/ This is usually predicated on the supply costs to the utility, which reflects any change in consumption rates. In this program, the consumer can access the electricity prices which can be changed from hour to hour based on the market price which allows consumers to manage the electricity bills by reducing their consumption at periods when prices are greater and extra power used amid an off-peak period when prices are less. Implementation of an RTP program requires automated interval metering, investment technology, communications, and billing systems so its cost is high. Because of this, only a limited number of the biggest industrial and commercial customers (>1 MW) use this program. Both RTP and CPP programs must provide backup generation to ensure service continuity, amplify performance and continue savings. During this stage, uncertainties emerge with regard to which programs fulfill most or all requirements for an optimized suppression in end user demand on the utility system and which program is more effective to be used in different customer fields, i.e. commercial or industrial. For example, in the USA, a survey carried out on the topic of DR programs, which focused more on commercial and industrial customers and the results acquired are depicted in Figure 11.5 [8]. It's clear that utilities didn't overwhelmingly use one program more than any other, however it can be noted that load response programs are more common relative to price response programs.

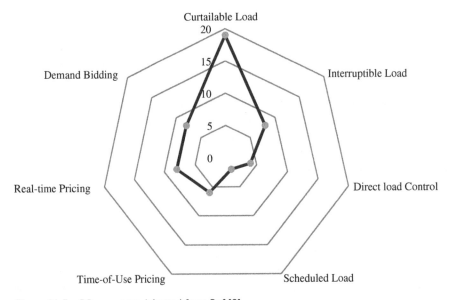

Figure 11.5 DR programs. Adapted from Ref [8].

11.4   End-User Engagement

End users will contribute a great deal in SG in the near future as a step to link between energy providers and end users (customers) to promote smart energy behavior which results in energy efficiency (measured by the required power being provided), flexibility (such as using any surplus power in high energy demand periods) and sustainable energy generation (by using renewable sources to generate power such as photovoltaic [PV] cells). This chapter presents a persistent and a unified view on involving end users in SGs and suggestions for end-user obligation. The main factor affecting market design for DR is end users. Enough information can be drawn from end users' feedback, and bidding behavior can help when developing pricing schemes and helping an indirect way of decision making. End-user engagement in the SG process changes from the "activation phase" when new participants begin engaging, then move to the "continuation phase" where interaction aimed at changing smart energy behavior to an improved level of "smartness" as shown in Figure 11.6 [11].

To move from the "activation phase" to the "continuation phase" two distinctive parts must be determined as enablers (reasons to tempt an end user to take part in DR programs) and barriers (reasons that cause end users to refrain from participating) which are presented in Figure 11.7. It's clear that, using SG technology will increase the comfort level for customers through more advanced ways of controlling loads and getting involved or taking part in the electricity market [12]. In addition, an opportunity presents itself to control energy consumption and gain incentives such as reducing the energy bill. And of course, using SG will have great benefits for the environment such as hindering production of greenhouse gases by implementing renewable energy sources into the grid. From the previous discussion, we can conclude the goals which must be achieved to attract participants to SG DR programs to move into the "continuation phase" which can be counted within the following points as success factors for end-user engagement [13, 14]:

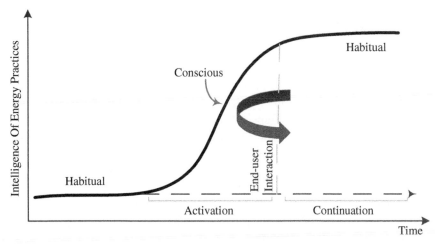

Figure 11.6   Clarification of the procedure of end-user-interaction defining the "activation" and "continuation" phases. Ref Num [11].

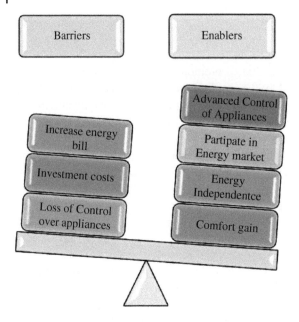

Figure 11.7 A stylized interpretation.

- Provide all the information relevant to each customer such as detailed billing information and present more information about SG program taking into consideration privacy.
- Provide attractive financial incentives and provide support and services.
- Building trust over time.
- Understand end users by taking into consideration direct and indirect feedback to result in actions.
- Guarantee a continuous information flow.
- Elicit and follow-up end-users' expectations on which to base engagement strategies.

11.5 Challenges of DR within Smart Grid

The challenges which face implementing DR programs in SG are wide ranging and have several axes [15]:

1) Electrical demand has a linear curve with the price of electricity which means any change in the electricity price will induce a proportional change in demand that means any added value of SG will lead to "demanding" a lot from customers which, in turn, means that utilities must increase interesting offers for customers. In the sense that, providing SG services to the end user will be a barrier to end user engagement in DR programs.
2) It is, as yet, largely unclear how utilities can analyze enablers and barriers for end users to identify targeted end-user groups and present a clear vision used in taking decisions, provide incentives and pricing schemes.
3) Provide communication pathways and marketing strategies to recruit end-users in smart energy projects.

4) SG has many stockholders cooperated to provide continuous service, the challenge is how to cooperate between all SG stockholders and end-user participants.
5) The system benefits of DR are not necessarily sufficient to cover the cost needed for control and communications infrastructure so research study for any redesign to market such a DR is needed to ensure the economic stability to the system.
6) Providing security for the information that flows to establish protection and reliability.
7) How to ensure consumers will not withdraw from DR programs. The system operator must prepare an additional plan to overcome this inconvenience. In addition, an extra penalty must be implemented on participants in this case.

11.6 Demand-Side Management

Demand Side Management (DSM) is a group of programs that connect together to achieve one main objective which is to provide customers with continuous and efficient energy in the long term with less cost. Nowadays, using SG helps utilities to implement DSM through providing the scale and scalability from all smart meter's readings which permits utilities to gather and evaluate usage information at intervals. Such real-time information is capable of being transferred to consumers in their premises allowing real-time feedback on consumption to achieve supply-demand balancing. DSM contains two main principles to achieve supply- demand balancing load shifting and energy efficiency. Load shifting through DR programs which work on transfer loads during high demand periods to off-peak periods. Two types of load shifting can be considered based on the shifting period. Shifting customer demand during 20–50 hours/year with the greatest demand is called critical peak shift and daily peak shift when changing customer demand in the timeframe of 24 hours every day with the greatest demand [16]. Energy efficiency and conservation through energy conservation programs makes customers decrease the total demand for electricity, if the reduction in overall demand for electricity is done through reducing the amount of electricity that the customer receives from utility called (energy conservation) and if the reduction of overall demand electricity is done through maintaining the amount of electricity that the customer received from utility called (energy efficiency). Figure 11.8 illustrates the major impact areas of DSM.

11.7 DSM Techniques

The concept of DR and DSM is most commonly utilized as synonymous or misused, sometimes the common techniques used for DSM are known as DR methods. However, DR and DSM are two distinct procedures, but they are capable of being used together in synchronicity. DR programs are reactive methods implemented to encourage consumers to decrease their energy consumption in the short time based on calls from utilities which are used either in system emergencies or at peak periods. Depending on the reduced electricity price for the off-peak period, this may cause shifting of all loads to that period, which causes another peak of demand, as well as service interruption. To overcome this, management strategies should be planned and implemented to manage a group of

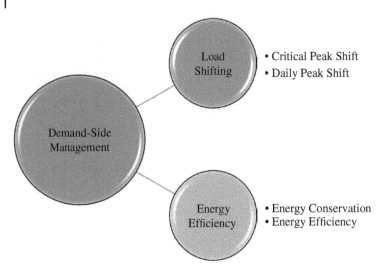

Figure 11.8 Major areas of dDSM.

users, ultimately, reducing energy consumption predicated on a system-wide perspective. DSM is a proactive approach used to manage and shift the load using the same amount of power with less cost per unit generation in power plants while maintaining efficient energy [17]. DSM techniques are applied for energy saving, reducing energy consumption, modifying the demand profile and load management. To achieve the goals of DSM, the load shapes of daily electricity demands can be considered as a pointer of energy consumption for all periods between peak and off-peak times which can be controlled through the following techniques:

- Peak clipping
- Valley filling
- Load shifting
- Strategic conservation
- Strategic load growth
- Flexible load building

Table 11.3 presents a brief description of each technique.

Before designing DSM models, there are some challenges that must be overcome. One of these challenges is the need to model customer behavior, optimize pricing, study fluctuating demand and develop decision-theoretic tools. All these challenges lead to the use of decision-theoretic approaches, for instance, game theory, optimization to appropriately model and examine the implementation of the DR method to achieve flexible communication with customers through SG.

- **Optimization**

With this approach, all consumers cooperate, managing their resources by reducing their energy consumption. Optimization models are used to decrease utility sharing. Although there is no conflict between users, energy prices or real-time tariffs vary with the total

Table 11.3 A brief description of each technique [18, 19].

Technique	Shape of technique	Description
Peak clipping		Decrease in peak demand.
Valley filling		Increased demand at off-peak.
Load shifting		Demand shifting to non-peak period.
Strategic conservation		Decrease of utility loads during a 24-hour period.
Strategic load growth		The rise of utility loads during 24 hours or less.
Flexible load building		Increased demand or a program that the customer can alter his energy consumption as needed.

number of users and customer participation, which are not the same, so decreasing the overall bill could be unfair. The two main methods of optimization in DSM are stochastic optimization and robust optimization. Stochastic optimization methods for problems use the uncertainty model with an association with the probabilistic model, but only for systems that have inaccurate distribution or uncertainty characterized by a group of values that do not depend on a probabilistic model. However, stochastic optimization is used for a long time but is considered as a limited model because of its reliance on historical statistics, and its failure to deal with hazards directly. In contrast, to a greater extent, robust optimization is simpler and more effective. Nowadays, most DSM methods with ambiguous data are designed with robust optimization.

- **Game theory**

This is a mathematical tool used to analyze relationships between players of the power grid and the utility who seek to maximize their own benefit through interaction with each other using game theory, we can identify solutions for conflict with cooperation between players. Game theory provides a framework to design the most suitable active side management model by studying the pricing scheme that provides incentives for the participants. Nowadays, game theory is the most preferred approach for DSM as the concept of game theory based on choosing the optimal behavior by undertaking a cost/benefits analysis for each option and selecting the most suitable with the least cost for all individuals. The game-theory model can decrease the entire cost of energy for each electricity payment on a day-to-day basis.

11.8 DSM Evaluation

DSM is an essential task used to meet the target of the electricity market. However, it is essential to develop systems responsible for evaluating the DSM model. This evaluation depends on systematic measurement of the program performance and whether that program produces valid and reliable results to the utility executives as well as utility and regulatory agencies who examine the quality of each program used. Utilities are capable of using the evaluation results as documented data for energy saving, load reduction, cost effectiveness which can be used to search for methods to enhance the design of future programs and increase the number of participants, thereby increasing energy savings and reducing the cost of the power produced. This evaluation is based on three main points that must be covered, the control of overall energy consumption, the optimal distribution of power and a lower cost through the adjustment of suitable electricity price policies. To measure achievements of any DSM program, two types of evaluation must be covered, process and impact evaluations as shown in Figure 11.9 Process evaluations examine the performance of the program during implementation [20]. Impact evaluations examine the effect of the program on the operational and any economic benefits.

DSM program evaluations are processed through four models [21]:

- Engineering
- Statistical
- Hybrid engineering/statistical
- Metering

Figure 11.9 Relationships between DSM Table 11.1.1 Programs. Adapted from Ref [19].

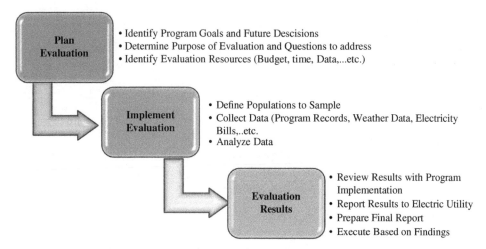

Figure 11.10 Process evaluation of DSM.

Different methods are applied using one of these models but some models use a combination of them which is considered the most effective evaluation method as it includes all aspects needed for evaluation process. The main concept of all these models is the collection of different data, analysis of the data and a summary of the costs and benefit functions of each factor, but the difference between one model and another is the data organization, processing, and the evaluation of collected data which mainly depends on different technologies. The system that can offer dynamic pricing instead of the flat rate price tariff for customers and demand stability can be evaluated as a suitable system for implementation.

The evaluation process is presented in Figure 11.10. Start by listing the sequence of activities associated with the DSM evaluation process. Identifying the program objectives is the first step in the evaluation process and focusing on the evaluation that can help in future decisions and affect the program to subsidize substantial energy or capacity resources. The evaluation should reflect the importance of the program through study of any change in budget, staff, time, or data. In addition, the possibility of examining alternatives through collecting all data about previous threats, methods used and their effect. All that means identifying all collected data. The second step is implementation of the evaluation through selecting some sample customers and collecting all data related to the program used such as any change in power demands, local weather data and records of utility bills. All results obtained must be merged, checked, and analyzed. The last step is presenting the evaluation results in report form revising the process with program staff and management. The DSM evaluation report provides the necessary information for program managers and staff to use for modification and improvement of current program operations.

11.9 Demand Response Applications

In all actuality, DR can be a key beneficiary of SGS. Using DR and DSM techniques will pursue inbuilding the demand and supply to improve the efficiency of energy consumption in SGs and ultimately integrate the user-side of electrical loads. In addition, coordination

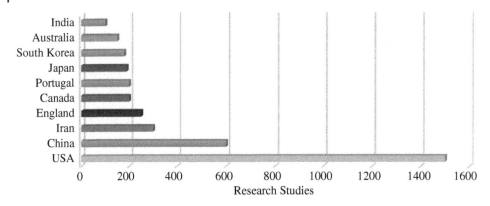

Figure 11.11 Major Research Studies worldwide. Ref [22]. Reproduced with permission from IEEE.

of consumers, electric cars, microgrids, and utility cooperation, are only possible through efficient DSM. That leads to one fact; demand-response plans enable organized management of the electricity supply and demand and are projected to play a vital role in SGs. Over the last decade, transition to SG concept is the evolutionary path to engage the customers in managing demand. SG is the most important application of DR techniques. For the last 10 years, a survey of the impact of DR applications on decrease of peak demand in different countries around the world is presented in Figure 11.11 [22].

As is clear, USA is first followed by China and Iran. A survey held in England mentions that every 5 cents energy decline at peak times led to a $2.7 billion reduction on the entire price of generation, transmission, and distribution [23]. The England Independent Operator (ISO-NE) delivers three types of DR application to its consumers: real-time demand, real-time pricing and one day-ahead DR [21]. In Japan, a 30% reduction of energy use is realized on a pilot DR project that is employed in 69 apartments [24]. According to the 2014 Europe SG project evaluation, the entire budget of DR-related projects is 400 million Euros [25]. In the USA, dynamic pricing rates were used to investigate the effect of DR applications on the decline of peak demand. Feldman and Lockhart [26] calculated a value reaching up to $9.7 billion by 2023 achieved by the global market for DR activating technologies [26, 27]. Such surveys are used to examine the potential of using DR applications in different countries at decreasing peak demand on the grid. Considering new smart appliances and applying the DR in a large scale under the SG concept will give a new outlook on generation and demand and on the operation of SG. However, the consumption of each customer needs to be measured, analyzed, and managed.

11.10 Summary

This chapter presents an overview of Demand Management as an important step to manage energy demand, giving customers a voice in changing their demand for electricity amid peak periods, and decreasing the energy they consume as a whole, while ultimately increasing grid efficiency through evaluation of the survivability of the system when failures occur or at peak periods. A brief description is presented for the main two columns of implementing demand management in SG DR and DSM. This study illustrates the benefits to be gained from implementing DR programs for both utilities and customers, the types of DR and the most used techniques in each type, in addition to end-use and facility goals for decreasing the load, as well as activating new technological selections. Utility costs and system advantages are two main concepts that exemplify the purpose of customers taking part and any barriers to participation. Studying (DSM) as a feature of the SG that entails any actions taken on the demand-side that vary the load profile on the generation side. This is followed by classifications of DSM into two broad categories: (i) reducing consumption; and (ii) shifting consumption also with a brief description of different techniques used to achieve DSM. This chapter discusses the importance of evaluation in shaping the effectiveness of DSM programs and determines whether it can or cannot achieve the goals of utilities and shows how the accuracy at which evaluations can decide on suggestions posed, estimations, and data turning this into valuable information to determine the best cost and most efficient processes for both utilities and customers.

References

1 Nadeem Javaid, Adnan Ahmed, Sohail Iqbal, and Mahmood Ashraf. "Day ahead real time pricing and critical peak pricing-based power scheduling for smart homes with different duty cycles." *Energies* 11, no. 6 (2018): 1464.

2 Houda Daki, Asmaa El Hannani, Abdelhak Aqqal, Abdelfattah Haidine, and Aziz Dahbi. "Big Data management in smart grid: concepts, requirements and implementation." *Journal of Big Data* 4, no. 1 (2017): 1–19.

3 Hussein Jumma Jabir, Jiashen Teh, Dahaman Ishak, and Hamza Abunima. "Impacts of demand-side management on electrical power systems: a review." *Energies* 11, no. 5 (2018): 1050.

4 US Department of Energy (2006). Benefits of demand response in electricity markets and recommendations for achieving them. *Tech. Rep. US Dept. Energy.*

5 Xiaoxing Lu, Kangping Li, Hanchen Xu, Fei Wang, Zhenyu Zhou, and Yagang Zhang. "Fundamentals and business model for resource aggregator of demand response in electricity markets." *Energy* (2020): 204 117885.

6 Pierre Pinson, and Henrik Madsen. "Benefits and challenges of electrical demand response: a critical review." *Renewable and Sustainable Energy Reviews* 39 (2014): 686–699.

7 Shafinejad, M.M., Saharkhiz, K.E. and Hedayati, S. (2016). Implement demand response various programs, operating reserve for Guilan Province's industrial customers and economic assessment of the project. *24th Iranian Conference on Electrical Engineering*, Shiraz, Iran (10–12 May 2016). IEEE.

8 Rockey Mountain Boulder (2006). Colorado for Southwest Energy Efficiency Project. Demand Response: An Introduction Overview of Programs, technologies, and lessons learned. http://large.stanford.edu/courses/2014/ph240/lin2/docs/2440_doc_1.pdf (accessed 15 February 2021).

9 Goldman, C. (2010). Coordination of energy efficiency and demand response. https://www.epa.gov/sites/production/files/2015-08/documents/ee_and_dr.pdf (accessed 15 Februaru 2021).

10 Adrian Bejan, Peter Vadász, and Detlev G. Kröger, eds. *Energy and the Environment*. Vol. 15. Springer Science & Business Media, 2012.

11 Valkering, P., Laes, E., Kessels, K. et al. (2014). How to engage end-users in smart energy behaviour?. *EPJ Web of Conferences*, Budapest, Hungary (27–30 October 2013). EDP Sciences.

12 Josue Campos do Prado, Wei Qiao, Liyan Qu, and Julio Romero Agüero. "The next-generation retail electricity market in the context of distributed energy resources: vision and integrating framework." *Energies* 12, no. 3 (2019): 491.

13 Kris Kessels, Carolien Kraan, Ludwig Karg, Simone Maggiore, Pieter Valkering, and Erik Laes. "Fostering residential demand response through dynamic pricing schemes: a behavioural review of smart grid pilots in Europe." *Sustainability* 8, no. 9 (2016): 929.

14 Ni, Zhen, and Avijit Das. "A new incentive-based optimization scheme for residential community with financial trade-offs." *IEEE Access* 6 (2018): 57802–57813.

15 Venkat Durvasulu, and Timothy M. Hansen. "Benefits of a demand response exchange participating in existing bulk-power markets." *Energies* 11, no. 12 (2018): 3361.

16 Brandon Davito, Humayun Tai, and Robert Uhlaner. "The smart grid and the promise of demand-side management." *McKinsey on Smart Grid* 3 (2010): 8–44.

17 Ahsan Raza Khan, Anzar Mahmood, Awais Safdar, Zafar A. Khan, and Naveed Ahmed Khan. "Load forecasting, dynamic pricing and DSM in smart grid: a review." *Renewable and Sustainable Energy Reviews* 54 (2016): 1311–1322.

18 Lokeshgupta, B., Sadhukhan, A. and Sivasubramani, S. (2017). Multi-objective optimization for demand side management in a smart grid environment. *7th International Conference on Power Systems*, Toronto, Canada (1–3 November 2017). IEEE.

19 Ankit Kumar Sharma, and Akash Saxena. "A demand side management control strategy using Whale optimization algorithm." *SN Applied Sciences* 1, no. 8 (2019): 870.

20 Peter Warren, "Demand-side policy: global evidence base and implementation patterns." *Energy & Environment* 29, no. 5 (2018): 706–731.

21 Chew, B., Feldman, B., Ghosh, D. and Surampudy, M. (2018). 2018 Utility Demand Response Market Snapshot. Smart Electric Power Alliance.

22 Elma, O. and Selamoğullari, U.S. (2017). An overview of demand response applications under smart grid concept. *4th international conference on electrical and electronic engineering*, Ankara, Turkey (8–10 April 2017). IEEE.

23 Economic Regulation Authority (2014). Inquiry into Microeconomic Reform in Western Australia: Final Report. https://www.parliament.wa.gov.au/publications/tabledpapers.nsf/displaypaper/3911797aff57f87c8fab25af48257d23001716d4/$file/1797.pdf (accessed 15 February 2021).

24 Jun Dong, Guiyuan Xue, and Rong Li. "Demand response in China: regulations, pilot projects and recommendations – a review." *Renewable and Sustainable Energy Reviews* 59 (2016): 13–27.

25 Zgajewski, T. (2015). Smart electricity grids: A very slow deployment in the EU. Egmont Paper No. 74. http://aei.pitt.edu/63582/1/74.pdf (accessed 15 February 2021).

26 Feldman, B. and Lockhart, B. (2014). *Automated Demand Response*. Navigant Research.

27 Hale, E.T., Bird, L.A., Padmanabhan, R. and Volpi, C.M. (2018). Potential roles for demand response in high-growth electric systems with increasing shares of renewable generation. https://www.nrel.gov/docs/fy19osti/70630.pdf (accessed 15 February 2021).

12

Business Models for the Smart Grid

The shift toward a sustainable global energy system will substantially depend on renewable energy sources and decentralized energy systems, which needs a profound reorganization of the energy sector. The traditional energy utility business model is dealing with a lot of stress because of this, and energy services are assumed to play a vital role for the energy shift. Thus, business model innovation is the main driver for the successful execution of the energy transition; the energy utilities transform themselves from commodity suppliers to service providers. This shift from a product-oriented, capital-intensive business model predicated upon tangible assets, in the direction of a service-oriented, expense-intensive business model that depends on intangible assets could potentially result in more managerial and organizational problems.

This chapter will shed light on the business model concept, its main components and how it can be used to analyze the impact of smart grid (SG) technology to create, deliver and capture a value for utility businesses. Then, the value chain for both traditional and smart energy industries is discussed. After that, different electricity markets are described. This will be followed by a review for previously proposed SG business models with their future levers. Finally, the chapter will highlight the potential of applying the blockchain technology in the electricity market.

12.1 The Business Model Concept

The business model concept is based on two theories of management: the resource-based view and transaction cost economics. The primary logic of business models is to find a new mixture of a firm's resources and abilities, then identify the most efficient method of organizing the transactions between these resources and abilities to produce value for the firm and its consumers. A business model can be understood as a structural guide that defines a firm's organizational and financial architecture. Also, it can be defined as the tool by which companies send value to consumers, enable consumers to pay for value and translate those payments into profit. A business model depends on four fundamental

Smart Grid and Enabling Technologies, First Edition. Shady S. Refaat, Omar Ellabban, Sertac Bayhan, Haitham Abu-Rub, Frede Blaabjerg, and Miroslav M. Begovic.
© 2021 John Wiley & Sons Ltd. Published 2021 by John Wiley & Sons Ltd.
Companion website: www.wiley.com/go/ellabban/smartgrid

aspects: value proposition, customer interface, infrastructure, and revenue model. These elements can be described by, [1, 2]:

- *Value proposition*: denotes the bundle of products and services that amount to a certain value with regard to the consumer and permits the company to receive revenue;
- *Customer interface*: entails the total interaction with the consumer. It includes consumer relationship, consumer segments, and distribution channels;
- *Infrastructure*: depicts the architecture of the company's value creation process. It entails assets, know-how, and partnerships;
- *Revenue model*: describes the relationship between costs to generate the value proposition and the revenues that are produced by delivering the value proposition to customers.

As illustrated in Figure 12.1, the four conceptualization elements of the business model are collaborating with one another, to ensure a full level of interaction. Furthermore, the notion of a business model could be utilized as a way to order and create categories or blueprints that will aid in the understanding of the business phenomena. Also, it helps managers to design, implement, operate, change and control their own businesses. In addition, business models open up fresh opportunities for new technologies to play a role in the markets and create value for them.

The business model framework, as illustrated in Figure 12.2, is used to study the influence of SG on electricity firms' business models. This framework distinguishes between three components: value creation, value delivery, and value capture. The value creation comprises the value proposition and customer relationships and is responsible for what consumers really require and if firms are capable of developing a proposition that meets this requirement. The value delivery entails the allocation of resources and capabilities to

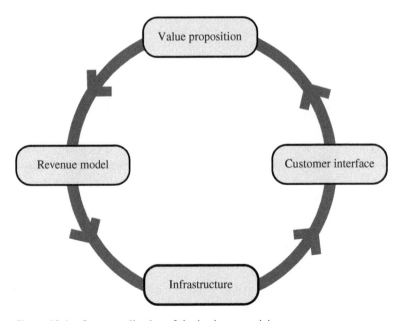

Figure 12.1   Conceptualization of the business model canvas.

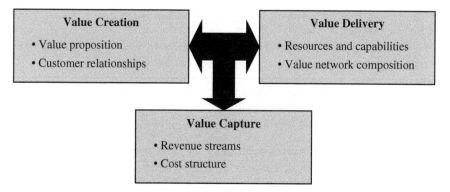

Figure 12.2 The business model framework used to understand the impact of SGs.

find a solution for organizing the company's delivery of the value to its consumers. The value capture refers to new revenue streams and the cost structure and takes into account the financial base of the business model and the way the firm makes customers pay for the cost generated and how profit is generated. Clearly, there are interdependencies among all components involved; signifying a novel technique for value capture cannot be achieved without considering the changes in value generation and transfer components, [3].

SG technology could be disturbing to power firms, because there could be substantial variations in the organization of the electricity market. The two-sided transmission of power and data has the potential to change the value chain of power generation, transmission, distribution, and supply. Reliance among distinct actors of the network could increase and consumers will require additional sources of value than electricity alone. Hence, SG has an impact on the business model innovation of these firms. Table 12.1 highlights the core influence of SGs on electricity firms' business model innovation, [3].

12.2 The Electricity Value Chain

To gain a full understanding of the current electricity industry, the entire industry value chain should be considered in any assessment. The traditional electricity value chain has within it the generation, transmission, distribution, retail and consumption guide from energy source to end use, as illustrated in Figure 12.1. Energy and information flow in a single direction, and all except the largest of consumers play an inactive part. The electricity value chain components are explained as [4–6]:

- Production of electricity implies the conversion of primary energy resources into electric power. The biggest share of electricity in industrialized countries is produced in large-scale power plants that depend on fossil fuels and nuclear energy and the production assets are mostly owned by a small group of utilities.
- Transmission includes the transport of electricity at high voltages across a long distance by the transmission grid. The transmission system operator (TSO) is in charge of balancing the power supply and demand in the respective region of operation. Considering the many TSOs taking part in the market, the transmission grid is normally a natural monopoly in its region.

Table 12.1 The core influence of SGs on electricity firms' business model innovation. Ref [3]. Reproduced with permission from ELSEVIER.

	Enabling impact	Constraining impact
Value creation	• Novel value suggestions predicated upon renewable energy and energy services; • Increase in demand for greater power quality; • Consumers' demand response supports network balancing.	• Empowering customers control and reducing electricity usage; • Customers generate their own electricity and become prosumers; • Customer engagement does not materialize.
Value delivery	• Enhancing optimization of the electricity network; • Enhancing marketing with the association with real-time data on electricity usage; • Leverage assets of specialized ICT or energy service providers.	• The value of conventional power plants erodes; • The risk that new entrants become competitors; • Increased complexity of the value network requires new capabilities.
Value capture	• New revenue streams with use of services and big data; • Opportunity to become a central actor in a multi-sided market; • Decreased costs from less grid maintenance and load shift.	• Reduced revenues from selling electricity; • Increased cost from investments in SG infrastructure; • Uncertainty about potential changes in revenues and costs.

- Distribution networks are built to distribute electrical power to the end consumers at low voltage level. The distribution network normally has only a few connection points to the transmission grid and other distributions networks are merely connected between each other by the transmission grid. The distribution network operator (DSO) is responsible for all aspects regarding the connection of end users to the grid.
- Retail is an administrative assignment that entails communication with the end consumer. Retailers buy power from producers, traders, or exchange and sell it to end consumers. Retail deals with the purchase of electricity, metering and billing.
- Consumption of energy is measured on the consumer-end of the meter.

In the case of the power industry, the traditional value model entails consumers getting reliable and universal power at acceptable rates, for which they deliver providers reciprocal value in the form of intermittent (usually monthly) revenue. Customers vastly interact with their utility when they switch on their lights and when they look at their utility bill. Customers usually have a very limited understanding of how they receive the energy generated (Figure 12.3).

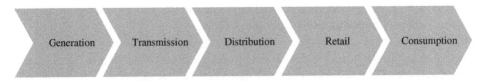

Figure 12.3 The traditional electricity value chain.

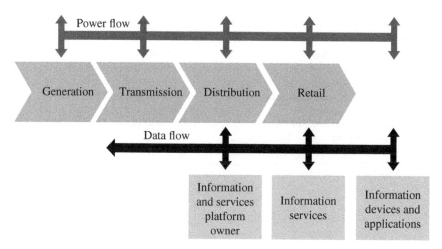

Figure 12.4 The emerging electricity value chain with both power and information flow.

The employment of SG technologies will increase complications to the network, transferring power and information in several directions and allowing a host of new participants and business models. Distributed energy resources (DER) including consumer-owned renewable production, plug-in electric vehicles, and energy storage will expand the value chain to consider assets operated at a close proximity to the end user. The end users, who have the potential to demonstrate a small combination of demand response, power or energy storage to the system, will additionally be a crucial aspect of the new value chain. All of these changes will lead to the emerging electricity value chain illustrated in Figure 12.4. Thus, utilities will repeatedly own consumers and their consumption, while new services and products will allow them to efficiently help consumers by changing and implementing new business abilities into the expanded value chain. Table 12.2 highlights the difference between the traditional and the forward thinking (smart) utilities, [7].

12.3 Electricity Markets

Normally, the power market is organized into financial and physical power markets as illustrated in Figure 12.5. Electricity markets possess retail and wholesale components. Retail markets consider the sales of electricity to customers; wholesale markets usually entail the sales of electricity between electric utilities and electricity traders before it is sold to customers. A big part of the wholesale market and specific retail markets are competitive, with prices established competitively. Where rates are chosen depending on the costs, market fundamentals should be considered here due to the fluctuation in supply and demand that will influence customers by affecting the price and reliability of electricity. The key supply aspects that influence costs entail fuel costs, capital costs, transmission capacity and restrictions and the functioning behavior of power plants. Demand (in terms of sharp changes and extremely high levels) also influence costs,

Table 12.2 Traditional and the smart utilities. Ref [7]. Reproduced with permission from Cognizant LLC.

Factor	Traditional utility	Smart utility
Business model	Easy, depends on steadily rising electricity sales, usually from an expanding asset base of centralized generation and common delivery infrastructure.	Complicated, integrated energy services helping different and developing consumer needs with an information-qualified infrastructure.
Services	Regimented communized services.	Spread, innovative service provider.
Electricity demand	Increasing.	Flattening with a potential decrease, except the implementation of new electric vehicles.
Capacity cost	Average price of new capacity steady or decreasing.	Average cost of new capacity (renewable) is greater than the traditional sources until achieving commercial maturity.
Utility objectives	Reliability, consumer service, inexpensive (decreasing costs), returns to shareholders.	Reliability, environmental quality, service quality, inexpensive (decreasing costs), returns to shareholders.
Customer's role	Passive	Active, possesses technology and incentives to control energy consumption and produce energy.

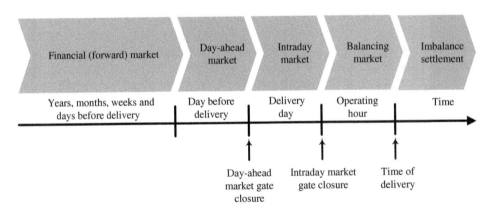

Figure 12.5 Structure of electricity trading for different time horizons.

particularly if less-efficient, more-expensive power plants should be switched on to serve load. The different markets can be defined as [8, 9]:

- *The day-ahead market* (the largest physical power market) is operated as a "spot market": sellers and buyers submit selling and buying bids to a market broker who, in turn, determines a price based on a set of rules that are known to everybody, bidding closes

12–36 hours before the delivery hour. Once the price is set, it is communicated to all players in the market, as well as to which extent their bids have or have not been accepted by the market. As consumption of electricity strongly varies through the day, the market price also varies. A price is typically set for pre-defined intervals (15 minutes to 1 hour) the day before the physical delivery of the electricity.

- *Intraday market* (in between market) is a trading market where modifications to trades made in the day-ahead market are normally done until 60 minutes before delivery. The market opens after the day-ahead market is closed.
- *Balancing markets*, electricity markets should consider a suitable mechanism to ensure a real-time equilibrium between consumption and production of power. The challenge of balancing (regulating power) is normally the task of TSOs. Hence, TSOs should be given suitable tools to handle rapid deviations from day-ahead (or hour-ahead) trading-based schedules.
- *The retail market* for electricity thus works in the same way as any other retail market. A large part of what is called the "deregulation" of the electricity industry has to do with ensuring the retail market is as open as possible. The retail electricity market allows end-use consumers to select the "energy choice," i.e., to select their supplier from competing retailers. Electricity retailers typically offer their customers a fixed price over a longer period, ranging from one month up to a few years. With monthly prices, the costs of electricity for the consumer follow the seasonal and long-term trends in electricity price on the wholesale market.

Both the wholesale and retail market are relatively reliant on each other as the wholesale price is the main part of the retail price of electricity. The reorganization of the electricity industry is assumed to initiate a more competitive wholesale market that will probably lead to the decline of wholesale prices. This will permit electricity retailers to establish decreased retail costs for their product and still remain financially feasible. Furthermore, SGs may penetrate distributed resources including DER and demand control resources into the industry-wide power system, introducing alterations to the decision-making frame of the present power industry. In the SG arena, there must be established communication among markets' domain and other entities in a SG. In addition, the interdependences among the market and generation entities are crucial due to the fact that matching generation with consumption depends on markets, as illustrated in Figure 12.6. Market participants are trading in wholesale and retail markets. Production companies take part in the wholesale market by selling large-scale amounts of energy. Retailers are companies that deliver power to end consumers via the distribution network and in the future, they could aggregate generation capacities of DER and make them available in the market [10]. A high percentage of retailers become involved in a trading organization to allow participation in the wholesale market. The emergence of many new market players will need regulatory alterations among regulators and distribution network service providers to ensure a fair and a transparent network access. Moreover, technical and commercial arrangements of network inter-connection are to transcend to effectively distribute suitable authority and responsibility [11, 12].

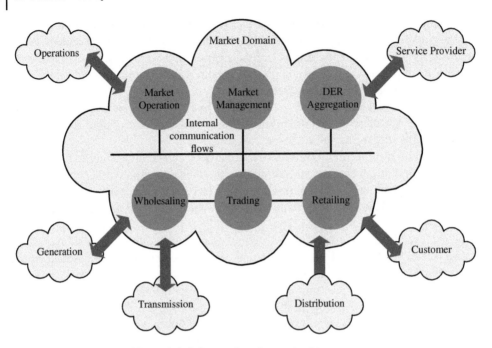

Figure 12.6 Market entities and their interactions in smart grid arena.

12.4 Review of the Previous Proposed Smart Grid Business Models

Advancing a new business model (BM) needs a deep comprehension of basic consumer needs, why competitors did not manage to meet those needs, as well as the technological and organizational trajectories. Also, the procedure of learning and modifying the BM possesses equal value to the designing aspect. Moreover, knowing the consumers and competitors' behavior varies from initial conjectures and makes employing a new BM move at a rapid pace. This section presents a review of different business models which have previously been proposed for the SG.

12.4.1 Timing-Based Business Model

The study in [13] examines the application of timing-based business models to increase the flexibility in power systems and it also shows how new market entrants and incumbent companies create and capture value by offering different flexibility along the whole grid (from supply to demand). These timing-based business models unite timing (as a core capability) with complementary supply and demand-based valuables. The linkage between timing as a core capability and the complementary valuable has consequences for energy policy to deliver the regulatory frame for such "coupled-service" business models and is a new challenge for energy policy. Four different business models have been suggested by uniting timing with: assets, e.g. the timing of large-scale power plants (power-plant optimization); a portfolio of small-scale distributed power plants (virtual

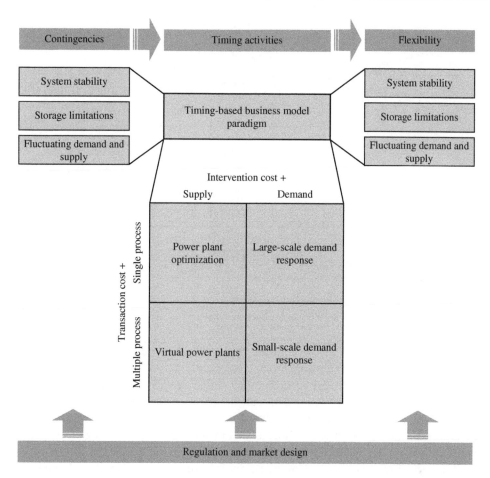

Figure 12.7 The timing-based business model and its created value for the SG [13]. Reproduced with permission from ELSEVIER.

power plant); controllable consumer behavior, e.g. large-scale and distributed, small-scale demand-response, as illustrated in Figure 12.7 [13].

12.4.2 Business Intelligence Model

Electricity utilities require a real-time business intelligence system to employ several data sources (smart meters and other SG devices,), extract and display Key Performance Indicators (KPIs) (using data mining), leverage existing investments, enhance scalability and security, and preserve resources and costs. As power companies evolve to implement and operate SGs, this will lead to variations in companies' information systems and also increase the need for real-time analytics. SG data analytics is the procedure carried out for collecting, aggregating, inspecting, cleaning, interpreting, visualizing, and modeling SGs, entailing data mining, which is responsible for modeling and knowledge discovery for predictive analytical reasons. SG analytics delivers information on the effectiveness of the energy ecosystem with the objective to sidestep power failures, enable decision making and increase revenue, [14].

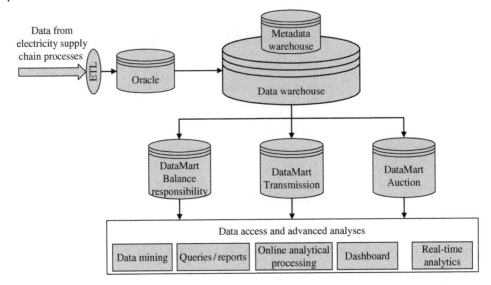

Figure 12.8 The business intelligence framework which can be applied to the SG. Reproduced with permission from Springer Nature.

A vast number of data can be produced from metering, monitoring and other SG procedures. The collected heterogeneous data is stored in a data warehouse after extract, transform and load (ETL) processes through, e.g. the Oracle (Object-Relational Database Management System). Also, these data can be accessed for different grid analytics through DataMart, (a DataMart is the access layer of the data warehouse environment). The problem is to form a sufficient dimensional model for taking note of data concerning the advancement and administration of the electricity market. Moreover, the formed data infrastructure aids in gathering disparate data that stems from several vendors' devices. The business intelligence framework, as illustrated in Figure 12.8, strives to improve the effectiveness and reliability of present procedures by enhance data and supply chain data flows, [15].

12.4.3 Business Models for Renewable Energy

Free and local availability, cleanliness, eco-friendliness, and sustainability of renewable energy resources (RERs) made some economists and policy makers acknowledge that achieving sustainable growth is the maximal consumption of RERs. Since the RE industry provides a profitable future with regard to security and rural area development aspects, a potential exists for private sector investments. From a private investor's standpoint, the RE industry is an entrepreneurial industry with technological, political and market uncertainties in which conventional evaluation of investment and market analysis are not easy. Investigations show that the main concerns of investors to begin RE investment are market and business aspects. Hence, researching dominant business structure in the RE industry is among the most crucial fields of study for diffusion of RE development [16–18].

There is research that addresses the RE business model from two standpoints: customer and utility. Two generic business models are mentioned in the literature: Customer-side renewable energy business models and utility-side renewable energy business models, both can be summarized as the following, [19]:

- Customer-side renewable energy business model, the renewable energy systems are located at the premises of the customer. Present technologies are photovoltaic (PV), solar thermal hot water, micro-combined heat and power systems (micro-CHP), geothermal heat pumps, and micro wind turbines. The size of the systems is normally between a few kilowatts and about 1 MW. The value proposition delivered by the utility are from a simple consulting services to a full-services package considering financing, ownership and operation of the asset.
- Utility-side renewable energy business model, where the projects are more superior than consumer-side projects ranging from 1 MW to few hundred MW. Typical technologies are onshore and offshore wind farms, large scale PV projects, biomass power plants, and solar thermal power plants. The value proposition in this business model is bulk generation of electrical power transferred to the grid. The electrical power is fed into the grid and transmitted to the customer by the traditional electricity value chain.

The different characteristics of the two business models are summarized in Table 12.3 [2]. Utility-side renewable energy business models are present, but consumer-side renewable energy business models are at an early stage of development, [2]. Utilizing the business model concept as an analytical tool to compare the implementation of customer-sited PV systems in three different countries: Germany, Japan and the United States has been discussed in [20]. Results show that the PV business models depend on the different national contextual environment, that includes the policy framework, transaction prices, the electricity market, the building sector and customer-associated factors. Moreover,

Table 12.3 Utility-side vs. customer-side business model. Ref [2]. Reproduces with permission from ELSEVIER.

	Utility-side business model	Customer-side business model
Value proposition	Bulk generation of electricity transferred to the grid	Customized solutions and energy related services.
Customer Interface	• Electricity as commodity; • Customer pays per unit.	• Consumers take part in energy production by hosting the production system and sharing advantages with the utility; • Long-term customer relationship.
Infrastructure	Low number of large-scale assets Centralized generation.	• Large number of small-scale assets; • Generation close to point of consumption.
Revenue Model	• Revenues through feed-in of electricity; • Economies of scale from large projects and project portfolios.	• Revenue from direct use, feed-in and/or from services; • High transaction costs.

policy makers and business managers are supposed to study the contextual environment to come up with the best business models and enhance their design [21].

12.4.4 Service-Oriented Business Models

Product-service system (PSS) is an amalgamation of products and services that are employed to fulfill certain demands of consumers. In the power utility domain, the Product-Service System has been discussed as a concept associated with energy efficiency. Reference [22] outlines three major service-oriented business models: Customer-owned product centered BMs, where the consumer possesses the product (renewable technologies or energy management tools) associated with power production or management; Third-party service centered BMs, where a third party delivers energy services to the consumer; and finally, energy community BMs, where resources are pooled and shared (sharing resources) among the community members. Value seized by offering solutions to complicated problems such as electricity peaks, or financial problems such as up-front price of renewable energy systems, including the expansion of renewable energy access and ensuring social acceptance of renewable energy diffusion.

12.4.5 Prosumer Business Models

When SG end users become prosumers, they evolve to become the most essential value makers within the SG. Prosumers are assumed to become the main actor in the advancement of the SG and they have the greatest services-based business creation potential in the energy market value chain. However, the improved characteristics of the end user side come at the expense of needing more technology to enable their usability, which includes advanced metering infrastructure, energy management systems, energy storage, distributed generation, and advanced communication networks infrastructure. When implementing the business model concept to the prosumer-centered energy market value chain, new business prospects are prone to spring up. Reference [4] discussed four prosumer-oriented business models and the characteristics of these models are summarized in Table 12.4 [5].

12.4.6 Integrated Energy Services Business Model

The possibility of implementing locally available Distributed Generation (DG) with Demand Response (DR) to provide flexibility to the grid is still unexploited in most electricity markets. A novel business model of Integrated Energy Services (IES) that recognizes these possibilities has been proposed in [23]. The model integrates DG and DR for industrial entities, with possible advantages for these companies and for energy suppliers and system operators, as illustrated in Figure 12.9. In the development of IES across Europe, three countries have been considered (Italy, the UK, and Spain) and have been investigated and the following factors have been highlighted as the main drivers for implementing the IES across Europe: rising electricity costs, flexibility services for the network, increasing experience with interruptible and direct load-control programs, and incentive schemes for combined heat and power (CHP) systems.

Table 12.4 Different prosumer-oriented business model characteristics. Adapted from Ref [5].

Business Model	Value Proposition	Costumer Interface	Infrastructure	Revenue Model
Energy service companies' prosumer-oriented business model	• Enhanced energy efficiency; • Decreased energy costs; • Energy performance contraction.	• Prosumer interactions management; • Prosumer segmentation; • Real-time media- or web-based communications.	• SG data management; • Grid monitoring.	• Energy savings; • Energy efficiency improvements; • Charge for performance/ service level offered.
Virtual Power Plant prosumer-oriented business model	• Flexibility of energy generation; • Granting a prosumer market access.	• Advanced systems for energy management.	• Distributed generation systems; • Electricity storage devices.	• Electricity sale • Energy consumption, production, storage based on real-time energy pricing.
Aggregator/ Retailer prosumer-oriented business model	• Operate and optimize energy consumption made by prosumers; • Demand response; • Flexible electricity tariffs according to momentary market conditions.	• Prosumers community; • Prosumer relationship management; • Automatic energy price information.	• Advanced metering infrastructure; • Automatic metering services (AMS).	• Real-time and critical peak pricing (RTP); • Time of use pricing (ToU).
Distributed system operator prosumer-oriented business model	• Security of supply and quality of service; • Choice of energy source; • System flexibility services; • Market facilitation.	• Active demand program; • Real-time media- or web-based communications; • In-home displays.	• Grid connection; • Smart metering systems; Local network services.	• Energy selling; • Static pricing; • Provision of connection services; Transmission/ distribution fees.

12.4.7 Future Business Model Levers

Utilities may select from a range of directions to move forward from where they are today. Multiple models could require to be employed to achieve diverse market needs or certain regulatory structures in many parts of the globe, or even jurisdictions. In any case, utilities

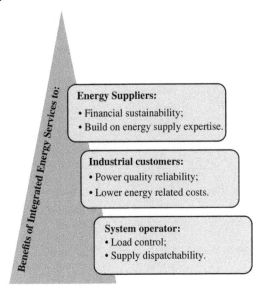

Figure 12.9 Benefits of integrated energy services.

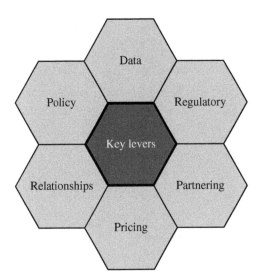

Figure 12.10 Future business model levers for the SG. Ref [24]. Used with permission from PriceWaterhouseCoopers - PWC.

will need to understand the way to leverage their present business position and the external market to improve their future competitive positioning. Companies possess many levers that could be implemented to improve their preparedness for the future and position themselves for prosperity, these levels as illustrated in Figure 12.10 are, [24]:

- Data, ownership of grid and consumer information can offer a knowledge platform that improves the value proposition.
- Policy, mandates and legislation are capable of forming the competitive field and improve the incumbent's opportunity to take part.
- Relationships: The usual interfaces of the incumbent with consumers, suppliers, stakeholders and other providers form a common platform.

- Pricing, changes in usual price-recognition models can offer higher chances for price recovery and extra flexibility to rate design.
- Partnering, improved creative and delivery abilities can expand the incumbent's present platforms and market reach.
- Regulatory, decisions can offer offensive (exploit) and defensive (preserve) methods to improve market involvement.

Companies can use these levers to improve their starting point for determining their future roles and for future market participation. Furthermore, to capture the SG business opportunity, SG players should deepen their understanding of where the value is in the developing SG and they should come up with a convincing model to purse this value.

12.5 Blockchain-Based Electricity Market

Blockchain is a type of distributed, electronic database (ledger) that can grasp any SG information and put the way on how the information should be updated. So, it is a secure, inexpensive, instantaneous and resilient mechanism for peer-to-peer (P2P) transaction platforms among unlimited users (or possibly devices) that uses decentralized storage to record all transaction data, with a resulting immutable transaction record. Its applications are projected to improve numerous industries, including P2P electricity trading. Blockchain technology improves the market role of customers and producers. It allows prosumers to buy and sell energy directly, with a high level of autonomy. Blockchain can establish coordination among local power generators and customers in their vicinity, providing distributed, real-time power markets. Also, it could establish the foundation for metering, billing, and clearing processes. Blockchain technology could radically change energy as we know it, by beginning with individual sectors first but eventually changing the energy market. Blockchain technology demonstrates several advancements, so the energy sector uses the opportunity of the blockchain to digitize and streamline administration of current procedures implemented and utilize the technology to offer new products and services where transparency, trust, and simplicity gives value. Blockchain addresses complex issues for the future digital energy system, such as, [25]:

- Interoperability, gives a common standard for data format, business processes, and communication protocols;
- Parallel products, executing software on the blockchain implies a common environment for products;
- Features, anyone can use; always on; immutability; real time; mass data; payment as a by-product;
- Trust, shared truth through 51% rule of all nodes in the system on public blockchain;
- Security, within the system all data is encrypted.

The electricity market can be reshaped based on Blockchain, which can be explained by the following scenario. This scenario entails two electricity producers and one electricity consumer which trade among each other regarding a blockchain. The producers publish exchange proposals of energy (in kWh) for currency (in USD) in a data stream, which works as a publishing board. The customers study the proposals and tries to determine its

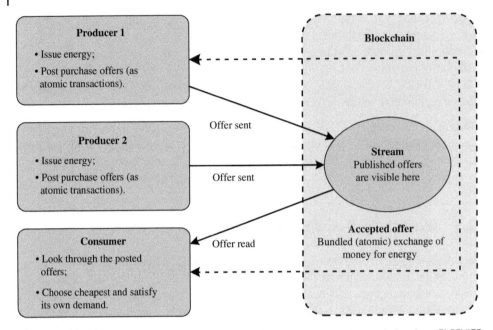

Figure 12.11 Blockchain based electricity market [25]. Reproduced with permission from ELSEVIER.

energy need at a low price. When a proposal is accepted, it is implemented as an atomic exchange (i.e. two simultaneous transactions are done and both should succeed or fail together). A typical trade proceeds as illustrated in Figure 12.11, [25]:

1) The producer nodes make and publish exchange proposals of kWh for USD in the stream "elec-market-open." The preparations need producers to lock enough energy assets and encode information about the exchange;
2) The customer node seeks the proposals regarding every publisher and decodes them;
3) The customer compares the offers and selects the one which minimizes energy cost;
4) The customer formulates a transaction matching the selected offer through the locking of sufficient funds and joining the selected offer with payment details. It then encodes this and submits the approved exchange to the chain;
5) Lastly, the customer makes sure that the payment was accepted by the chain.

12.6 Conclusion

Previously, operating an integrated utility from production through to consumer supply was easily understood as the utility managed the whole value chain. Presently, utilities are subjected to a unique alteration in their operating environment, which needs a wide remaking of business models. The utility business model is currently questioned by the rapid employment of distributed renewable energy sources, moreover to, decrease electricity demand, promote smarter distribution and regulation, and more motivated consumers and competitors. Utilities require to invest their efforts on distributed and integrative business models and increase consumer- and service-oriented methods.

References

1 Richter, M. (2013). German utilities and distributed PV: how to overcome barriers to business model innovation. *Renewable Energy* 55: 456–466.

2 Richter, M. (2013). Business model innovation for sustainable energy: German utilities and renewable energy. *Energy Policy* 62: 1226–1237.

3 Shomali, A. and Pinkse, J. (2016). The consequences of smart grids for the business model of electricity firms. *Journal of Cleaner Production* 112: 3830–3384.

4 Boons, F. and Lüdeke-Freund, F. (2013). Business models for sustainable innovation: state-of-the-art and steps towards a research agenda. *Journal of Cleaner Production* 45: 9–19.

5 Rodríguez-Molina, J., Martínez-Núñez, M., Martínez, J.-F., and Pérez-Aguiar, W. (2014). Business models in the smart grid: challenges, opportunities and proposals for prosumer profitability. *Energies* 7: 6142–6171.

6 Richter, M. (2012). Utilities' business models for renewable energy: a review. *Renewable and Sustainable Energy Reviews* 16 (5): 2483–2493.

7 Nair, S. (2013). Building a Thriving and Extended Utilities Value Chain. https://www.cognizant.com/InsightsWhitepapers/Building-a-Thriving-and-Extended-Utilities-Value-Chain.pdf (accessed 15 February 2021).

8 Kim, S., Hur, S., and Chae, Y. (2010). Smart grid and its implications for electricity market design. *Journal of Electrical Engineering & Technology* 5 (1): 1–7.

9 Knežević, G., Fekete, K. and Nikolovski, S. (2010). Applying agent-based modeling to Electricity Market simulation. *33rd International Convention MIPRO*, Opatija, Croatia (24–28 May 2010). MIPRO.

10 Babic, J. and Podobnik, V. (2016). A review of agent-based modelling of electricity markets in future energy eco-systems. *International Multidisciplinary Conference on Computer and Energy Science*, Split, Croatia (13–15 July 2016). IEEE.

11 Bollen, M.H.J. (2011). *The Smart Grid: Adapting the Power System to New Challenges*. Morgan & Claypool.

12 Bohi, D.R. and Palmer, K.L. (1996). The efficiency of wholesale vs. retail competition in electricity. *The Electricity Journal* 9 (8): 12–20.

13 Helms, T., Loock, M., and Bohnsack, R. (2016). Timing-based business models for flexibility creation in the electric power sector. *Energy Policy* 92: 348–358.

14 Lukić, J., Radenković, M., Despotović-Zrakić, M. et al. (2016). A hybrid approach to building a multi-dimensional business intelligence system for electricity grid operators. *Utilities Policy* 41: 95–106.

15 Lukić, J., Radenković, M., Despotović-Zrakić, M. 3 et al. (2017). Supply chain intelligence for electricity markets: a smart grid perspective. *Information Systems Frontiers* 10: 91–107.

16 Aslani, A. (2014). Private sector investment in renewable energy utilisation: strategic analysis of stakeholder perspectives in developing countries. *International Journal of Sustainable Energy* 33 (1): 112–124.

17 Schleicher-Tappeser, R. (2012). How renewables will change electricity markets in the next five years. *Energy Policy* 48: 64–75.

18 Aslani, A. and Mohaghar, A. (2013). Business structure in renewable energy industry: key areas. *Renewable and Sustainable Energy Reviews* 27: 569–575.

19 Mario Richter, "Utilities' Business Models for Renewable Energy: Evidence from Germany", World Renewable Energy Congress 2011, pp. 2385–2392, 8–13 May 2011, Sweden.

20 Strupeit, L. and Palm, A. (2016). Overcoming barriers to renewable energy diffusion: business models for customer-sited solar photovoltaics in Japan, Germany and the United States. *Journal of Cleaner Production* 123: 124–136.

21 Tantau, A. and Staiger, R. (2020). Evolving business models in the renewable energy. In: *Sustainable Business: Concepts, Methodologies, Tools, and Applications* (ed. Information Resources Management Association), 395–413. IGI Global https://books.google.com.qa/books?id=kJvLDwAAQBAJ&pg=PP1&hl=ar&source=gbs_selected_pages&cad=2#v=onepage&q&f=false.

22 Hamwi, M. and Lizarralde, I. (2017). A review of business models towards service-oriented electricity systems. *Procedia College International pour la Recherche en Productique* 64: 109–114.

23 Gaspari, M., Lorenzoni, A., Frías, P., and Reneses, J. (2017). Integrated energy services for the industrial sector: an innovative model for sustainable electricity supply. *Utilities Policy* 45: 118–127.

24 PWC (2017). The business models of the future – PwC. https://www.pwc.com/gx/en/utilities/publications/assets/pwc-future-utility-business-models.pdf (accessed 15 February 2021).

25 Sikorski, J.J., Haughton, J., and Kraft, M. (2017). Blockchain technology in the chemical industry: machine-to-machine electricity market. *Applied Energy* 195: 234–246.

13

Smart Grid Customers' Acceptance and Engagement

Insofar as societies rely on energy and its management, radical modification in energy-related technologies, such as the Smart Grid (SG), are expected to come in huge forms of social disruption. As one might expect, there are numerous claims regarding the risks that lie ahead as traditional energy resources become insufficient, as populations rise, and as expectations soar. Consequently, future societies will experience innovations and big transformations to enhance their energy consumption. The SG can employ different sets of electricity resources, which entails big power plants and distributed renewable energy resources, electric energy storage, demand response (DR), and electric vehicles (EVs). According to numerous visions for the SG, customers will come to play a vital role in the energy systems of tomorrow. To successfully build a SG, it is necessary to ensure customers' acceptance. The ultimate employment of the SG is reliant on the end users' approval of SG products and services. However, motivating the residential space in the SG is still a challenge. This chapter aims to give energy systems researchers and decision makers with a good insight into the fundamental drivers of customer acceptance of the SG and the logical steps for their engagement to apply the SG technology and make it feasible in a timely manner. The content of this chapter is based on the authors' publication [1].

13.1   Introduction

Electric power system (EPS) is recognized as the main artificial system that encompasses the most area. Yet, it is an extremely weak system as it could collapse easily in the event of big disturbance, disaster, etc. The traditional EPS is built and utilized for one-way power transmission, from centralized, large-scale, and fossil fuel-based power stations, to the customers, as illustrated in Figure 13.1. Moreover, such centrally produced electric power could not be stored, leaving it as the ultimate "just-in-time" commodity. Limited supply and demand balancing could have dire implications on the power system, including blackouts or large system collapse [2]. Furthermore, the rising growth in demand as population increases with changes of living standards and quality, mean the world's electric power system faces many challenges and issues. This entails aging infrastructure, implementation

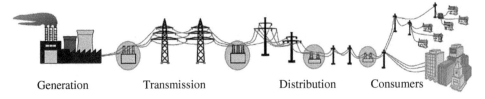

Generation   Transmission   Distribution   Consumers

Figure 13.1 Traditional electric power system structure.

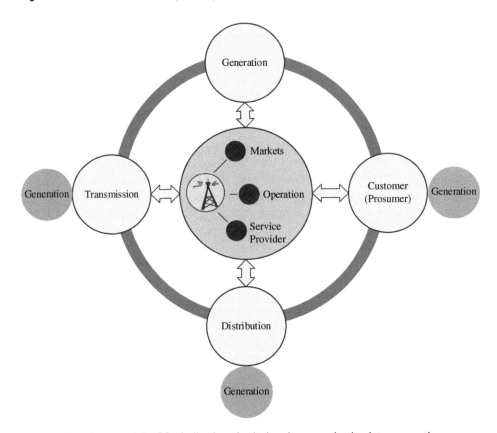

Figure 13.2 Concept of the SG – indicating physical and communication interconnections.

of rising quantities of stochastic in nature renewable energy sources (RES) and electric vehicles, the necessity to enhance the security of supply, and the necessity to decrease greenhouse gas emissions [3].

The SG is a modern electric power grid infrastructure presented for enhanced efficiency, reliability and security, with gradual integration of many RES and energy storage devices, by smart energy management algorithms, automated control, and modern information and communications technologies. It manages the requirements and capabilities of all power generators, grid operators, end-users, and electricity market stakeholders to function all elements of the system as efficiently as possible, while minimizing the costs and environmental impacts with maximum system stability, reliability and resiliency [3–12].

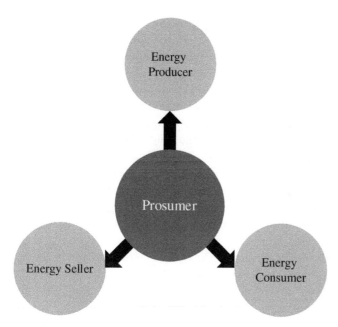

Figure 13.3 Concept of prosumer on SG.

Figure 13.4 Building blocks of the SG [22]. Reproduced with permission from ELSEVIER.

Figure 13.2 illustrates the new concept of SGs and the functional relationship among different subsystems and technologies. The bulk generation, transmission, distribution, and customers being electricity producers and consumers (prosumers) are physically connected and linked via communication systems with the markets, operations, and service providers [13]. Prosumers not only consume energy but also share surplus energy produced by renewables with grid and/or with other customers in a community, as illustrated in Figure 13.3, furthermore, prosumers are not only an essential stakeholder of the future SGs but in addition have a crucial role in peak demand management [14].

The primary aim of the advancement of the SG is on the technology side, which is one among four other building blocks of the SG, as illustrated in Figure 13.4. Little attention is given to involving individual electricity consumers, which could contribute to the foundation block for the SG integration, [15, 16]. The building sector is responsible for 31% of total final energy consumption and for 29% of CO_2 emissions into the atmosphere, hence, buildings represent a huge opportunity for decreasing global energy consumption and ultimately reducing climate gas emissions [17]. Thus, the residential sector is an essential asset when it comes to balancing electricity demand and supply, but this is reliant on households' (consumers) acceptance of the SG technology to increase the outcomes of SG technology implementation. Present technology follows robust, but questionable, assumptions of expected social acceptance of the basic aspects of these SGs. Consumers' reservations regarding SG technology could impede the employment of this technology and, as a result, the development of the SG. Hence, to appropriately develop and effectively implement the SG technology to accomplish its goals, research is required to attain a better comprehension of what makes customers accept the SG technology [18–20]. The ongoing challenges of electric utilities for customer engagement entail: overcoming low levels of SG and electricity system understanding, considering customer concerns regarding the SG and mentioning the advantages and opportunities of the integration of the SG to customers without creating unrealistic expectations [21].

This chapter will give an explanation on the drivers of customer acceptance of the SG and the logical steps for motivating the customer to promote this technology. This overview will support energy systems researchers and decision makers to better understand customers behavior for a faster inclusion of them. This chapter will begin by explaining the consumer as one of the SG domains. Then, understanding the SG consumer will be mentioned, followed by SG consumer acceptance, then consumer engagement in the SG will be addressed. Finally, challenges for SG customer engagement and policy recommendations will be explained.

13.2　Customer as One of the Smart Grid Domains

For a better understanding of the SG in its customer context, it is essential to categorize the electric system paradigm into three interconnected pillars that directly influence the customer end of the system, as illustrated in Figure 13.5, these pillars are, [5, 23]:

1) Smart Customer: the set of technologies which allow customers to track and manage their energy consumption.
2) Smart Utility: the utility which enhances its operational efficiency and optimizes revenue realization by employing monitoring, control and pricing and reaching out to its consumers with innovative energy management programs.
3) Smart Market: The structure of the market that permits implementation of the technologies, decision logics, and information at both consumer and utility ends to establish an economically efficient solution in the new dynamic paradigm that the digital technology has created.

A SG is responsive to consumer, utility, and energy market needs by complete and dynamic implementation of systems and processes. A SG also allows the utilities to deliver, and consumers to receive, reliable, cost effective, environmentally friendly energy, and services. Figure 13.6 shows a conceptual model of the SG, including seven main functional areas known as domains with information flowing between these domains, and the flow of electricity from power sources by transmission and distribution systems to the consumers.

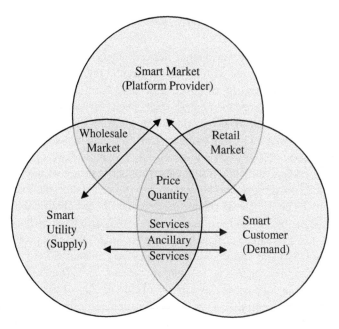

Figure 13.5 The three pillars of the SG: smart marker; smart utility and smart customer.

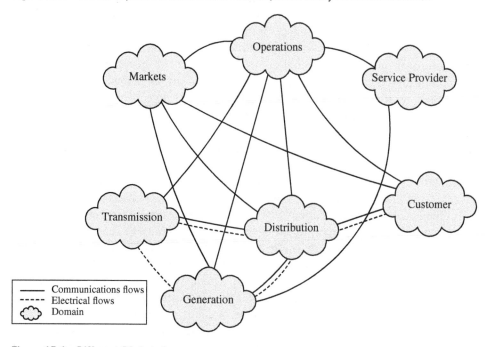

Figure 13.6 Different SG domains.

The consumer is then the stakeholder that the whole grid was made to support and it is the domain where electricity is consumed. The functional requirements within the consumers' premises, entailing the capability to produce, store, monitor, and manage the electricity usage of consumers, [24–29].

13.3 Understanding the Smart Grid Customer

Individuals around the world are aware of climate change, but customers still do not fully comprehend the correlation of electricity usage with its environmental influence; merely 42% of them realize that. Yet, the large majority of customers, 75% claim, overestimate, they comprehend enough regarding the way to manage their electricity consumption. Although, merely 28% have heard about electricity management programs. Additionally, among customers who have heard of electricity management programs, the majority, 58%, still have no knowledge about whether or not their own electricity providers deliver these programs and merely 9% have participated in one, this small enrollment rate shows a high inertia even among well-informed customers. Smaller energy bill (88%), smaller environmental impact (66%) and improved comfort level (better management of the heating and cooling) (51%) are the core three factors which most urge customers to participate in electricity management programs. Yet, a rise in electricity bill (46%), provider making a profit from personal electricity saving (41%) and additional utility access to personal electricity consumption data (32%) are the top three factors which demotivate customers from utilizing electricity management programs. More positively, customers are motivated to understand electricity management programs, and they are open to take them into account as an alternative and probably register for a program where electricity providers could not remotely restrict the use of any home appliances [30–35].

Different surveys have been carried out around the globe to get an insight regarding the preparedness of energy customers for the SGs. In February 2010, 68% of Americans had no knowledge of SGs. However, approximately 57% of US adults were aware of the magnitude of their energy consumption. In addition, 67% of them mentioned that they would limit their consumption if they possessed the knowledge required to do so, and 75% of them wanted to monitor and manage their electricity consumption [20]. In May 2011, US customers were asked about their SG awareness; 41% mentioned that they were aware, while 59% said they were not. Furthermore, when probed about their knowledge level regarding the SG, 69.9% mentioned that they possessed no knowledge, 26.1% mentioned that they were slightly knowledgeable, and 4.1% mentioned that they were relatively knowledgeable about SG. Moreover, when asked to rate the possible advantages of the SG, respondents took into account the following: save money (81.3%), more efficient electricity utilization (78.2%), decreasing electricity consumption (73.2%), create new jobs (67.4%), preserving the environment better (63.9%), increase US security (62.6%), more usage of domestic energy (58.3%), decrease the rate at which power outages occur (61.7%), and improve coordination with utility when outages take place (52.0%). Additionally, they were asked if they agreed or disagreed with some possible negative implications of the SG, 83% mentioned that the cost to construct an SG was of great concern, 79% mentioned that the price of smart meters is of concern, and 65% mentioned that the lack of privacy regarding electricity consumption to a SG is of concern [36]. In December 2011, approximately 51% of the US customers mentioned that they had never heard about SG and 21% mentioned that they had heard about it but had no knowledge about its concept or use [37]. In 2012, a survey of Canadian customers reported a 27% (21% basic and 6% complete understanding) understanding of the SG [38, 39]. In January 2014, 51% of UK adults could be encouraged to participate in a community energy project if it would save them money, relative to merely 17% who would do so to decrease carbon emission, [40]. In 2015, almost three-quarters of Canadians expressed interest in participating in programs that help utilities manage demand; furthermore, 84% support a plan for a smarter grid and 82% indicate a keen interest in active participation in purchasing SG devices, [41]. In May 2016, the Smart Grid Consumer Collaborative (SGCC) determined that up to 68% of consumers were interested in participating in a smart thermostat program (smart appliances example), this program acts as a gateway-opportunity to engage consumers and offer them information about additional smart energy products and services, [42].

13.4 Smart Grid Customer Acceptance

Although consumers are the central players in SGs, they are occasionally not getting essential considerations in technology design, and in some cases, their enthusiasm for taking part in the SG are not fully valued. Undeniably, any technological advancement must be thought of following two topics: one associated with the technological tools and the other entailing investigations on who must accept, adopt, and utilize it. As for each innovation, which includes alterations in users' life habits, one should know if and how such innovation could be accepted from possible consumers [43].

Energy behavior is either investment or habitual behavior. The former usually entails the implementation of a new technology, maybe the purchase of a new appliance. Habitual behavior is routine behavior such as switching off the lights when leaving a room. The customer energy behavior could be comprehended by investigating the interactions between cognitive norms (e.g. beliefs, understandings), material culture (e.g. technologies, building form) and energy practices (e.g. activities, processes). As indicated in Figure 13.7, these aspects of behavior are relatively interactive. Cognitive norms affect people's selection of technologies and the practices they accept. Material culture has a

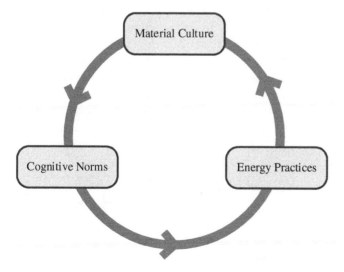

Figure 13.7 The energy cultures conceptual framework.

solid impact on cognitive norms and on the range of people's potential energy practices. Energy practices show how technologies are utilized, and somewhat shape people's beliefs and understandings [44–52].

To effectively model a change in customers' behavior regarding SG initiatives, a comprehension of how an alteration in existing information results in a knowledge shift and associates a number of influencers to make a change in behavior in the individual actor is required. Present theories of how people behave aim to associate possible influencers entailing their personal norms and predispositions, social effect, habits and emotions. Well-known theoretical frameworks for individual-level innovation and adoption studies entail the Theory of Reasoned Action (TRA), the Theory of Planned Behavior (TPB), the Technology Acceptance Model (TAM), and the Value-based Adoption Model (VAM). This section briefly outlines these theories.

The TRA, illustrated in Figure 13.8, has been utilized for studies about customers' integration to innovative technology. Social psychology described TRA as universal interpretations of distinct human behaviors, and mentioned that behavior was described by two typical factors, attitude and subject norm. The factors of belief and evaluation mostly had an impact on the attitude, while the norm belief and the encouragement to follow shaped the subject norm. The TRA mentioned that the individual performance of a given behavior is mainly determined through a person's intention to implement that behavior. Intention is itself a result of the amalgamation of attitudes toward a behavior. That is the positive or negative evaluation of the behavior and its expected outcomes and subjective norms, which are the social pressures applied on an individual due to their perceptions of what other people think they must do and their disposition to conform with these [53, 54].

The TRA has some limitations. The first limitation is the substantial risk of inconsistency between attitude and norm, as norms can be shaped as attitude and vice versa. Secondly, action is free to act without any limitation when one intends. Yet, practically, numerous other external factors like environment, time, limited ability, etc. might be the hurdle in executing the perceived behavior. To address the limitations of the TRA, the TPB, as shown in Figure 13.9, considers an extra dimension of perceived behavioral control as the determinant of behavioral intention. The TPB enlarges the boundaries of the TRA, a purely volitional control, by entailing a belief factor that considers the possession of requisite resources and chances to execute a certain behavior. By adding perceived behavioral control component, TPB can explain the connection among behavioral intention and actual behavior [55–57]. In [58], TPB is implemented to get an insight about the customer behavior for SG integration where the impact of individual's Resistance to Change (RC), as

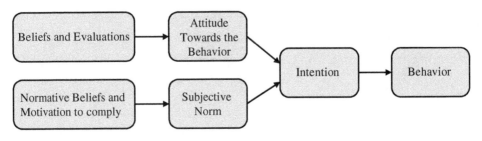

Figure 13.8 Theory of reasoned action.

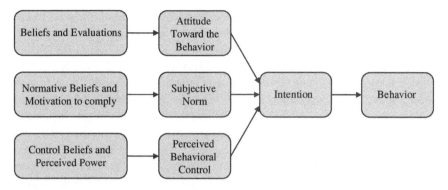

Figure 13.9 Theory of planned behavior.

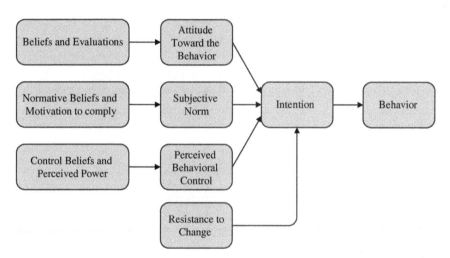

Figure 13.10 Theory of planned behavior model with resistance to change variable.

indicated in Figure 13.10, examined as an added variable, which leads to behavioral intention. This is mainly predicated upon the notion that customers with high RC could have a different propensity in forming intention to implement SG technologies and habits, relative to low RC customers.

The TAM has been extensively implemented in research as a model to predict and describe the customers' acceptance of a new technology [59] and lately has been utilized to study customers' acceptance of SG products and services [60–64]. TAM is utilized to give a basis for tracing the effect of external variables on internal beliefs, attitudes and intentions. It proposes that perceived ease of utilization and perceived usefulness are the two most essential aspects in exemplifying system utilization. Perceived usefulness is described as the level of belief that to utilize a certain technology will improve the attainment of valued goals. And, perceived ease-of-use is described as the level to which utilization of that certain technology is understood to be easy and effortless. These beliefs are expected to determine a person's attitude regarding the utilization of the new technology, with the attitude toward utilizing and the perceived usefulness interacting to change consumers' behavioral intention to use. The actual use is impacted by a consumers' behavioral intention in the same way as perceived usefulness is affected by perceived ease of use, as illustrated in Figure 13.11, [18].

Furthermore, to Perceived Usefulness and Perceived Ease of Use of the SG shown in TAM, Perceived Risk is known as a main factor impacting the Intention to accept the SG. Perceived Risk influences Intention to Use and has a jointly influencing relationship with Perceived Usefulness, as shown in Figure 13.12 [65]. To implement the TAM to investigate customers' acceptance of SG technology, the external variables set impacting their acceptance of the SG should be chosen, as shown in Figure 13.13. Perceived Compatibility and Perceived Understanding have been chosen as external variables that impact the Perceived Ease of Use. Perceived Compatibility denotes the level to which the way of utilizing a new technology is consistent with those of the present technology, as the higher the compatibility a new innovative technology has with present technology, the sooner it is accepted. Moreover, Perceived Understanding is as follows; the greater the understanding of the SG the customers possess, the higher the acceptance of SG technologies is. Furthermore, three external variables were set as effects on the TAM's Perceived Usefulness variable. The first is Perceived Electricity Rate Saving, which should have the most direct influence on Perceived Usefulness. The second variable is Perceived Eco-environment. As the global interest in climate change rises, the

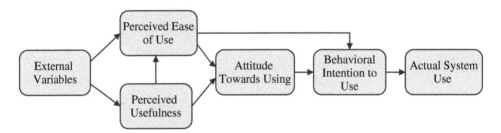

Figure 13.11 Technology acceptance model.

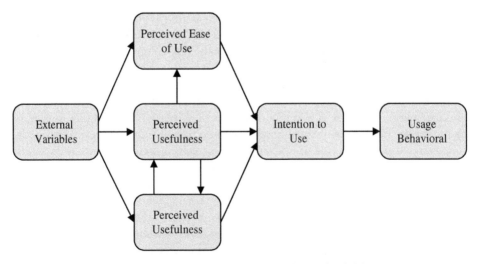

Figure 13.12 Technology acceptance model integrated with perceived risk.

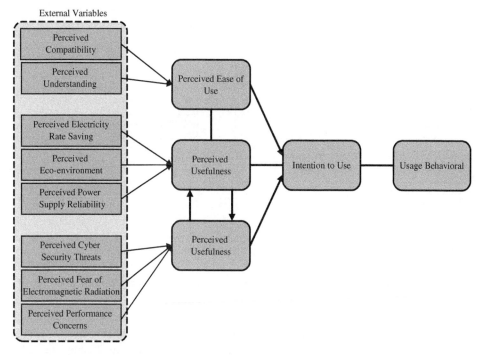

Figure 13.13 Technology acceptance model for SG with external variables set.

customers' environmental protection consciousness also increases. The third variable is Perceived Power Supply Reliability. With regard to the most crucial possible benefits of the SG is the decrease of power loss and a smooth recovery from power failure. Moreover, Perceived Risk could be impacted through: Perceived Cyber Security Threats, Perceived Fear of Electromagnetic Radiation and Perceived Performance Concerns [65, 66].

Using the TAM for SGs to clarify consumer acceptance might not consider the whole picture [67]. The TAM was established and primarily utilized for examining the technology integration in an organizational context, where technology adopters are merely technology users, without any financial burdens, so they have less attitude. Yet, adopters of SG technologies are individuals with a dual role as technology and service consumers. Hence, the integration of SG products and services is predicated upon personal reasons and the adopters carry the financial costs themselves. Therefore, the perceived value, which infers a "trade-off" between costs and benefits, is of paramount importance for consumer acceptance of SG technologies. Generally, if the perceived value of SG is high (i.e. the benefits are perceived to be more than the sacrifices), then customers are likely to utilize it. The VAM [68], specifically perceived benefits, perceived costs, and perceived value, has numerous benefits relative to the TAM when describing technology adoption and hence consumer acceptance of the technology. VAM is developed on two fundamental constructs to describe perceived value, specifically benefits and sacrifices, as illustrated in Figure 13.14. The benefits entail usefulness and enjoyment, while sacrifice considers technicality and perceived fee. Thus, adoption intention is determined by the evaluated value, a detailed comparison of benefits and sacrifices [69, 70].

There are many methods to improve the number of customer acceptances of SG. First, the development of SG technologies must aim for the simplicity of user interface by, for

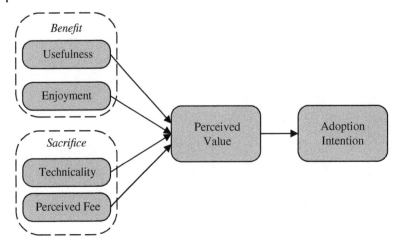

Figure 13.14 Value-based adoption model of technology.

instance, clear visuals, intuitive settings, easy steps for inputting preferences, and preprogrammed default settings. Second, explaining the advantages of SG, entailing long-term financial benefits, is crucial to its social marketing and political success. Giving customers information regarding money saved, and quantity of emissions avoided, motivates environmentally-friendly preference-setting and maintenance of preset defaults. Finally, the utilization of pre-implementation pilots to enhance observability and ultimately diminish customer concerns regarding smart electricity technology.

Public acceptance of the SG is an essential goal to its deployment. Social acceptance of the SG is the acceptance of all layers and sectors of the society to all institutional alterations required for its employment and it has three dimensions: socio-political, community, and market. Figure 13.15 describes these dimensions with the core issues associated with them,

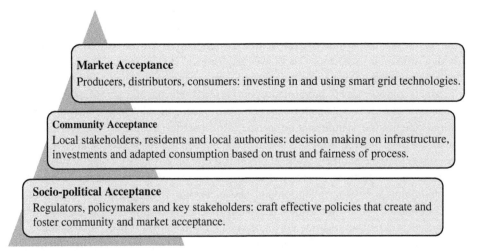

Figure 13.15 The three dimensions of the SG social acceptance, which are: market, community and socio-political acceptance.

showing that each form of acceptance is not enough on its own to promote SGs; merely environments where national social and political frameworks meet community interests and market drivers will experience SGs rapidly implemented [71–76].

13.5 Customer Engagement in the Smart Grid

Although utility-customer relationships were one-sided, utilities usually allocated most of their resources to energy generation, distribution, and ratemaking (setting the rates or prices that they will charge their consumers). No longer are utilities merely energy suppliers, they are transforming into becoming energy management solutions partners. Utilities must convert themselves from energy suppliers to energy service advisors. The challenge of accomplishing more customer motivation with SG technology is among behavioral changes. Smart energy behavior entails two different types of behavior: habitual and one-shot. Energy related practices, for example, washing, cooking, heating, etc. could usually be considered habitual. Yet, behaviors regarding a change of practices, for instance, choosing if participating in a SG project and/or to buy smart appliances, are one-shot behavior. There are many factors that customers must take into account when choosing if participating in a SG program or not and they could be classified into enablers and barriers. Table 13.1 shows an outline of the number of potential enablers and barriers of consumer engagement in SG [77–80].

Motivating a sustainable energy behavior change is a psychologically, socially, and culturally complex problem. Technology adoption and diffusion theories aim to describe how and why innovations come about, enter into use, and become widespread. The innovation decision process entails five stages, as illustrated in Figure 13.16, [81–83]: (i) knowledge, which shows when a customer becomes aware of the innovation and obtains information regarding how it works; (ii) persuasion, which shows the stage when the customer forms an opinion regarding the innovation (favorable or unfavorable); (iii) decision, which shows the stage where the customer becomes involved in activities that result in a decision whether to adopt or refuse the innovation; (iv) implementation, which is when the customer begins to utilize the innovation; and finally (v) confirmation, which is when the customer aims to support a decision made in favor of integration, but where this decision could be changed if given opposing information. These stages usually follow each other in a time-ordered manner. Moreover, the rate of adoption is described as the relative speed with which an innovation is integrated by members of a social system. There are five aspects that explains the rate of adoption, as shown in Figure 13.17: Perceived attributes of innovation, type of innovation decision, communication channels, the social system, and the efforts of the change agent. One could see that the perceived attributes of innovation play a substantial role in the rate of adoption of an innovation [84].

The Transtheoretical Model (TTM) explains the step-by-step process of behavior change and the Fogg Behavior Model describes the influences that result in behavior change. The TTM, as illustrated in Figure 13.18, also known as the "Stages of Change" Model, is a recognized theory of behavioral change processes. It shows that intentional behavior change is a process taking place in a series of stages, five main stages, instead of a single event. The stages progress is as follows: pre-contemplation, contemplation, preparation, action,

Table 13.1 Possible enablers and barriers of end-user engagement in SG projects [77].

Category	Enablers	Barriers
Comfort	• Comfort (gain)	• Comfort (loss)
Control	• More energy independence ("energy autarky"); • Extended possibilities to participate in the electricity market; • More advanced control of appliances, e.g. using mobile devices.	• Loss of control over appliances.
Environment	• Environmental benefits.	
Finance	• Financial or in-kind incentives; • Energy bill reduction.	• Investment costs; • Increased energy bill.
Knowledge & Information	• More transparent and frequent billing; • Detailed knowledge about electricity use.	• Unclear information regarding the SG program (technologies / incentives / cost plans); • Limited competences to handle new technologies or to discuss with energy suppliers; • Limited awareness regarding the goal of using SGs and its opportunities; • Anticipated risks such as detrimental health impact.
Security	• Improved reliability of energy supply.	• Privacy and security concerns.
Social process	• Role models; • Customer testimonial; • Community feelings; • Competition; • Fun.	• Free rider effects; • Job losses.

maintenance. Progression by these stages needs communication, support, and guidance. Pre-contemplation is when individuals are unaware of an issue or do not care to employ a solution. The contemplation stage is when the individual is aware of the issue and is deciding whether or not to try a change. When the individuals choose to change behavior, they arrive at the preparation stage where they get ready for the change and then take action. The action stage is when individuals are implementing the target action. The maintenance stage is when the individuals implement the new routine. Relapse is when individuals fall back into the old habits. This could occur during any of the stages. When relapse takes place, the reason should be evaluated and the individuals should be motivated to continue with the change [85]. SG should deliver the anticipated benefits, so that people participate to some extent with at least three behavioral changes: motivation, ability and triggers, as

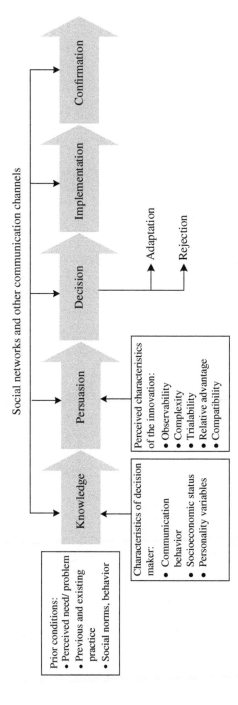

Figure 13.16 The innovation decision process in the SG arena.

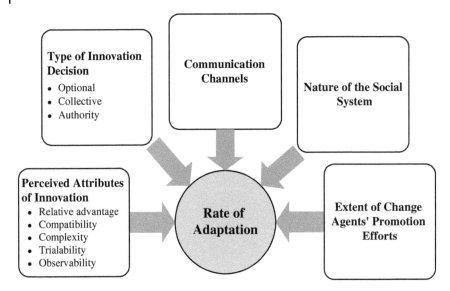

Figure 13.17 Factors affecting the rate of innovation adoption toward SG.

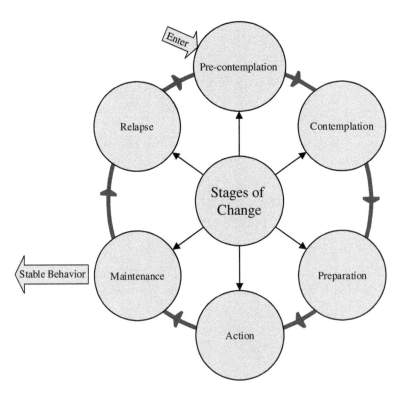

Figure 13.18 Transtheoretical model – the process of change toward SG.

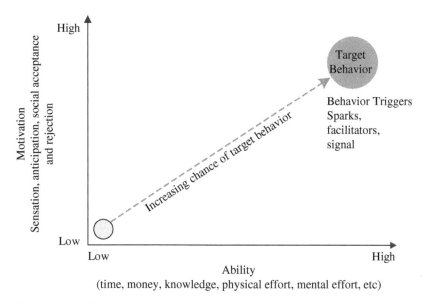

Figure 13.19 The Fogg behavior model has three factors: motivation, ability, and triggers.

shown in Figure 13.19. The individuals should have enough motivation, sufficient ability, and an effective trigger at the same time for a behavior change to take place. A notification alone will never fill all three roles. Yet, a notification could be the trigger that makes the target behavior happen. A motivator and the ability to do something should be in place before a trigger will have any impact [86].

Allowing customers to take part in controlling their electricity use more efficiently will be an essential success factor for utilities that aim to embrace the opportunities of the SG. Customer outreach and education could support utilities to refrain from the "trough of disillusionment," as shown Figure 13.20, and substantially speed up customer acceptance and implementation of SG technologies [87]. International SG programs are utilizing numerous channels to encourage their customers. Press releases, media events and face-to-face communication are widely used for this purpose. However, more strategies include newsletters, reports, awards, phone calls, hotlines, mail shots, the internet, television, advertisements, and the social media, where, the consumer is taken through a journey of ever-increasing awareness, knowledge, trust, and involvement.

Positive involvement of customers with electricity is an essential success factor for materializing the possible gains of the SG and the limited involvement is worrying. Energy providers should aim to form different relationships with customers by focusing on three essential areas, as illustrated in Figure 13.21:

- Knowing the customer; energy customers' values and preferences vary. What is apparent is that customers fall into segments predicated upon non-conventional criteria, for example, their attitudes and values about the environment and their individual sense of responsibility to make a difference.
- Reaching the customer; current interactions between energy providers and customers usually focus on billing or customer service. Marketing, sales and channel

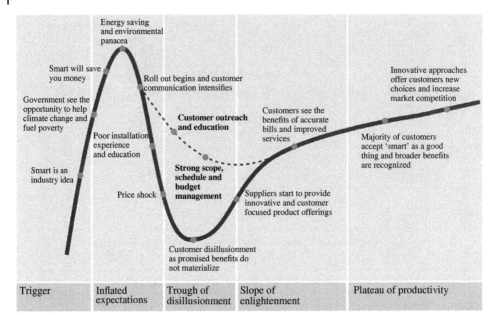

Figure 13.20 SG expectation cycle [88]. Reproduces with permission from PriceWaterhouseCoopers - PWC.

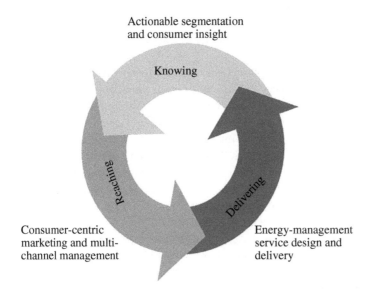

Figure 13.21 Energy providers-consumer new relationships focus areas in SG.

management for non-conventional customer-facing tasks need skills that organizations might not have yet. Consequently, trials to engage with customers regarding energy efficiency, conservation and smart metering often fall flat. Determining the right messages and developing the right third-party partnerships to encourage integration will be hard tasks.

Delivering energy efficiency and conservation experience; customers have a tendency to have different preferences and responses to programs predicted upon individual price sensitivity, levels of self-control vs. external control, and environmental influence.

13.6 Challenges for Consumer Engagement, Policy Recommendation and Research Agenda

The main challenges entail privacy issues, Radio Frequency (RF) safety, and potential rate rises. Each one of these challenges will be reviewed [89]. The privacy issue has been raised by numerous customers as one of the most substantial problems in the implementation of the SG. The SG, by the smart meter, has the capability to capture thorough electricity consumption in every home down to 15-minute intervals over a number of months. This information is sufficient to come up with a profile that includes the personal habits of the customer, for instance, cleaning, cooking, sleeping, or absence from the home. Moreover, it is not known who this information belongs to, the customer or the utility. It could be relatively valuable in the hands of marketing firms or criminals. To solve privacy challenges and protect the customer, it is possible to design privacy directly into the SG by making it the default (Privacy by Design) in all physical, administrative and technological aspects of the SG [90]. RF Safety influence of the RF signals transported by the smart meter is an additional concern raised by customer groups. This resistance is predicated upon safety concerns that considers RF signals being transported during the day in the home. Yet, cell phones and microwaves both produce more RF signals than smart meters, in addition, numerous internet wireless routers which broadcasts RF signals 24 hours a day throughout the home without customer backlash. The main difference is choice, customers select each of these RF devices. Yet, they are required to utilize RF-based smart meters. RF exposure from a smart meter is far below-and more infrequent-than other common electric devices. The smart meters broadcast their signal for less than a total of 15 minutes each day. Furthermore, the electric panel and wall behind the smart meter block much of the RF signal from entering the home, so exposure to the radio waves is less. Experts determined that it could take 30 years of living with a smart meter to take in the same RF exposure that a cell phone user takes in just one day [91]. There is a belief within many customer advocacy organizations that the SG will drive rate rise because of increased utility costs for smart meters installations and the related energy management systems. The utilities have a long-term time horizon and depreciation model spanning 20 years, which permits the influence on customers to be decreased. Yet, this information is not shared with the customer, hence, feeding the customer belief that rates would rise [91].

In addressing customer-related obstacles, the following policies are recommended for more customer engagement [92]:

- Give customers with access to usage, pricing and carbon mix data for use in energy monitoring applications.
- Provide customers with uniform and consistent privacy policies.
- Coordinate SG stakeholders in a sustained customer awareness and education campaign. Provide incentives to assist with the purchase of consumer SG devices.
- Policymakers and regulators must evaluate the best means of ensuring that customers gain useful information and education about SG technologies and options.

Research must concentrate on social and business aspects of SG developments. The essential aspects of this research goal are as follows:

- Coming up with additional socially embedded visions on SGs and the benefits it holds; this should not merely be the responsibility of "experts," but should include all suitable actors;
- A change in the goal on advancing SG elements and systems regarding the benefits it will offer, taking energy use practices as the starting point;
- Forming and examining the innovative user-centered business strategies and ecosystems; there are existing pilot projects that test new business strategies, yet they are very technology driven.

13.7 Conclusion

SGs are one of the largest developments in the power utilities industry worldwide in modern times. They provide the potential to substantially decrease grid inefficiency, allow more interactive demand management, better implement distributed power sources into the grid and facilitate new uses for electric power. SG is the new electric energy of which the customers play a vital role. SG motivates and engages consumers, who are an essential part of the electric power system. Customer support is essential for the transformation to SG from the planning stages to full integration. Once SG is established, customer attitudes and willingness to adopt certain technologies will affect the impact of SG.

This chapter has given energy systems researchers and decision makers an insight into the fundamental drivers of customer acceptance of the SG. Yet, for the customer to accept and embrace the integration of the SG, utilities and policymakers should communicate the advantages effectively to the public. For consumer buy-in to SG delivering knowledge is of paramount importance, the more customers learn about what is happening, the more motivated they are toward it. Consumer engagement within the electric power industry is an evolving, ongoing process that is just on the beginning to emerge. It is more of a culture, a method of doing business where your consumers are at the heart and have choices. It needs the utility to constantly listen, educate, interact, and give access to important information that provides those choices. Lastly, the SG customer is informed, adjusting the method they utilize and purchase electricity. They have choices, incentives, and disincentives to adjust their purchasing patterns and behavior. Furthermore, the emerging SG represents a core and large-scale shift away from educational and social norm change initiatives and toward implemented technology and automation.

References

1 Ellabban, O. and Abu-Rub, H. (2016). Smart grid customers' acceptance and engagement: an overview. *Renewable and Sustainable Energy Reviews* 65: 1285–1298.
2 US Department of Energy (2015). An Assessment of Energy Technologies and Research Opportunities", Quadrennial Technology Review. https://www.energy.gov/sites/prod/files/2015/09/f26/Quadrennial-Technology-Review-2015_0.pdf (accessed 15 February 2021).

3 IEA (2011).Smart Grids: Technology Roadmap. https://www.iea.org/reports/technology-roadmap-smart-grids (accessed 15 February 2021).

4 Hossain, M.S., Madlool, N.A., Rahim, N.A. et al. (July 2016). Role of smart grid in renewable energy: an overview. *Renewable and Sustainable Energy Reviews* 60: 1168–1184.

5 Ardito, L., Procaccianti, G., Menga, G., and Morisio, M. (2013). Smart grid technologies in Europe: an overview. *Energies* 6: 251–281.

6 Fadaeenejad, M., Saberian, A.M., Fadaee, M. et al. (January 2014). The present and future of smart power grid in developing countries. *Renewable and Sustainable Energy Reviews* 29: 828–834.

7 Brown, M.A. (April 2014). Enhancing efficiency and renewables with smart grid technologies and policies. *Futures* 58: 21–33.

8 Niesten, E. and Alkemade, F. (January 2016). How is value created and captured in smart grids? A review of the literature and an analysis of pilot projects. *Renewable and Sustainable Energy Reviews* 53: 629–638.

9 Eltigani, D. and Masri, S. (December 2015). Challenges of integrating renewable energy sources to smart grids: a review. *Renewable and Sustainable Energy Reviews* 52: 770–780.

10 El-hawary, M.E. (2014). The smart grid – state-of-the-art and future trends. *Electric Power Components and Systems* 42 (3–4): 239–250.

11 Mikalauskas, I. (2015). Economic, social and environmental benefits of smart grids. *European Journal of Interdisciplinary Studies* 7 (2): 19–28.

12 NREL (2015).The Role of Smart Grid in Integrating Renewable Energy. https://www.nrel.gov/docs/fy15osti/63919.pdf (accessed 15 February 2021).

13 Ribeiro, P.F., Polinder, H., and Verkerk, M.J. (fall 2012). Planning and designing smart grids: philosophical considerations. *IEEE Technology and Society Magazine* 31 (3): 34–43.

14 Zafar, R., Mahmood, A., Razzaq, S. et al. (2017). *Prosumer Based Energy Management and Sharing in Smart Grid*. Renewable and Sustainable Energy Reviews.

15 Ürge-Vorsatz, D., Cabeza, L.F., Serrano, S. et al. (January 2015). Heating and cooling energy trends and drivers in buildings. *Renewable and Sustainable Energy Reviews* 41: 85–98.

16 Manic, M., Wijayasekara, D., Amarasinghe, K., and Rodriguez-Andina, J.J. (Spring 2016). Building energy management systems: the age of intelligent and adaptive buildings. *IEEE Industrial Electronics Magazine* 10 (1): 25–39.

17 Skjølsvold, T.M. and Ryghaug, M. (2015). Embedding smart energy technology in built environments: a comparative study of four smart grid demonstration projects. *Indoor and Built Environment* 24 (7): 878–890.

18 Broman Toft, M., Schuitema, G., and Thøgersen, J. (1 December 2014). Responsible technology acceptance: model development and application to consumer acceptance of smart grid technology. *Applied Energy* 134: 392–400.

19 Wolsink, M. (January 2012). The research agenda on social acceptance of distributed generation in smart grids: renewable as common pool resources. *Renewable and Sustainable Energy Reviews* 16 (1): 822–835.

20 Ponce, P., Polasko, K., and Molina, A. (July 2016). End user perceptions toward smart grid technology: acceptance, adoption, risks, and trust. *Renewable and Sustainable Energy Reviews* 60: 587–598.

21 IndEco Strategic Consulting Inc. (2013). Smart Grid Consumer Engagement: lessons from North American utilities. https://www.indeco.com/ideas/smart-grid-consumer-engagement/ (accessed 15 February 2021).

22 Rahman, S. (2015). Challenges & Opportunities in Renewable Energy Role of the Smart Grid. *IEEE ICCIT Conference*, Dhaka, Bangladesh (21–23 December 2015). IEEE.

23 Tabors, R.D., Parker, G. and Caramanis, M.C. (2010). Development of the Smart Grid: Missing Elements in the Policy Process. *43rd Hawaii International Conference on System Sciences*, Hawaii, USA (5–8 January 2010). IEEE.

24 National Institute of Standards and Technology (2014). NIST Framework and Roadmap for Smart Grid Interoperability Standards, Release 3.0. Special Publication (NIST SP) 1108r3.

25 de Haan, J.E.S., Nguyen, P.H., Kling, W.L. and Ribeiro, P.F. (2011). Social interaction interface for performance analysis of smart grids. *IEEE First International Workshop on Smart Grid Modeling and Simulation*, Brussels, Belgium (17 October 2011). IEEE.

26 Rodríguez-Molina, J., Martínez-Núñez, M., Martínez, J.-F., and Pérez-Aguiar, W. (2014). Business models in the smart grid: challenges, opportunities and proposals for prosumer profitability. *Energies* 7 (9): 6142–6171.

27 Suleiman, H., Ahmed, K.A., Zafar, N. et al. (June 2012). Inter-domain analysis of smart grid domain dependencies using domain-link matrices. *IEEE Transactions on Smart Grid* 3 (2): 692–709.

28 Fang, X., Misra, S., Xue, G., and Yang, D. (2012). Smart grid – the new and improved power grid: a survey. *IEEE Communications Surveys and Tutorials* 14 (4): 944–980.

29 Tuballa, M.L. and Abundo, M.L. (June 2016). A review of the development of smart grid technologies. *Renewable and Sustainable Energy Reviews* 59: 710–725.

30 Cooke, D. (October 2011). *Empowering Customer Choice in Electricity Markets*. International Energy Agency (IEA).

31 Wijaya, T.K. (2015). Pervasive Data Analytics for Sustainable Energy Systems. PhD thesis. École Polytechnique Fédérale De Lausanne.

32 Bigerna, S., Bollino, C.A., and Micheli, S. (May 2016). Socio-economic acceptability for smart grid development – a comprehensive review. *Journal of Cleaner Production* 131 (2016): 399–409.

33 Accenture (2013). *The New Energy Consumer Handbook*. Accenture Energy Consumer Services.

34 Michael Valocchi and John Juliano, "Knowledge is power: driving smarter energy usage through consumer education", IBM Institute for Business Value, pp. 1–16 January 2012.

35 Swaminathan, S. and Ting, T. (January–February 2013). Customer focused approach to implementing smart grid applications for publicly owned utilities. *The Electricity Journal* 26 (1): 79–83.

36 Zpryme Research & Consulting, LLC (2011). The New Energy Consumer. https://zpryme.com/reports/the-new-energy-consumer/ (accessed 15 February 2021).

37 Smart Grid Consumer Collaborative (2012). Consumer Pulse and Segmentation Research Program-Wave 2. www.smartgridcc.org (acessed 15 February 2021).

38 SmartGrid Canada (2012). The Canadian Consumer and Smart Grids – A Research Report. SmartGrid Canada.

39 IndEco Strategic Consulting Inc. (2013). Smart Grid Consumer Engagement: lessons from North American utilities. https://www.indeco.com/pdfs/smart-grid-consumer-engagement.pdf (accessed 15 February 2021).

40 Department for Energy and Climate Change (2014). Community Energy Strategy: Full Report. https://assets.publishing.service.gov.uk/government/uploads/system/uploads/attachment_data/file/275169/20140126Community_Energy_Strategy.pdf (accessed 15 February 2021).

41 Hiscock, Jennifer. "Smart grid in Canada 2014.",pp.1-40 (2014). https://www.nrcan.gc.ca/sites/www.nrcan.gc.ca/files/canmetenergy/files/pubs/SmartGrid_e_acc.pdf.

42 The Smart Grid Consumer Collaborative(2016). Smart Thermostats —The Next Wave in Consumer Empowerment. https://smartenergycc.org/smart-thermostats-the-next-wave-in-consumer-empowerment-white-paper/ (accessed 15 February 2021).

43 Sintov, N.D. and Schultz, P.W. (April 2015). Unlocking the potential of smart grid technologies with behavioral science. *Frontiers in Psychology* 6: 1–8.

44 Stephenson, J., Barton, B., Carrington, G. et al. (October 2010). Energy cultures: a framework for understanding energy behaviours. *Energy Policy* 38 (10): 6120–6129.

45 Lopes, M.A.R., Antunes, C.H., Janda, K.B. et al. (March 2016). The potential of energy behaviours in a smart(er) grid: policy implications from a Portuguese exploratory study. *Energy Policy* 90: 233–245.

46 Barrios-O'Neill, D. and Schuitema, G. (February 2016). Online engagement for sustainable energy projects: a systematic review and framework for integration. *Renewable and Sustainable Energy Reviews* 54: 1611–162154.

47 Stern, P.C. (March 2014). Individual and household interactions with energy systems: toward integrated understanding. *Energy Research & Social Science* 1: 41–48.

48 Kowsari, R. and Zerriffi, H. (December 2011). Three dimensional energy profile: a conceptual framework for assessing household energy use. *Energy Policy* 39 (12): 7505–7517.

49 van der Schoor, T. and Scholtens, B. (March 2015). Power to the people: local community initiatives and the transition to sustainable energy. *Renewable and Sustainable Energy Reviews* 43: 666–675.

50 Lopes, M.A.R., Antunes, C.H., and Martins, N. (August 2012). Energy behaviours as promoters of energy efficiency: a 21st century review. *Renewable and Sustainable Energy Reviews* 16 (6): 4095–4104.

51 Snape, J.R. (2011). Understanding energy behaviours and transitions through the lens of a smart grid agent based model. ECEEE Summer Study Conference.

52 Maréchal, K. (15 March 2010). Not irrational but habitual: the importance of "behavioural lock-in" in energy consumption. *Ecological Economics* 69 (5): 1104–1114.

53 Christaktopoulos, A. and Makrygiannis, G. (2012). Consumer Attitudes towards the Benefits provided by Smart Grid – a Case Study of Smart Grid in Sweden. Master thesis. Mälardalen University.

54 Hsu, J.-Y. and Yen, H.-L. (2012). Customers' adoption factors and willingness to pay for home energy information management system in Taiwan. *International Proceedings of Computer Science and Information Technology* 45: 11–16.

55 Han, H., (Jane) Hsu, L.-T., and Sheu, C. (June 2010). Application of the theory of planned behavior to green hotel choice: testing the effect of environmental friendly activities. *Tourism Management* 31 (3): 325–334.

56 Mengolini, A. and Vasiljevska, J. (2013). The Social Dimension of Smart Grids. Joint Research Centre – Institute for Energy and Transport. http://ses.jrc.ec.europa.eu (accessed 15 February 2021).

57 Huijts, N.M.A., Molin, E.J.E., and Steg, L. (January 2012). Psychological factors influencing sustainable energy technology acceptance: a review-based comprehensive framework. *Renewable and Sustainable Energy Reviews* 16 (1): 525–531.

58 Perri, C. and Corvello, V. (2015). Smart energy consumers: an empirical investigation on the intention to adopt innovative consumption behavior. *International Journal of Social, Behavioral, Educational, Economic, Business and Industrial Engineering* 9 (9): 3030–3042.

59 Davis, F.D. (Sep., 1989). Perceived usefulness, perceived ease of use, and user acceptance of information technology. *MIS Quarterly* 13 (3): 319–340.

60 Yunus, A.R., Abu, F., Jabar, J., and Ahmad, A. (2015). Empowering smart customer to participate in electricity supply system. *Australian Journal of Basic and Applied Sciences* 9 (4): 110–114.

61 Park, C.K., Kim, H.J. and Kim, Y.S. (2012). An empirical study of the smart grid Technology Acceptance Model in Korea. *IEEE International Energy Conference and Exhibition*, Florence, Italy (9–12 September 2012). IEEE.

62 Orillaza, J.R.C., Orillaza, L.B. and Barra, A.L. (2014). Consumer acceptance of prepaid metering and consumer preferences. *IEEE PES Asia-Pacific Power and Energy Engineering Conference*, Kerala, India (21–23 November 2021). IEEE.

63 Abu, F., Yunus, A., Majid, I. et al. (2014). Technology acceptance model (TAM): empowering smart customer to participate in electricity supply system. *Journal of Technology Management and Technopreneurship* 2 (1): 85–94.

64 Stragier, J., Hauttekeete, L. and De Marez, L. (2010). Introducing Smart grids in residential contexts: Consumers' perception of smart household appliances. *IEEE Conference on Innovative Technologies for an Efficient and Reliable Electricity Supply*, Waltham, USA (27–29 September 2010). IEEE.

65 Park, C.-K., Kim, H.-J., and Kim, Y.-S. (September 2014). A study of factors enhancing smart grid consumer engagement. *Energy Policy* 72: 211–218.

66 Guerreiro, S., Batel, S., Lima, M.L., and Moreira, S. (December 2015). Making energy visible: sociopsychological aspects associated with the use of smart meters. *Energy Efficiency* 8 (6): 1149–1167.

67 Curtius, C., Künzel, K., and Loock, M. (Feb. 2012). Generic customer segments and business models for smart grids. *International Journal of Marketing* 51 (2–3): 63–74.

68 Kim, H.-W., Chan, H.C., and Gupta, S. (February 2007). Value-based adoption of mobile internet: an empirical investigation. *Decision Support Systems* 43 (1): 111–126.

69 Lin, T.-C., Wu, S., Hsu, J.S.-C., and Chou, Y.-C. (December 2012). The integration of value-based adoption and expectation–confirmation models: an example of IPTV continuance intention. *Decision Support Systems* 54 (1): 63–75.

70 Kwon, H.-K. and Seo, K.-K. (August 2013). Application of value-based adoption model to analyze SaaS adoption behavior in Korean B2B cloud market. *International Journal of Advancements in Computing Technology(IJACT)* 5 (12): 368–373.

71 Wüstenhagen, R., Wolsink, M., and Bürer, M.J. (May 2007). Social acceptance of renewable energy innovation: an introduction to the concept. *Energy Policy* 35 (5): 2683–2691.

72 Wolsink, M. (January 2012). The research agenda on social acceptance of distributed generation in smart grids: renewable as common pool resources. *Renewable and Sustainable Energy Reviews* 16 (1): 822–835.

73 Sovacool, B.K. and Ratan, P.L. (September 2012). Conceptualizing the acceptance of wind and solar electricity. *Renewable and Sustainable Energy Reviews* 16 (7): 5268–5279.

74 Enevoldsen, P. and Sovacool, B.K. (January 2016). Examining the social acceptance of wind energy: practical guidelines for onshore wind project development in France. *Renewable and Sustainable Energy Reviews* 53: 178–184.

75 Hanger, S., Komendantova, N., Schinke, B. et al. (April 2016). Community acceptance of large-scale solar energy installations in developing countries: evidence from Morocco. *Energy Research & Social Science* 14: 80–89.

76 Mengolini, A. and Vasiljevska, J. (2013). The social dimension of smart grids. *Joint Research Centre of the European Commission*: 1–45.

77 Valkering, P., Laes, E., Kessels, K. et al. (2014). How to engage end-users in smart energy behavior. *EPJ Web of Conferences* 79: 04003.

78 Luthra, S., Kumar, S., Kharb, R. et al. (May 2014). Adoption of smart grid technologies: an analysis of interactions among barriers. *Renewable and Sustainable Energy Reviews* 33: 554–565.

79 Moreno-Munoz, A., Bellido-Outeirino, F.J., Siano, P., and Gomez-Nieto, M.A. (January 2016). Mobile social media for smart grids customer engagement: emerging trends and challenges. *Renewable and Sustainable Energy Reviews* 53: 1611–1616.

80 Skopik, F. (March 2014). The social smart grid: dealing with constrained energy resources through social coordination. *Journal of Systems and Software* 89: 3–18.

81 Rogers, E.M. (2003). *Diffusion of Innovations*, 5e. New York, USA: Free Press.

82 Wilson, C. and Dowlatabadi, H. (June 2007). Models of decision making and residential energy use. *Annual Review of Environment and Resources* 32: 169–203.

83 Broman Toft, M. (2014). Consumer Adoption of Sustainable Energy Technology – The Case of Smart Grid Technology. PhD thesis. Aarhus University.

84 Schabram, S. (2013). Customer-Centric Business Models for Smart Grid Development. Master thesis. Uppsala University.

85 He, H.A., Greenberg, S. and Huang, E.M. (2010). One Size Does Not Fit All: Applying the Transtheoretical Model to Energy Feedback Technology Design. *Conference on Human Factors in Computing Systems*, Atlanta, USA (10–15 April 2010). SIGCHI.

86 Fogg, B.J. (2009). A behavior model for persuasive design. *4th International Conference on Persuasive Technology*, Claremont, USA (26–29 April 2009).

87 van der Zanden, G.-J. (2011). The Smart Grid in Europe: The Impact of Consumer Engagement on the Value of the European Smart Grid. Master of science thesis. Lund University.

88 "Smart from the start", PricewaterhouseCoopers, https://www.pwc.com/gx/en/utilities/energy-supply/pdf/managing-smart-grid-programmes.pdf, 2010.

89 Gupta, A. (2012). Consumer adoption challenges to the smart grid. *Journal of Service Science* 5 (2): 79–86.

90 Cavoukian, A., Polonetsky, J., and Wolf, C. (2010). Smart privacy for the smart grid: embedding privacy into the design of electricity conservation. *Springer* 3: 275–294.

91 Wamsted, D.J. (March / April 2012). Clarity for customers. *Electric Perspectives* 37 (2): 22–36.

92 TIA (2011). Smart Grid Policy Roadmap: Consumer Focused and Technology Driven. https://www.tiaonline.org/wp-content/uploads/2018/05/Smart_Grid_Policy_Roadmap_-_Consumer_Focused_and_Technology_Driven.pdf (accessed 15 February 2021).

14

Cloud Computing for Smart Grid

Amira Mohammed and Dabeeruddin Syed

Future power grids are predicted to transfer into Smart Grid (SG) through the use of advanced communication and information technologies. A vast range of computing resources and storage are to be offered to store and use the big SG data. Cloud Computing (CC) can offer all these computing and storage resources. CC is considered the next-generation computing paradigm because of its advantages in network access, massive computation services, storage capacities, and various application opportunities including a SG.

14.1 Introduction

CC has attracted both academia and industry attention because it is a potential computing paradigm, enabling global access to shared pools of computing resources that include computer networks, servers, storage, applications, and services that must be accessed easily with minimum efforts [1]. SG applications can be categorized into fundamental and emerging applications. The fundamental applications denote energy management strategies, reliability models, security, and privacy, in addition to promoting demand-side management (DSM). Emerging applications include the deployment of electric vehicles (EVs) and mobile charging stations. SG is characterized by automated energy delivery, monitoring, and consumption with players from utilities, the market, and customers. A reliable and efficient communication system and applications are vital for SGs to control the entire grid operations in a reliable and efficient manner while analyzing huge chunks of customers' data. Thus, CC can provide a more resilient and smarter approach to manage the power grid. SG computational requirements will be able to benefit from the CC model with adaptable resources and services distributed in the network, and from the available parallel processing capabilities. The CC concept is considered to be an effective tool for SGs although it faces few challenges related to security and reliability.

In this chapter, we will discuss the fundamental relationship between SG and CC services. The architectural principles of CC services will be discussed. The chapter will provide a sufficient understanding of various forms of CC. The chapter will introduce CC characteristics and applications for the SG. We will also examine the benefits and drawbacks of those characteristics associated with SGs. Opportunities and difficulties of using CC in SGs will be discussed. Furthermore, the major categories of data security challenges of CC will be discussed.

14.2   Overview of Cloud Computing for Smart Grid

CC is a promising technology with convenient functionality to integrate electrical power system resources through internal networks. This integration improves robustness, storage capacity, and load balance.

Information and communication layers of the SG are critical for the physical components to communicate with each other and to establish a versatile, scalable, resilient, and efficient power grid. The communication system layer is structured in a centralized and/or distributed structure. Initially, the data is collected by smart meters and this data is sent to the collector which communicates it through access points in the communication network. Then, the data is transported to the service provider or the distributed or centralized control center [2]. The main challenge of the SG is managing various types of intelligent devices such as power assets, recorders, sensors, measuring units, and smart meters also, processing this huge amount of data received from different devices is another challenge. CC is a great opportunity to address these challenges, which provides tools, approaches, and computational resources on demand that provide various benefits such as energy-saving, cost-saving, agility, scalability, and flexibility. Figure 14.1 describes the relationship between the SG, big data, and CC technologies.

To guarantee a robust, reliable, efficient grid with secure communication, supply and demand balancing must be adopted. CC could be utilized to execute the communication process among substations, loads, and the main power grid. As the data intensity is high so scalable platforms are required to deal with different SG applications and to improve the reliability, scalability, privacy, and robustness of this communication to construct and employ the functional SG architecture. For that, CC platforms can be utilized as a proper solution. The crucial features of the CC model are presented in Figure 14.2 [3].

The crucial features of the CC model were described by the National Institute of Standards and Technology (NIST) as follows [4–6]:

- On-demand self-service: A customer has the chance to individually save computing resources such as time and network storage automatically without the need of individuals to communicate with each service provider.
- Broad network access: Resources are accessible across the network. It could be opened up through typical procedures used by client platforms such as tablets, laptops, and mobile phones.
- Resource pooling: A multi-tenant model utilized to assist several customers using a vast range of computing resources. The consumer possesses zero control or knowledge regarding the precise location of the offered resources.
- Rapid elasticity: CC has a flexible essence of storage and memory devices. It is capable of extending or condensing itself automatically depending on the workload from the users, as required.

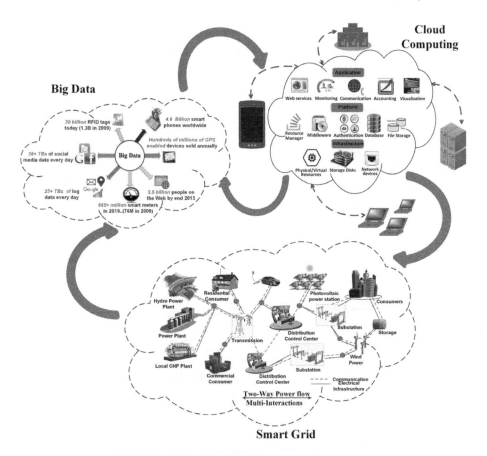

Figure 14.1 The relationship between SG, cloud computing, and big data.

- Measured service: CC automatically control and manage resources and offers a gauging platform to consumers. Cost-optimization procedures are delivered to consumers, allowing them to provide and purchase for merely the used resources.

From the previous characteristics, it's concluded that CC can offer great advantages for organizations and citizens by enhancing the system flexibility and achieve cost reduction if the previous essential characteristics are taking into consideration when building the CC model. However, some concerns must be considered such as data privacy and security. Therefore, proper protection measures to organizations' and customers' data should be guaranteed.

14.3 Cloud Computing Service Models

CC is described as a model developed on-demand network to enable wide access of users and enterprises to a shared pool of adjustable resources including servers, storage devices, network connectivity devices, CPUs, and memory with minimal management. CC allows users to store and process large data to achieve coherence between users and computational resources, that serves as a utility provider.

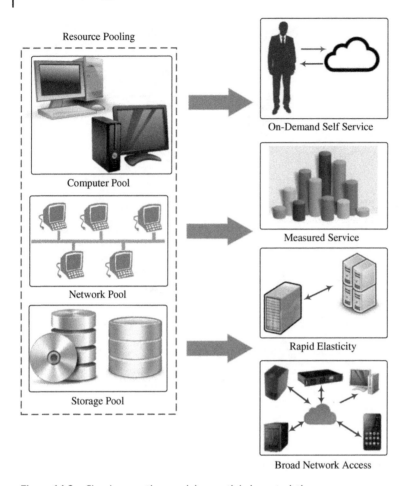

Figure 14.2 Cloud computing model essential characteristics.

CC can provide three main layers of services as illustrated in Figure 14.3 [7]:

- Infrastructure as a Service (IaaS)
- Platform as a Service (PaaS)
- Software as a Service (SaaS)

14.3.1 Infrastructure as a Service (IaaS)

The infrastructure service model is the layer responsible for providing a scalable infrastructure to consumers and for supplying the higher layers with the needed information to develop the cloud software. The IaaS layer is physical hardware as it comprises all the hardware needed (servers, nodes, networking, data center, storage, etc.) to make CC possible [8]. This platform provides access to the cloud using different user interfaces, for instance, web service applications, programing interfaces, and command-line interfaces.

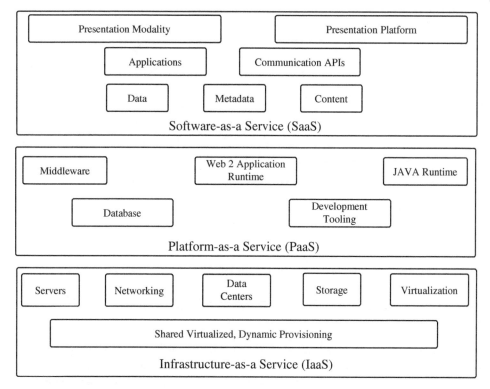

Figure 14.3 Cloud computing layers.

14.3.2 Platform-as-a-Service (PaaS)

The platform layer is the middle layer of the Cloud that offers a platform for users to build and execute their applications. The PaaS layer is responsible for the runtime pursuance of customers through creating applications and software and web tools by developers and programmers. PaaS is representing the software layer which is predicated upon the IaaS layer and is implemented to create the highest layer of the Cloud [9].

14.3.3 Software-as-a-Service (SaaS)

The Software layer is the top layer of the Cloud, built on the top of the Platform and Infrastructure layer. Also, the layer that is used by customers and is responsible for delivering the application programs, software, and web tools to the end-users. This layer can be accessible by computer, smartphone, or tablet. The three cloud-service layers are combined to offer CC in three deployment models: public clouds, private clouds, and hybrid clouds as illustrated in Figure 14.4 [10]. Choosing the type of CC will be based on the characteristics of data and the user application.

Public Cloud
Public cloud provides services and infrastructure as a public service provided off-site over the internet. The Public Cloud has some benefits to users such as reducing the capital cost

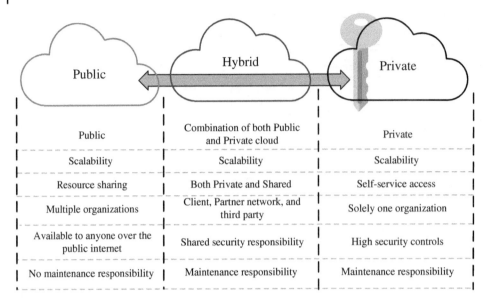

Public	Hybrid	Private
Public	Combination of both Public and Private cloud	Private
Scalability	Scalability	Scalability
Resource sharing	Both Private and Shared	Self-service access
Multiple organizations	Client, Partner network, and third party	Solely one organization
Available to anyone over the public internet	Shared security responsibility	High security controls
No maintenance responsibility	Maintenance responsibility	Maintenance responsibility

Figure 14.4 Three forms of the clouds; Public, Private, or Hybrid clouds.

by providing an infrastructure and offering a high level of efficiency in sharing resources [11]. Also, it offers cost-saving by using CC as a utility. In this case, the customer only pays for their use. The public Cloud is available to everyone through an internet connection; therefore, it needs special security measures. Choosing the public Cloud depends on the application such as:

- When the applications are used by lots of people in a standardized form such as e-mail.
- In case of collaborative projects to share data and meetings.
- In the case of test and develop application codes.
- When incremental capacity is needed to increase computer capacity at peak periods.
- Employing ad-hoc software development projects through a (PaaS) software.

Private cloud

This type is also named internal cloud as it's designed for restricted use by one organization. Building and managing the private cloud can be done by the organization or by special providers [12]. The private cloud offers comparable advantages to Public computing such as scalability, self-service, superior performance, easy customization, and compliance, however, the private cloud conducted the services by means of a proprietary architecture or based on a third-party provider, for example, Amazon Web Services (AWS) or Microsoft Azure which maintain all resources used by customers to be accessed over the internet providing a high level of accuracy. A private cloud provides a high degree of control, reliability, and security and can be considered the obvious choice in some cases such as:

- When all the services and infrastructure are safeguarded by a private network.
- Huge companies have the opportunity to build and control their cloud data centers.

Hybrid Cloud

The hybrid cloud model is an amalgamation of public and private cloud models trying to gain the advantage of each model and address the limitations of the public and private cloud models. The main benefits of Hybrid Cloud are control (each organization can ensure privacy to its infrastructure), flexibility (take the opportunity of adding resources for the public cloud when required), and cost-effectiveness (the ability to scale for the extra computing power only when needed).

This model divides the service infrastructure into two groups, one group runs under private clouds while the other group runs under public clouds. However, hybrid clouds are considered flexible models [13]. Using multiple security is needed to ensure secure communication between two different models with a different security level. Hybrid cloud is the best in some applications such as:

- Using (SaaS) application with some security considerations such as using a virtual private network (VPN) for additional security.
- Used in different market projects through implementation of the public cloud to interact with clients but, at the same time, keep all customers' data secured using a private cloud.

These deployment models explained previously are used based on their main purpose in SG applications. Studies suggest that the public cloud is a reasonable choice for SG utilities that require a standardized service and less complexity. For the application which needs high-security issues, the private cloud is the more effective model to be used. Hybrid cloud is used for SG applications that need a scalability aspect added to it by providing a central role to Cloud Service Providers (CSPs) to provide benefits of both public and private clouds.

14.4 Cloud Computing Architecture

The section presents a few basic CC architectures. Each of these architecture models presents a fundamental implementation and characteristic of CC environments.

14.4.1 Workload Distribution Architecture

The computational resources on cloud can be horizontally scaled by the addition of identical resources and a load balancer between them to manage the load on the resources. Having a load balancer will avoid over-utilization or under-utilization of the computational resources. However, the efficiency of such a distribution architecture depends upon the load balancer algorithms and run-time logic. Figure 14.5 represents the workload distribution architecture with horizonal scaling of computational resources and the server requests managed by the load balancer [14].

14.4.2 Cloud Bursting Architecture

Cloud burst architecture enables the computational resources to be actively added or changed based on the demands of the users [15]. With automatic cloud bursting architecture, an organization, which uses the private environment for their usual capacity requests,

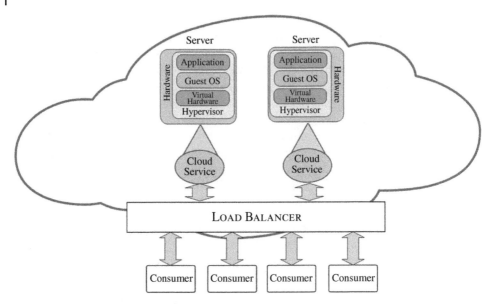

Figure 14.5 Workload distribution architecture.

can rent the computational resources from cloud vendors to meet the higher-usage requests on their private environment as and when required. An automatic mechanism should be in place for the organization to determine the load status on their servers. When a cloud status algorithm detects an overload in terms of service requests, a request for a new Virtual Machine (VM) is put in place. Figure 14.6 illustrates the cloud bursting architecture for CC.

14.4.3 Dynamic Scalable Architecture

Dynamic Scalable Architecture envisions the on-demand scaling capability of computational resources based on the user requests [16]. There is an automatic load monitor balancer which works with a scaling algorithm. If the user requests are above the capacity threshold of the current cloud environment, then new computational resources are added to the workload processing. The architecture is capable of providing horizontal scaling (cloud instances with more computational resources are added), vertical scaling (the processing capacity of the current cloud instance is increased) or dynamic relocation (the current instance resource is moved to a host higher capacity threshold). Figure 14.7 depicts the dynamic scalable architecture used in CC.

14.4.4 Elastic Resource Capacity Architecture

The architecture is typically associated with virtual servers and the dynamic allocation of computational resources to the server from a resource pool of CPUs and memory [17]. The load monitor and balancer run an intelligent script that analyzes the workload of the requests and the analysis results are then sent to the hypervisor which can then reclaim CPUs and memory from the resource pool to meet the requests. Figure 14.8 represents the

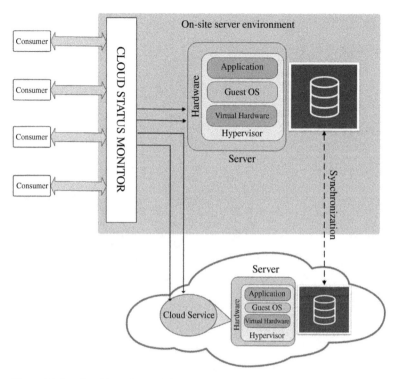

Figure 14.6 Cloud bursting architecture.

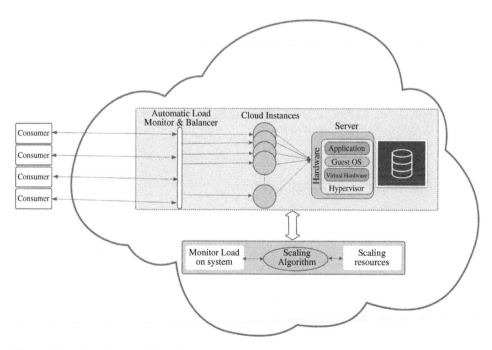

Figure 14.7 Dynamic scalable architecture.

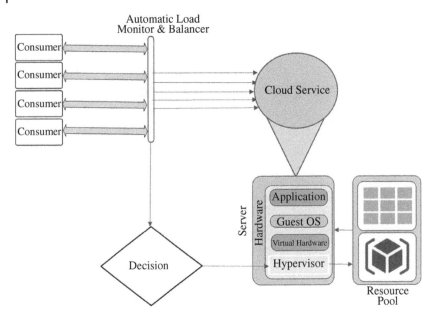

Figure 14.8 Elastic resource capacity architecture.

elastic resource capacity architecture in which hypervisor calls for increased resources from a pool based on the decision made on user presented workload.

14.4.5 Resource Pooling Architecture

The resource pooling architecture is based on the idea of grouping identical computational resources together into one or more pools [18]. One or more of the resource pools in the architecture are maintained by a system so that these are synchronized at any instant in time. Different types of pools include physical server pools, virtual server pools, CPU pools, memory (RAM) pools, storage pools, etc.

14.5 Cloud Computing Applications

Researchers around the globe are working to gain benefits from implementing CC in different fields of the power system. Coupling SG networks with CC can be considered the most important CC application for supporting SG operations [19]. Such coupling would enhance the abilities of electricity generation, transmission, and distribution [20]. Figure 14.9 presents the coupling of CC with SG in various sectors [21].

To integrate SG with CC technology, the Cloud must be chosen as private, public or hybrid based on the requirements and applications. These applications based on power usage, monitoring, data storage, energy management, and processing which affect SG performance, and power dispatching in the SG as depicted in Figure 14.10 [22].

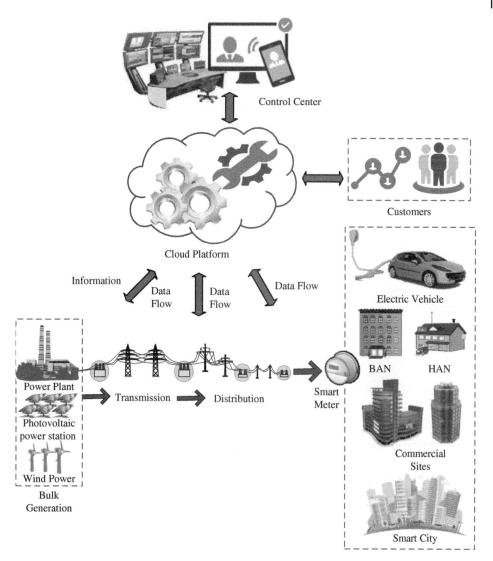

Figure 14.9 Cloud computing platform coupled with SG.

14.5.1 Cloud Applications for SG Performance

Cloud platforms can deal with data related to different SG areas to improve the grid performance such as the data related to customer applications (dynamic DR system, accurate information related to energy consumption, and dynamic pricing which helps to optimize and control power usage). In addition, CC provides the sharing of the data and resources in the same pool to lower the cost instead of adding extra devices to increase system capacity. Cloud is a single platform capable of supporting IT-related operation in SG which, in turn, leads to fast managing of the SG data and assets [23]. CC has the ability recover quickly in case of disturbances by using all the available data

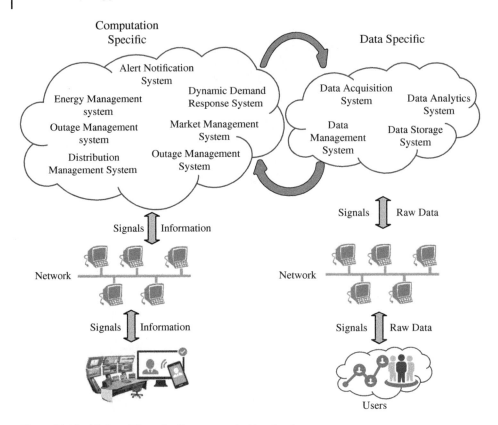

Figure 14.10 Various SG applications supported by cloud.

which increases SG reliability and prevents connection losses. This helps in improving grid performance and efficiency while ensuring high security by adopting private cloud platforms. The cloud can improve the SG capability in integrating different renewable sources and smart homes [24].

14.5.2 Cloud Applications for Energy Management

Energy management is a significant consideration in SGs. The integration of cloud applications in the SG is one of the most used ways to achieve better energy management. The cloud application could be utilized for energy storage and data storage. Figure 14.11 presents a comparison between two energy management scenarios; with and without cloud application. Figure 14.11a presents the conventional SG without cloud in which the link between grid components and utility providers is done through the communication network only. Conversely, the second scenario in Figure 14.11b represents the SG coupled with cloud applications [25]. In this scenario, SG components communicate directly via the cloud and share the required data by using cloud services. This decreases the needed time for taking the necessary decisions. In addition, it provides customers with real-time information such as energy usage and pricing information.

(a)

(b)

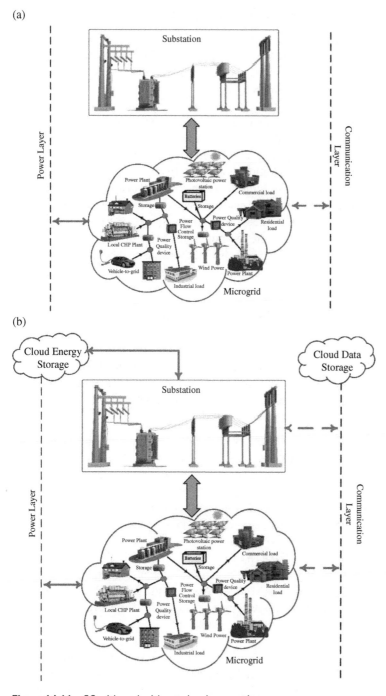

Figure 14.11   SG with and without cloud computing.

14.5.3 Cloud Computing-Based Power Dispatching in SG

SG construction with the intelligent energy and communication network with the consumers, operation, management, and maintenance represents continuous cost increase which is considered the main reason to hamper SG construction. For that reason, different technology centers search for some applications to aid in an accurate, definitive, and efficient energy management solution in real-time. CC, as a valid and intelligent tool, can be the most suitable way to manage this huge amount of data, achieve economic benefits, and decrease the management costs through using the same resources for various functions. This may help in cutting down the repeated investment which, in turn, decreases the management cost and complexity [26]. In addition, it provides the interaction process between the customers information (smart meter information, customer satisfaction), and utilities information (power demand, fuel cost, and weather data) to match the power generation against consumers' demand. CC will process this information based on the economic power dispatching model and share the output with both utility and customers. Implementing CC in SG is an economic approach for utilities since they need to optimize the investment in the communication and computing facilities.

14.6 Cloud Computing Characteristics in Improving Smart Grid

CC has many characteristics of interest for improving SG applications. Figure 14.12 presents the CC characteristics which positively affect the SG in four main areas: security, reliability, scalability, and performance. The characteristics are detailed as follows:

- **Agility**
  Any reconstruction of the SG infrastructure will be managed by CC. Agility enables the rapid provisioning of computational resources with growing and expanding needs. It encompasses the concept of adaptability, balancing, elasticity, scalability, and coordination.
- **Reliability**
  CC provides a disaster recovery strategy that increases system reliability. Fault-tolerant concepts such as isolation, redundancy, etc. can be implemented in CC architecture for increased reliability.
- **Application Programming Interface (API)**
  A cloud API is coordinated with a cloud infrastructure to assign computing, storage, and network resources for SG applications/services. Also, cloud APIs enable the development of services and applications for resource assignment. These APIs are mainly responsible for the users' ability to ubiquitous access.
- **Low cost**
  Different models of CC are provided such as the public cloud or delivery cloud model which imparts lower cost for SG customers by choosing a suitable model to achieve the requirements of different applications. Additionally, the pay-as-you-go model of CC allows users to pay only for the services and resources they have utilized. Furthermore, the utilization of remote servers removes the requirement for in-house servers, storage devices, application requirements and avoids the overhead costs of software updates, etc.

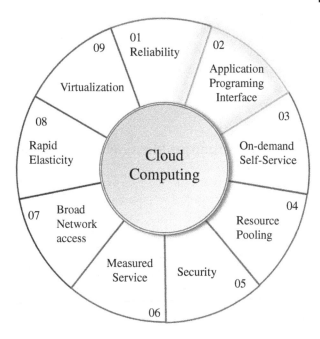

Figure 14.12 The cloud computing characteristics.

- **Availability**
  All data is available to customers by accessing the system via the internet from any device anywhere through the cloud feature, which is device and location independent. Availability is very critical for all enterprise services which demand 99.99% availability. The availability is maintained through one of the techniques including infrastructure availability, middleware availability, application availability, restarting failed application instances if recovery is possible, requests bypassing the failed application instances, etc.
- **Maintenance and virtualization**
  Maintenance and virtualization are the main features of CC. Utilizing computer resources allows to serve as other resources without any need for new installations for running the applications and performing computations. Also, all the data and information are shared between servers and users through CC storage devices which eases the maintenance process for any part of the grid.
- **Multi-tenancy characteristic**
  One software can serve multiple applications and many customers, while each customer has the capability to modify a few aspects of the application which is economical due to the fact that software development and maintenance charges are joint.
- **Performance**
  In the CC platform, web services are provided which increases system performance. Performance is also attributed to the fact that CSP give specialized cloud and application services with high efficiency, velocity, and performance when compared to a single-company data center. Further, the on-demand type of service yields high flexibility in upscaling and downscaling computational resources in a short time and without any additional installation cost.

- **Security**

  Also, the use of private cloud platforms will provide information security and prevent information losses. Even if public or hybrid cloud platforms are employed, data is secured behind an enterprise-level firewall, protected against Distributed Denial-of-service (DDoS) and other cyber-attacks, and the client can benefit from the security expertise of the cloud vendors.

14.7 Opportunities and Challenges of Cloud Computing in Smart Grid

Implementing CC technology in performing the SG has great interest from both the power industry and the research area. The CC provides low-cost computing relative to all the models used and ensures fast response in the event of power outages or failures. The potential scalable capacity of cloud services allows for enhanced protection in case of attacks, helps the dynamic management of demand and supply curves, and provides unlimited data storage to save consumer data. All these attractive features of CC are a great gain to the SG [27]. This chapter presents a brief description of the benefits and challenges of integrating CC technology within the SG.

14.7.1 Opportunities to Apply CC in SG

14.7.1.1 Scalability

In a SG environment, fast changes in SG applications and components are deployed on a large scale which means new and many devices are added such as home appliances, smart meters, micro-grids, substations, sensor nodes, communication-network devices, and small and variable energy resources. Also, deploying huge storage devices are needed to accommodate all those collected data. This provokes the challenge of how to set up and offer data storage capacity for all new SG applications with such multiple devices. The principle for effective management of a massively distributed energy system based on small and variable energy resources is to solve the scalability challenge by installing enough storage devices and distributing them to cover the needs of the SG. This can easily be achieved by utilizing CC technology [28]. CC has autoscaling and load-balancing support features. This allows for the Cloud platform to add computational, storage, and other resources as and when required making the cloud-based applications more scalable. Therefore, these applications can add more instances to manage growing workloads and the addition of more processing data. Consequently, SG utilities can respond to market and technology changes immediately and when they have an intensive flow of data.

14.7.1.2 Cost Efficiency

Moving toward a secure and renewable energy-based electric system is the target of electric utilities that could be achieved through switching the traditional power grid into an SG network. SG, as a bidirectional power and information flow, exchanges all the information to manage and control the power generation, distribution, and consumption. Robust information exchange with low cost can be provided through the internet via cloud platforms and with a dedicated network utilized for the SG [29].

14.7.1.3   Central Data Storage

CC provides not only distributed but also a single-cloud data center with a lower cost relative to any used computing hardware to store and process SG huge amount of data with high performance and a shorter time [30]. Also, it supports shared information among users after meeting privacy requirements, and thereby, a shared communication platform provided by the cloud could be adopted to avoid the employment of numerous middleware software and interfaces to open the data by SG systems.

14.7.1.4   Real-Time Response

Vast sizes of data in SGs are to be processed in real-time for energy management and control which can be done through CC with its distributed data processing centers. This delivers a scalable load balancing technology in the SG and gives rapid reaction against unpredicted failures such as network device misconfiguration, software fault in the memory management component, Web Services [31]. From the previous discussion, it's clear that the employment of CC can be beneficial for different areas of the SG. However, there are some barriers to the implementation of CC technology in SGs. Figure 14.13 presents both prospects and difficulties of CC for SGs [28, 32].

14.7.2   Challenges of Applying Cloud Computing for SGs

Up until now, SG utilities possess frightening thoughts about CC's authenticity. This is enough to build a lot of challenges that must be considered. The major difficulties of implementing CC for SG applications can be explained as follows:

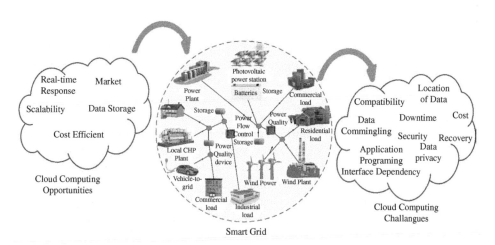

Figure 14.13   Cloud computing opportunities and challenges for smart grid.

14.7.2.1 Location of Data

Data sources of SG applications are placed at different and multiple locations. It is crucial that the data is optimally, structurally, and efficiently integrated, stored, and processed to extract business value from it. This is a major challenge in the data management and processing in the SG. Therefore, determining the data location through a CSP is crucial for SG applications processing and security [33]. A coherent communication architecture is to be built on a cloud platform to gather information from heterogenous recorders, devices, equipment, sensors, etc. in the SG.

14.7.2.2 Data Commingling

CC enables different users to access and store multi-applications in a resource pool which can be considered mixing of data from CSP. Data that is commingled can present a security vulnerability which is an important factor needed in SG applications [34]. Data of multiple users is stored in the same database and the architecture is heavily reliant on software for data isolation and separation. A security breach on the cloud provider would expose the data of utility along with other clients owing to a centralized database management system in the multi-tenant architecture. Hence, high regulation is required for security and there needs to be a clear separation between the data of different clients.

14.7.2.3 Application Programming Interfaces Dependency

CC applications applied by CSP are reliant on utility Application Programming Interfaces (APIs). These APIs are employed to communicate data and information between cloud applications and systems. It is important to thwart the risks of security breaches and to ensure compliance during data communication through APIs. Conventionally, developer registration, protocols such as HTTPS, etc. are employed to protect against security breaches. However, the high dependency of cloud applications on APIs makes it difficult for SG services in CC to migrate from one CSP to another. Also, the migration would require a long time.

14.7.2.4 Compatibility

The most critical challenge with CC is meeting the SG auditing compliance requirements. Location of data and inefficient security policy are two factors that cause difficulties for CC to become compatible with auditing requirements in the SG [35].

14.7.2.5 Inefficient Cloud Security Policy

Data Security is among the most difficult problems when implementing CC in SG. Data security is based on the level of security provided by CSP. Unaffected cloud security policy may affect the information security in the grid which is considered a reason for avoiding companies from attracting benefits of the cloud or causing disagreements between SG utilities. CC is related to information technology which allows users to store and process data to achieve coherence between users and utilities in an economic way through three service provider layers. CSP layers attract significant amounts of threats in (i) service models, which include IaaS, PaaS and SaaS, (ii) deployment models, which include private, public and hybrid Cloud, and (iii) network issues and any internet-based services. Figure 14.14 illustrates the three major categories of data security challenges [36].

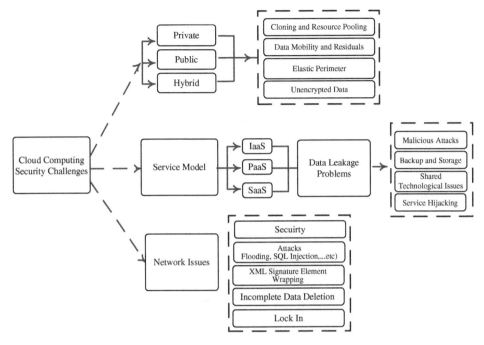

Figure 14.14 Major categories of data security challenges.

A) **Security challenges related to deployment models**

There are many security issues associated with the deployment of the three fundamental types of private, public, and hybrid models. One of the challenges is related to illegal access as a result of sharing data in the same network which is called resource pooling. This could result in data security threats that include data leakage, data remnants, and inconsistent data [37]. Among the most vital attributes of CC is multitenancy which permits more than one user to run the applications on the same infrastructure which adds more security risk to the public cloud such as illegal access to the data. Identity management in CC allows the user to access its private data and distribute it to various services across the network. This feature in the cloud is considered a drawback of interoperability stemming from distinct identity protocols and codes which need more security procedures such as providing a password or fingerprint for achieving a higher level of security.

B) **Security challenges related to service models**

All cloud service models are utilized in real-time over the cloud [38]. That means the users should depend greatly on the service provider for security reasons. For some models such as in PaaS, the cloud providers make available few controls to the users but there is no guarantee with regard to the data protection of customers. This offers proper security and compliance as mentioned below.

Data leakage or data deletion affects security, integrity, and breaches. One solution to this problem is storing copies of redundant data instead of storing the actual copies of the data and provide a link to access this data which is called deduplication with a limitation on the number of user uploads per time window. One of the most known challenges is the shared technological problems such as when IaaS users transport their services in a flexible way by contributing the infrastructure to gain sharing technologies and advantages. However, this structure may not deliver robust isolation properties which leads to security complications on the cloud [39].

C) **Network issues**

Cloud mainly deals with the internet network and remote computers or servers which save all the information for executing different applications. The network structure of the cloud deals with different types of attacks and threats such as browser security, Denial of Service (DoS) attacks, locks-in, and incomplete data deletion. To successfully use CC technology, the structure of the network must be upgraded to develop security standards to affirm data's confidentiality, integrity, and availability.

14.8 Multiple Perspectives for Cloud Implementation

The deployment of CC in SG applications has multiple standpoints which can be grouped as organizational, technical, economic, and political standpoints. Table 14.1 illustrates the most effective criteria associated with each perspective. These standpoints are also described below.

1) **Organizational perspective**

Organizational perspectives are the needed criteria that must be taken into account from organizations that use CC in SG applications. The criteria are collected from several contexts such as security, privacy, culture, and strategy. The following are descriptions of the criteria.

- Security: Any access or transmission to utilities' data must be authorized to provide information security.
- Privacy: The shared data must be reasonable. Organizations must guarantee that their cloud providers employ technical and administrative controls to guard their data.

Table 14.1 The criteria of cloud computing prospective.

Organizational	Technical	Economical	Social	Political
Security	Real-time	Security Reinforcement	Incentives	Policies
Privacy	Consistency	Flexibility	Privacy	Code/Standards Compliance
Culture	Performance Latency	Deployment Model	Complementary Technologies	–
Strategy	Technology Roadmap	Network Deployment Model	–	
Subject-matter knowledge	Fault-Tolerant	Scalability	Training level requires	–
	Interoperability			–

- Subject matter knowledge: Each organization must have the required knowledge to implement or deal with IT systems.
- Culture: How to control the overall process by involving third parties to reduce the cost of control and maintenance.
- Strategy: Each organization must implement its own approach for managing and maintaining the system.

2) **Technical Perspective**

The most common technical characteristics needed for the use of CC for SG applications are the ability to fit the required resources dynamically. The network must be able to implement that by public internet or any dedicated network through choosing the best model based on the needed application and the ability to implement security reinforcement rather than using new technologies to boost the information in the SG. The following are descriptions of the criteria.

- Real-time response: The system's ability to respond in less time.
- Consistency: The ability of CC to provide stable and accurate services.
- Fault-tolerant: Resiliency against disasters when one or more of system components fails.
- Performance latency / Availability: Average time cloud services are available.
- Interoperability: The capability of a cloud system to transport, exchange data with other cloud and IT systems or utilization of data.
- Technology roadmap: A strategy that achieves short- and long-term development goals. This means that CC vendors' plans are considered to enhance the status quo of technology taken into account in these criteria. This describes the main forces that have the potential to form the future of technology.
- Technology maturity: CC technology preparedness level for implementation in several environments [40–42]

3) **Social perspective**

The criteria required to be taken into account from the society's viewpoint are categorized as follows:

- Incentives that encourage users to adopt CC services, which is known as end-user motivation.
- Privacy of the data shared through the cloud is one of the most important aspects of data security to prevent any misuse of the data.
- Complementary technologies availability: Availability of the technology needed to allow the consumers to associate and utilize the cloud services.
- Availability of training needed to allow end-users to access cloud services.

4) **Economical perspective**

Implementing CC in the SG has different economic criteria that should be considered such as scalability which means the ability of the CC system to expand for future development without high expenses. CC systems must provide immediate access to hardware without upfront capital investment for users and should ensure system flexibility to fit different requirements without additional expenses which are known as system flexibility. One of the most important perspectives is choosing the cloud service model (private, public, or hybrid) which should be used based on the required function, for example, infrastructure as a service (Iaas), platform as a service (Paas), and software as a service (Saas). Security reinforcement must be taken into consideration by developing existing technologies to improve the current status of the cloud to allow differentiation between vendors to meet the required security level.

5) **Political perspective**

Political perspectives mean all the associated criteria from studying the government policies' that could affect CC has the chance to reinforce or deteriorate the benefits of cloud services for SGs. Also, a government R&D framework is needed which relates to the support that can be presented by governments such as grants to help implement the cloud technology for SG and to develop the used technology [43]. The effect of codes and standards that judge the system quality must be taken into consideration to ensure the reliability of power systems using CC. These codes and standards are important to adopt CC systems in SGs and to study the effect of its implementation on the reliability of the power system. For example, compliance with North American Electric Reliability Corporation – Critical Infrastructure Protection (NERC CIP) standards which ensure the reliability of bulk power systems and protect all those who use it from cyber threats would be a challenge [44]. All corporations and institutional entities that work using bulk power systems should meet all the regulations set by NERC in order to stay in operation. These regulations are mandated and compelled by the Federal Energy Regulatory Commission (FERC) [45].

14.9 Conclusion

A review on incorporating CC into the SG architecture was discussed in this chapter with a goal of achieving a reliable and effective energy distribution system. Energy management strategies in SGs can be accomplished with the help of the cloud, rather than among end-user's devices. This architecture offers larger cost-effective computational resources including memory and storage to examine computing mechanisms for energy management. The implementation of CC in SGs is an envisioned strategy to be used for transforming the SG architecture. The development of standards and policies to ensure utilization of this opportunity is needed at different levels.

References

1 Kapil, D., Tyagi, P., Kumar, S. and Tamta, V.P. (2017). Cloud computing: Overview and research issues. *International Conference on Green Informatics*, Fuzhou, China (15–17 August 2017). IEEE.

2 Le, T.N., Chin, W.-L., Truong, D.K. et al. (2016). Advanced metering infrastructure based on smart meters in smart grid. In: *Smart Metering Technology and Services-Inspirations for Energy Utilities* (ed. M. Eissa). IntechOpen.

3 Ali, M., Khan, S.U., and Vasilakos, A.V. (2015). Security in cloud computing: opportunities and challenges. *Information Sciences* 305: 357–383.

4 Bohn, R.B., Lee, C.A. and Michel, M. (2020). The NIST cloud federation reference architecture. Special Publication (NIST SP) 500-332.

5 Bhan, R., Ahmad, M.S., Jain, M. et al. (2019). VM Availability in Presence of Malicious Attacks in Open-Source Cloud. *9th International Conference on Cloud Computing, Data Science & Engineering*, Noida, India (10–11 January 2019). IEEE.

6 Naveen, P., Danquah, M., Sidhu, A. and Abu-Siada, A. (2016). Cloud computing for energy management in smart grid-an application survey. *IOP Conference Series: Materials Science and Engineering*, Miri, Malaysia (6–8 November 2015). IOP.

7 Jelassi, M., Ghazel, C. and Saïdane, L.A. (2017). A survey on quality of service in cloud computing. *3rd International Conference on Frontiers of Signal Processing*, Paris, France (6–8 September 2017). IEEE.

8 Shahzadi, S., Iqbal, M., Qayyum, Z.U. and Dagiuklas, T. (2017). Infrastructure as a service (IaaS): A comparative performance analysis of open-source cloud platforms. *IEEE 22nd International Workshop on Computer Aided Modeling and Design of Communication Links and Networks*, Lund, Sweden (19–21 June 2017). IEEE.

9 Mensah-Bonsu, E., Kwarteng, I.O. and Asiedu-Larbi, E. (2019). SECURITY CHALLENEGES OF CLOUD COMPUTING IN GHANA. https://www.academia. edu/40271946/SECURITY_CHALLENEGES_OF_CLOUD_COMPUTING_IN_GHANA (accessed 15 February 2021).

10 Radu, L.-D. (2017). Green cloud computing: a literature survey. *Symmetry* 9 (12): 295.

11 Gonzalez, N.M., Carvalho, T.C.M.d.B., and Miers, C.C. (2017). Cloud resource management: towards efficient execution of large-scale scientific applications and workflows on complex infrastructures. *Journal of Cloud Computing* 6 (1): 13.

12 Hsin Tse, L., Kao, C.H., Po Hsuan, W. and Yi, H.L. (2014). Towards a hosted private cloud storage solution for application service provider. *2014 International Conference on Cloud Computing and Internet of Things*, Changchun, China (13–14 December 2014). IEEE.

13 Kurdi, H., Enazi, M. and Al Faries, A. (2013). Evaluating firewall models for hybrid clouds. *European Modelling Symposium*, Manchester, UK (20–22 November 2013). IEEE.

14 Erl, T., Puttini, R., and Mahmood, Z. (2013). *Cloud Computing: Concepts, Technology, & Architecture*. Pearson Education.

15 Wu, H., Ren, S., Garzoglio, G. et al. (2013). Automatic cloud bursting under fermicloud. *International Conference on Parallel and Distributed Systems*, Seoul, South Korea (15–18 December 2013). IEEE.

16 Chieu, T.C., Mohindra, A., Karve, A.A. and Segal, A. (2009). Dynamic scaling of web applications in a virtualized cloud computing environment. *IEEE International Conference on e-Business Engineering*, Macau, China (21–23 October 2009). IEEE.

17 Park, J-.W., Yeom, J.K., Jo, J. and Hahm, J. (2014). Elastic resource provisioning to expand the capacity of cluster in hybrid computing infrastructure. *IEEE/ACM 7th International Conference on Utility and Cloud Computing*, London, UK (8–11 December 2014). IEEE.

18 Zhu, Zhuangdi, Alex X. Liu, Fan Zhang, and Fei Chen. "FPGA resource pooling in cloud computing." IEEE Transactions on Cloud Computing (2018).

19 Dong, X.D. (2019). Cloud computing application to manage smart grid system. *Science* 4 (3): 369–374.

20 Simmhan, Y., Aman, S., Cao, B. et al. (2011). An informatics approach to demand response optimization in smart grids. *Natural Gas* 31: 60.

21 Kaur, K. and Kumar, N. (2014). Smart grid with cloud computing: Architecture, security issues and defense mechanism. *9th International Conference on Industrial and Information Systems*, Gwalior, India (15–17 December 2014). IEEE.

22 Nachiket Kulkarni, S.V., Lalitha, N.L., and Deokar, S.A. (2019). Real time control and monitoring of grid power systems using cloud computing. *International Journal of Electrical & Computer Engineering* 9 (2): 2088–8708.

23 Ezhilarasi, S. (2016). DDTA-DDoS defense techniques and attributes to integrate Smart Grid and Cloud. *Online International Conference on Green Engineering and Technologies*, Coimbatore, India (19 November 2016). IEEE.

24 Schaefer, J.L., Siluk, J.C.M., de Carvalho, P.S. et al. (2020). Management challenges and opportunities for energy cloud development and diffusion. *Energies* 13 (16): 4048.

25 Bera, S., Misra, S., and Rodrigues, J.J.P.C. (2014). Cloud computing applications for smart grid: a survey. *IEEE Transactions on Parallel and Distributed Systems* 26 (5): 1477–1494.

26 Haug, K., Candel, Kretschmer, T., and Strobel, T. (2016). Cloud adaptiveness within industry sectors–measurement and observations. *Telecommunications Policy* 40 (4): 291–306.

27 Bagherzadeh, L., Shahinzadeh, H., Shayeghi, H. et al. (2020). Integration of Cloud Computing and IoT (CloudIoT) in Smart Grids: Benefits, Challenges, and Solutions. *International Conference on Computational Intelligence for Smart Power System and Sustainable Energy*, Odisha, India (29–31 July 2020). IEEE.

28 Yigit, M., Gungor, V.C., and Baktir, S. (2014). Cloud computing for smart grid applications. *Computer Networks* 70: 312–329.

29 Ward, R.M., Schmieder, R., Highnam, G., and Mittelman, D. (2013). Big data challenges and opportunities in high-throughput sequencing. *Systems Biomedicine* 1: 0–1.

30 Iosup, A., Ostermann, S., Yigitbasi, M. et al. (2011). Performance analysis of cloud computing services for many tasks scientific computing. *IEEE Transactions on Parallel and Distributed Systems* 22: 931–945.

31 Fang, X., Misra, S., Xue, G., and Yang, D. (2012). Managing smart grid information in the cloud: opportunities, model, and applications. *IEEE Network* 26: 32–38.

32 Cioara, T., Anghel, I., Salomie, I. et al. (2019). Exploiting data centres energy flexibility in smart cities: business scenarios. *Information Sciences* 476: 392–412.

33 Deng, W., Liu, F., Jin, H. et al. (2013). Harnessing renewable energy in cloud datacenters: opportunities and challenges. *IEEE Network* 28 (1): 48–55.

34 Hurwitz, J.S., Kaufman, M., Halper, F., and Kirsch, D. (2012). *Hybrid Cloud for Dummies*. Wiley.

35 Hasan, N. and Ahmed, M.R. (2013). Cloud computing: opportunities and challenges. *Journal of Modern Science and Technology* 1: 34–36.

36 Faheem, M., Shah, S.B.H., Butt, R.A. et al. (2018). Smart grid communication and information technologies in the perspective of industry 4.0: opportunities and challenges. *Computer Science Review* 30: 1–30.

37 Parekh, D.H. and Sridaran, R. (2013). An analysis of security challenges in cloud computing. *International Journal of Advanced Computer Science and Applications* 4 (1): 38–46.

38 Jensen, M., Schwenk, J., Gruschka, N., and Iacono, L. On technical security issues in cloud computing, in: IEEE international conference on Cloud Computing, CLOUD'09. *IEEE* 2009: 109–116.

39 Krishna, B.H., Kiran, S., Murali, G., and Reddy, R.P.K. (2016). Security issues in service model of cloud computing environment. *Procedia Computer Science* 87: 246–251.

40 Armbrust, M., Fox, A., Griffith, R. et al. (2010). A view of cloud computing. *Communications of the ACM* 53 (4): 50–58.

41 Wang, L., Von Laszewski, G., Younge, A. et al. (2010). Cloud computing: a perspective study. *New Generation Computing* 28 (2): 137–146.

42 Syed, D., Zainab, A., Refaat, S.S. et al. Smart grid big data analytics: survey of technologies, techniques, and applications. *IEEE Access* https://doi.org/10.1109/ACCESS.2020.3041178.

43 Bakken, D. *Smart Grids: Clouds, Communications, Open Source, and Automation*, 2014. CRC Press.

44 Montanari, M., Cook, L.T. and Campbell, R.H. (2012). Multi-organization policy-based monitoring. *IEEE International Symposium on Policies for Distributed Systems and Networks*, North Carolina, USA (16–18 July 2012). IEEE.

45 Chaichi, N., Lavoie, J., Zarrin, S. et al. (2015). A comprehensive assessment of cloud computing for smart grid applications: A multi-perspectives framework. *Portland International Conference on Management of Engineering and Technology*, Portland, USA (2–6 August 2015). IEEE.

15

On the Pivotal Role of Artificial Intelligence Toward the Evolution of Smart Grids

A Review of Advanced Methodologies and Applications

Mohamed Massaoudi, Shady S. Refaat and Haitham Abu-Rub

Significant attention is paid to Smart Grids (SG) and Artificial intelligence (AI) from the energy community as a means to improve the power delivery service using information and communication technologies (ICT). SG combines the most efficient self-monitoring, control, and management frameworks in power systems. Critical decision-making reflects the essential role of AI in various parts of power systems. Due to the growing energy demand, state-of-the-art machine learning (ML) techniques play an essential role in SG energy management. The next-generation grid infrastructure should be protected against cyber-attacks while maintaining a high level of flexibility to achieve considerable financial benefits. AI lies at the core of the SG paradigm to ascertain its robustness and effectiveness. This chapter addresses the status of AI as a central element in SG while focusing on the recent progress of research on ML techniques to pave the future work in the SG area.

15.1 Introduction

Due to the continuous industrial revolution and world population growth, the electrical grid faces many operational challenges in terms of power scalability, reliability, and autonomous control [1]. For the time being, the power grid transition for greater efficiency and flexibility enhancement reveals a strong consensus. Meanwhile, the SG paradigm provides the most effective solution to overcome the major issues in the outdated grid such as poor adaptation to outliers and heterogeneous sources [2]. The SG concept revolutionizes the traditional utility grid with high reliability, eco-friendly impact, human safety, and infrastructure protection. This interactive power system paves the way for customers and suppliers to exchange their benefits from a bidirectional coordination platform for electricity and information. The SG framework provides a performance toolkit based on ICT [3, 4]. The SG paradigm supervises and promotes grid operations at a high level of expertise [5].

Currently, the distributed renewable energy market is the fastest-growing sector with an estimation of surpassing traditional sources in 2050 [6]. For their smooth and mature penetration, the SG requires a robust and intelligent coordination platform between the different elements of power systems. Traditional automation, based on instructive tasks and

Smart Grid and Enabling Technologies, First Edition. Shady S. Refaat, Omar Ellabban, Sertac Bayhan, Haitham Abu-Rub, Frede Blaabjerg, and Miroslav M. Begovic.
© 2021 John Wiley & Sons Ltd. Published 2021 by John Wiley & Sons Ltd.
Companion website: www.wiley.com/go/ellabban/smartgrid

traditional methods for tedious routine operations between the utility grid parts, is ineffective in dealing with unexpected situations and sustainability problems. The use of conventional approaches for electrical operations through deterministic programming mean the power flow issues remain difficult to control due to the heterogeneous multi-agent practitioners for the "energy mix" generation of the grid [2]. Furthermore, the conventional automation techniques require manual monitoring restoration and operation regulation leading to frequent problems and downtimes especially with the incorporation of Renewable Energy Sources (RES). The SG's underpinnings tend to automatically communicate with different electrical components and deduce the future behavior of each section using extensive calculations in which AI has the main share of their effective deployment.

AI systems have attributed to the realization of Sustainable Development Goals (SDGs) including the vital integration of RES. Furthermore, the decarbonization of fossil fuel systems via the deployment of Electric Vehicles (EVs) using Vehicle to Grid (V2G) and Grid to Vehicle (G2V) concepts requires smart infrastructure. This infrastructure supports a real-time demand response (DR), charge scheduling, fair energy provision, congestion management, and market bidding optimization in which AI methodologies are mandatory for their implementation [7]. Therefore, power systems-based AI gain tremendous interest from engineers and scientists to make use of their huge potential for the SG concept. The SG enormously requires intelligent management to allow diverse RES for bulk penetration to grid-connected systems and ensures well-organized generation scheduling and effective management control. AI-based technologies work under the SG umbrella to provide the correct decision-making and guarantee self-controlled, pervasive networks with broadband ICT. The fusion of AI and SG presents a powerful tool for handling sophisticated problems. When the automation is inefficient in interpreting non-linear and narrow behavior, computer reasoning is competently able to carry out complex tasks with high expertise. Cognitive scientists define AI as the fusion of science and engineering for the benefit of intelligent computational programming [2]. The AI applications include learning, reasoning, and taking actions to face complicated problems such as electric operations.

Nowadays, Computational Intelligence (CI) illustrates the core factor for the digital automation of SG systems taking advantage of the existing infrastructure in the level of hardware, software, and data availability [8]. In energy systems, AI applications include prediction, system modeling and control, and energy management. Predictive models investigate the power system behavior to generate forecasts about the system variability for a specific time horizon. Predictive Modelling (PM) techniques are applied to demand, renewable power prediction, fault detection prediction, etc. The ML concept involves Supervised Learning (SL) models such as Neural Networks and Genetic Algorithms (GAs) [9]. While system modeling consists of defining a symbolic function for a transformation process. For example, ES efficiently interferes with the model simulation using knowledge structure and logic control [9]. Furthermore, system control and energy management employ robust AI algorithms to secure normal operations of power systems.

Doubtless, SG, and AI have aroused substantial attention at a fast pace in the last decade. Furthermore, technology evolution creates a significant emergence for SG. For example, 5G network technology becomes available on many continents around the globe. Many countries take advantage of this technology to increase the size of the data collected and use it for behavioral understanding and predictive applications. From another side, the potential of the computational work witnesses a huge improvement. In particular, the recent waves of quantum computing that deeply integrate Internet Of Things (IoT) concept in electric systems allow AI to be upgraded to AI2.0. Therefore, the improvement of AI techniques is accelerated to meet the high-performance requirements and promote the usage of smart meters and controllers to improve and monitor the power quality in the SG.

The rising interest in AI techniques underpinning the SG area intensified the need for a taxonomic review to summarize the most recent development in SG application areas. However, the existing body of knowledge reported so far in available published papers only focus on the improvement of a specific technique or application for sub-areas of SG. Pioneering review articles for AI & SG applications are reported in Table 15.1 [6, 10–14].

Table 15.1 List of the ML-based AI review papers for 2020.

Ref.	Keywords	Scope	Outcomes
[10]	AI, design, intelligent controller, predictive maintenance, power electronic systems, prognostics and health management	AI applications for power electronics	AI has a huge impact on power electronic systems. AI, particularly, expert systems, FL, metaheuristic methods, and machine learning are deeply investigated due to their importance in the design, control, and maintenance of power systems. Although the increasing interest of AI in electrical systems, the practical implementations of AI are scarcely available.
[11]	RES, Forecasting models; ANN, EM, ML models, Energy planning	renewable energy and electricity prediction models applied to SG	Short-term forecasting is deeply investigated for power system applications compared to the long-and medium-term forecasting. Despite the recent development of the forecasting models, traditional models still have a competitive performance in terms of accuracy and robustness.
[12]	SG; Grid control; Grid reliability; RES; Conventional grid	SG concept and its reliability in presence of RES integration	The major concerns of the grid reliability include the frequency deviation, overloads, synchronization, voltage collapse, and control measures.

(Continued)

Table 15.1 (Continued)

Ref.	Keywords	Scope	Outcomes
[13]	ML, Smart dispatch, Deep neural networks, Parallel systems	Characteristics and challenges for smart dispatch systems, development of AI methods, and parallel learning in power systems	ML models are essential to smart dispatching and generation control. However, the robustness, reliability and, accuracy still under improvement. Significant attention is paid to dispatching robot technology using the new generation AI is meaningfully presented. Big data is fundamental for SG deployment.
[14]	Distributed energy resources; distributed grid intelligence; demand response management,	The role of AI for the integration of RES, demand response, energy response, grid management and security	The SG management highly requires AI techniques. The SG should pay more attention to self-learning, complete automation, self-healing, and skilled workforce proprieties of the SG paradigm.
[6]	Smart city, Intelligent transportation System, Smart grids, Cyber-security	Applications of AI in smart city with the development of ICT	Big data and 5G technology empower the development of Smart cities. The standardization of SGs communication infrastructure is of utmost importance to increase the power grid efficiency.

From Table 15.1, it can be concluded that this research strategy leads to losing sight of significance in tracing the development line of the energy field. The major motivation for writing this study comes from providing a unifying overview of the AI methods related to the SG field and demonstrated with real-world applications. To the best of the authors' knowledge, this study offers the largest high-level review to date from a broad perspective on the status of AI as a basic ingredient in SG infrastructure with the advances of research on this area. Particularly, this study includes the major applications of AI techniques from a compendious standpoint functionality of nearly 450 scientific peer-reviewed papers from scientific research engines.

15.2 Research Methodology and Systematic Review Protocol

The collection of mainstream research papers on SG/AI from Web of Science, Scopus, IEEE Xplore, Science Direct, and Google scholar were conducted as the largest databases of peer-reviewed articles. Only peer-reviewed articles written in English, providing experimental results, and having a unique identifier from the aforementioned databases were taken into consideration including reviews, research articles, patent reports, and conference proceedings. The adopted methodology for conducting this chapter used a combination of keywords categorized into three main groups, specifically, AI, smart grid (SG), and prediction. The keywords are employed as follows:

- Artificial intelligence: "Artificial intelligence," "Deep learning," "Forecasting," "Machine learning," "Support vector machine, Artificial neural network," "Computational intelligence," Fuzzy logic, "Hybrid model," "Metaheuristics," "Edge computing," "Big data," "Data Science."
- Smart grid: "Smart grid," "Smart cities," "Load demand," "Electricity price," "Electric vehicles," "Renewable energy," "Cyber-physical systems," "Photovoltaics," "Wind energy," "MPPT," "Fault detection," "Grid resilience."
- Prediction: "Forecasting," "Prediction," "Optimization," "Predictive maintenance," "Model predictive control," "Classification," "Regression," "Clustering," "Optimization," "Time series analysis."

The keyword combinations are employed in the systematic review protocol to provide a broad overview of research trends in AI techniques in SG concept as shown in Figure 15.1.

The keyword combinations are employed in the systematic review protocol to provide a broad overview of research trends in AI techniques in SG concept.

This study contributes to the existing research papers by answering the following questions:

- Q1: What are the motivations for the transition of the old grid to SG?
- Q2: What are the most recent AI techniques deployed to SG applications?
- Q3: Which metrics have been used to evaluate the ML models' accuracy?
- Q4: What specific limitations are found in the employment of a particular technique?
- Q5: What are the shortcomings and future research directions of AI techniques in SG?

The answers to these questions take into consideration multiple sources of information to ensure the accuracy and objectivity of the main findings of this research work. The search methodology focuses on recent research articles from 2015 to 2020 to identify the comprehensive statues of the AI applications on SG. The filtering process results in 450

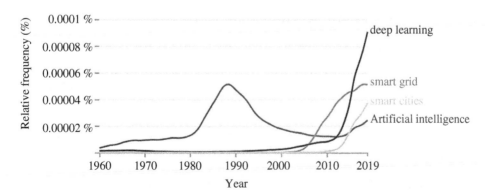

Figure 15.1 Frequency of use of terms AI, SG, DL, and smart cities in books from 1960 to 2020 (*Source:* Data from GoogleBooks Ngram Viewer, 2020).

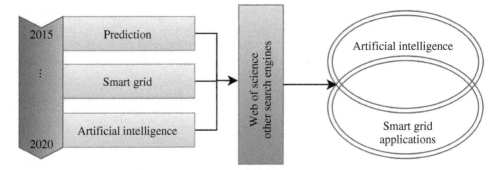

Figure 15.2 Search methodology based on keyword combinations.

research papers selected based on their relevance by reading the title, abstract and conclusion, and full text. The filtered articles are tabulated and unified to facilitate the comparative analysis and assessments according to the prediction horizon, applications, used data, error measures, AI classes, experimental setup, etc. The following criteria were applied: (i) SG and power systems are considered. (ii) The feasibility analysis of the forecasting models is given high importance in the selection process. (iii) the evaluation of the forecasting models emphasizes the use of scale-free metrics. (iv) Future directions and perspectives take into consideration the latest research articles to give a general standpoint of the current statue of SG-based AI and future work.

Figure 15.2 presents a timescale variation on the frequency of use of terms AI, SG, DL, and smart cities in scientific books from Google Books Ngram Viewer.

It can be seen from Figure 15.2 that the popularity of SG significantly increased from 2005. Smart cities approximately follow the SG behavior while Deep Learning (DL) achieves an exponential peak since 2013. This high correlation between the SG and smart cities lies in their complementarity. However, they must take advantage of the increasing trend of DL and AI to promote their applicability in real-world policies and move from the conceptional stage. Further, it can be noted a relative decline of AI in recent years due to the appearance of DL which significantly contributes to the AI research area by all means.

15.3 Century-Old Grid and Smart Grid Transition

The scarcity of conventional energy resources in the near future and their increasing threats to the environment extremely require the transition toward RES [15]. Several milestones have been reached that are indicative of a vital need for RES in the modern economy due to industrial expansion and technological development. Thus, nations around the globe are promoting the use of RES deployment [16]. The ongoing increase of renewable plants reveals technical issues on their deployment in the current power grid for their high sensitivity to weather parameters such as temperature, wind speed, and irradiation. Traditionally, the utility grid concept is not perfectly adapted to emerge such sources in unidirectional energy flow due to partial observability and uncertainty issues. Furthermore, the high disruption of RES and their sensibility to weather conditions made the control of the mixed flow of energy a difficult task. The outdated grid faces a major collapse due to RES reliance. The bidirectional energy flow requires advanced technology to meet renewable incorporation requirements.

Similarly, with global modernization and development of nations across the globe, domestic electricity consumption has dramatically increased from 14 TWh in one year to reach 997 TWh in 2018 [17]. Yearly, the load demand is soaring by 3.5% with expectations to reach a 50% increase in 2050 [17]. The residual, industrial, and transportation sectors are the most dominant sectors for the current heavy consumption. Nevertheless, electricity production mostly uses gas turbines and vapor turbines as sources of energy. Following the ongoing increase in electricity demand, the extinction of these sources is significantly accelerated. The traditional utility grid has four phases, namely, generation, distribution, transmission, and supply. The transmission lines have a unidirectional energy flow where the electricity is delivered to customers via transportation lines. The high voltage in this phase over long distances is stepped down via transformers in the distribution lines. The distribution phase supplies the energy provided to customers in the final step. The four phases are linked sequentially to build an archaic system.

Although the SG technology presents a potentially powerful paradigm to electrical operations, the complexity of SG requires advanced automation tools to control power systems. The bulk penetration of RES into the electrical grid leads to unstable and volatile power generation. This volatility requires AI-based models to address the uncertainty and intermittency of renewable power systems [18]. Furthermore, the wide development of Advanced Metering Infrastructures (AMI) and Wide Area Monitoring Systems (WAMS) intensifies the necessity of AI-based techniques to deal with the massive data produced.

Taking as an example the demand forecasting, stakeholders must provide an effective solution to carry out the load variations at different levels and meet the market demands [19]. However, generating accurate load forecasting (LF) is a challenging task due to the existence of several uncertain factors. These factors include social activities, economic indicators, and seasonal effects. The accuracy of LF from AI techniques has a strong effect on the operational cost and the control strategies for power systems [20]. Furthermore, the flexibility of electricity prices in deregulated markets is ensured by the AI technique to optimize the real-time pricing and billing in usage time.

Researchers have long sought to investigate the different elements that increase the profitability of the grid utility. AI methods provide the essential techniques to identify the power consumption behavior by users [11]. Using this identification ability, the consumers could ensure the reasonable pricing of the energy system and optimize their profitability from the electrical grid. AI tools make use of the power consumption analysis to prevent consumers from abnormal behavior detection. One of the basic goals for a sustainable SG system is to smooth the integration of RES in the electrical grid. Several challenges should be taken into consideration to prevent grid failure. Alternative energy forecasting is very beneficial to enhance the economic operation of the power system.

AI paradigm could provide fault tolerance and self-healing system to control electrical systems. The safety of equipment against every environmental threat could be conducted using AI. Fault diagnosis and AI techniques could enlarge the life cycle of the equipment by the early treatment of the operating system before the complete failure. Furthermore, the power network security protection is assured by the AI techniques to prevent the system from any potential threats. Outages caused by electric Distribution and Transmission failures occurred frequently over these last years [21]. Blackouts can occur due to cyber security attacks and device failures from environmental conditions and sudden accidents where the branch of a tree falling to the ground causes a blackout of all the neighboring houses [22]. In an uncertain and complex environment, the SG system continuously deals with deep information flow leading to possible attacks on the physical power grid. AI ensures quick detection of malware and intrusion. Furthermore, AI methods draw the best strategies for cyber security protection to protect energy systems.

In an SG, AI deployment contributes to the energy systems through budget optimization, self-automation, and efficient management. Particularly, the most noticeable impact of AI deployment in power systems lies in higher improvement in terms of prediction accuracy, powerful security, and high performance. AI models include different aspects based on the type of information used for data representation mapping. These aspects are applied to different classes according to the data provided and the problem complexity. ML classes include SL, Semi-Supervised Learning (SSL), Unsupervised Learning (UL), and Reinforcement Learning (RL) as illustrated in Figure 15.3.

According to Figure 15.3, the application of ML for SG includes classification, regression, clustering, dimensionality reduction, and control applications. The classification techniques are regarded as the identification process of different correlations for feature vectors to distinct output classes. While the regression techniques determine the exact value of labeled data from the data representation [23]. The regression accuracy is measured through error metrics [23]. Dimensionality reduction is employed to optimize feature numbers for the multi-labeled system. This optimization is conducted to identify the most suitable data representations for a specific task and avoid the Curse of Dimensionality

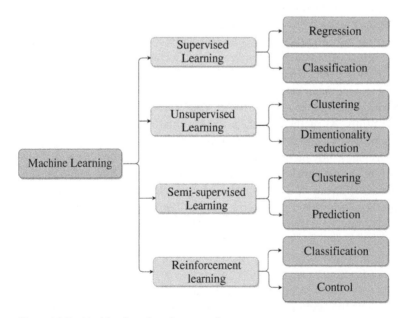

Figure 15.3 Machine learning classes and usage.

(COD) [24]. The COD occurs when the prediction system is overfitted [24]. The clustering defines the grouping of database samples to classes [25]. It could be used as a preprocessing stage for the enhancement of classification and prediction accuracy results. For UL, the clustering consists of grouping the inputs to sub-classes according to specific criteria [26].

AI contributes to the power grid by autonomous and fast equipment control. The communication of the system elements provides flexibility to the grid and ensures an optimal load supply balance. The grid integrity via AI ensures the required protection from cyber threats. The generated data provide a reliable source with AMI systems. The received information is used to tune and enhance AI techniques. The data management and control platform are carried out through AI. Great importance has been addressed to AI techniques due to their crucial role in grid transformation. The information-sharing framework assigns core importance to intelligence since it presents the key success with the support of edge computing [27].

In light of the inherent necessity of advanced tools, the system adopts some fundamental aspects. Those aspects consist of digitalization, flexibility, resiliency, and intelligence [12]. The efficiency of such a system will improve the penetration of alternative energy sources since the stability of the outdated grid is threatened by the intermittent weather change. With AI, precautions are taken for any slight perturbance to select the next suitable decision making. However, the transition to future grids cannot be achieved without the high connectivity of advanced power electronics and devices with each other through high-performance computing and monitoring techniques. With daily operations, a huge amount of data is generated and stored in the system. The role of an intelligent grid system is to manage this data and make it useful for monitoring and control. The static hierarchical grid requires AI techniques for a safe and stable transition.

15.4 Review of AI Methods

Recently, AI aroused greater attraction at a far faster pace than even a decade ago. This appears quite obviously when regarding the massive number of technologies where CI fingerprint is meaningfully marked [28]. In the energy field, AI interferes in modeling, simulation, and prediction to cope with energy sources heterogeneity and operation complexity [23]. CI contributes to power systems with smart programming and advanced methodologies that emerge the SG-enabling strategies. Fault diagnosis, maintenance scheduling, restoration, remedial control, substation monitoring, and management extremely rely on AI to work efficiently and meet the requirements for safe and secured power system operations [28]. Similarly, the deployment of AMI increases the amount of data collected exponentially from the system operations. AMI architecture incorporates ICT technologies, smart meters, and data management platforms [29]. Smart meter data analytics deployment overtakes consumption measurement to provide a massive database from a further usage perspective. Using AI techniques, the collection of this rich information allows the SG paradigm to cope with people's lifestyles and decrease their electric bills. The following subsection depicts the commonly applied AI methods to the SG system and the key elements of the evaluation procedure.

15.4.1 Commonly Deployed Methods

The research community takes advantage of the widespread popularity of AI methods to resolve the operational problems of power systems. In the AI domain, the hybridization approach offers insights to improve the existing models. From a quantitative investigation, hybrid models are commonly implemented in the energy systems as a clue for higher robustness and precision according to a fair assessment basis [30]. Figure 15.4 illustrates the frequently used ML models for SG linked to their most suitable platforms for deployment according to the number of citations from the Scopus engine.

According to Figure 15.4, The most overwhelming models in real-world applications include Support Vector Machines (SVM), Principal Component Analysis (PCA), Gradient Boosted Machines (GBM), K-Nearest Neighbors (KNN), Decision Trees (DT), DL, Random Forest (RF), GA, Ensemble Methods (EM), Convolutional Neural Network (CNN), Auto-Regressive Integrated Moving Average (ARIMA), and Generative Adversarial Networks (GAN). For the sake of following the share of ML methods usage for SG systems limited to 2019–2020, several popular ML models have been coupled with "Smart grid" and searched on google scholar research engine. As a result, a pictorial representation of the AI techniques correlated to SG with their frequency of use is illustrated in Figure 15.5.

According to Figure 15.5, Artificial Neural Networks (ANN) and Swarm intelligence are still a hot topic of research for scientists due to their mature development and high efficiency. The frequent use of ML models is explained by the increased popularity of open-sourced platforms dedicated to facilitating their implementation such as Keras [31]. On the other side, every method is preferred for a specific application of the SG area to perform a custom task better than the rest of the models due to a special inner mechanism. For example, DL takes advantage of its well-designed architecture to do feature engineering by default. For a flow and cohesion section, this part is devoted to tackling the most relevant

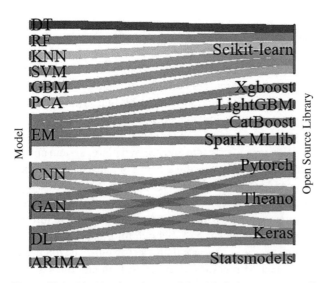

Figure 15.4 Machine learning models with their open-source libraries.

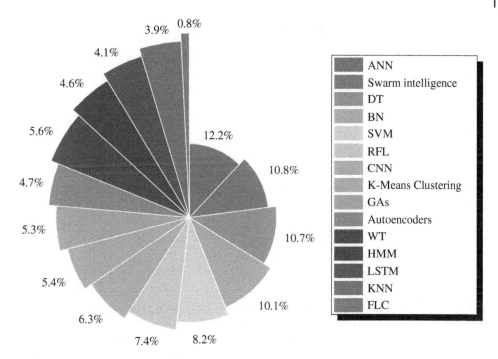

Figure 15.5 Pictorial representation of the AI Models applied to SG in 2019–2020 (*Source:* Data from Google scholar, 2020). Timescale evolution of Artificial Neural Network.

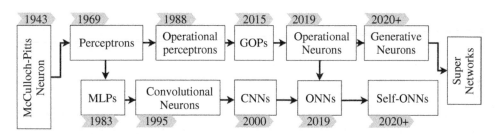

Figure 15.6 Timescale evolution of Artificial Neural Networks with Operational Neuron Networks (ONNs) Generalized Operational Perceptrons (GOPs), +: under improvement. Super Networks belongs to Super AI which meets the technological singularity (These concepts are purely speculative at the time of writing this review and may not exist in the future) Adapted from [33, 34].

AI techniques used in SG operations. These techniques consist of ANN, Fuzzy Logic, Ensemble methods (EM), DL, ES, SVM, and hybrid models.

15.4.1.1 Artificial Neural Networks-Based (ANN)

Conceptually, the ANN model is an interconnection of linked processing units called neurons to mimic the present understanding of the human brain and its associated nervous system [32]. Over the past decades, ANN has achieved a significant evolution from the simple neuron model to more complex ANN structures as shown in Figure 15.6.

The information processing passes through different computational layers, namely, input, hidden, and output layers. The computational stages are forward propagation in the input layers, weight adjustment in the hidden layers, and backpropagation in the output layers. The N vector inputs pass through the activation function $\phi(.)$ with initial weights w and bias b values of each node. The weighs w_K and the bias b_K of elementary processors called perceptrons are adjusted through the loss function between the ground truth and the prediction values in every back-propagation.

The ANN model is perfectly tailored for monitoring and control of non-linear energy systems as it copes with the high variability of weather conditions and the heterogeneity of energy sources [35]. ANN reveals several features that promote its implementation for non-linear problems of single- and multi-complex tasks such as distributed representation and computation, mapping capabilities, powerful generalization, and high-speed information processing [35]. There are numerous other advantages of this concept especially in higher-dimensional settings, still, the loophole lies in algorithm complexity, lengthy-training time for large ANN, and the higher computational burden [35].

The authors in [35] implement the ANN model for the short-term demand prediction of a microgrid supply system using one-year's historical data. The backpropagation neural network (BP) is proposed to generate a Mean Absolute Percentage Error MAPE = 2.12% [35]. The simulation results yield that increasing the complexity of ANN does not necessarily produce a greater forecasting accuracy [35]. The authors in [36] demonstrate that ANN can forecast global horizontal irradiance and global temporal irradiance with a high exactitude. In the authors' work [36], the forecasting horizon was set to five minutes ahead while the historical database was built from seven meteorological data centers. The authors in [37] use the ANN classifier for fault detection diagnostics in DC microgrids. Their work aims to stabilize the current, frequency, and phasor information for better integration of RES. The use of Wavelet Transform (WT) processing stands for capturing the characteristic changes during the faults. Localization of the faults is determined by the ANN classifier for fast detection and isolation of the faults in the system.

15.4.1.2 Fuzzy Logic-Based

Fuzzy Logic Control (FLC) is the development of the fuzzy set theory and its associated techniques. This supervisory technique mimics human cognition and approximate reasoning [38]. FLC involves uncertainty in engineering by attaching degrees of certainty contrary to binary logic. The Fuzzy control system involves four elements, particularly, fuzzification module, knowledge base, inference engine, and defuzzification module as shown in Figure 15.7 [38].

Due to the imprisoning robustness of the FLC model and its great generalization capabilities, it has been widely integrated with energy systems [39]. Paper [40] proposed a smart meter with an autonomous energy management system using FL. The implementation of FL system enables a saving of 34% of energy consumption during the peak period. However, the proposed controller is very sensitive to weather variations and user's stochastic behavior.

Authors in [41] propose a trust management model-based fuzzy system. The proposed model is designed to secure the SG network against compromising lightweight devices by measuring the trustworthiness of devices. The trustworthiness of smart devices is the FLC system that has been deployed to facilitate RES integration and monitoring of electricity

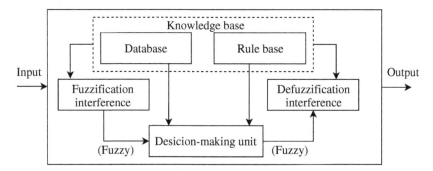

Figure 15.7 Fuzzy inference system structure.

distribution and production between different elements in the grid. The simulation results lead to a classification accuracy of 80%. In [42], the authors introduce an Interval Type-2 Fuzzy Logic System (IT2FLS) based Maximum Power Point Tracker (MPPT) to maximize the Photovoltaic (PV) generation. The proposed model not only tracks the maximum PV power but also pays attention to results uncertainty. The IT2FLS model contributes to the stability of the MPPT model while optimizing the PV power generation compared to Perturbe and Observe algorithm.

According to the reviewed articles, it has been demonstrated that Fuzzy systems have a flexible and suitable architecture where the fuzzy rules could be changed to serve any logical system. Therefore, the FLC model provides great efficiency in modeling and controlling dynamical non-linear systems that require a high level of expertise in deriving mathematical parametric equations. Furthermore, FLC exhibits several features including ease of implementation and the ability to cope with imprecise, incomplete, or uncertain information. However, the shortcomings of FL include its dependence on the problem formulation in terms that it does not always provide a satisfactory performance due to misleading fuzzy rules and membership functions parameters. Thus, careful attention must be paid to fuzzy rules selection with all membership combinations. Moreover, as FLC is widely applied to control systems, the evaluation process is costly due to the fees of hardware implementation.

15.4.1.3 Ensemble Methods-Based

Model assembling targets building a strong model from base learners with higher generalization ability, overfitting avoidance, fault tolerance, and adaptability to solve complex problems. Ensemble learning techniques include two major classes based on the way of building base learners: competitive and cooperative ensembles [43]. Competitive ensembles use multiple base predictors to constrict a powerful ensemble through averaging the prediction of single-base predictors. The prediction output could take into account purring the base predictors. The weak predictors' diversity is maintained by employing different parameters, initial conditions, kernel diversity, or data diversity [43]. On the other side, for cooperative ensembles, the prediction task is partitioned into distinguished sub-folds. Then, the cooperative model computes each sub-fold individually. The prediction results are averaged to generate the final forecast output.

Cooperative ensemble forecasting includes two classes: Pre-processing and post-processing. In the pre-processing, the subsets of the cooperating ensembles are divided without taking into account the data characteristics. Wavelet Decomposition and Empirical Mode Decomposition (EMD) are two popular time-series examples for pre-processing [44]. While in the post-processing, the data linearity is taken into account. The cooperative ensembles attribute for each base models the most suitable subsets according to their potential capabilities to carry out specific data characteristics. For example, ARIMA can be used as a linear model and associated with a non-linear model such as Generalized Auto-Regressive Conditional Heteroscedasticity (GARCH) model, SVM, and ANN. Ensemble learning techniques present efficient meta-algorithms for forecasting problems related to SG systems [44].

The goal of using EM in energy systems consists of avoiding class imbalance, COD, and concept drift. This goal is conducted using two frameworks according to the dependency of inducers in the training stage [45]. The dependent training uses transfer learning to correlate between the actual inducer and the following inducers and acquire knowledge. On the other side, independent training separates the base learners to be trained individually. For example, Bootstrap Aggregation (Bagging) such as RF and DT, stacking, and blending approaches are the most well-known independent frameworks. While Boosting is one of the commonly used dependent frameworks in practical applications for its high efficiency including Adaptative Boosting (AdaBoost), Categorical boosting (Catboost), and Gradient Boosting Machines (GBM) [46]. Bagging is an intelligent algorithm that employs a parallel generation of individual estimations to optimize the variance. Although Bagging approach demonstrates its prediction ability in high-variance data, studies approve that the risk of overfitting is high. Therefore, it is recommended to define a threshold when the learning rate stops improving. While using the Boosting approach, the accuracy is increasingly improved via sequential generation and weighted average.

Boosting is advantageous in the sense that the optimization accords more importance on the bias rather than the variance. The stacking approach consists of creating multiple prediction levels from heterogeneous models [47]. The output result is generated through a combination of meta-learner training folds and using an out-of-fold learning technique. Those folds are utilized as a base learner and the information is transferred via the folds training. The boosting method prevents the system from overfitting and generates a high-quality prediction [48]. However, the Stacking restraints resume in time-consuming with a huge computational burden requirement. Blending is quite similar to Stacking in terms that it uses the fold-training technique but the validation set is made only from the training folds. The holdout set accelerates the training process to simplify the learning process [49].

A novel predictive method for energy price classification was proposed in [50]. The proposed method does not identify the exact value of electricity price but it classifies the values to a specific threshold. Their work was extended to include an investigation of Classification and Regression Trees (CART) and C5.0 models. The combination of EM provides high accuracy of 99.3%. Zhang et al. used a combination of two EM, specifically, RF and XGBoost to provide generalization capabilities for fault detection of wind turbines [51]. RF model is implemented to rank the features according to their contribution to the domain knowledge (DK) while XGBoost trains the model from three features.

15.4.1.4 Deep Learning-Based

DL is a promising branch of ML with breathtaking potential and widespread adoption from the research community and big tech companies. As a novel wave of ANN research, DL takes advantage of multiple processing layers, improved parallel calculation, and excellent representation ability to boost the performance of computational models using massive high dimensional data [52, 53]. Neural network architectures with different configurations and structures such as Recurrent Neural Network (RNN), CNN, GAN, Deep Belief Network (DBN), and auto-encoders are the typical DL models to solve complex problems in the area of SG and power electronics [54, 55]. For instance, the Long Short-Term Memory (LSTM) model dramatically improved the state-of-the-art in time series prediction due to its strong ability to learn long-tailed temporal dependencies using memory units and customized gates [55]. These gates tend to reduce the vanishing gradient from RNN, where the information is lost in-depth when it backpropagates. However, RNN-based structures including LSTM and Gated Recurrent Unit (GRU) have limited potential in learning spatial features. Unlike LSTM networks, CNNs analyze hidden patterns using pooling layers for scaling, shared weights for memory reduction, and filters for capturing the semantic correlations by convolution operations in multiple-dimensional data. Thus, CNN architecture acquires a strong potential in understanding spatial features [56]. Despite the CNN potential, CNN model suffers from its disability in capturing special features [56]. GAN has gained tremendous interest as a mainstream super-resolution model. GAN uses a generator and discriminator neural network to create synthetic data that follows the same distribution from the original data set [57]. More specifically, the generator network mimics the data distribution using noise vectors to confuse the discriminator in differentiating between the fake image and real image. The role of the discriminator is trained to distinguish the generated fake image created by the generator from the original image following the two-player zero-sum game [58]. Despite the large popularity of GANs, the assessment of GANs is quite challenging and often limited to computer vision and image recognition applications.

In [59], the authors employed an improved DBN for short-term load forecasting (STLF) using meteorological data, load demand data, and demand-side management data. The improvement methodology consists of the virtue of processing units, specifically, the Hankel matrix and gray relational analysis for correlation analysis, Gauss-Bernoulli restricted Boltzmann machines (RBMs) for identifying the probability density of data, Bernoulli-Bernoulli RBMs for processing binary data, and mixed pre-training and GA for parameter optimization. The proposed model in [59] is found outperforming other benchmarks with a MAPE = 6.07%. From the obtained results, a general conclusion can be drawn that giving higher importance to demand-side management (DSM) data and the electricity price data enhances prediction accuracy of the proposed model. However, the proposed methodology requires four different kinds of data to operate. This condition is difficult to satisfy in practical industrial applications. Capsule network (CN) has been used in [60] for fault detection and diagnosis. CNs are an advanced form of CNN comprising a group of vector neurons, primary capsules layer, convolution layer, and digit capsules layer. Capsules in CN identify spatial patterns between the lower-level entities and apply a dynamic routing algorithm to recognize these relationships. From the authors' work in [60], a weight-shared capsules network is employed to further supplement the generalization capability of the original CN. The proposed model is found perfectly tailored for an automobile transmission. Nevertheless, the proposed model is dependent on the quality of labeled data to perform fault diagnosis of machinery. This data is difficult to acquire, especially when the faults happen rarely.

In [57], the authors used GAN and CNN for a day ahead of PV power forecasting. Wasserstein Generative Adversarial Networks with Gradient Penalty (WGANGP) is proposed to classify 33 meteorological weather types for generating synthetic training data. From the simulation results, it has been found that weather classification plays an important role in determining the most suitable PV power forecasting technique. In [61], the authors proposed a DL approach for interval wind forecasting. The proposed model employs CNN, and GRU models to learn Spatio-temporal representations for probabilistic forecasting. From the loss vs. epoch variation, it is worth mentioning that the convergence of the model was reached in the first 10 iterations which may lead to serious over-fitting problems. In [62], the authors proposed an interval probability distribution learning model for wind forecasting. The proposed model, namely, deep mixture density neural network, employs RBMs, rough set theory, and fuzzy sets to capture the uncertainties from the wind data. It has been found that fuzzy type II inference system could supplement the DL models' robustness leading to a Root Mean Square Error RMSE = 0.419 (m/s) for 10-minute ahead forecasting. Despite the strong competitiveness of the proposed forecasting system, a serious hyper-parameter optimization and high computational power are extremely required in the proposed model application in real operational conditions. To overcome the mentioned problems for wind power forecasting, a Spiking neural network (SNN) was introduced as one of the third generation ANNs [63]. SNNs are considered as efficient models for temporal coding in neurons where the neurons interact with other nodes through excitatory or inhibitory spikes. Furthermore, SNNs can support huge parallel processing using neuron clusters which significantly accelerates the execution time [63]. The proposed forecasting system in [63] associates SNN with a group search optimizer for automatic hyperparameter tuning to perform probabilistic forecasting with excellent performance and a short training time of 1.31 seconds.

Despite the inherent characteristics of DL, overfitting persists as a challenging problem. Overfitting happens with inadequate data size leading to poor generalization potential [64]. Regularization techniques such as dropout and batch normalization tackle the overfitting problem by using shared weights or network capacity and internal co-variate shift reduction. Data augmentation (DA) can be mentioned as an efficient technique to reduce overfitting caused by data scarcity. Furthermore, DL models require extensive work on hyperparameter optimization and model structure. Although DL is deeply investigated in power electronics, the reported limitations impede the latter concept from becoming a canonical approach in the industry. Table 15.2 presents a list of the mainstream DL methods with their advantages and disadvantages of SG applications.

15.4.1.5 Expert Systems-Based

The Domain Expert (DE) concept is a deep understanding of a specific field with strong skills and backgrounds of what is currently known and what not everyone can do. ES presents an intelligent computer program that resolve difficult problems. This architecture deploys artificial human expertise using a knowledge-based ES [76]. The DK, firstly introduced in 1970, presents a well-established prediction method [76]. ES architecture is illustrated in Figure 15.8.

Regarding Figure 15.8, the ES is divided into three stages. The first stage acquires the information from experts and transcribes it into rules, frames, and semantic nets with

Table 15.2 Mainstream DL architectures with their advantages, limitations, and example applications.

Model	Reference	Advantages	Limitations	Exemplary Applications
GRU	[55, 65, 66]	• Simple and effective gated structure • Convenience to solve complex problems • Fast training • Change tracking over time • Preserve the characteristics of long dependencies • Alleviate the vanishing gradient problems • Able to capture abstract temporal features • High performance for sequential data	• Poor spatial features representation • Difficult implementation • Difficult parameter tuning	Health monitoring/ PV power forecasting
SAE[a]	[67]	• Do not need unlabeled data for training • No need for un label led data for training • Better feature descriptions process • Automatic dimensionality reduction • Excellent performance for high-level feature abstractions and representations • Easier to follow the reduced loss/cost function • Easy-to-implement	• Extensive processing time and fine tuning • Hyperparameter optimization is difficult • The relevance of information is not considered • Training can suffer from vanishing of the errors	Fault Diagnoses
DBN	[59, 68]	• Unsupervised training • High effectiveness for feature extractions	• Disable to process multi-dimensional • Training can be very slow and inefficient	Load Forecasting/ fault detection
RBM[b]	[69]	• Ability to learn a probability distribution over inputs • Incorporate top-down feedback • Efficient inference with narrow inputs	• Difficulties in the determination of reconstructed • Data-driving probabilities over visible and hidden layers • High time complexity required for the inference • Extensive parameter optimization is for big data	Wind speed forecasting

(Continued)

Table 15.2 (Continued)

Model	Reference	Advantages	Limitations	Exemplary Applications
CN	[70–72]	• Overcome the deficiency of max-pooling • The parallel architecture reduces the gradient explosion • Less training time which allows handling the massive data from large power grids	• Not well-applied in experimental setup • Can only learn problems where the negative input represents the same class as the input itself	fault detection
GNN[c]	[73, 74]	• Preserves the principal topological relations between node adjacencies • Generalize over arbitrary topologies, routing schemes and traffic intensity • Able to model graph-structured information • Achieve relational reasoning and combinatorial generalization over information structured as graphs • Perfectly tailored for different topologies without retraining	• Poor performance with different class labels • Low accuracy with dissimilar features • Inability to detect and count graph substructures-informative for downstream tasks	Distributed Economic Dispatch
DRN[d]	[75]	• Ability to skip connection to avoid vanishing and exploding gradients • High performance with spatio-temporal dependencies	• Problem of diminishing feature reuse • Long training time especially for large number of layers	Load forecasting

[a] Slacked auto-encoders
[b] Restricted Boltzmann Machines
[c] Graph Neural Network
[d] Deep Residual Network.

boolean logic. The source of information includes human experts, datasets, or scientific papers in the field of interest [9]. This means that the original dataset includes factual, heuristic, or queries background with contributing to the DK by a computer recognizable Boolean rule. In the next stage, acquired knowledge is applied to a real case study. The Interface Engine (IE) is a coordination framework that links the user to the DK via the input and output interface. This coordination allows the ES to interact with the environment which leads to generating conclusions and explanations. In the last stage, IE also

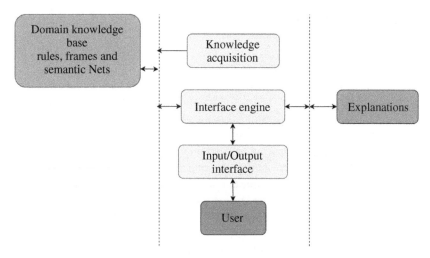

Figure 15.8 Flow diagram of expert system.

provides a convenient explanation for the output decision via the data used and the rules applied. Furthermore, the improvement of KB is achieved through additional rules and new situations that contribute to the KB and update the existing information acquired. The architecture of ES demonstrated its suitability as it copes with the complexity of SG concept [77–79].

Authors in [80] introduced an advanced method for the control of nonlinear systems using expert knowledge and the deep fuzzy neural network (DFNN). The proposed schema passes by an offline pretraining using DFNN and online post-training for corrective actions. The DK is encrypted into the rule-based to enable online learning. The knowledge expertise of adaptive learning-based control improves the achieved results by 50%. In [81], the mainstream expert control systems were reviewed for the control of wind turbines. The proposed model takes into consideration the dynamic operations of wind energy (WE) systems and the control response approaches to generate its proper rules. A general conclusion can be drawn that the efficient management of the environmental variables and the control rules for the use of the stored knowledge expertise is the core factor for the proper solutions. Belief Rule-Based Expert Systems based DL (BRBES-DL) was proposed in [82] to carry out the uncertainty analysis related to power generation. A deep processing layer has been integrated as an alternative to the activation weight function of the original BRBES to enhance the prediction performance of the underlying system under uncertainty. Despite the high potential of the BRBES-DL model, it lacks universality to be limited only to supervised regression problems.

The authors in [83] present an electric load prediction model on a receiver Operating Characteristic Curve (ROC). The used model in this study is a fuzzy expert system which has a high performance of 87%. The use of ES as one of the AI approaches provides a user-friendly interface that facilitates the interconnection with the environment. However, the vital need of experts to establish scaling rules make ES difficult to implement in some meticulous situations. Furthermore, the correlated rules of ES are highly dependent on the system of interest which leads to a poor generalization potential [10]. Thus, the usage of ES loses sight of significance in recent years contrary to ML which relies on CI more than knowledge expertise to solve particular tasks.

15.4.1.6 Support Vector Machines-Based

It is well documented in the literature that SVM model demonstrates its usefulness in data analysis, prediction, and optimization tasks. SVM architecture consists of an optimal boundary for the separation and identification of feature samples using multiple transformations of the database and a structural risk minimization principle. This algorithm applies statistical learning theory by creating a threshold margin for clustering the dataset; SVM achieves better results with SL especially with small data. Authors in [84] combine an improved SVM and Extreme learning Machine (ELM) for STLF. The hyperparameters tuning for ELM and improved SVM is conducted by GA and Grid Search (GS) method respectively. Although the high performance of improved SVM-GS (93.25%), the ELM-GA proved its superiority with an overall accuracy of 96.42%.

Authors in [85] proposed a novel algorithm based on SVM and improved moth-flame optimization for PV power forecasting. The proposed method consists of optimizing the parameters of SVM for a better adaptation to weather conditions. The proposed algorithm succeeds to achieve high-quality results with a coefficient of determination $R^2 = 99.62\%$ for the rainy weather. In [86], the hybridization of the SVM and WT decomposition is introduced for short-term power prediction. The proposed method aims to study the DA impact on the hybrid model accuracy in the time-frequency domain. The WT-SVM is compared to a hybrid radial basis function (RBF)-SVM model using one wind farm from Texas. WT-SVM demonstrated its performance superiority in terms of RMSE and Mean Relative Error (MRE). An uncertainty analysis is conducted to measure the potential risk factor in the prediction of WT-SVM model. The WT-SVM model performs a 90% confidence degree which demonstrates the high efficiency of the proposed model.

15.4.1.7 Hybrid Models-Based

Previous studies, to date, have primarily been focused on AI model' hybridization as a potentially viable approach since it gives individual models a massive upgrade in the majority of cases. Conceptually, hybrid techniques include two or more heterogeneous predictors. The weakness of a single predictor in capturing individual and group behavior from data is compensated by the rest of the models. The evolution of AI passes through the fusion of techniques and is a key solution to overcome narrow tasks in which conventional ML techniques generate poor accuracy. A variety of hybrid methods uses optimizers and predictors in the forecasting engine to minimize the errors. The hybridization of ML techniques illustrates a promising approach, especially with narrow datasets. Figure 15.9 presents a variety of hybrid techniques dedicated to SG applications.

These techniques outperform conventional models and significantly ameliorate model uncertainty. There is extensive interest in these types according to the number of papers that dramatically increased over the last 10 years. The authors in [87] depict the LF problem using an Adaptive Neuro-Fuzzy Inference System (ANFIS) and compared with the FL model. A yearly dataset is widely used for the training part. ANFIS is a supervised adaptive Multi-Layer Perceptron (MLP). It combines the self-learning capability from Neural Networks (NN) with the linguistic expression function of FL. The mechanism of this

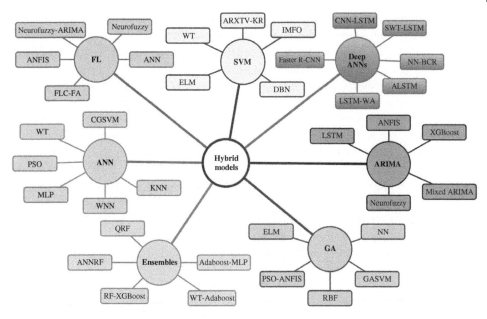

Figure 15.9 Representation of hybrid models applied to SG.

architecture consists of converting the series data into fuzzy inputs. In the inference system, these fuzzified data build the IF-THEN rules, the implication and defuzzification take place to indicate output weighting factors from the preceding part and then convert the fuzzy rules into crisp values. The proposed method in [87] provides accurate hourly load forecasts for the electric consumption data of Turkey using a monthly three years dataset from 2009 to 2011. Although both algorithms are very efficient in load prediction, ANFIS model outperforms FL [87]. However, the high complexity results in long computational time. Some authors in [88] integrated an ANFIS model but with a two-stage feature selection with a hybrid Backtracking Search Algorithm (BSA). The proposed model deals with an electricity price prediction. BSA presents an efficient method for membership function tuning. The feature selection is held through a multi-objective binary-valued BSA to create a various subset from the dataset. Then, ANFIS model assesses the importance of each dataset.

Paper [89] proposed a hybrid method for WE forecasting. The proposed method combines a DBN and Support Vector Regression (SVR). DBN is a supervised DL technique having l-layers and parabolic hidden cells [89]. The construction of layers is made by RBM. The idea of this fusion is supported by the fact that the wind consists of two levels: hourly average wind power level and turbulent residual error level. Thus, each predictor is concerned with a single level. Despite the transient stochastic variation, DBN-SVR model generates quite accurate results with an error MAE equal to 0.014. Furthermore, the proposed model outstrips individual models' accuracy (SVR and DBN). However, the proposed method is time-consuming [89]. In a similar vein, the authors in [90] combine LSTM and CNN for power demand forecasting.

The hybrid approach namely (c,1)-LSTM-CNN proved their high performance IN various areas. For the energy field, this hybrid model uses dual-stage attention and the inputs are passed through LSTM cells for features extraction with two hidden layers. Then, CNN model uses these data processed to make predictions. Compared with S2S LST, ARIMA, and (c,1)-LSTM. The model shows a better performance according to MAPE and RMSE scores. However, the suggested model is high complexity for just a short horizon in which more sampler techniques can generate for the same range of accuracy. Paper [91] combines WT and ensemble learning types illustrated in the AdaBoost model. This research work tackles the problem of short-term wind speed forecasting. WT is deployed to transform the data into numerous scales and derive a divided subset. Overall, AI methods are divided into seven learning groups with different degrees of complexity according to their distinctive architectures as shown in Figure 15.10.

According to Figure 15.10, The AI learning models include Statistical learning, Ensemble learning, Swarm intelligence, ML and RL, Bayesian learning, DL, and Hybrid models. The learning diversity refers essentially to the nature of rules deployed, parametric functions, the information processing stages, learning styles, and application domains. A clear understanding of the AI taxonomy mapping enables users to select the most suitable algorithm

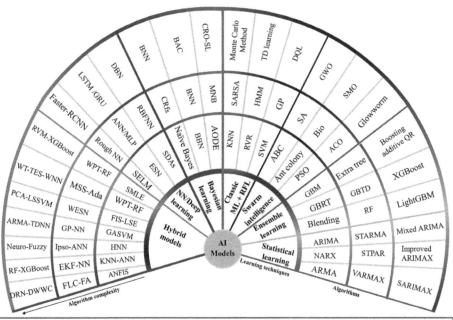

Abbreviations: *ACO= Ant Colony Optimization, ABC = Ant Bee Colony, Ada= Adaptive boosting, AODE= Aggregating One-Dependence Estimators classifier, ANFIS= Adaptive neuro-Fuzzy Inference System, ANN=Artificial Neural Network, ARIMA= AutoRegressive Integrated Moving Average, ARIMAX= AutoRegressive Integrated Moving Average process with eXogenous inputs, ARMA= AutoRegressive Moving Average, BAC= Bayesian Actor-Critic, BBN= Bayesian Belief Network, Bio= Biological swarm chasing algorithm, BNN= Bayesian Neural Network, CRfs= Conditional random fields, CRO-SL= Coral Reefs Optimization algorithm with Substrate Layers, CNN-WT= Convolutional Neural Network-Wavelet Transform, , DCNN= Deep Convolutional Neural Network, DQL= Deep Q-Learning, DBN= Deep Belief Networks, DRN= Deep Residual Networks, DWT= Discrete Wavelet Transform, DWWC= Dynamically Weighted Wavelet Coefficients, ELM= Extreme Learning Machine, ESN= Echo State Networks FA= Firefly Algorithm, Faster R-CNN= Faster Region-based Convolutional Neural Network, FCRBM= Factored Conditional Restricted Boltzmann Machine, FLC = Fuzzy Logic Controller, FIS= Fuzzy Inference System, GA= Genetic Algorithm, , GARCH=Generalized Autoregressive Conditional Heteroskedastic, GASVM= Genetic Functionality Support Vector Machine, Glowworm= Glowworm swarm based optimization, GBM= Gradient Boosting Machine, GBRT= Gradient Boosted Regression Tree, GBTD= Gradient Boosting Theft Detector, GP= Gaussian Process, GRU= Gated Recurrent Unit, GWO= Grey Wolf Optimization, HMM= Hidden Markov Models, IPSO= Improved Particle Swarm Optimization, KNN= K-Nearest Neighbors, , LightGBM = Light Gradient Boosting Method, LSSVM= Least Squares Support Vector Machines, LSTM= Long Short-Term Memory, LES= Least-Squares Estimation, NN= Neural Networks, PDRNN= Pooling-based Deep Recurrent Neural Network, SD-EMD= Similar Days-selection-Empirical Mode Decomposition, Neuro-Fuzzy= Artificial neural networks-Fuzzy logic, LES= Linear exponential smoothing, MLP= MultiLayer Perceptron, MSS= Model Structure Selection, HNN= Hybrid Neural Network, MNB= Multinomial Naïve Bayes, NARX=Temporal difference learning, NB=Naïve bayes, NN=Neural Networks, PCA= Principal Component Analysis, PSO= Particle Swarm Optimization, QR=Quantile Regression, RBFNN=Radial Basis Function Neural Networks, RF=Random Forest, RFL= Reinforcement learning, RVR= Relevance Vector Regression, SARSA= State-Action-Reward-State-Action, SA=Simulated Annealing, SARIMAX= Seasonal AutoRegressive Integrated Moving Average process with eXogenous inputs, SDAs= Stacked Denoising Autoencoders, , SELM= Stacked extreme learning machine, SMEL= Stacking Heterogeneous Ensemble Learning model, SMO= Spider Monkey Optimization, SMLE= Semi-parametric Maximum Likelihood Estimator, STARMA= Space Time AutoRegressive Moving Average, STPAR= Smooth transition periodic autoregressive, SVM= Support Vector Machines, TES=Triple Exponential Smoothing, TD learning= Temporal Difference learning, TDNN= Time Delay Neural Network, VARMAX= Vector AutoRegressive Moving Average with eXogenous inputs, VMD= Variational Mode Decomposition, WNN= Weighted Nearest-Neighbor, WPT= Wavelet Packet Transform, WT= Wavelet Transform, XGBoost= Extreme Gradient Boosting,.*

Figure 15.10   Representation of hybrid models applied to SG.

for their applications and thus, improving the prediction system performance. It must be noted that all algorithms mentioned in the flowchart are taken from real applications on power systems and SG. On the other side, the short-term prediction is investigated rather intensively. The efficiency of these methods is limited for further steps ahead in medium- and long-term predictions. Table 15.3 illustrates the recent emerging AI-based SG applications. These methods are classified into the year of publication, author, model, output, and problem type.

Table 15.3 List of hybrid models for smart energy applications.

Year	Author/ Reference	Model	Output	Problem type
2019	Xu et al. [92]	Hybrid NN- BCR	Electric load	Regression
2018	Jiang et al. [93]	SVR With Hybrid parameters Optimization	Electric load	Regression
2018	Rafiei et al. [94]	Hybrid GELM, Wavelet preprocessing and bootstrapping	Electric load	Regression
2018	Karmacharya et al. [95]	Hybrid DWT-ANN	Fault Location	Classification
2019	Zhen et al. [96]	PCPOW model	type of Sky images	Classification
2018	Chen et al. [97]	Unsupervised Feature Learning and CSAE model	Fault detection	Classification
2018	Shafiullah et al. [98]	Hybrid ST-FFNN	faults detection	Classification
2019	Mishra et al. [99]	FDOST-DT	Fault Classification	Classification
2018	Li et al. [100]	Hybrid WPD-Boost-ENN -WPF	Wind speed	Regression
2018	Song et al. [101]	CEEMDAN-GWO combined model	Wind speed	Regression
2019	Park et al. [102]	CLD-based linear prediction with LSTM	Electric load	Regression
2020	Incremona et al. [103]	LASSO-FFT algorithm	Electric load	Regression
2019	Yan et al. [104]	Hybrid SWT-LSTM	Energy Consumption	Regression
2019	Chu et al. [105]	LSTM-DQN-based scheduling algorithm	state of battery	Regression
2018	Ghadimi et al. [106]	Hybrid MSFE	Electricity price	Regression
2019	Ahmad et al. [107]	hybrid ANN-based DALF model	Electric load	Regression
2019	Ahmad et al. [108]	Hybrid ANNSE	state of the distribution network	Regression
2019	Naz et al. [109]	ELR and ERELM	Electric Load and Price	Classification
2019	Álvarez et al. [110]	PSF- funHDDC	Electricity Demand	Clustering

(Continued)

Table 15.3 (Continued)

2018	Kuo *et al.* [111]	CNN-LSTM	Electricity prices	Regression
2019	Bissing *et al.* [112]	MLR- ARIMA- Holt–Winters model	Energy Price	Regression
2019	Jesus *et al.* [113]	nonlinear regression and SVM	electricity price	Regression
2019	Aimal [114]	ECNN- EKNN	Electricity Load Forecasting	Classification
2019	Hafeez *et al.* [115]	FCRBM-ELF	Load Forecasting	Regression
2018	Gonzàlez [38]	Hilbertian ARMAX Model	Electricity Price Forecasting	Regression
2018	Dong *et al.* [116]	SVRCCS model	Electric Load Forecasting	Regression
2019	Wang *et al.* [117]	CQRA	Electric Load Forecasting	Regression
2019	Ouyang *et al.* [118]	DBN	power load	Regression
2019	Yu *et al.* [119]	GRUNN-DTW	Electric load	Regression
2019	Zhang *et al.* [120]	improved QRNN	Electric load	Regression

With regard to Table 15.3, it can be seen as an increasing focus on assembling and hybridization of methods implemented. However, ES and metaheuristic algorithms in AI have lost their popularity due to their high complexity. Prediction, optimization, and control are the dominant sorts of using CI due to the availability and abundance of high-resolution data in most cases. Therefore, the prediction system can learn solely from concept examples. For particular problems, scare data limits the profitable use of ML due to high acquisition cost or low data storage efficiency. These problems include anomaly detection, predictive control, diagnosis, time series classification, etc. DA is an effective approach to overcome poor data quality [100]. DA uses basic approaches such as Fast Fourier transform and WT to convert the data into the time domain, frequency domain, or time-frequency domain [121]. Moreover, several advanced methods were proposed in the literature to increase the size of the limited labeled data using decomposition methods, model methods, and learning methods [122]. Table 15.4 present the advantages and limitations of all the presented methods.

All these algorithms have their strengths and weakness in terms of hyperparameter settings, data exploration of the computational burden. Therefore, metaheuristic optimization techniques were adopted to support AI algorithms or provide an alternative for specific optimization tasks due to their efficiency and scalability for various complex applications including the design of the perfect architectures of DL approaches. In Table 15.5, a list of the most recent metaheuristic algorithms is presented with their limitation and advantages and their exemplary applications on energy systems.

Table 15.4 Comparative study of the proposed methods.

Model	Strengths	Limitations
ANN	• Ability to approximate continuously differentiable • functions • Ability to deliver high-quality results • Fault tolerance ability • Provide a distributed memory for computation	• Easy to overfitting • Local minima can reduce the model performance • Less accuracy • The complex structure of a large-scale power grid makes ANN less reliable • Convergence can be slow • Sensitivity to the initial randomization of weight matrices. • Unexplained behavior of the network • Improper initial weights may affect the learning convergence speed and make learning fall into local optima • More hidden layers in ANNs may lead to weak training results
FL models	• High Flexibility and consistency • Efficient computation • Universal approximation capability • Accurate and practical • Handle quantitative/qualitative information and uncertainties with high exactitude	• Hard to define the optimal fuzzy rule base • High complexity • A large number of parameters to tune • Can be affected by the curse of dimensionality
Ensemble Models	• Easy to understand • Non-parametric • Better generalization capability of multiple models • A high degree of freedom in the bias/variance trade-off	• Previous knowledge of system • May overfit data • It can get stuck in local minima • The training cost is too high • Computationally expensive • More Memory constrains
DL models	• offer unsupervised feature engineering • High robustness and reliability • Strong generalization capability and large datasets training/Robust against noise successful applications in many domains	• Requires a large amount of training data labels • Previous knowledge of system • Results can be incomprehensible • Increasingly hard to train as the number of layers increases • Memory intensive • High time-consuming • Difficulties in the determination of proper network structure • Hardware Dependence • Extensive hyperparameter optimization

(Continued)

Table 15.4 (Continued)

Model	Strengths	Limitations
Expert systems	• Consistent solutions for complex problems • Logical understanding and reasonable explanations • Robust and fast adaptability to sudden disturbances • High reliability	• Lack of common sense • High implementation and maintenance cost • Difficulty in creating inference rules • Time-consuming • Poor universality • Require continuous updating • Difficult knowledge acquisition
SVM	• Memory-efficient • More effective in high dimensional spaces • Excellent generalization ability • High scalability • Can address high dimensional data even with relatively small training samples	• Cannot perform automatic gene selection • Difficulties of the kernel function selection • Low convergence speed with large data sets • Lack of efficiency in handling • multi-class scenarios • High computational cost • Sensitive to noise
Hybrid models	• Better generalization capabilities • High accuracy • Overcoming the limitations of each method used alone	• More complex structure • High computational burden • Difficult to implement

Table 15.5 Comparative study of metaheuristic algorithms for SG applications.

Algorithm	Ref.	Advantages	Weakness	Exemplary Application
Grey wolf optimizer	[123]	• High generalization potential • Impressive characteristics • Competitive results • Good convergence ability • Widely used in neural network training • High optimization precision	• Easy to fall into local optimum with high-dimensional data • Imbalance between exploration and exploitation • Slow convergence	MPPT optimization

Table 15.5 (Continued)

Algorithm	Ref.	Advantages	Weakness	Exemplary Application
Salp Swarm optimizer	[124]	• Equipped with fewer parameters and operators • Simple and easy to implement • Effective in various applications	• Low precision • Low optimization dimension • Slow convergence speed • Can fall into the local stagnation • Feature selection problem in the packaging mode • The mathematical model has a poor transition between exploitation and exploration	Wind Power Prediction
Binary/ Chaotic bat optimizer	[125–127]	• Competitive accuracy and effectiveness • Less parameters tuning • Fast convergence • Able to handle a large number of variables • Excellent global search ability • Overcoming the local optimum entrapment	• Low Computational efficacy • Can present an insufficiency at exploration phase • Can suffer from lack of memory of best solution	Energy management
krill herd optimizer	[128–130]	• Excellent flexibility • High computationally efficiency • Simple • Fast in Finding the global search regions	• Low performance in performing local search • Large number of iterations with global search • The krill follows the same direction leading to local optimum solution	Fault diagnosis
Marine predators opitmizer	[131–133]	• Rapid and accurate convergence • Outstanding exploration ability • High local minimal avoidance ability • Extreme balance between the exploration and the exploitation	• multi-thresholding problems • Could fail to reach a better position than the local stored one	PV power optimization
Whale optimizer	[134–136]	Overcoming the local optimum entrapment Simplicity Strong optimization ability Selection of a few parameters Strong search strategy More efficient in storage space	Low precision in complex problems Slow convergence speed Difficult adjustment of internal parameters Problem of determining the optimal threshold Random location update mechanism Low population diversity Local optimal solution due to Premature convergence	Supply side management

Table 15.5 (Continued)

Algorithm	Ref.	Advantages	Weakness	Exemplary Application
Dragonfly optimizer	[137–139]	• Simplicity • High efficiency • Fast settling time • Fast convergence rate • Low peak over-shoot	• Low premature convergence • Poor balancing local and global search capability • Poor local exploitation ability	Wind Speed Forecasting
Ant Lion optimizer	[140]	• Less parameters adjustment • Higher accuracy • Not constrained by the problem • High robustness and reliability • High accuracy and convergence efficiency • Rapid response • Reduced computation requirements • Less parameter tuning	• Unbalanced exploration and exploitation problem • Easy to fall into the local optimal solution • Four-bar truss design problem • Easy to appear premature phenomenon leading to local optimum solution	power rescheduling

15.4.2 Machine Learning Model Evaluation

In SG systems, the evaluation of the AI techniques is given high importance due to the uncertainty of the heterogeneous parameters to the grid. The volatility of the SG environment may cause a huge loss of accuracy which threatens the whole SG system operations. The assessment of ML models is conducted using a diversity of score metrics [141], etc. Score metrics are devoted to revealing a key idea on the model performance for unseen data in the testing phase. The growing interest in the accuracy measures is demonstrated by a large number of ML techniques proposed. However, adopting the most convenient score function according to data characteristics and model proprieties is non-trivial for ML techniques. For the classification type, confusion matrix and classification accuracy are commonly used due to their simplicity and generalization capabilities for such tasks. However, for regression problems, the score functions diversity requires a better understanding of the forecasting system. Table 15.6 presents the mathematical equations of the most popular accuracy metrics from regression task techniques [141]. These methods were found to be the most suited indicators to truly reflect the performance of ML techniques.

y_i, y_i, and $\bar{y}$ denote the ground truth, the forecasted value, and the normalization factor respectively. n and k are the forecast horizon and the kth sample respectively. Performance metrics are employed for the model selection stage to judge the suitability of the model face to a specific task, the training stage to follow the learning structure, or in the tuning and

Table 15.6 Most popular error measures for regression tasks.

Name	Abbreviation	Equation
Mean Absolute Error	MAE	$MAE = \dfrac{1}{N}\sum_{i=0}^{N-1}\|y_i - \hat{y}_i\|$
Root Mean Squared Error	RMSE	$RMSE = \sqrt{\dfrac{1}{N}\sum_{i=0}^{N-1}(y_i - \hat{y}_i)^2}$
Mean Absolute Percentage Error	MAPE	$MAPE = \dfrac{100(\%)}{N}\sum_{i=0}^{N-1}\left\|\dfrac{\hat{y}_i - y_i}{y_i}\right\|$
Coefficient of determination	R^2	$R^2 = 1 - \dfrac{\sum_{N=0}^{N-1}(\hat{y}_i - y_i)^2}{\sum_{N=0}^{N-1}(\overline{y}_i - y_i)^2}, \overline{y} = \sum_{i=0}^{N-1} y_i$
Root Mean Squared Logarithmic Error	RMSLE	$RMSLE = \sqrt{\dfrac{1}{N}\sum_{i=0}^{N-1}\left(\log_e(y_i+1) - \log_e(\hat{y}_i+1)\right)^2}$
Adjusted coefficient of determination	Adjusted R^2	$Adjusted\ R^2 = 1 - \dfrac{(1-R^2)(N-1)}{N-k-1}$
Mean Squared Logarithmic Error	MSLE	$MSLE = \dfrac{1}{N}\sum_{i=0}^{N-1}\left(\log_e(y_i+1) - \log_e(\hat{y}_i+1)\right)^2$
Median absolute error	MdAE	$MdAE = \underset{i=1,...,N}{Md}\left(\|y_i - \hat{y}_i\|\right)$
Neyman Chi-Square Distance	NCSD	$NCSD = \sum_{i=0}^{N-1}\dfrac{\hat{y}_i - y_i}{y_i}$
Standard deviation	σ	$\sigma = \sqrt{\dfrac{\sum_{i=0}^{N-1}(y_i - \overline{y})}{N}}$
Normalized Root Mean Squared Error	nRMSE	$nRMSE = \dfrac{\sqrt{\dfrac{1}{N}\sum_{i=0}^{N-1}(y_i - \overline{y}_i)^2}}{\overline{y}_i}$
Mean Relative Error	MRE	$MRE = \dfrac{1}{N}\left(\sqrt{\sum_{i=0}^{N-1}\|y_i - \hat{y}_i\|(y_i + \hat{y}_i)}\right)\times 100\%$
Mean Bias Error	MBE	$MBE = \sqrt{\dfrac{\sum_{i=0}^{N-1}(\hat{y}_i - y_i)}{N}}$

(Continued)

Table 15.6 (Continued)

Name	Abbreviation	Equation
Mean absolute scaled error	MASE	$MASE = \sqrt{mean\left(q_t\right)^2}; q_t = \dfrac{\left|\hat{y}_t - y_t\right|}{\dfrac{1}{N-1}\sum_{i=2}^{N}\left|y_i - y_i - 1\right|}$
Reference RMSE	RRMSE	$RRMSE = \dfrac{RMSE}{RMSE_{reference}}$
Median relative absolute error	MdRAE	$MdRAE = median\left|r_t\right|; r_t = \dfrac{\left|\hat{y}_t - y_t\right|}{y_t - y_t - 1}$

Table 15.7 Classes of score metrics.

Class	Description	Advantages	Shortcomings	Error Examples
Absolute errors	based on absolute error calculation.	• Popularity • Simplicity	• Scale dependencies. • High sensibility to outliers. • Low reliability with different datasets.	RMSE
Percentage errors	based on percentage error ratio	• High Reliability	• Biased error measures • High sensibility to outliers. • Possible errors due the division by zero • Asymmetry problems	MAPE
Scaled errors	based on the q bias	• Symmetrical • Not effected by outliers	• Possible errors due the division by 0	MASE
Symmetric errors	based on the ratio of the absolute error value	• Less calculation errors • Reliability	• Non-symmetric values • Possible errors due the division by 0 • Less usage	Symmetric MAPE
Relative errors	based on the bias rt	• High Reliability	• Possible errors due the division by 0	MdRAE
Relative measures	based on reference errors	• Ease of Interpretability	• Possible errors due the division by 0 • Less reliability	RRMSE
Other error measures	customized to the data	• Scale dependencies reduction	• Possible errors due the division by 0 • High sensibility to outliers.	NRMSE

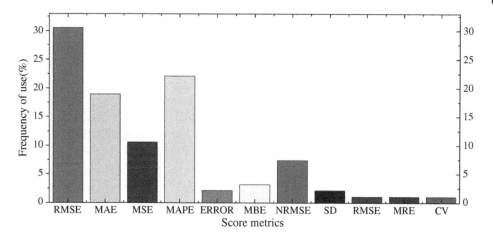

Figure 15.11 Commonly used score metrics for regression.

adjustment stage to enhance the model efficiency. Authors in [141] classified the score metrics into seven-folds according to the nature of the data. These folds include measures based on percentage errors, absolute forecasting errors, scaled errors, symmetric errors, measures based on relative errors, relative measures, and other error measures as shown in Table 15.7.

From Table 15.7, some innovative score metrics were proposed for the prevention of some mathematical errors such as the zero-denominator error. For this reason, the authors in [142] suggest the use of RMSE which becomes the most reliable approach for forecasting techniques assessment. The widely used ML score metrics for regression evaluation in the recent research articles are ranked according to their use for SG applications in Figure 15.11.

From Figure 15.11, it can be concluded that the RMSE is the most reliable and widely used score metric followed by MAPE and MAE. Regardless of percentage errors, generally the score metrics the smaller the better. Many researchers aim to make standard criteria for effectiveness measures. Nevertheless, the selection of error metrics depends essentially on the nature of the data and the type of application deployed. If the task estimated is classification then it's preferable to use the ratio of the correct estimated samples by the forecasting horizon. On the other side, for regression, data exploration lies at the core of the goodness of prediction models. Therefore, the choice of which one is computed should first be given greater consideration for the zero-denominator error (RMSE, MAPE, Mean Absolute Scaled Error (MASE)). The Cross-Validation (CV) method is also an effective method for avoiding biased data.

This method employs trained data from a part of the dataset to make predictions of another fold. Despite the efficiency of CV, it's time-consuming. For multiple types of regression techniques, the best method for assessment uses the R^2 error with automatic model selection. In [143], the authors introduce two categories for score metrics selection which are Absolute Accuracy Measures (AAM) and Accuracy measures relative to another Method (AMRAM). AAM uses its mathematical function for calculation such as RMSE and MAE. They are commonly used because they are simple and straightforward to interpret. AMRAM models involve the Percentage of times Better. This includes the percentage of time that one method is more accurate than another. Despite all the work done, it must be mentioned that it is necessary to find a conventional method to assess the regression techniques for a better universality in the interpretability of the prediction performance.

15.5 Major Applications of AI in Smart Grid

For energy systems, AI techniques provide potential benefits to SG applications including the full automation of decision- making. Toward that end, power systems require logical reasoning and metaheuristic search methods to effortlessly solve narrow problems. On the other side, the forecasting horizon plays an essential role in assigning the ML method to its suitable application. Tables 15.8–15.10 list the recent AI methods for SG applications for short-, medium-, and long-term horizons.

Regarding Tables 15.8–15.10, the AI applications for SG systems include predictive modeling, power quality, power system transients, and grid security [164]. However, the majority of AI techniques are oriented for the short-term horizon. In this section, a variety of successful applications of AI techniques in the SG system have been briefly explained with examples. In the interest of brevity, the selected applications cover some of the frequently discussed applications including LF, Alternative Energy Forecasting (AEF), maximum power point (MPP) optimization, fault diagnosis, Cybersecurity awareness, EV integration, and electricity price forecasting.

15.5.1 Load Forecasting

LF is the essence of strategic management operations in SG as it offers an optimization solution for operation costs, optimum load supply tradeoff, and power quality [47]. LF is the analysis of historical data related to environmental conditions, charge graph behavior, and historical consumption. The output of LF is an accurate prediction of the microgrid's consumption to determine preventive actions for stable operating conditions. The time horizons include short, medium-, and long-term predictions. The exact quantity of supply the load needs are ensured using LF techniques. An accurate prediction facilitates the coordination of SG, and thus, guarantees the budget-saving commitment and grid stability.

In [120], the authors' associated Quantile Regression with Neural Network for STLF. This hybrid technique is found more accurate than single models. For the validation procedure, the historical database included six successive years of an hourly electric load from GEFcom2014 competition and ambient temperature from 2001 to 2011

Table 15.8 Short-term SG-based AI applications.

Index	Ref.	Forecasting horizon	Involved models	Application	Best accuracy	Dataset location
1	[144]	10 min	WPD[a], VMD[b], KELM, IGWO[c]	Wind Power	$R^2 = 99{,}27\%$	La Haute Borne, France
2	[145]	15-min	Conv-LSTM[d], MWO[e], BPNN, CEN	Wind Speed	MAPE = 2,62%	a wind farm in Jiangsu province
3	[146]	1 s	SCM[f]	Voltage Stability Assessment	Accuracy = 98%	Hong Kong power grid
4	[147]	week ahead, day ahead	KF[g], WNN[h], ANN	Electric Load	MAPE = 2,15%	Egypt & Canada
5	[148]	day ahead	RNN[i]	Electricity price	MAPE = 5,63%	ISO New England dataset
6	[149]	hourly day-ahead	GCA[j], SDAE[k], RANN[h]	Electricity price	MAPE = 7,66%	data of Ontario, Canada"
7	[150]	5 min	AC[l], DDPG[m]	energy consumption	$R^2 = 98\%$	Henan, China
8	[151]	Multi-step	EWT[n], RL, LSTM, DBN, ESN	Wind Speed	MAPE = 8,17%	Xinjiang, China
9	[152]	1 hour	GAN, KNN	PV generation	MAE = 0,10	GEFCom 2014 solar track
10	[59]	Multistep up to 4 h	DBN, RBM	Electrical load	MAPE = 4,50%	Tianjin, China

[a] Wavelet Packet Decomposition.
[b] Variable Mode Decomposition.
[c] Improved Grey Wolf Optimization.
[d] Convolutional LSTM.
[e] Modified Whale Optimization.
[f] Shapelet Classification Method.
[g] Kalman Filtering.
[h] Wavelet Neural Network.
[i] Rough ANN.
[j] Grey Correlation Analysis.
[k] Stacked Denoising Auto-Encoders.
[l] Asynchronous Advantage Actor-Critic.
[m] Deep Deterministic Policy Gradient.
[n] Empirical Wavelet Transform.

taken from 25 weather stations. The proposed model in [120] targets memory capacity and time-consuming tradeoff minimization. The hybrid model architecture is slightly different from QRNN by tuning the weights and bias before the complete data is fed to the prediction system. Moreover, a batch layer is added to accelerate the training process.

Table 15.9 Medium-term SG-based AI applications.

Index	Ref.	Forecasting horizon	Involved models	Application	Best accuracy	Dataset location
1	[153]	48 h	BO[a], SVR[b], ENN, BiLSTM, ELM, BFGS[c]	Wind power forecasting	RMSE = 16,39 MW	Canada
2	[154]	one month	AR[d], NPAR[e], STAR[f], and ARIMA	Electricity Consumption	MAPE = 4,83%	Pakistan grid
3	[155]	one month	ELM	Electricity price	MAPE = 84,35%	New York grid
4	[156]	one month, half year, One year	cycle-based LSTM, TD-CNN[g]	Electric load	MRE = 4,2%	load of HangzhoU
5	[157]	one month, One year	Generalized Additive model	Electricity Price	APE = 1,9%	Turkish grid
6	[158]	one month	GKR[h], RFE[i], KNN	Wind Speed and Power	RMSE = 11,37 MW	ISO-New England data
7	[159]	one week	AR, MA[j]	Power demand	MAPE = 4,46%	PJM, United States

[a] Butterfly Optimization.
[b] Support Vector Regression.
[c] Royden, Fletcher, Goldfarb, and Shanno Quasi-Newton back-propagation.
[d] Auto-Regressive.
[e] Non-parametric Auto-Regressive.
[f] Smooth Transition AutoRegressive.
[g] time-dependency CNN.
[h] Gaussian Kernel regression model.
[i] Random Feature Expansion
[j] Moving Average.

Table 15.10 Long-term SG-based AI applications.

Index	Ref.	Forecasting horizon	Involved models	Application	Best accuracy	Dataset location
1	[160]	1 year	FB[a]	Electricity consumption	MAPE = 1,61%	China
2	[161]	5 years	least-square SVM, ARIMA	Electricity consumption	MAPE = 1,02%	Turkey
3	[162]	2 years	MARS[b], ANN, LR	Electric Load	MAPE = 3,6%	Turkish grid
4	[163]	1 year	NARX[c]	Wind speed	MAE = 0,8	KT. Malaysia

[a] Fuzzy Bayesian theory.
[b] multivariate adaptive regression splines.
[c] Non-linear Auto-Regressive exogenous.

Authors in [75] implemented Deep Residual Networks (DRN) for STLF. The DRN model investigates the DK indicators to enhance the prediction quality. Paper [165] implemented a DL framework based on Long-Short Term Memory and Whole selected Appliances. The aforementioned paper showed that the number of appliances boosts the accuracy of their proposed model [165]. A Hybrid Kalman Filters for load prediction in the Very Short-Term Load Forecasting (VSTLF) has been comprehensively introduced in [166]. The model architecture presents not only a combination of a two-level Wavelet neural network method, but it also provides prediction interval estimates online via Kalman filters (WNNHKF). The proposed model is a validation of real-world datasets from ISO, New England.In [167], the authors proposed model a Holographic Ensemble Forecasting Method (HEFM) for STLF. The framework built consists of four steps presented as follows: Firstly, the dataset selection is deployed to pick the most relevant information. Three years of historical data includes weather proprieties, week type data, load data, week type data from Guangdong, China. In the second step, the obtained dataset is combined during the sampling space level. The third step consists of defining and tuning the non-linear heterogeneous prediction techniques, namely, Least-Squares Support-Vector Machines (LSSVM), iGBRT, and Back Propagation Neural Network (BPNN). Initially, the multimodal hyperparameters are randomly pre-defined. Finally, the Bootstrap predictor is online trained and the Holographic Ensemble Forecasting (HEF) is processed to obtain the final decision. HEF is a combination of information about online second learning. Using the ISO New England dataset, the model generates a MAPE error of 1.95% and outperforms the Artificial Neural Network RF model. The authors in [168] used a Kernel Online Extreme Learning Machine (K-ELM). The multimodal output deploys Cholesky decomposition to produce its final output. The simulation results of K-ELM model demonstrate its efficiency with superior accuracy compared to GNN.

A novel Causal Markov Elman Network model (CMEN) has been comprehensively depicted in [169]. The model imposes a selective collection of feature dataset that has the most contribution to the DK with a State-of-the-Art causal engine. This procedure allows the system to select the best predictive performant model for the electricity LF framework. To validate the proposed model, 222 neighboring houses were utilized for one year of load acquisition with a time step of 30 minutes. The comparative simulation between SVM, RNN, DNN, and CMEN proves that CMEN has greater accuracy than the rest of the models.

15.5.2 Alternative Energy Forecasting

Due to the high volatility of weather conditions, energy generation is intermittent and unstable. To exemplify, harnessing the energy from wind and solar energy has a major demerit due to the fluctuation of irradiance and the wind speed [170]. The sensitivity of the environmental situation threatens grid stability and causes unsatisfactory penetration to the public grid. In this context, renewable energy forecasting contributes to the power systems by balancing the energy fed to the load and augmenting the efficiency and reliability of RES [171, 172].

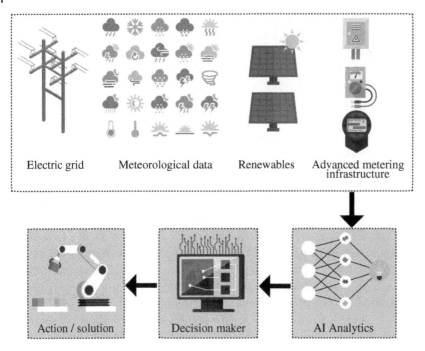

Figure 15.12 Forecasting procedure.

Regarding Figure 15.12, the forecasting procedure for renewable energy systems passes through four stages. Firstly, AMI capture'the weather and power variation. In the second stage, the registered dataset is analyzed using AI techniques. The pattern investigation conducts accurate forecasts which are directly linked to decision making in the third stage. The final output illustrates the preventive solution for weather disturbance. AEF attracts scientists and engineers for the goal of maximizing the profitability of the inexhaustible flows of energy. The AI paradigm ensures a reliable unit commitment between the grid and the load. SG requires AEF to manage the ongoing energy distribution and secure a smooth alternative energy integration. The following section covers PV and WE prediction-based AI.

15.5.2.1 Photovoltaic Energy
PV energy is the dominant type of alternative source of energy with a share of 37% [173]. The generation of solar energy from PV modules comes from the conversion of irradiation into electricity. It has many benefits regarding its clean generation and fast replenishment. In addition, the operation phase is noise-free since there are no mechanical moving parts. For those factors, PV energy has the support community with a total installation of 516 GW by the end of 2018 and estimated to touch 332 GWh in 2020 [173]. However, the electricity generation directly employs the sunlight which is unstable during the day. PV power forecasting protects the SG from abnormal behavior using AI techniques [174]. New forecasting techniques aim to cover the weak points of the old models in terms of accuracy, computational work, and time consumption [175]. The weather

information for data-based approaches includes temperature, humidity, wind speed and direction, solar direction, and precipitation. Using ML models, the extraction of the most relevant indicators from the aforementioned list is conducted through a variety of feature importance approaches. This step minimizes the database and improves the quality of the results.

The authors in [176] combined GA, Particle Swarm Optimization (PSO), and ANFIS to create a multimodal predictor. The hybrid proposed model uses a Gaussian Process Regressor (GPR) to optimize the feature number via the binary GA. Then, model relativity ANFIS is optimized to generate the final output. The predictor is validated using Goldwind microgrid database from Beijing for one day ahead. The proposed model outperforms the benchmark models.

In [177], a Weighted Gaussian Process regression (WGPR) is introduced to predict the PV power from weather conditions. The proposed method deals with the high-level heteroscedasticity of relevant inputs with outlier values [178]. The methodology is to include the nonlinear correlation measurement and Weighted Local Outlier Factor (W-LOF). The samples are given weights according to the Density-based Local Outlier Factor (D-LOF). The forecast horizon remains accurate for 15 minutes. The proposed method outperforms the typical predictors including LSSVM, ANN, GPR with an RMSE of 186.51 W. A DL technique based on Long Short-Term Memory Neural Network (LSTM) and Attention mechanism (ALSTM) has been proposed in [179]. This framework consists of an association of two LSTM cells. The features employed include the PV module temperature and the PV power, respectively, with a threshold for the significant features according to their relevance to the PV power. However, this method is not efficient when the time horizon is longer.

In [180], a Deep CNN Model for short-term prediction has been proposed. The used data is the ambient temperature, the historical PV power, and the irradiation. Feature extraction and mapping leads deploy a CNN convolutional layers to build a DL model named PVPNet. The registered RMSE is 163.15 W for 24 hours of forecasts. For the assessment of the used dataset in an hour ahead of PV power forecasting, the authors in [181] used SVR, Lasso, and polynomial regression fed by various feature selection schemes. The database used for the simulation belongs to Virginia Tech Research Center (VTRC) in Arlington. The dataset includes weather temperature, solar irradiance, module temperature, wind velocity, and PV output power. It has been shown that SVR is the most accurate predictor on the list. However, the accuracy decreased with more unnecessary added features, such as the temperature and the humidity.

Furthermore, indirect methods also presented efficient tools for PV prediction [182]. The temperature and irradiation prediction could lead to accurate PV power prediction [183]. In this context, the authors' work in [184] implemented LSTM NN targeting short-term solar irradiance prediction. The authors in [185] employ the DNN model to predict the Canadian weather. The authors in [186] introduced a novel ML model using a time-based regional analog for one-day ahead of PV power forecasting. It can be concluded from the reported literature that long-term forecasting methods number is very limited due to the high volatility of weather conditions and the sensibility of PV plants from every environmental change.

15.5.2.2 Wind Power

To date, WE has become one of the fastest-growing industries in the energy market. WE achieve a significant increase every year reaching 591 in 2018 with an annual growth of 51 GW/year [187]. However, WE is characterized by a stochastic variation. The narrow behavior of weather parameter limits WE penetration to power systems. Therefore, wind power forecasting presents a key solution to increase its integration SG systems.

In the energy management context, authors in [188] proposed the LSTM DNN method for wind-speed prediction. The contribution of reported work is a deep novel feature extraction approach developed based on stacked denoising autoencoders and batch normalization before the dataset is entered into the predictor. With a threshold of outliers' samples using a density-based spatial clustering of application with noise (DBSCAN), the data has a better cleaning process. Paper [189] proposed a Reduced SVM (RSVM). In the proposed model, a dimensionality-reduction step is made using PCA to extract the most informative features from the dataset. A new ML ensemble-based wind forecasting has been introduced in [190]. The authors in [191] implemented Markov Chain models for a horizon of six hours ahead. Paper [192] created a segmentation from the theory of optimal stratification. The short-term prediction is the goal of the aforementioned proposed method. Other notable techniques used for PV power forecasting is illustrated in [193], where a novel statistical technique has been comprehensively explained. The proposed technique deploys AdaBoost Numerical weather predictions (NWPs) data clustering and Reference Wind Masts (RWMs) correlation. The dataset transition from wind-to-wind and wind-to-power is processed using ANN/SVM models. The final output deploys Model Output Statistics (MOS) a MOS and weighted average combination. A real data-driven case from Turkey confirms the effectiveness of this approach for 48-hour prediction.

15.5.3 Electrical Vehicles Integration Based AI

With the recent waves of electrification, EVs have drawn a lot of attention in the research and industrial community due to their non-fossil fuel characteristic and their socio-economic impact on stakeholders and users [194, 195]. EVs are also used for DSM. EVs can be equipped with electric engines power powered by high-density lithium-ion (Li-Ion) batteries, Lead-Acid batteries, Nickel-Metal Hydride Batteries, or Ultra-capacitors. However, the integration of EVs in SG is a challenging problem due to its temporal demand dependencies and wide interaction with various elements.

The state-of-charge prediction of Li-Ion battery in EVs has been investigated [196]. The adopted prediction technique uses an improved deep neural network which is configured based on empirically established heuristics. This methodology of parameter sittings is highly dependent on the nature of the structured data. In such cases, automatic tuning is fundamental to give better generalization potential to the proposed model. In [197], an integrated time series model based-LSTM has been implemented for dynamic prediction of the brake pressure of EVs to overcome the limitation of the existing model predictive control approaches. The data from signals and system states are stored from the controller area network bus. Next, the proposed system predicts the velocity with EV states to accurately forecast the braking pressure [197]. The simulation results show that the proposed technique is very promising, achieving an accuracy of RMSE = 0.189 MPa for one-step prediction. However, due to the sensitivity of the autonomous emergency brake system (AEB), risk uncertainty, and sensitivity analysis are extremely required for the safety of drivers [197].

Authors in [198] proposed a self-charging strategy to manage the EVs charge schedule using DL techniques. A concrete Li-ion battery was used to validate the model performance and meet the load demand balance. However, the limitation of state of charge estimation based on ML lies in the high dependency of the battery types and models. In [199], an effective prioritized deep deterministic policy gradient model for the pricing of EV charging is proposed. The EV pricing and its flexibility potential have been examined using the proposed deep RL technique which shows a profitability enhancement for the EV aggregator. Despite the advantage of using this deep RL model, the model generalizations with realistic variability of the exogenous inputs require further investigations.

15.5.4 MPPT-Based AI

Due to the non-monotonic Power-Voltage curve for RES generation, the energy generated from solar modules and wind turbines is unsteady due to weather parameter variations. The meteorological conditions perturb the energy generated during the operation time. The major factors that directly influence the generation of electricity from wind and solar energy are sun-oriented irradiance, temperature, wind speed, and cloud shading. The weather parameters have fast–varying behavior. The impact of this variation return with the difficulty of maximum energy generation from RES regarding the environmental conditions. The weather sensibility causes multiple Local Maximum Power Point (LMPP) for the alternative generators. Missing these points in energy production causes a significant loss for consumers and grid utility.

In order to dynamically set the optimum PV power, search optimization algorithms are deployed. MPP algorithms ensure optimum efficiency by maximizing energy production. A MPPT is an electrical device that leads the PV system to constantly deliver as much energy as possible. AI tends to cope with the PV effect and meet with high PV module efficiency through the deployment of robust controllers [200]. Tracking the MPP requires efficient metaheuristic search algorithms. Nevertheless, finding a suitable MPPT is a challenging task [201, 202]. The number of tools is significantly increased these years with the progress of ML and swarm intelligence [203]. MPPT model selection is conducted according to the following criteria:

- Performance efficiency
- Convergence speed
- Ease of implementation
- Cost.

MPPT techniques have been used in the real-time implementation from online and offline training. As shown in Figure 15.13, Several MPPT techniques are proposed in the literature including ANN, FLC, Chaos theory models, Evolutionary computing probability, and non-linear ML modes.

Authors in [204] presented a hybridization of the Overall Distribution algorithm and PSO (OD-PSO) based MPPT. The proposed method aims to track the Global Maximum

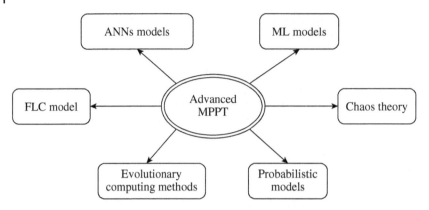

Figure 15.13 Flowchart of advanced MPPT algorithms.

Power Point (GMP) under partial shading conditions (PSC). This sophisticated approach offers a fast-tracking of GMP without prior deep knowledge of the PV system. Compared with the benchmarks, the proposed model has a superior efficiency of 98.29% with the shortest tracking time of 1.64 seconds. The authors in [173] proposed an adaptive RF technique based-MPPT for Permanent-Magnet Synchronous Generator (PMSG). The goal aims to track the PMSG maximum power. Padmanaban et al. proposed a hybrid approach for PV MPP trackers. ANFIS and Ant Bee Colony Optimization (ACO) are associated to build ANFIS-ABC technique. The proposed model was validated on real-time implementation via Dspace (DS1104). ABC is dedicated to determining the membership function.

The authors in [205] introduced a hybrid GA neural network-RBF network (BP- RBF network). For boosting the efficiency of the hybrid model architecture, GA carries out the tuning data centers of RBF. The experimental results showed that the proposed algorithm is capable of adapting well to weather changes such as temperature variation and shading conditions. An efficient hybrid method for tracking the MPP has been proposed in [206]. The mechanism of the proposed method is a combination of Coarse Gaussian SVM (CGSVM) and ANN models to construct the ANN-CGSVM. CGSVM as a data mining technique is used as a classifier to generate a fitness function using irradiance (G) and Temperature (T) as inputs, and the PV current (Ipv) as output. It should be pointed out that CGSVM training does not require much data (only 19 samples in the study). Then, the fitness function is used to predict new training samples. At this stage, ANN predicts the response values from the newly generated data. Finally, the error value is calculated as the difference between the PV operating current and the response values and passed to a Proportional–Integral–Derivative (PID) controller for tuning and determining the duty cycle. Simulation results showed that the proposed technique succeeded in tracking the MPP and extracting the PV energy with acceptable efficiency. However, the model assessment has concluded that although the suggested method has a fast convergence speed, ANFIS, ANN-PO, and ANN- CGSVM methods achieve the greatest accuracy performance. The low performance lies in the lack of training samples given to CGSVM.

In [207], the authors incorporated a Memetic RL (MRL) in the optimum PV generation via the MPPT search tool. RL, the used method, is based on a Q-learning algorithm. Under PSC, the proposed model tends to carry out the maximization of photoelectric conversion.

The action-reward mechanism with multiple groups of agents presents an efficient paradigm for nonlinear systems since the learning rate is significantly improved during operation. Furthermore, the methodology incorporates a memetic computing platform, which is a key factor for the determination of local and global power point. The MRL-based MPPT proceeds as follows: The learning model is built using the feedback reward-action making via multiple small-scale knowledge matrices. This approach leads the system to communicate with the environment and carry out analysis with conclusions [208]. The second stage is the exploration and exploitation balance of agents. The model in this stage classifies agents via the knowledge metrics. The final step is the global information exchange where the complex reassembly is collected with regard to the agents' importance. The information exchange is made through a Shuffled frog-lapping algorithm. In the proposed case study, the irradiance and temperature are the only indicators used for the fitness and knowledge metrics calculation. A comparative study of MRL with the state-of-the-art methods, including Incremental Inductance (INC), PSO, GAs, Artificial Bee Colony (ABC), Cuckoo Search Algorithm (CSA), Teaching-Learning-Based Optimization (TLBO), and Grey Wolf Optimizer (GWO) shows that the proposed method has the best performance in energy generation under PSC due to the high searching potential of memetic computing RL framework [209]. Nevertheless, the limitation of this algorithm lies in power fluctuations caused by its high convergence stability.

In [210], a Bayesian Networks (BN) based MPPT has been proposed to lead the system toward optimum PV production. BN is implemented to accelerate the tracking time through Bayesian rules. BN is a directed acyclic probabilistic graph with conditional dependencies and unique arbitrary variables for each neuron. BN applies to the MPP search with the voltage and current being the inputs. BN lies at the core of the optimal current (I) and voltage (V) direction. The dataset consists of I-V values and the open-circuit voltage of the module (VOCM). The architecture of the proposed MPPT technique takes place once the dataset is acquired by the BN. The proposed technique provides the output node (S) as a result of the computational work to lead the system for the next transfer point orientation. Within the right orientation, the output node is fed to the INC algorithm. The INC model generates a back loop to BN with the actual (Impp,Vmpp). Moreover, this model creates an error signal with respect to INC equality $(I/V + dI/dV = 0)$ of the PV slope and the optimal reference value. This equation is optimized using PID controller via Ziegler-Nichols tuning method to reduce the steady-state oscillations. Compared to the Whale Optimization Algorithm-Differential Evolution (WODE) method, the BN MPPT is outperforming in terms of greater efficiency and quicker response. However, the reliability of BN-MPPT highly depends on the tuning of the PID controller which can lead to overshoots and oscillations during the peak hours of daylight.

From the cited examples, soft computing techniques are contributing to MPPT with the necessary flexibility and efficiency which could not be achieved by conventional methods. Indeed, the continuous environmental changes require such robustness in MPPT systems. The PV cells are associated in series and parallel to increase the voltage or current, respectively. The association of these connections together to build the PV module faces a problem when these series cells do not generate the same energy due to several environmental factors. To clarify, PSC are a stochastic parameter that affects PV plants generation according to the sky positions. This factor not only causes a significant reduction of power generation, but it also has a destructive impact that decreases the lifetime of PV panels. The heavy investments are damaged regarding the equipment aging acceleration. As a result of PSC, hot-spots appear on the surface of the PV modules generating overheating in a local cell. This phenomenon is of paramount importance due to the sunlight temperature distribution which can cause a faster degradation of the PV panels. The hot-spotting effect happens if the low current PV cells in a string of numerous high short-circuited current. The PV conversion produces heat in this case instead of electricity. Thus, advanced MPPTs are supposed to focus on dealing with such an issue. A bench of metaheuristic optimization techniques inspired by the collector behavior of the natural elements has proven of high efficiency with the PSC [211].

From this perspective, advanced algorithms are presented to increase the efficiency of sustainable energy systems such as solar, and wind sources. Efficient MPPTs boost energy generation and improve the power quality encouraging the integration of RES. For grid-connected, isolated, or stand-alone PV systems, MPPT is the essence of guaranteeing the security and safety of the equipment. On the other hand, advanced algorithms pose a problem of high computational work. This factor reflects the expensive cost of these electronic devices. A compromise between the complexity-efficiency should be taken into consideration to improve the utilization of these techniques.

15.5.5 Fault Diagnosis-Based AI

For SG systems, the maintenance expenses of energy systems are costly. Some parts of these systems become irredeemable when the maintenance action is delayed. Moreover, the manual diagnosis is time-consuming and depends greatly on the proficiency and expertise of infield men. Situational awareness for fast-acting changes is the key solution for providing a stable, and highly distributed energy resource penetration. Fault diagnosis plays an important role in safe energy operations in SG paradigm [212]. For instance, paper [213] summarizes the possible failures for PV systems. The possible failures are classified into three levels: first failure (1F), second failure (2F), and order failure (OF) [213]. Figure 15.14 illustrates the classification of the fault of PV systems.

Fault detection and prediction in SG ensure a secure dispatchability, long-lasting equipment, and efficient monitoring [214].

One of the methods to enhance fault detection via ML techniques is to improve the AMI. Wang et al. proposed an ensemble method for the aggregated load with (M) sub-profiles [215]. This method contradicts the traditional ensemble model where the number of clusters is maximized through optimizing the number combinations. The proposed algorithm consists of building an accurate predictor by associating a group of weak estimators to generate multiple predictions. The sub-profiles are hierarchically clustered for K number, where K = [k1, k2, . . ., kN], and the prediction output is obtained from each grouped load individually. Then, the weights (w) for N predictions are calculated and combined to generate the final forecast. The application of this method uses a smart Metering Electricity Customer Behavior Trails (CBTs) in Ireland. The simulation gives an accuracy of 4.05% in terms of MAPE and outperforms the individual forecasts. The authors in [216] proposed an efficient technique for predictive PV malfunctions. The proposed

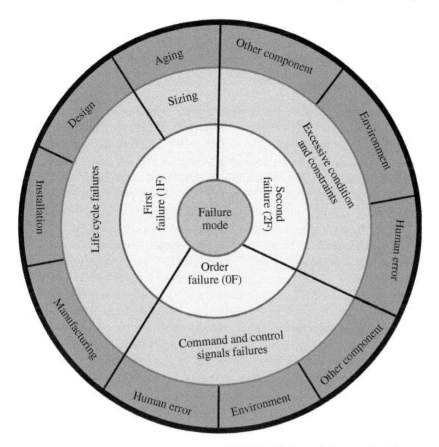

Figure 15.14 Classification of faults in power systems.

method consists of associating LSTM cells for feature extraction and Softmax function for fault classification. The classifier function is presented as follows [216]:

$$i_t = \sigma\left(w_i\left[h_{t-1}, x_t\right] + b_i\right) \tag{15.1}$$

$$f_t = \sigma\left(w_f\left[h_{t-1}, x_t\right] + b_f\right) \tag{15.2}$$

$$o_t = \sigma\left(w_o\left[h_{t-1}, x_t\right] + b_o\right) \tag{15.3}$$

Where i_t, f_t, o_t denote the input, forget, and output gates, respectively, σ is the sigmoid function, w_i is the Wight for the representative gate (x) neurons, h_{t-1} is the input at the current timestamp, and b_x presents the Biases for the respective gates (x). The collected data and proposed model are based on simulations. It must be noted that the homogeneity of data samples in fault detection problems causes a lower learning rate and remarkably lessens prediction accuracy. The noiseless data is carried out by adopting a virtual database with the addition of a fake noise for the training process. The score accuracy of the suggested approach ranges between 97.66% and 99.23% for Non-Condensable (NC), Hot Spot Fault (HSF), and Line-to-Line Fault (LLF).

15.5.6 AI and Cyber SG Security

In any SCs operation, the network stores the new data via cloud storage. The size of the generated data from the grid operations is huge and increases each day in the order of thousands of zettabytes (1021) [217]. Currently, the security of Big data became a burden, especially with the evolution of ICT technologies that increase the information size. The acquired information requires high computational potential with complicated structures collected from two-way communication SGs. Advanced data analytics are required to make use of this data for quick decision-making and strategic planning. However, cyber security attacks affect the data through encryption and physical threats [218, 219]. An SG cyber-physical system is a hybridization of physical network systems and cyber systems [220]. The physical system transfers the real word information to the cyber system using Digital to Analog Converters (DAC). The received data is analyzed to generate actions using Analog to Digital converters (ADC). The structure of the SG cyber-physical system is shown in Figure 15.15.

Nowadays, the worldwide vulnerability in SG against networking threats is transformed into heavy investments of a Trillion dollars. Since ICT pilots everything on the grid, any unfamiliar access significantly affects the whole system. ML techniques are introduced to overcome this weakness and establish a mature and safe SG system [221–224]. The attacks frequently occur in the transmission lines with a rate of 62% [225]. For a reliable electricity grid, a high-level security platform is distributed over the whole system sections to retain integrity and confidentiality during operating conditions. The data register must be shared only with the authorities to keep the information authentic, and safe from any epidemic modification. Numerous defensive techniques were proposed against the attack threats including Reject On Negative Impact (RONI), adversarial training, Defense distillation, Ensemble method, Differential privacy, and Homomorphic encryption [226]. Cyber

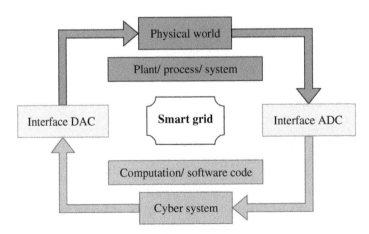

Figure 15.15 Structure of the cyber-physical system.

Security standards promote the identification risk assessment requirement and algorithm efficiency. The authors in [227] proposed a Cyber-SAVs framework. This framework provides data management and establishes the cyber trust theory via the Operational Trust Indicator (OIT) feature. In other words, the system detects the false alarm, affected sensor, missed detection, and productivity trust. From further investigation, ML models could cause vulnerabilities from malicious worms such as poisoning, evasion, impersonation, and inversion attacks to the training data [228]. The security strategies classify the threats according to the impact on classifiers, security violations, and attack specificity [229]. The cited attacks target the training data and cause a mal performance with low accuracy. Thus, the protection of ML models from unwanted hacks is given prime importance for SG.

15.5.7 Electricity Price Forecasting

To date, the current grid utilities are authorized to fix the tariff-rate of electricity depending on industrial, commercial, and domestic usages [230]. The prices are uniform and consider morning, afternoon, and night times. The transition toward SG liberates the energy market with the integration of numerous electricity providers. Therefore, prices are customized and the end-user chooses the most affordable electricity from the restructured markets [230]. In a pool-based open power market, the electricity price is the key solution for optimum benefits with the load demand. The deregulation of the market competitively requires the tariff rate prediction to guide bidders for efficient plans. The investors and industrial partners need extremely accurate electricity price forecasting to cope with their operations strategies and thus, maximize their profit [230]. However, the electric pricing is intermittent depending essentially on the spikes of the supply and loads. Figure 15.16 highlights the major factors that have a high impact on the energy trade-off tariff.

From Figure 15.16, the electric bills are essentially affected by load uncertainties, time indicators, volatility in crude oil prices, generation outages, transmission congestions,

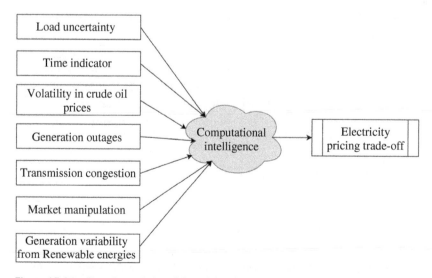

Figure 15.16 Flowchart of electricity pricing factors.

market manipulations, and unstable renewable sources generation. AI links all aforementioned parameters to provide a continuous spot-price trend prediction over a time horizon. This is conducted by the deep investigation and pattern analysis of several elements that impact the pricing [231].

Forecasting energy prices is not an easy task due to the nonlinearity of energy market elements and their high sensibility causing instability pricing trade-off. AI provides an efficient tool for optimal cost-effectiveness and greater scheduling. In [232], the authors introduced a CNN model optimized by GA for locational marginal price (LMP) forecasting, GA was employed for hyperparameter optimization of the CNN architecture. For one day ahead forecasts, The GA-CNN model outperformed several benchmarks including LSTM, SVM, and the original CNN for all the seasons. However, the proposed GA-CNN lacks a comparative study with the mainstream optimization algorithms such as GWO, Ant Lion optimizer, and whale optimizer (see Table 15.5). This comparison is necessary as GA is relatively old and less efficient than the cited optimizers. In [233], an ensemble approach was employed to predict the LMP using three distinguished elements, specifically, the system energy price, Transmission/congestion cost, and cost of marginal components. Although the results seem to be very promising, the proposed model lacks generalization and requires a high-quality data set to perform as expected. From the reported work, it has been noted that LMP signals are frequently employed as a means to transfer the estimated electricity price in advance to customers and manufacturers.

The authors in [234] proposed a hybrid dual-stage model for electricity spot price prediction. This approach involves an Auto-Regressive Time-Varying with exogenous Variables (ARXTV), Kernel regression (KR), WT, and SVM. The methodology of this combination proceeds as follows. The dataset is modified on a novel methodology through WT. Using this data, an elementary forecast is generated using ARXTV. On the other side, electricity spikes are classified using SVM and estimated using KR. The proposed methodology in [234] has greater performance than individual predictors. However, it generates less accurate results compared with ANN. The simulation results lead to the conclusion that the complexity does not contribute a meaningful performance.

Paper [235] introduced a hybrid method based on a Relevance vector machine (RVM), XGboost, Elastic Net, Gaussian RBF, and Polynomial Kernels (PK). The hourly data was collected from the English electricity market in order to validate the proposed approach. The short-term LMP is forecasted using the following as inputs: the previous day change in Day-Ahead (DA) LMP, previous hour DA Demand, previous Day change in DA Demand, time indicator, hourly Dry Bulb temperature, daily average Dry bulb temperature, and crude oil prices., The proposed model in [235] has a very fast training stage (only 88 seconds) and high accuracy in terms of low error measures.

15.6 Challenges and Future Scope

In light of the current research studies reported in the study, to our knowledge, the main challenges of the AI methods in SG are reported as follows:

- The time horizon in the majority of the research work is oriented for short-term prediction. However, the development of medium- and long-term forecasting is quite limited.
- The massive data accumulation requires an extremely highly developed infrastructure to analyze data from different endpoints in power systems. Cloud computing and distributed collaboration platforms for scalable ML and DL are still weak. Despite multiple high performing platforms such as AWS and SAS, SG big data analytics still lacks a unified solution to analyze petabytes of unstructured data across different formats.
- Probabilistic forecasting is not widely investigated for SG applications. Quantifying variability and uncertainty are vital for a dynamic environment and unavoidable deviations of point forecasting.
- The applications of AI-based algorithms are frequently assessed in normal conditions where the model is performing optimally. However, real-world experiments prove that ML models' accuracy significantly decreases with a perturbing environment.
- Although data-driven methods have been integrated into variant sub-areas in SG, some fault detection problems do not have sufficient historical data which makes SL poorly performing.
- UL has scarcely been investigated. Despite the massive data generated from different sources, poor data availability is still a major issue for some faults in power systems. UL is highly important in resolving this problem by the identification of the fault without the need for labeled data. This explains the ongoing industrial interest and investment toward dynamic control and fault detection-based UL.
- Cyber-physical SG systems have a massive data flow which raises the issue of how the security of the grid in the long term is ensured. The deep investigation of cyber-physical threats over the SG platform is relatively scarce.
- Hybrid models were very competitive for prediction problems of the grid. Therefore, the number of papers that addressed the hybridization of AI techniques has increasingly progressed.
- Regional prediction techniques are rarely investigated in research papers where the model could perform well in any other direction.
- Online experimental validation of the forecasting methods is still marginally investigated with conformed characteristics of the industrial applications. Broadly speaking, the development of the prediction model loses sight of significance for the algorithm adaptability of AI in the SG paradigm.
- Little work has been done on transfer learning of ML models for SG applications. Transfer learning could improve the quality of prediction results by combining multiple sources of information with a lack of data availability. Furthermore, transfer learning significantly reduces the computational time for complex problems through the use of pre-trained models. Despite these advantages, not much attention is paid to transfer learning in the SG paradigm.
- Smart cities and infrastructure resilience which AI sustains has not received enough attention from the reviewed literature. Big data processing and IOT technologies using AI techniques are increasingly integrated into the livability of the urban fabric. However, Big data velocity, volume, and variety through smart cities need to be addressed by AI techniques. As such, urban governance, intelligent transportation systems, and digitized infrastructure are dependent on AI, Big Data, and ML for compliance with SDGS.

With regard to the reviewed work from previous studies, it can be observed that DL has acclaimed significant attention in the recent few years due to the high availability of IOT devices and smart meters. However, the complex ANN architectures are computationally expensive and time-consuming. Therefore, parallel and distributed computing presents a highly active area of research to reduce training time and computational complexity in order to have better adaptability to industrial applications of SG.

AI technology becomes one of the top policy trends due to its significant economic, social, and, environmental contributions. Over the last few years, many countries such as China and the USA prepared a strategic roadmap to speed up AI diffusion on a larger scale, especially for the energy market. As a result of the extensive investments of this high technological expertise, the significant increase in the share of scientific papers in the AI-based SG landscape has been noticed. From the World Economic Forum (WEF), 20% of the market jobs could have the fingerprint of AI techniques creating 133 million jobs by 2022. Despite the recent waves of AI democratization on an international level, the adaptation of such high technology is now at a low level of spread in developed countries. This is due to the high financial cost of the redefinition of the traditional grid infrastructure and adding the necessary flexibility to be perfectly tailored for the SG paradigm.

On the other side, a growing interest in cloud computing has been found to manage the acquisition of high-dimension data from multi-agent systems. Excellent parallel performance in speedup and scale-up is vitally needed to achieve rational allocation and data sharing of resources of the SG paradigm. Ultimately, the multi-dimensional and multi-level coordinated SG requires a robust generalized AI where all the automation work is curried by ML and nature-inspired algorithms. The key insights of a high performing SG paradigm are essentially linked to the development of AI tools to draw the operation guidelines for future energy systems.

15.7 Conclusion

Amidst the fourth industrial revolution, AI-as-a-service enables smooth penetration of advanced techniques across SG through a variety of applications. This review study outlines the common emerging AI-driven edge computing for SG operations. The vast majority of case studies from the literature demonstrated that hybridization and assembling straight AI techniques lead to better accuracy and high robustness. From the reviewed articles, it has been found that the current development of AI deployment paves the way for large SG expansion. However, the AI progress faces several challenging points including significant computational burden, higher complexity, and scalability. Seven recent applications of SG-based AI were depicted including LF, power forecasting, Maximum Power Point Tracking, fault detection, electricity price prediction, EV integration, and cyber security awareness. For their safe and mature implementation, Big data is extremely needed as a powerful framework for data exchange in energy systems. On the other side, the cyber security attacks increasingly menace SG resilience which requires further investigation.

References

1 Sun, Q., Zhang, N., You, S., and Wang, J. (2019). The dual control with consideration of security operation and economic efficiency for energy hub. *IEEE Transactions on Smart Grid* 10 (6): 5930–5941.

2 Vinuesa, R., Azizpour, H., Leite, I. et al. (2020). The role of artificial intelligence in achieving the sustainable development goals. *Nature Communications* 11 (1): 1–10.

3 Albano, M., Ferreira, L.L., and Pinho, L.M. (2015). Convergence of smart grid ict architectures for the last mile. *IEEE Transactions on Industrial Informatics* 11 (1): 187–197. https://doi.org/10.1109/TII.2014.2379436.

4 Müller, S.C., Georg, H., Nutaro, J.J. et al. (2016). Interfacing power system and ict simulators: challenges, state-of-the-art, and case studies. *IEEE Transactions on Smart Grid* 9 (1): 14–24.

5 You, M., Zhang, X., Zheng, G. et al. (2020). A versatile software-defined smart grid testbed: artificial intelligence enhanced real-time co-evaluation of ict systems and power systems. *IEEE Access* 8: 88 651–88 663.

6 Ullah, Z., Al-Turjman, F., Mostarda, L., and Gagliardi, R. (2020). Applications of artificial intelligence and machine learning in smart cities. *Computer Communications* 154: 313–323.

7 Rigas, E.S., Ramchurn, S.D., and Bassiliades, N. (2015). Managing electric vehicles in the smart grid using artificial intelligence: a survey. *IEEE Transactions on Intelligent Transportation Systems* 16 (4): 1619–1635. https://doi.org/10.1109/TITS.2014.2376873.

8 Sun, Q. and Yang, L. (2019). From independence to interconnection – a review of ai technology applied in energy systems. *CSEE Journal of Power and Energy Systems* 5 (1): 21–34.

9 Bose, B.K. (2017). Artificial intelligence techniques in smart grid and renewable energy systems – some example applications. *Proceedings of the IEEE* 105 (11): 2262–2273.

10 Zhao, S., Blaabjerg, F., and Wang, H. (2020). An overview of artificial intelligence applications for power electronics. *IEEE Transactions on Power Electronics*, Early access 36: 4633–4658.

11 Ahmad, T., Zhang, H., and Yan, B. (2020). A review on renewable energy and electricity requirement forecasting models for smart grid and buildings. *Sustainable Cities and Society* 55: 102052.

12 Ourahou, M., Ayrir, W., Hassouni, B.E., and Haddi, A. (2020). Review on smart grid control and reliability in presence of renewable energies: challenges and prospects. *Mathematics and Computers in Simulation* 167: 19–31.

13 Yin, L., Gao, Q., Zhao, L. et al. (2020). A review of machine learning for new generation smart dispatch in power systems. *Engineering Applications of Artificial Intelligence* 88: 103372.

14 Ali, S.S. and Choi, B.J. (2020). State-of-the-art artificial intelligence techniques for distributed smart grids: a review. *Electronics* 9 (6): 1030.

15 Brockway, P.E., Owen, A., Brand-Correa, L.I., and Hardt, L. (2019). Estimation of global final-stage energy-return-on-investment for fossil fuels with comparison to renewable energy sources. *Nature Energy* 4 (7): 612–621.

16 Bogdanov, D., Farfan, J., Sadovskaia, K. et al. (2019). Radical transformation pathway towards sustainable electricity via evolutionary steps. *Nature Communications* 10 (1): 1–16.

17 EIA (2019).International Energy Outlook. https://www.eia.gov/outlooks/ieo/pdf/ieo2019.pdf (accessed 12 February 2021).

18 Mohamed-Seghir, M., Krama, A., Refaat, S.S. et al. (2020). Artificial intelligence-based weighting factor autotuning for model predictive control of grid-tied packed u-cell inverter. *Energies* 13 (12): 3107.

19 Al Mamun, A., Sohel, M., Mohammad, N. et al. (2020). A comprehensive review of the load forecasting techniques using single and hybrid predictive models. *IEEE Access* 8: 134 911–134 939.

20 Syed, D., Refaat, S., Abu-Rub, H. et al. (2019). Averaging ensembles model for forecasting of short-term load in smart grids. *IEEE International Conference on Big Data*, Los Angeles, USA (9–12 December 2019). IEEE.

21 D'silva, S., Shadmand, M., Bayhan, S., and Abu-Rub, H. (2020). Towards grid of microgrids: seamless transition between grid-connected and islanded modes of operation. *IEEE Open Journal of the Industrial Electronics Society* 1: 66–81. https://doi.org/10.1109/OJIES.2020.2988618.

22 Liang, G., Weller, S.R., Zhao, J. et al. (2016). The 2015 Ukraine blackout: implications for false data injection attacks. *IEEE Transactions on Power Systems* 32 (4): 3317–3318.

23 Mosavi, A., Salimi, M., Faizollahzadeh Ardabili, S. et al. (2019). State of the art of machine learning models in energy systems, a systematic review. *Energies* 12 (7): 1301.

24 Kobak, D. and Berens, P. (2019). The art of using t-sne for single-cell transcriptomics. *Nature Communications* 10 (1): 1–14.

25 Huang, D., Wang, C.-D., and Lai, J.-H. (2017). Locally weighted ensemble clustering. *IEEE Transactions on Cybernetics* 48 (5): 1460–1473.

26 Fahy, C., Yang, S., and Gongora, M. (2018). Ant colony stream clustering: a fast density clustering algorithm for dynamic data streams. *IEEE Transactions on Cybernetics* 49 (6): 2215–2228.

27 Chen, J. and Ran, X. (2019). Deep learning with edge computing: a review. *Proceedings of the IEEE* 107 (8): 1655–1674.

28 Wirtz, B.W., Weyerer, J.C., and Geyer, C. (2019). Artificial intelligence and the public sector – applications and challenges. *International Journal of Public Administration* 42 (7): 596–615.

29 Boudet, H.S. (2019). Public perceptions of and responses to new energy technologies. *Nature Energy* 4 (6): 446–455.

30 Fallah, S.N., Deo, R.C., Shojafar, M. et al. (2018). Computational intelligence approaches for energy load forecasting in smart energy management grids: state of the art, future challenges, and research directions. *Energies* 11 (3): 596.

31 Chollet, F. (2018). *Deep Learning mit Python und Keras: Das Praxis-Handbuch vom Entwickler der Keras-Bibliothek*. MITP-Verlags GmbH & Co. KG.

32 Hasson, U., Nastase, S.A., and Goldstein, A. (2020). Direct fit to nature: an evolutionary perspective on biological and artificial neural networks. *Neuron* 105 (3): 416–434.

33 Veniat, T. and Denoyer, L. (2018). Learning time/memory-efficient deep architectures with budgeted super networks. *IEEE Conference on Computer Vision and Pattern Recognition*, Salt Lake City, USA (18–23 June 2018). IEEE.

34 Hurlburt, G. (2017). Superintelligence: Myth or pressing reality? *IEEE IT Professional* 19 (1): 6–11.

35 Ayoub, N., Musharavati, F., Pokharel, S. and Gabbar, H.A. (2018). Ann model for energy demand and supply forecasting in a hybrid energy supply system. *IEEE International Conference on Smart Energy Grid Engineering*, Oshawa, Canada (12–15 August 2018). IEEE.

36 Notton, G., Voyant, C., Fouilloy, A. et al. (2019). Some applications of ann to solar radiation estimation and forecasting for energy applications. *Applied Sciences* 9 (1): 209.

37 Jayamaha, D., Lidula, N., and Rajapakse, A.D. (2019). Wavelet-multi resolution analysis based ann architecture for fault detection and localization in dc microgrids. *IEEE Access* 7: 145 371–145 384.

38 Hannan, M.A., Ghani, Z.A., Hoque, M.M. et al. (2019). Fuzzy logic inverter controller in photovoltaic applications: issues and recommendations. *IEEE Access* 7: 24 934–24 955.

39 Rizwan, M., Jamil, M., Kirmani, S., and Kothari, D. (2014). Fuzzy logic based modeling and estimation of global solar energy using meteorological parameters. *Energy* 70: 685–691.

40 Chojecki, A., Rodak, M., Ambroziak, A., and Borkowski, P. (2020). Energy management system for residential buildings based on fuzzy logic: design and implementation in smart-meter. *IET Smart Grid* 3 (2): 254–266.

41 Reza, S.S., Mahbub, T.N., Islam, M.M. et al. (2019). Assuring cybersecurity in smart grid networks by fuzzy-logic based trust management model. *IEEE International Conference on Robotics, Automation, Artificial-intelligence and Internet-of-Things*, Dhaka, Bangladesh (29 November–1 December 2019). IEEE.

42 Hong, Y.-Y. and Buay, P.M.P. (2020). Robust design of type-2 fuzzy logic-based maximum power point tracking for photovoltaics. *Sustainable Energy Technologies and Assessments* 38: 100669.

43 Ren, Y., Suganthan, P., and Srikanth, N. (2015). Ensemble methods for wind and solar power forecasting – a state-of-the-art review. *Renewable and Sustainable Energy Reviews* 50: 82–91.

44 Huang, X., Shi, J., Gao, B. et al. (2019). Forecasting hourly solar irradiance using hybrid wavelet transformation and elman model in smart grid. *IEEE Access* 7: 139 909–139 923.

45 Sagi, O. and Rokach, L. (2018). Ensemble learning: A survey. *Wiley Interdisciplinary Reviews: Data Mining and Knowledge Discovery* 8 (4): e1249.

46 Persson, C., Bacher, P., Shiga, T., and Madsen, H. (2017). Multi-site solar power forecasting using gradient boosted regression trees. *Solar Energy* 150: 423–436.

47 Massaoudi, M., Refaat, S.S., Chihi, I. et al. (2020). A novel stacked generalization ensemble-based hybrid lgbm-xgb-mlp model for short-term load forecasting. *Energy* 214: 118874.

48 Zhou, H., Zhang, Y., Yang, L. and Liu, Q. (2018). Short-term photovoltaic power forecasting based on stacking-svm. *9th International Conference on Information Technology in Medicine and Education*, Hangzhou, China (19–21 October 2018). IEEE.

49 Li, S., Wang, P., and Goel, L. (2015). A novel wavelet-based ensemble method for short-term load forecasting with hybrid neural networks and feature selection. *IEEE Transactions on Power Systems* 31 (3): 1788–1798.

50 Reston Filho, J., Affonso, C. and de Oliveira, R. (2015). Energy price classification in north brazilian market using decision tree. *12th International Conference on the European Energy Market*, Lisbon, Portugal (19–22 May 2015). IEEE.

51 Zhang, D., Qian, L., Mao, B. et al. (2018). A data-driven design for fault detection of wind turbines using random forests and xgboost. *IEEE Access* 6: 21 020–21 031.

52 LeCun, Y., Bengio, Y., and Hinton, G. (2015). Deep learning. *Nature* 521 (7553): 436–444.

53 Wang, H., Amagata, D., Makeawa, T. et al. (2020). A dnn-based cross-domain recommender system for alleviating cold-start problem in e-commerce. *IEEE Open Journal of the Industrial Electronics Society* 1: 194–206.

54 Wang, H., Lei, Z., Zhang, X. et al. (2019). A review of deep learning for renewable energy forecasting. *Energy Conversion and Management* 198: 111799.

55 Massaoudi, M., Chihi, I., Sidhom, L. M. et al. (2019). Performance evaluation of deep recurrent neural networks architectures: Application to pv power forecasting. *2nd International Conference on Smart Grid and Renewable Energy*, Doha, Qatar (19–21 November 2019). IEEE.

56 Massaoudi, M., Refaat, S.S., Abu-Rub, H. et al. (2020). Pls-cnn-bilstm: an end-to-end algorithm-based savitzky–golay smoothing and evolution strategy for load forecasting. *Energies* 13 (20): 5464.

57 Wang, F., Zhang, Z., Liu, C. et al. (2019). Generative adversarial networks and convolutional neural networks based weather classification model for day ahead short-term photovoltaic power forecasting. *Energy Conversion and Management* 181: 443–462.

58 Wang, K., Gou, C., Duan, Y. et al. (2017). Generative adversarial networks: introduction and outlook. *IEEE/CAA Journal of Automatica Sinica* 4 (4): 588–598.

59 Kong, X., Li, C., Zheng, F., and Wang, C. (2019). Improved deep belief network for short-term load forecasting considering demand-side management. *IEEE Transactions on Power Systems* 35 (2): 1531–1538.

60 Huang, R., Li, J., Wang, S. et al. (2020). A robust weight-shared capsule network for intelligent machinery fault diagnosis. *IEEE Transactions on Industrial Informatics* 16: 6466–6475.

61 Afrasiabi, M., Mohammadi, M., Rastegar, M., and Afrasiabi, S. (2020). Advanced deep learning approach for probabilistic wind speed forecasting. *IEEE Transactions on Industrial Informatics* 17: 720–727.

62 Khodayar, M., Wang, J., and Manthouri, M. (2018). Interval deep generative neural network for wind speed forecasting. *IEEE Transactions on Smart Grid* 10 (4): 3974–3989.

63 Wang, H., Xue, W., Liu, Y. et al. (2020). Probabilistic wind power forecasting based on spiking neural network. *Energy* 196: 117072.

64 Srivastava, N., Hinton, G., Krizhevsky, A. et al. (2014). Dropout: a simple way to prevent neural networks from overfitting. *The Journal of Machine Learning Research* 15 (1): 1929–1958.

65 Zhao, R., Wang, D., Yan, R. et al. (2017). Machine health monitoring using local feature-based gated recurrent unit networks. *IEEE Transactions on Industrial Electronics* 65 (2): 1539–1548.

66 Dong, M. and Grumbach, L. (2019). A hybrid distribution feeder long-term load forecasting method based on sequence prediction. *IEEE Transactions on Smart Grid* 11 (1): 470–482.

67 Gao, W., Wai, R.-J., and Chen, S.-Q. (2020). Novel pv fault diagnoses via sae and improved multi-grained cascade forest with string voltage and currents measures. *IEEE Access* 8: 133 144–133 160.

68 Karimi, M., Majidi, M., MirSaeedi, H. et al. (2019). A novel application of deep belief networks in learning partial discharge patterns for classifying corona, surface, and internal discharges. *IEEE Transactions on Industrial Electronics* 67 (4): 3277–3287.

69 Zhang, C.-Y., Chen, C.P., Gan, M., and Chen, L. (2015). Predictive deep boltzmann machine for multiperiod wind speed forecasting. *IEEE Transactions on Sustainable Energy* 6 (4): 1416–1425.

70 Chen, L., Qin, N., Dai, X., and Huang, D. (2020). Fault diagnosis of high-speed train bogie based on capsule network. *IEEE Transactions on Instrumentation and Measurement* 69: 6203–6211.

71 Huang, R., Li, J., Li, W., and Cui, L. (2019). Deep ensemble capsule network for intelligent compound fault diagnosis using multisensory data. *IEEE Transactions on Instrumentation and Measurement* 69 (5): 2304–2314.

72 Zhu, Z., Peng, G., Chen, Y., and Gao, H. (2019). A convolutional neural network based on a capsule network with strong generalization for bearing fault diagnosis. *Neurocomputing* 323: 62–75.

73 Li, J., Li, X., and He, D. (2019). A directed acyclic graph network combined with cnn and lstm for remaining useful life prediction. *IEEE Access* 7: 75 464–75 475.

74 Lu, Q., Liao, X., Li, H., and Huang, T. (2020). Achieving acceleration for distributed economic dispatch in smart grids over directed networks. *IEEE Transactions on Network Science and Engineering* 7: 1988–1999.

75 Chen, K., Chen, K., Wang, Q. et al. (2018). Short-term load forecasting with deep residual networks. *IEEE Transactions on Smart Grid* 10 (4): 3943–3952.

76 Wagner, W.P. (2017). Trends in expert system development: a longitudinal content analysis of over thirty years of expert system case studies. *Expert Systems with Applications* 76: 85–96.

77 Khan, A.R., Mahmood, A., Safdar, A. et al. (2016). Load forecasting, dynamic pricing and dsm in smart grid: a review. *Renewable and Sustainable Energy Reviews* 54: 1311–1322.

78 Chavero-Navarrete, E., Trejo-Perea, M., Jáuregui-Correa, J.C. et al. (2019). Expert control systems for maximum power point tracking in a wind turbine with pmsg: state of the art. *Applied Sciences* 9 (12): 2469.

79 Velusamy, D. and Pugalendhi, G. (2020). Water cycle algorithm tuned fuzzy expert system for trusted routing in smart grid communication network. *IEEE Transactions on Fuzzy Systems* 28 (6): 1167–1177.

80 Sarabakha, A. and Kayacan, E. (2019). Online deep fuzzy learning for control of nonlinear systems using expert knowledge. *IEEE Transactions on Fuzzy Systems* 28: 1492–1503.

81 Navarrete, E.C., Perea, M.T., Correa, J.J. et al. (2019). Expert control systems implemented in a pitch control of wind turbine: a review. *IEEE Access* 7: 13 241–13 259.

82 Islam, R.U., Hossain, M.S., and Andersson, K. (2020). A deep learning inspired belief rule-based expert system. *IEEE Access* 8: 190 637–190 651.

83 Sanjani, M.A. (2015). The prediction of increase or decrease of electricity cost using fuzzy expert systems. *4th Iranian Joint Congress on Fuzzy and Intelligent Systems*, Zahedan, Iran (9–11 September 2015). IEEE.

84 Ahmad, W., Ayub, N., Ali, T. et al. (2020). Towards short term electricity load forecasting using improved support vector machine and extreme learning machine. *Energies* 13 (11): 2907.

85 Lin, G.-Q., Li, L.-L., Tseng, M.-L. et al. (2020). An improved moth-flame optimization algorithm for support vector machine prediction of photovoltaic power generation. *Journal of Cleaner Production* 253: 119966.

86 Liu, Y., Shi, J., Yang, Y., and Lee, W.-J. (2012). Short-term wind-power prediction based on wavelet transform – support vector machine and statistic-characteristics analysis. *IEEE Transactions on Industry Applications* 48 (4): 1136–1141.

87 Çevik, H.H. and Çunkas, M. (2015). Short-term load forecasting using fuzzy logic and anfis. *Neural Computing and Applications* 26 (6): 1355–1367.

88 Pourdaryaei, A., Mokhlis, H., Illias, H.A. et al. (2019). Short-term electricity price forecasting via hybrid backtracking search algorithm and anfis approach. *IEEE Access* 7: 77 674–77 691.

89 Tao, Y. and Chen, H. (2016). A hybrid wind power prediction method. *IEEE Power and Energy Society General Meeting*, Boston, USA (17–21 July 2016). IEEE.

90 Kim, M., Choi, W., Jeon, Y., and Liu, L. (2019). A hybrid neural network model for power demand forecasting. *Energies* 12 (5): 931.

91 Shao, H., Deng, X., and Cui, F. (2016). Short-term wind speed forecasting using the wavelet decomposition and adaboost technique in wind farm of East China. *IET Generation, Transmission and Distribution* 10 (11): 2585–2592.

92 Xu, F.Y., Cun, X., Yan, M. et al. (2018). Power market load forecasting on neural network with beneficial correlated regularization. *IEEE Transactions on Industrial Informatics* 14 (11): 5050–5059.

93 Jiang, H., Zhang, Y., Muljadi, E. et al. (2016). A short-term and high-resolution distribution system load forecasting approach using support vector regression with hybrid parameters optimization. *IEEE Transactions on Smart Grid* 9 (4): 3341–3350.

94 Rafiei, M., Niknam, T., Aghaei, J. et al. (2018). Probabilistic load forecasting using an improved wavelet neural network trained by generalized extreme learning machine. *IEEE Transactions on Smart Grid* 9 (6): 6961–6971.

95 Karmacharya, I.M. and Gokaraju, R. (2017). Fault location in ungrounded photovoltaic system using wavelets and ann. *IEEE Transactions on Power Delivery* 33 (2): 549–559.

96 Zhen, Z., Pang, S., Wang, F. et al. (2019). Pattern classification and pso optimal weights based sky images cloud motion speed calculation method for solar pv power forecasting. *IEEE Transactions on Industry Applications* 55 (4): 3331–3342.

97 Chen, K., Hu, J., and He, J. (2016). Detection and classification of transmission line faults based on unsupervised feature learning and convolutional sparse autoencoder. *IEEE Transactions on Smart Grid* 9 (3): 1748–1758.

98 Shafiullah, M. and Abido, M. (2018). S-transform based ffnn approach for distribution grids fault detection and classification. *IEEE Access* 6: 8080–8088.

99 Mishra, P.K., Yadav, A., and Pazoki, M. (2019). Fdost-based fault classification scheme for fixed series compensated transmission system. *IEEE Systems Journal* 13 (3): 3316–3325.

100 Li, Y., Shi, H., Han, F. et al. (2019). Smart wind speed forecasting approach using various boosting algorithms, big multi-step forecasting strategy. *Renewable Energy* 135: 540–553.

101 Song, J., Wang, J., and Lu, H. (2018). A novel combined model based on advanced optimization algorithm for short-term wind speed forecasting. *Applied Energy* 215: 643–658.

102 Park, K., Yoon, S., and Hwang, E. (2019). Hybrid load forecasting for mixed-use complex based on the characteristic load decomposition by pilot signals. *IEEE Access* 7: 12 297–12 306.

103 Incremona, A. and De Nicolao, G. (2019). Spectral characterization of the multi-seasonal component of the italian electric load: a lasso-fft approach. *IEEE Control Systems Letters* 4 (1): 187–192.

104 Yan, K., Li, W., Ji, Z. et al. (2019). A hybrid lstm neural network for energy consumption forecasting of individual households. *IEEE Access* 7: 157 633–157 642.

105 Chu, M., Li, H., Liao, X., and Cui, S. (2018). Reinforcement learning-based multiaccess control and battery prediction with energy harvesting in iot systems. *IEEE Internet of Things Journal* 6 (2): 2009–2020.

106 Ghadimi, N., Akbarimajd, A., Shayeghi, H., and Abedinia, O. (2018). A new prediction model based on multi-block forecast engine in smart grid. *Journal of Ambient Intelligence and Humanized Computing* 9 (6): 1873–1888.

107 Ahmad, A., Javaid, N., Mateen, A. et al. (2019). Short-term load forecasting in smart grids: an intelligent modular approach. *Energies* 12 (1): 164.

108 Ahmad, F., Tariq, M., and Farooq, A. (2019). A novel ann-based distribution network state estimator. *International Journal of Electrical Power & Energy Systems* 107: 200–212.

109 Naz, A., Javed, M.U., Javaid, N. et al. (2019). Short-term electric load and price forecasting using enhanced extreme learning machine optimization in smart grids. *Energies* 12 (5): 866.

110 Martínez-Álvarez, F., Schmutz, A., Asencio-Cortés, G., and Jacques, J. (2019). A novel hybrid algorithm to forecast functional time series based on pattern sequence similarity with application to electricity demand. *Energies* 12 (1): 94.

111 Kuo, P.-H. and Huang, C.-J. (2018). An electricity price forecasting model by hybrid structured deep neural networks. *Sustainability* 10 (4): 1280.

112 Bissing, D., Klein, M.T., Chinnathambi, R.A. et al. (2019). A hybrid regression model for day-ahead energy price forecasting. *IEEE Access* 7: 36 833–36 842.

113 Lago, J., De Ridder, F., and De Schutter, B. (2018). Forecasting spot electricity prices: deep learning approaches and empirical comparison of traditional algorithms. *Applied Energy* 221: 386–405.

114 Aimal, S., Javaid, N., Islam, T. et al. (2019). An efficient cnn and knn data analytics for electricity load forecasting in the smart grid. *Workshops of the International Conference on Advanced Information Networking and Applications*, Matsue, Japan (27–29 March 2019). Springer.

115 Hafeez, G., Javaid, N., Riaz, M. et al. (2019). An innovative model based on fcrbm for load forecasting in the smart grid. *International Conference on Innovative Mobile and Internet Services in Ubiquitous Computing*, Sydney, Australia (3–5 July 2019). Springer.

116 Dong, Y., Zhang, Z., and Hong, W.-C. (2018). A hybrid seasonal mechanism with a chaotic cuckoo search algorithm with a support vector regression model for electric load forecasting. *Energies* 11 (4): 1009.

117 Wang, Y., Zhang, N., Tan, Y. et al. (2018). Combining probabilistic load forecasts. *IEEE Transactions on Smart Grid* 10 (4): 3664–3674.

118 Ouyang, T., He, Y., Li, H. et al. (2019). Modeling and forecasting short-term power load with copula model and deep belief network. *IEEE Transactions on Emerging Topics in Computational Intelligence* 3 (2): 127–136.

119 Yu, Z., Niu, Z., Tang, W., and Wu, Q. (2019). Deep learning for daily peak load forecasting–a novel gated recurrent neural network combining dynamic time warping. *IEEE Access* 7: 17 184–17 194.

120 Zhang, W., Quan, H., and Srinivasan, D. (2018). An improved quantile regression neural network for probabilistic load forecasting. *IEEE Transactions on Smart Grid* 10 (4): 4425–4434.

121 Wen, Q., Gao, J., Song, X. et al. (2019). Robuststl: A robust seasonal-trend decomposition algorithm for long time series. *AAAI Conference on Artificial Intelligence*, Hawaii, USA (27 January–1 February 2019). AAAI.

122 Hu, Z., Tan, B., Salakhutdinov, R.R. et al. (2019). Learning data manipulation for augmentation and weighting. *Advances in Neural Information Processing Systems*, Vancouver, Canada (8–14 December 2019). NIPS.

123 Guo, K., Cui, L., Mao, M. et al. (2020). An improved gray wolf optimizer mppt algorithm for pv system with bfbic converter under partial shading. *IEEE Access* 8: 103 476–103 490.

124 Tan, L., Han, J., and Zhang, H. (2020). Ultra-short-term wind power prediction by salp swarm algorithm-based optimizing extreme learning machine. *IEEE Access* 8: 44 470–44 484.

125 Taher, S.A., Fatemi, S., and Honarfar, O. (2019). Optimal reconfiguration of distribution network for power loss reduction and reliability improvement using bat algorithm. *Soft Computing Journal* 8 (1): 0–0.

126 Liang, H., Liu, Y., Shen, Y. et al. (2018). A hybrid bat algorithm for economic dispatch with random wind power. *IEEE Transactions on Power Systems* 33 (5): 5052–5061.

127 Mugemanyi, S., Qu, Z., Rugema, F.X. et al. (2020). Optimal reactive power dispatch using chaotic bat algorithm. *IEEE Access* 8: 65 830–65 867.

128 Rostami, M.-A., Kavousi-Fard, A., and Niknam, T. (2015). Expected cost minimization of smart grids with plug-in hybrid electric vehicles using optimal distribution feeder reconfiguration. *IEEE Transactions on Industrial Informatics* 11 (2): 388–397.

129 Prasad, S. and Kumar, D.V. (2017). Optimal allocation of measurement devices for distribution state estimation using multiobjective hybrid pso–krill herd algorithm. *IEEE Transactions on Instrumentation and Measurement* 66 (8): 2022–2035.

130 Zhang, Y., Li, X., Zheng, H. et al. (2019). A fault diagnosis model of power transformers based on dissolved gas analysis features selection and improved krill herd algorithm optimized support vector machine. *IEEE Access* 7: 102 803–102 811.

131 Yousri, D., Babu, T.S., Beshr, E. et al. (2020). A robust strategy based on marine predators algorithm for large scale photovoltaic array reconfiguration to mitigate the partial shading effect on the performance of pv system. *IEEE Access* 8: 112 407–112 426.

132 Yousri, D., Abd Elaziz, M., Oliva, D. et al. (2020). Reliable applied objective for identifying simple and detailed photovoltaic models using modern metaheuristics: comparative study. *Energy Conversion and Management* 223: 113279.

133 Soliman, M.A., Hasanien, H.M., and Alkuhayli, A. (2020). Marine predators algorithm for parameters identification of triple-diode photovoltaic models. *IEEE Access* 8: 155 832–155 842.

134 Qiao, W., Huang, K., Azimi, M., and Han, S. (2019). A novel hybrid prediction model for hourly gas consumption in supply side based on improved whale optimization algorithm and relevance vector machine. *IEEE Access* 7: 88 218–88 230.

135 Elazab, O.S., Hasanien, H.M., Elgendy, M.A., and Abdeen, A.M. (2018). Parameters estimation of single-and multiple-diode photovoltaic model using whale optimization algorithm. *IET Renewable Power Generation* 12 (15): 1755–1761.

136 Mirjalili, S. and Lewis, A. (2016). The whale optimization algorithm. *Advances in Engineering Software* 95: 51–67.

137 Bo, H., Niu, X., and Wang, J. (2019). Wind speed forecasting system based on the variational mode decomposition strategy and immune selection multi-objective dragonfly optimization algorithm. *IEEE Access* 7: 178 063–178 081.

138 Shi, Z., Liang, H., and Dinavahi, V. (2017). Direct interval forecast of uncertain wind power based on recurrent neural networks. *IEEE Transactions on Sustainable Energy* 9 (3): 1177–1187.

139 Cui, X., Li, Y., Fan, J. et al. (2020). A hybrid improved dragonfly algorithm for feature selection. *IEEE Access* 8: 155 619–155 629.

140 Ibrahim, I.A., Hossain, M., and Duck, B.C. (2019). An optimized offline random forests-based model for ultra-short-term prediction of pv characteristics. *IEEE Transactions on Industrial Informatics* 16 (1): 202–214.

141 Botchkarev, A. (2019). A new typology design of performance metrics to measure errors in machine learning regression algorithms. *Interdisciplinary Journal of Information, Knowledge and Management* 14: 45–76.

142 Roy, K., Das, R.N., Ambure, P., and Aher, R.B. (2016). Be aware of error measures. Further studies on validation of predictive qsar models. *Chemometrics and Intelligent Laboratory Systems* 152: 18–33.

143 Syntetos, A.A. and Boylan, J.E. (2005). The accuracy of intermittent demand estimates. *International Journal of Forecasting* 21 (2): 303–314.

144 Han, Y. and Tong, X. (2020). Multi-step short-term wind power prediction based on three-level decomposition and improved grey wolf optimization. *IEEE Access* 8: 67 124–67 136.

145 Chen, G., Li, L., Zhang, Z., and Li, S. (2020). Short-term wind speed forecasting with principle-subordinate predictor based on conv-lstm and improved bpnn. *IEEE Access* 8: 67 955–67 973.

146 Zhu, L., Lu, C., Kamwa, I., and Zeng, H. (2018). Spatial-temporal feature learning in smart grids: A case study on short-term voltage stability assessment. *IEEE Transactions on Industrial Informatics* 16: 1470–1482.

147 Aly, H.H. (2020). A proposed intelligent short-term load forecasting hybrid models of ann, wnn and kf based on clustering techniques for smart grid. *Electric Power Systems Research* 182: 106191.

148 Zhang, C., Li, R., Shi, H., and Li, F. (2020). Deep learning for day-ahead electricity price forecasting. *IET Smart Grid* 3 (4): 462–469.

149 Jahangir, H., Tayarani, H., Baghali, S. et al. (2019). A novel electricity price forecasting approach based on dimension reduction strategy and rough artificial neural networks. *IEEE Transactions on Industrial Informatics* 16 (4): 2369–2381.

150 Liu, T., Tan, Z., Xu, C. et al. (2020). Study on deep reinforcement learning techniques for building energy consumption forecasting. *Energy and Buildings* 208: 109675.

151 Liu, H., Yu, C., Wu, H. et al. (2020). A new hybrid ensemble deep reinforcement learning model for wind speed short term forecasting. *Energy* 202: 117794.

152 Zhang, W., Luo, Y., Zhang, Y., and Srinivasan, D. (2020). Solargan: multivariate solar data imputation using generative adversarial network. *IEEE Transactions on Sustainable Energy* 12: 743–746.

153 Chen, C. and Liu, H. (2020). Medium-term wind power forecasting based on multi-resolution multi-learner ensemble and adaptive model selection. *Energy Conversion and Management* 206: 112492.

154 Shah, I., Iftikhar, H., and Ali, S. (2020). Modeling and forecasting medium-term electricity consumption using component estimation technique. *Forecasting* 2 (2): 163–179.

155 Parhizkari, L., Najafi, A., and Golshan, M. (2020). Medium-term electricity price forecasting using extreme learning machine. *Journal of Energy Management and Technology* 4 (2): 20–27.

156 Han, L., Peng, Y., Li, Y. et al. (2018). Enhanced deep networks for short-term and medium-term load forecasting. *IEEE Access* 7: 4045–4055.

157 Ilseven, E. and Göl, M. (2017). Hydro-optimization-based medium-term price forecasting considering demand and supply uncertainty. *IEEE Transactions on Power Systems* 33 (4): 4074–4083.

158 Yousuf, M.U., Al-Bahadly, I., and Avci, E. (2019). Current perspective on the accuracy of deterministic wind speed and power forecasting. *IEEE Access* 7: 159 547–159 564.

159 Boroojeni, K.G., Amini, M.H., Bahrami, S. et al. (2017). A novel multi-time-scale modeling for electric power demand forecasting: from short-term to medium-term horizon. *Electric Power Systems Research* 142: 58–73.

160 Tang, L., Wang, X., Wang, X. et al. (2019). Long-term electricity consumption forecasting based on expert prediction and fuzzy bayesian theory. *Energy* 167: 1144–1154.

161 Kaytez, F. (2020). A hybrid approach based on autoregressive integrated moving average and least-square support vector machine for long-term forecasting of net electricity consumption. *Energy* 197: 117200.

162 Nalcaci, G., Özmen, A., and Weber, G.W. (2019). Long-term load forecasting: models based on mars, ann and lr methods. *Central European Journal of Operations Research* 27 (4): 1033–1049.

163 Azad, H.B., Mekhilef, S., and Ganapathy, V.G. (2015). Long-term wind speed forecasting and general pattern recognition using neural networks. *IEEE Transactions on Sustainable Energy* 5 (2): 546–553.

164 Iorkyase, E.T., Tachtatzis, C., Glover, I.A. et al. (2019). Improving rf-based partial discharge localization via machine learning ensemble method. *IEEE Transactions on Power Delivery* 34 (4): 1478–1489.

165 Xia, M., Wang, K., Song, W. et al. (2020). Non-intrusive load disaggregation based on composite deep long short-term memory network. *Expert Systems with Applications* 160: 113669.

166 Guan, C., Luh, P.B., Michel, L.D., and Chi, Z. (2013). Hybrid kalman filters for very short-term load forecasting and prediction interval estimation. *IEEE Transactions on Power Systems* 28 (4): 3806–3817.

167 Zhou, M. and Jin, M. (2017). Holographic ensemble forecasting method for short-term power load. *IEEE Transactions on Smart Grid* 10 (1): 425–434.

168 Cheng, Y., Jin, L. and Hou, K. (2018). Short-term power load forecasting based on improved online elm-k. *International Conference on Control, Automation and Information Sciences*, Hangzhou, China (24–27 October 2018). IEEE.

169 Sriram, L.M.K., Gilanifar, M., Zhou, Y. et al. (2018). Causal markov elman network for load forecasting in multinetwork systems. *IEEE Transactions on Industrial Electronics* 66 (2): 1434–1442.

170 Córdova, S., Rudnick, H., Lorca, and Martínez, V. (2018). An efficient forecasting-optimization scheme for the intraday unit commitment process under significant wind and solar power. *IEEE Transactions on Sustainable Energy* 9 (4): 1899–1909. https://doi.org/10.1109/TSTE.2018.2818979.

171 Wu, Z., Li, Q., and Xia, X. (2021). Multi-timescale forecast of solar irradiance based on multi-task learning and echo state network approaches. *IEEE Transactions on Industrial Informatics* 17 (1): 300–310. https://doi.org/10.1109/TII.2020.2987096.

172 Muñoz, M.A., Morales, J.M., and Pineda, S. (2020). Feature-driven improvement of renewable energy forecasting and trading. *IEEE Transactions on Power Systems* 35 (5): 3753–3763. https://doi.org/10.1109/TPWRS.2020.2975246.

173 Jäger-Waldau, A. (2019). Snapshot of photovoltaics—february 2019. *Energies* 12 (5): 769.

174 Akhter, M.N., Mekhilef, S., Mokhlis, H., and Shah, N.M. (2019). Review on forecasting of photovoltaic power generation based on machine learning and metaheuristic techniques. *IET Renewable Power Generation* 13 (7): 1009–1023.

175 Massaoudi, M., Refaat, S.S., Abu-Rub, H. et al. (2020). A hybrid bayesian ridge regression-cwt-catboost model for pv power fore- casting. *IEEE Kansas Power and Energy Conference*, Manhattan, USA (13–14 April 2020). IEEE.

176 Semero, Y.K., Zhang, J., and Zheng, D. (2018). Pv power forecasting using an integrated ga-pso-anfis approach and gaussian process regression-based feature selection strategy. *CSEE Journal of Power and Energy Systems* 4 (2): 210–218.

177 Sheng, H., Xiao, J., Cheng, Y. et al. (2017). Short-term solar power forecasting based on weighted gaussian process regression. *IEEE Transactions on Industrial Electronics* 65 (1): 300–308.

178 Manjili, Y.S., Vega, R., and Jamshidi, M.M. (2018). Data-analytic-based adaptive solar energy forecasting framework. *IEEE Systems Journal* 12 (1): 285–296. https://doi.org/10.1109/JSYST.2017.2769483.

179 Zhou, H., Zhang, Y., Yang, L. et al. (2019). Short-term photovoltaic power forecasting based on long short term memory neural network and attention mechanism. *IEEE Access* 7: 78 063–78 074. https://doi.org/10.1109/ACCESS.2019.2923006.

180 Huang, C.-J. and Kuo, P.-H. (2019). Multiple-input deep convolutional neural network model for short-term photovoltaic power forecasting. *IEEE Access* 7: 74 822–74 834.

181 Alfadda, A., Adhikari, R., Kuzlu, M. and Rahman, S. (2017). Hour-ahead solar pv power forecasting using svr based approach. *IEEE Power & Energy Society Innovative Smart Grid Technologies Conference*, Arlington, USA (23–26 April 2017). IEEE.

182 Ahmed, R., Sreeram, V., Mishra, Y., and Arif, M. (2020). A review and evaluation of the state-of-the-art in pv solar power forecasting: techniques and optimization. *Renewable and Sustainable Energy Reviews* 124: 109792.

183 Massaoudi, M., Chihi, I., Sidhom, L. et al. (2019). Medium and long-term parametric temperature forecasting using real meteorological data. *IECON 2019-45th Annual Conference of the IEEE Industrial Electronics Society*, Lisbon, Portugal (14–17 October 2019). IEEE.

184 Yu, Y., Cao, J., and Zhu, J. (2019). An lstm short-term solar irradiance forecasting under complicated weather conditions. *IEEE Access* 7: 145 651–145 666.

185 Alzahrani, A., Shamsi, P., Dagli, C., and Ferdowsi, M. (2017). Solar irradiance forecasting using deep neural networks. *Procedia Computer Science* 114: 304–313.

186 Zhang, X., Li, Y., Lu, S. et al. (2018). A solar time-based analog ensemble method for regional solar power forecasting. *IEEE Transactions on Sustainable Energy* 10 (1): 268–279.

187 Murdock, H.E., Gibb, D., André, T. et al. (2019). Renewables 2019 global status report. https://www.ren21.net/wp-content/uploads/2019/05/gsr_2019_full_report_en.pdf (accessed 15 February 2021).

188 Wu, Y.-X., Wu, Q.-B., and Zhu, J.-Q. (2019). Data-driven wind speed forecasting using deep feature extraction and lstm. *IET Renewable Power Generation* 13 (12): 2062–2069.

189 Kong, X., Liu, X., Shi, R., and Lee, K.Y. (2015). Wind speed prediction using reduced support vector machines with feature selection. *Neurocomputing* 169: 449–456.

190 Du, P. (2018). Ensemble machine learning-based wind forecasting to combine nwp output with data from weather station. *IEEE Transactions on Sustainable Energy* 10 (4): 2133–2141.

191 Verma, S.M., Reddy, V., Verma, K. and Kumar, R. (2018). Markov models based short term forecasting of wind speed for estimating day-ahead wind power. *International Conference on Power, Energy, Control and Transmission Systems*, Chennai, India (22–23 February 2018). IEEE.

192 Wu, Y.-K., Su, P.-E., and Hong, J.-S. (2016). Stratification-based wind power forecasting in a high-penetration wind power system using a hybrid model. *IEEE Transactions on Industry Applications* 52 (3).

193 Buhan, S., Özkazanç, Y., and Çadırcı, I. (2016). Wind pattern recognition and reference wind mast data correlations with nwp for improved wind-electric power forecasts. *IEEE Transactions on Industrial Informatics* 12 (3): 991–1004.

194 Ahmad, F., Alam, M.S., Shariff, S.M., and Krishnamurthy, M. (2019). A cost-efficient approach to ev charging station integrated community microgrid: a case study of indian power market. *IEEE Transactions on Transportation Electrification* 5 (1): 200–214.

195 Shamami, M.S., Alam, M.S., Ahmad, F. et al. (2020). Artificial intelligence-based performance optimization of electric vehicle-to-home (v2h) energy management system. *SAE International Journal of Sustainable Transportation, Energy, Environment, & Policy* 1, no. 13-01-02-0007: 115–125.

196 How, D.N., Hannan, M.A., Lipu, M.S.H. et al. (2020). State-of-charge estimation of li-ion battery in electric vehicles: a deep neural network approach. *IEEE Transactions on Industry Applications* 56 (5): 5565–5574.

197 Xing, Y. and Lv, C. (2019). Dynamic state estimation for the advanced brake system of electric vehicles by using deep recurrent neural networks. *IEEE Transactions on Industrial Electronics* 67: 9536–9547.

198 López, K.L., Gagné, C., and Gardner, M.-A. (2018). Demand-side management using deep learning for smart charging of electric vehicles. *IEEE Transactions on Smart Grid* 10 (3): 2683–2691.

199 Qiu, D., Ye, Y., Papadaskalopoulos, D., and Strbac, G. (2020). A deep reinforcement learning method for pricing electric vehicles with discrete charging levels. *IEEE Transactions on Industry Applications* 56: 5901–5912.

200 Kumar, N., Singh, B., and Panigrahi, B.K. (2019). Pnklmf-based neural network control and learning-based hc mppt technique for multiobjective grid integrated solar pv based distributed generating system. *IEEE Transactions on Industrial Informatics* 15 (6): 3732–3742.

201 Yousri, D., Babu, T.S., Allam, D. et al. (2019). A novel chaotic flower pollination algorithm for global maximum power point tracking for photovoltaic system under partial shading conditions. *IEEE Access* 7: 121 432–121 445.

202 Ram, J.P., Pillai, D.S., Rajasekar, N., and Strachan, S.M. (2019). Detection and identification of global maximum power point operation in solar pv applications using a hybrid elpso-p&o tracking technique. *IEEE Journal of Emerging and Selected Topics in Power Electronics* 8 (2): 1361–1374.

203 Bendib, B., Belmili, H., and Krim, F. (2015). A survey of the most used mppt methods: conventional and advanced algorithms applied for photovoltaic systems. *Renewable and Sustainable Energy Reviews* 45: 637–648.

204 Li, H., Yang, D., Su, W. et al. (2018). An overall distribution particle swarm optimization mppt algorithm for photovoltaic system under partial shading. *IEEE Transactions on Industrial Electronics* 66 (1): 265–275.

205 Liu, N., Zhang, J., Zhao, S. et al. (2017). A novel mppt method based on large variance ga-rbf-bp. *Chinese Automation Congress*, Jinan, China (20–22 October 2017). IEEE.

206 Farayola, A.M., Hasan, A.N., and Ali, A. (2018). Efficient photovoltaic mppt system using coarse gaussian support vector machine and artificial neural network techniques. *International Journal of Innovative Computing Information and Control (IJICIC)* 14 (1).

207 Zhang, X., Li, S., He, T. et al. (2019). Memetic reinforcement learning-based maximum power point tracking design for pv systems under partial shading condition. *Energy* 174: 1079–1090.

208 Agostinelli, F., Hocquet, G., Singh, S., and Baldi, P. (2018). From reinforcement learning to deep reinforcement learning: an overview. In: *Braverman Readings in Machine Learning. Key Ideas From Inception to Current State* (eds. L. Rozonoer, B. Mirkin and I. Muchnik), 298–328. Springer.

209 Mohanty, S., Subudhi, B., and Ray, P.K. (2015). A new mppt design using grey wolf optimization technique for photovoltaic system under partial shading conditions. *IEEE Transactions on Sustainable Energy* 7 (1): 181–188.

210 Keyrouz, F. (2018). Enhanced bayesian based mppt controller for pv systems. *IEEE Power and Energy Technology Systems Journal* 5 (1): 11–17.

211 Belhachat, F. and Larbes, C. (2018). A review of global maximum power point tracking techniques of photovoltaic system under partial shading conditions. *Renewable and Sustainable Energy Reviews* 92: 513–553.

212 Abu-Rub, O.H., Khan, Q. and Refaat, S.S. (2019). Multi-level defects classifica- tion of partial discharge activity in electric power cables using machine learning. *International Conference on Artificial Intelligence*, Long Beach, USA (9–15 June 2019). ICAI.

213 Triki-Lahiani, A., Abdelghani, A.B.-B., and Slama-Belkhodja, I. (2018). Fault detection and monitoring systems for photovoltaic installations: a review. *Renewable and Sustainable Energy Reviews* 82: 2680–2692.

214 Andresen, C.A., Torsaeter, B.N., Haugdal, H. and Uhlen, K. (2018). Fault detection and prediction in smart grids. *IEEE 9th International Workshop on Applied Measurements for Power Systems*, Bologna, Italy (26–28 September 2018). IEEE.

215 Wang, Y., Chen, Q., Sun, M. et al. (2018). An ensemble forecasting method for the aggregated load with subprofiles. *IEEE Transactions on Smart Grid* 9 (4): 3906–3908.

216 Appiah, A.Y., Zhang, X., Ayawli, B.B.K., and Kyeremeh, F. (2019). Long short-term memory networks based automatic feature extraction for photovoltaic array fault diagnosis. *IEEE Access* 7: 30 089–30 101.

217 Emani, C.K., Cullot, N., and Nicolle, C. (2015). Understandable big data: a survey. *Computer Science Review* 17: 70–81.

218 Syed, D., Refaat, S.S., and Bouhali, O. (2020). Privacy preservation of data-driven models in smart grids using homomorphic encryption. *Information* 11 (7): 357.

219 Zainab, A., Refaat, S.S., and Bouhali, O. (2020). Ensemble-based spam detection in smart home iot devices time series data using machine learning techniques. *Information* 11 (7): 344.

220 Yu, X. and Xue, Y. (2016). Smart grids: a cyber-physical systems perspective. *Proceedings of the IEEE* 104 (5): 1058–1070.

221 Yu, Z.-H. and Chin, W.-L. (2015). Blind false data injection attack using pca approximation method in smart grid. *IEEE Transactions on Smart Grid* 6 (3): 1219–1226.

222 Musleh, A.S., Chen, G., and Dong, Z.Y. (2019). A survey on the detection algorithms for false data injection attacks in smart grids. *IEEE Transactions on Smart Grid* 11 (3): 2218–2234.

223 Kurt, M.N., Ogundijo, O., Li, C., and Wang, X. (2018). Online cyber-attack detection in smart grid: a reinforcement learning approach. *IEEE Transactions on Smart Grid* 10 (5): 5174–5185.

224 Karimipour, H., Dehghantanha, A., Parizi, R.M. et al. (2019). A deep and scalable unsupervised machine learning system for cyber-attack detection in large-scale smart grids. *IEEE Access* 7: 80 778–80 788.

225 Hossain, E., Khan, I., Un-Noor, F. et al. (2019). Application of big data and machine learning in smart grid, and associated security concerns: a review. *IEEE Access* 7: 13 960–13 988.

226 Fredrikson, M., Jha, S. and Ristenpart, T. (2015). Model inversion attacks that exploit confidence information and basic countermeasures. *22nd ACM SIGSAC Conference on Computer and Communications Security*, Denver USA (12–16 October 2015). ACM.

227 Matuszak, W.J., DiPippo, L. and Sun, Y.L. (2013). CyberSAVe - Situational awareness visualization for cybersecurity of smart grid systems. *ACM International Conference Proceeding Series*, Atlanta, USA (October 2013). SIGACT.

228 Zhao, M., An, B., Gao, W. and Zhang, T. (2017). Efficient label contamination attacks against black-box learning models. *International Joint Conference on Artificial Intelligence*, Melbourne, Australia (19–25 August 2017). IJCAI.

229 Wu, J., Ota, K., Dong, M. et al. (2018). Big data analysis-based security situational awareness for smart grid. *IEEE Transactions on Big Data* 4 (3): 408–417.

230 Nowotarski, J. and Weron, R. (2018). Recent advances in electricity price forecasting: a review of probabilistic forecasting. *Renewable and Sustainable Energy Reviews* 81: 1548–1568.

231 Hong, T. and Fan, S. (2016). Probabilistic electric load forecasting: a tutorial review. *International Journal of Forecasting* 32 (3): 914–938.

232 Hong, Y.-Y., Taylar, J.V., and Fajardo, A.C. (2020). Locational marginal price forecasting using deep learning network optimized by mapping-based genetic algorithm. *IEEE Access* 8: 91 975–91 988.

233 Zheng, K., Wang, Y., Liu, K., and Chen, Q. (2020). Locational marginal price forecasting: A componential and ensemble approach. *IEEE Transactions on Smart Grid* 11: 4555–4564.

234 Vu, D.H., Muttaqi, K.M., Agalgaonkar, A.P., and Bouzerdoum, A. (2019). Short-term forecasting of electricity spot prices containing random spikes using a time-varying autoregressive model combined with kernel regression. *IEEE Transactions on Industrial Informatics* 15 (9): 5378–5388.

235 Agrawal, R.K., Muchahary, F., and Tripathi, M.M. (2019). Ensemble of relevance vector machines and boosted trees for electricity price forecasting. *Applied Energy* 250: 540–548.

16

Simulation Tools for Validation of Smart Grid

Smart grids (SGs) are considered as complex systems due to interactions among the traditional power grid, active power electronic converters, and communication medium. Furthermore, a high penetration level of renewable energy resources in the SG requires automation and information technologies to ensure smooth and optimal operation in such cyber-physical energy systems. As a result of these requirements, the verification methods on the system level play a critical role in planning and developing SG systems. Although the laboratory-based testing technique seems a good option for the small-scale system, this technique does not suitable for large systems. Recently, simulation-based verification tools become more popular due to their advanced specifications and ability to operate on various levels of the SG. This chapter discusses the current state of simulation-based approaches including multi-domain simulation, co-simulation, and real-time simulation (RTS) and hardware-in-the-loop (HIL) for the SG. Furthermore, some SG planning and analysis software are summarized with their advantages and disadvantages.

16.1 Introduction

The future power system will contain large numbers of power electronics-based distribution generation (DG) units and dedicated communication networks so as to ensure proper control. These newly introduced components and communication systems in the traditional power system increase the complexity of the existing system. Furthermore, to ensure reliable operation and optimization of such cyber-physical power systems, cutting-edge automation and information technologies along with advanced control techniques and data analytics methods play an important role in the future power system and need to be considered in the system design phase [1]. For that reason, the validation and verification on the system level today becomes much more important during the whole engineering and deployment process as the future power system require more complex analyses than is currently performed [2].

Simulation is an approximate imitation of the operation of process or system; that represents its operation over time [3]. Simulation in the field of power systems plays crucial

Smart Grid and Enabling Technologies, First Edition. Shady S. Refaat, Omar Ellabban, Sertac Bayhan, Haitham Abu-Rub, Frede Blaabjerg, and Miroslav M. Begovic.
© 2021 John Wiley & Sons Ltd. Published 2021 by John Wiley & Sons Ltd.
Companion website: www.wiley.com/go/ellabban/smartgrid

role to comprehend system behavior not only under normal operating conditions but also in emergency situations. There are mainly two types of simulation technologies, namely, physical simulation and computer-based simulation. Physical simulation refers to simulation where physical objects are substituted for the real thing (small-scale lab setups, HIL test platforms, etc.) whereas computer-based simulation refers to simulation where all components of the system were modeled and run on the computer. Physical simulations are not suitable for the power system since this kind of simulation needs investment and laboratory space as well as safety requirements for the lab-scale testbed. On the other hand, computer-based simulation tools are cost-effective and flexible to replicate complex systems without being involved in actual architecture [4].

This chapter presents the current state of simulation-based validation approaches for the power systems and addresses SG validation needs. The simulation-based validation approaches will be discussed in three main groups, namely, multi-domain simulation, co-simulation, and RTS. Furthermore, most popular SG planning and analysis tools will be summarized.

16.2 Simulation Approaches

Simulation-based evaluation software has become a more valuable tool for power system planning, designing, and analysis since the introduction of more sophisticated computation solutions. Although simulation approaches are mainly focusing on power system and load flow calculation in traditional power systems, SG simulation studies need to consider a broader scope and handle more levels of complexity. There is much available software in the market for simulation of the SG and these are based on different approaches. In general, the simulation approaches can be divided into three main fields.

- Multi-domain simulation,
- Co-simulation, and
- RTS and HIL.

Each of these simulation approaches has its own advantages and disadvantages. A summary of all these approaches is given below.

16.2.1 Multi-Domain Simulation

Complex systems are generally composed of sub-systems that are involved in multiple disciplines such as electronics, control, software, machinery, etc. On the other hand, most of the traditional simulation software only offers a single field of modeling and simulation. This type of software is not suitable for SG simulation since the SG concept involves integrated communications, sensing and measurements, advanced control, etc. For that reason, recently, some software has been developed for multiple domain modeling and simulation based on interface and high-level architecture [5, 6]. The multi-domain simulation of complex system design not only requires sub-models in all related areas but also needs coordination in the overall system level. There are several simulation tools for multi-domain simulation. Some of the popular ones will be introduced briefly below.

Matlab/Simulink is a graphic programing language that is mainly used for modeling, simulating, and analyzing multi-domain systems. Simulink offers a friendly user interface and a bunch of customizable block libraries to build various models in a short time. Furthermore, the Simulink environment has very tight integration with the rest of the Matlab environment. In other words, Simulink can either feed the Matlab -m file or can be driven by the Matlab -m file. Simulink is widely used in various control applications and digital signal processing for multi-domain simulation and model-based design [7].

Modelica is another multi-domain simulation tool, which is an open-source software, object-oriented, and equation-based language to conveniently model large, complex, and heterogeneous physical systems [8]. Modelica is suitable for mechatronic models in robotics, automotive, and aerospace applications involving mechanical, electrical, hydraulic control and state-machine subsystems, process-oriented applications and generation and distribution of electric power. The models developed by Modelica can be imported conveniently into Simulink using export features of Modelica simulation environments. More information about Modelica can be found in [8].

MapleSim is the Modelica-based multi-domain simulation tool developed by Maplesoft [9]. MapleSim is an advanced system-level modeling and simulation tool that applies modern techniques to dramatically reduce model development time, provide greater insight into system behavior, and to produce fast, high-fidelity simulations. MapleSim, similar to Simulink and Modelica, uses symbolic and numeric mathematical engines to generate model equations, run simulations, and perform analyses [9].

Ptolemy II is also considered a multi-domain simulation tool, which is open-source software backing experimentation with actor-oriented design. Actors in this framework can be defined as software components that execute concurrently and communicate through messages sent via interconnected ports [10]. The model in Ptolemy II is comprised of a hierarchical interconnection of actors. The more information about Ptolemy II can be found in [10, 11].

16.2.2 Co-Simulation

Co-simulation is a technique where two different simulation environments are combined to create a collaborative system with data exchange between the two simulators [12]. In other words, co-simulation uses already well-established simulation tools to model and simulate each subsystem. The main advantage of the co-simulation is that each subsystem is modeled and simulated by an appropriate solver and data exchange among other subsystem solvers is possible. Hence, co-simulation allows simulating multiple systems with different time steps, at the same time. Furthermore, unlike multi-domain simulation, co-simulation enables simulating the large-scale systems as the computation burden is shared between simulators.

A conceptual framework of co-simulation is depicted in Figure 16.1 [12]. It can be seen that simulation of the power system dynamic is modeled and simulated in the "Simulator A" (OpenDSS, PSLF, etc.) by using continuous time step whereas simulation of communication network is modeled and simulated in the "Simulator B" (NS2, Omnet++, etc.) by using discrete time steps. In co-simulation, time synchronization needs to be used between

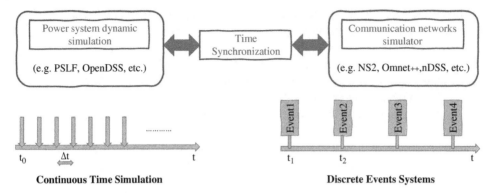

Continuous Time Simulation   **Discrete Events Systems**

Figure 16.1 A conceptual structure of co-simulation [12]. Reproduces with permission from ELSEVIER

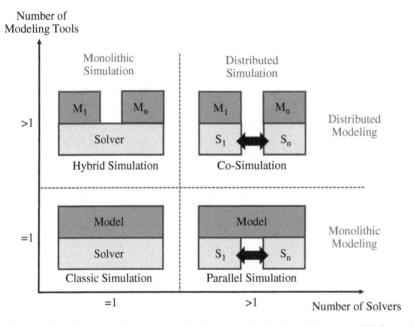

Figure 16.2 Distinction between co-simulation and other simulation types [1]. Reproduced with permission from Springer Nature

continuous time simulation and discrete time simulation systems, as shown in Figure 16.1. It is worth mentioning that the models in the co-simulation are developed and implemented independently [1].

Figure 16.2 shows the distinction between co-simulation and other simulation types. It can be seen that "classic" simulation uses a model and one solver in the same system. On the other hand, "parallel" simulation allows the employment of more than one solver to simulate a model. Both "classic" and "parallel" simulations use a single model which is called monolithic modeling. Another simulation type that is called "hybrid" simulation uses a single solver to simulate multiple models. In co-simulation, model and simulation are distributed into the subsystems, therefore, more than one model and one solver are used to simulate the considered system. It is clear that co-simulation brings considerable benefits while simulating large-scale systems. The major benefit of co-simulation is the separation of the modeling and simulation processes.

In general, the co-simulation structure is divided into two groups depends on the coupling strategies – namely individual and framework coupling – among subsystems. The co-simulation structures are illustrated in Figure 16.3. In the specific co-simulation structure, each sub-system is modeled and simulated by appropriate simulation software and each simulation software is connected to others through the interfaces that allows direct data exchange as shown in Figure 16.3a. On the other hand, generic co-simulation structure consists of co-simulation framework that is responsible for data exchange and temporal synchronization of several models as shown in Figure 16.3b. In this structure, the data exchange is an indirect way and a common time frame for this exchange is ensured by the synchronization algorithm. The main advantage of generic co-simulation is reducing the possibility of coupling errors as an individual coupling between simulation software is not necessary unlike the specific co-simulation structure depicted in Figure 16.3a.

Besides the well-established simulation tools and co-simulation frameworks, interfaces play a critical role to ensure the success of the co-simulation. The Functional Mockup Interface (FMI) is the one of the popular standard to be used in computer simulations to develop complex cyber-physical systems [13]. The FMI defines a container and an interface to exchange dynamic models using a combination of XML files, binaries, and C code zipped into a single file. Activities from systems modeling, simulation, validation, and test can be covered with the FMI-based approach. The FMI is supported by more than one hundred tools such as Dymola, SimulationX, Matlab, Python, PowerFactory, TRNSYS, etc. One of the advantages of the FMI is allowing users to develop their models and simulation tools accessible in the form of so-called Functional Mock-up Units (FMU) that is, simulation component compliant with the FMI specification [1].

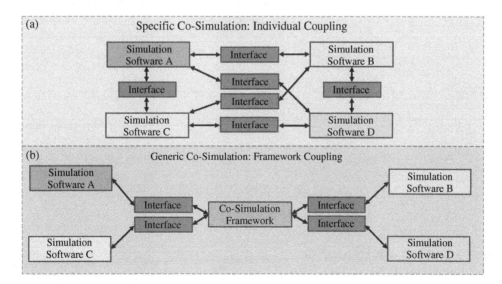

Figure 16.3 Specific and generic co-simulation structures.

16.2.3 Real-Time Simulation and Hardware-in-the-Loop

An RTS mimics the behavior of a physical system by running its computer-based model at the same rate as the actual wall clock time [14]. It means that, in the RTS, when the simulation takes a particular time, the same time has passed in the real world. In the RTS, instead of a variable step size, the simulation is executed in a fixed step size since time moves forward in an equal duration of time. The time required to solve the system equations in the simulation model should be less than the pre-defined fixed step. Otherwise, an overrun occurs. In simple words, RTS must produce the internal variables and output within the same length of time as its physical counterpart would. In general, the RTS is used in high-speed simulations, especially when closed-loop testing required.

The RTS plays an important role in testing and validation of the new systems before implementation of these systems in the field. To validate such systems, the RTS is commonly integrated with hardware that represents as a physical system. This kind of simulation is known as HIL. The HIL provides an effective platform that reduces the risk associated with performing tests and, at the same time, reduces the cost and time required for performing validation of the system. Typically, there are two kinds of HIL test platforms, namely controller HIL (CHIL) and power HIL (PHIL), depending on the properties of the interface between the real-time simulator and the hardware.

The CHIL provides advanced simulation platforms to test and validate the dynamic and transient performances of various controllers such as protective relays, power electronic control boards, and rotating machine controllers. In some test and validation studies, actual power devices need to be tested. Therefore, in such simulation studies, power devices such as power converters, photovoltaic (PV) emulators, etc. should be integrated in HIL tests. Such simulation is therefore called PHIL simulation [15]. It is worth mentioning that the PHIL simulation requires power amplification as the hardware in this test absorbs and/or sinks real power. Figure 16.4 depicts the overall structures of the CHIL and PHIL.

To model and simulate the controllers for the power electronics in different applications or the power systems – from large-scale high-voltage transmission systems to small-scale

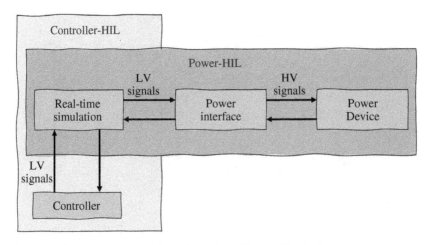

Figure 16.4 The structure of Controller-HIL and Power-HIL platforms.

low-voltage distribution systems – various simulation tools are available to use in a real-time simulator. For example, different real-time simulators are available in the market, for example, xPC Target and RT-LAB for mechatronic systems and using Matlab/Simulink, eFPGASIM and eDRIVESIM for power electronic simulation and eMEGASIM, HYPERSIM and RTDS for power grid RTS [16].

16.3   Review of Smart Grid Planning and Analysis Tools

Planning and analysis of the SG are becoming more complex as the components in the power system have grown in size and the number of interconnections has increased. To assist the SG planning and analysis, digital computers and advanced developed computer programs are used. Such programs include power-flow, stability, short-circuit, and transient software [17].

- The power-flow study is a numerical analysis of the flow of electric power in the power systems, be it the distribution, industrial, or transmission networks. The power-flow analysis tool is the basic tool that is available in almost all power system software. Although this tool is basic, the output of this analysis is very useful for the planning, designing, and operation of any electrical power systems. The power-flow software utilizes sparse matrix methods and multiple solution algorithms that include Newton–Raphson, Gauss–seidel, and fast decoupling techniques.
- Stability analyses tools are used to verify the performance of the power systems under disturbance conditions. These analyses are vital to determine whether generators remain in synchronism in case of disturbances. The disturbances can be a result of a sudden loss of generation or power line, by a sudden load increases or decreases, or by short circuit and switching operations. The stability program combines power-flow equations and machine-dynamic equations to compute the angular swings of machines during disturbances [17].
- The faults in the power system are very common and these faults need specific attention when the power system is designed. A short-circuits tool is used to analyze various of faults that include symmetrical and unsymmetrical faults (single line-to-ground, two line-to-ground, and line-to-line). The results of these analyses can be used to select circuit breakers for fault interruption, select relays that detect faults and control circuit breakers, and determine relay settings [17].
- Power system transients that are caused by utility switching operations or lightnings have the potential to damage the equipment or disrupt the operation. Power system transient studies are performed using electromagnetic transient programs, such as EMTP-RV, PSCAD/EMTDC, and ATP. Occasionally, the simulations are performed in conjunction with system monitoring for model validation and identification of important customer power quality problems [18]. The planning engineers utilize the results of transient studies to determine insulation requirements for the power lines, transformers, and other grid elements, and to select surge arresters that protect the assets against transient over voltages [17].

To sum up, SG planning and analysis tools should have functionalities to conduct steady-state, time series, dynamic, protection, and transient analyses. On the other hand, all these

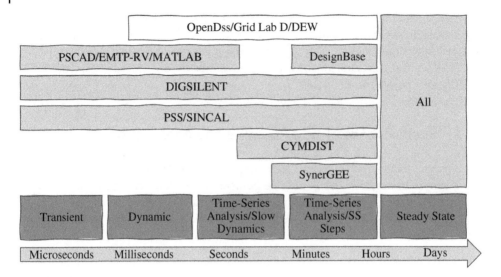

Figure 16.5 Summary of basic analysis fields in each tool.

analyses, unfortunately, cannot be offered only by one tool. For that reason, the basic functionality of some planning and analysis tools are summarized in this section. Figure 16.5 summarizes the capabilities of the software packages in the areas of basic functionality reviewed in this section.

16.3.1 PSCAD

PSCAD (Power Systems Computer Aided Design) is a powerful and flexible graphic user interface to the world-renowned, EMTDC electromagnetic transient simulation engine. PSCAD enables the user to schematically construct a circuit, run a simulation, analyze the results, and manage the data in a completely integrated, graphic environment. Online plotting functions, controls and meters are also included, enabling the user to alter system parameters during a simulation run, and thereby view the effects while the simulation is in progress [19].

PSCAD comes complete with a library of pre-programed and tested simulation models, ranging from simple passive elements and control functions, to more complex models, such as electric machines, full-on flexible alternating current transmission system (FACTS) devices, transmission lines and cables. If a required model does not exist, PSCAD provides avenues for building custom models [19].

Top features of PSCAD include its Python-based automation library that allows users to call PSCAD functions via custom scripts. Using the library, users can load a custom simulation model as well as workspaces and projects. Batch operations are also possible, enabling users to simultaneously run multiple simulations. The following are examples of studies routinely conducted using PSCAD [19]:

- Contingency studies of AC networks consisting of rotating machines, exciters, governors, turbines, transformers, transmission lines, cables, and loads.

- Relay coordination.
- Transformer saturation effects.
- Insulation coordination of transformers, breakers, and arrestors.
- Impulse testing of transformers.
- Sub-synchronous resonance (SSR) studies of networks with machines, transmission lines and High-voltage direct current (HVDC) systems.
- Evaluation of filter design and harmonic analysis.
- Control system design and coordination of FACTS and HVDC; including static synchronous compensator (STATCOM), VSC, and cycloconverters.
- Optimal design of controller parameters.
- Investigation of new circuit and control concepts.
- Lightning strikes, faults, or breaker operations.
- Steep-front and fast-front studies.
- Electric naval vessel design.
- Investigation of the pulsing effects of diesel engines and wind turbines on electric networks.

16.3.2 PowerWorld Simulator

PowerWorld Simulator is a power system simulation package designed from the ground up to be user-friendly and highly interactive. This simulator has the power for serious engineering analysis, but it is also so interactive and graphical that it can be used to explain power system operations to non-technical audiences [20].

The simulator consists of a number of integrated products. At its core, is a comprehensive, robust Power Flow Solution engine capable of efficiently solving systems of up to 100 000 buses. This makes Simulator quite useful as a stand-alone power flow analysis package. Unlike other commercially available power flow packages, however, Simulator allows the user to visualize the system through the use of full-color animated single-line diagrams complete with zooming and panning capability. System models can either be modified on the fly or built from scratch using Simulator's full-featured graphic case editor. Transmission lines can be switched in (or out) of service, new transmission or generation can be added, and new transactions can be established, all with a few mouse clicks. Simulator's extensive use of graphics and animation greatly increases the user's understanding of system characteristics, problems, and constraints, as well as how to remedy them.

The base package also contains all necessary tools to perform integrated economic dispatch, area transaction economic analysis, power transfer distribution factor computation, short-circuit analysis, and contingency analysis. All of the above features and tools are easily accessible through a consistent and colorful visual interface. These features are so well integrated that you will be up and running within minutes of installation. In addition to the features of the base Simulator package, various add-on tools are available [20]. A brief introduction to each follows:

- **Voltage Adequacy and Stability Tool (PVQV):** The purpose of the PVQV add-on is to allow the user to analyze the voltage stability characteristics of a system. After the PVQV simulation is complete, the user can graph various system parameters.

- **Optimal Power Flow Tool** (OPF): The purpose of an OPF is to minimize an objective (or cost) function. In Simulator OPF the Linear Programming OPF algorithm (LP OPF) determines the optimal solution by iterating between solving a standard power flow and solving a linear program to change the system controls thereby removing any limit violations.
- **Security Constrained Optimal Power Flow Tool** (SCOPF): The OPF tool minimizes an objective function (usually total operation cost) by changing different system controls while meeting power balance constraints and enforcing base case operating limits. The SCOPF tool takes it one step further by considering contingencies that may arise during system operation and ensuring that, in addition to minimizing the objective function, no unmanageable contingency violations occur.
- **Optimal Power Flow Reserves** (OPFR): Simulator OPF Reserves is the tool used to simulate Ancillary Services Reserve Markets.
- **Available Transfer Capability Analysis Tool** (ATC): ATC analysis determines the maximum MW transfer possible between two parts of a power system without violating any limits.
- **PowerWorld Simulator Automation Server (SimAuto):** SimAuto provides PowerWorld customers the ability to access PowerWorld Simulator functionality within a program written externally by the user.

16.3.3 ETAP

Another simulation software for electrical network modeling and analyzing is ETAP which stands for Electrical Transient Analyzer Program. This simulation software offers a comprehensive suite of integrated power system enterprise solutions to create a "digital twin platform" for the considered power system to analyze its dynamic, transients, and protection performance [21]. Briefly, ETAP power engineering software utilizes an electrical digital twin so that electrical engineers and operators can perform the following studies in offline or online mode:

- Load flow or power flow study.
- Short circuit or fault analysis.
- Protective device coordination, discrimination, or selectivity.
- Transient or dynamic stability.
- Substation design and analysis.
- Harmonic or power quality analysis.
- Reliability.
- Optimal power flow.
- Power system stabilizer tuning.
- Optimal capacitor placement.
- Motor starting and acceleration analysis.
- Voltage stability analysis.
- Arc flash hazard assessment.
- Ground loop impedance calculation.
- Battery modeling and simulation.

16.3.4 DIgSILENT PowerFactory

The calculation program PowerFactory, as written by DIgSILENT, is a computer-aided engineering tool for the analysis of transmission, distribution, and industrial electrical power systems. It has been designed as an advanced, integrated and interactive software package dedicated to electrical power systems and control analysis in order to achieve the main objectives of planning and operation optimization.

"DIgSILENT" is an acronym for "DIgital SImuLation of Electrical NeTworks". DIgSILENT was the world's first power system analysis software with an integrated graphic single-line interface. That interactive single-line diagram included drawing functions, editing capabilities and all relevant static and dynamic calculation features. The accuracy and validity of results obtained with PowerFactory has been confirmed in a large number of implementations, by organizations involved in planning and operation of power systems throughout the world [22]. To address users' power system analysis requirements, PowerFactory was designed as an integrated engineering tool to provide a comprehensive suite of power system analysis functions within a single executable program. Key features include:

- PowerFactory core functions: definition, modification, and organization of cases; core numerical routines; output and documentation functions.
- Integrated interactive single-line graphic and data case handling.
- Power system element and base-case database.
- Integrated calculation functions (e.g. line and machine parameter calculation based on geometrical or nameplate information).
- Power system network configuration with interactive or on-line SCADA access.
- Generic interface for computer-based mapping systems.

Use of a single database, with the required data for all equipment within a power system (e.g. line data, generator data, protection data, harmonic data, controller data), means that PowerFactory can easily execute all power simulation functions within a single program environment – functions such as load-flow, short-circuit calculation, harmonic analysis, protection coordination, stability calculation, and modal analysis [22].

16.3.5 OpenDSS

OpenDSS stands for Open Distribution Simulation Software that was released by Electric Power Research Institute (EPRI) to provide a free, open-source distribution system simulator for the power engineers. The main aim of this simulator is to make it accessible for the users and encourage the users to contribute to the models that were developed [23].

Although OpenDSS is an open-source simulation software, it provides wide-ranging tool for the distribution system planning and analysis including distributed generation integration, annual load and generation analysis, power flow analysis, etc. Power flow and harmonic analyses, dynamic and fault studies can be performed by OpenDSS. This simulator does not handle the communication layer; however, this simulator can be used in co-simulation with a communication simulator [23].

16.3.6 GridLab-D

GridLab-D is another open-source power system simulation tool that provides valuable information about the considered distribution system to users. Furthermore, GridLab-D is flexible, agent-based simulation engine that allows users to explore a wide variety of analyses or even constructing new scenarios during runtime. It uses Gauss–Seidel method for power flow analysis and has detailed models of power and transmission lines, transformers, voltage regulators, fuses, switches, shunt capacitor banks, etc. [23]. The main advantage of this simulation tool is the developed advanced algorithm that simultaneously coordinates the state of millions of independent devices, each of which is described by multiple differential equations. The more information about this simulator can be found in [24].

16.3.7 Conclusion

The current state of simulation-based approaches including multi-domain simulation, co-simulation, and RTS and HIL for the SG are described in this chapter. Furthermore, some SG planning and analysis software – PSCAD, PowerWorld Simulator, ETAP, DIgSILENT PowerFactory, OpenDSS, GridLab-D – are summarized with their advantages and disadvantages.

References

1 Steinbrink, C. Lehnhoff, S., Rohjans, S. et al. (2017). Simulation based Validation of Smart Grids - Status Quo and Future Research Trends. *8th International Conference on. Industrial Applications of Holonic and Multi-Agent Systems*, Lyon, France (28–30 August). Springer.

2 Palensky, P., van der Meer, A., Lopez, C. et al. (2017). Co-simulation of intelligent power systems: fundamentals, software architecture, numeric, and coupling. *IEEE Industrial Electronics Magazine* 11 (1): 34–50.

3 Banks, J., Carson, J., Nelson, B., and Nicol, D. (2001). *Discrete-Event System Simulation*, 3. ISBN 978-0-13-088702-3. Prentice Hall.

4 Razaq, A., Pranggono, B., Tianfield, H. and Yue, H. (2015). Simulating smart grid: Co-simulation of power and communication network. *50th International Universities Power Engineering Conference*, Stoke-on-Trent, UK (1–4 September 2015). IEEE.

5 Chirapongsananurak, P. and Santoso, S. (2016). Distribution system multi-domain simulation tool for wind power analysis. *IEEE Power and Energy Society General Meeting*, Boston, USA (17–21 July 2016). IEEE.

6 Jun, L., Guochen, W. and Yanyan, L. (2016). Multi-Domain Modeling Based on Modelica. MATEC Web Conferences, Xian, China (3 October 2016). EDP Sciences.

7 MathWorks (2021). Model based design: from concept to code, Matlab/Simulink. https://www.mathworks.com/products/simulink.html (accessed 12 February 2021).

8 Modelica(2017). A Unified Object-Oriented Language for Systems Modeling. Version 3.4. https://www.modelica.org/documents/ModelicaSpec34.pdf (accessed 12 February 2021).

9 MapleSim3 – Multi-Domain Modeling and Simulation, https://www.maplesoft.com/products/maplesim/academic/.

10 Eker, J., Janneck, J., Lee, E.A. et al. (January 2003). Taming heterogeneity – the Ptolemy approach. *Proceedings of the IEEE* 91 (1): 127–144.

11 Lee, E.A., Brooks, C., Feng, T.H. et al. (2010). Ptolemy II 8.0.1: An open-source software framework supporting experimentation with actor-oriented design. University of California.

12 Bhor, D., Angappan, K., and Svalingam, K.M. (2016). Network and power-grid co-simulation framework for smart grid wide-area monitoring networks. *Journal of Network and Computer Applications* 59: 274–284.

13 Awais, M.U., Palensky, P., Elsheikh, A. et al. (2013). The high level architecture RTI as a master to the functional mock-up interface components. *International Conference on Computing, Networking and Communications*, San Diego, USA (28–31 January 2013). IEEE.

14 ERIC-LAB (2000). Real Time Simulation. http://www.eric-lab.eu/drupal/node/4 (accessed 12 February 2021).

15 Steurer, M., Bogdan, F., Ren, W. et al. (2007).Controller and Power Hardware-In-Loop Methods for Accelerating Renewable Energy Integration. *IEEE Power Engineering Society General Meeting*, Tampa, USA (24–28 June 2007). IEEE.

16 Dufour, C., Andrade, C. and Bélanger, J. (2010). Real-Time Simulation Technologies in Education: a Link to Modern Engineering Methods and Practices. *11th International Conference on Engineering and Technology Education*, Ilhéus, Brazil (7–10 March 2010). IEEE.

17 Duncan Glover, J., Sarma, M.S., and Overbye, T.J. (2012). *Power System Analysis and Design*, 5e. Cengage Learning.

18 Grebe, T. (2016). Power System Transient Studies using EMTP-RV. *Electric Energy T&D Magazine* (May/June).

19 PSCAD (2018). PSCAD User's Guide v4.6. https://hvdc.ca/knowledge-base/read,article/160/pscad-user-s-guide-v4-6/v: (accessed 12 February 2021).

20 Power World Corporation (2012). PowerWorld Simulator Version 16. Interactive power system simlation, analysis and visualization. https://www.powerworld.com/files/Simulator16_Help_Printed.pdf (accessed 12 February 2021).

21 Mohammed, T., Malika, A., Bouzidi, B., and Toufik, T. (2013). Electrical network modelling and simulation tools: the state of the art. *Journal of Electrical and Control Engineering* 5: 3.

22 DigSilent GmbH(2012). DIgSILENT Power Factory User Manual. https://www.academia.edu/35776168/DIgSILENT_PowerFactory_User_Manual_2017 (accessed 12 February 2021).

23 Bush, S.F. (2014). *Smart Grid: Communication-Enabled Intelligence for the Electric Power Grid*. Wiley.

24 GridLab-D (2017). A Unique Tool to Design the Smart Grid. https://www.gridlabd.org/index.stm (accessed 12 February 2021).

17

Smart Grid Standards and Interoperability

To accelerate the smart grid (SG) concept and integrate this concept in the main utility grid with its considerable benefits, many issues must be handled before the SG becomes a major player of the main utility grid. One of the important issues with the SG is standards. Although many standards have already been introduced for the distributed generators, metering, and communication medium, more standards are still needed to integrate SG components into the main grid. These standards include generation sources, smart home appliances, and energy management systems (EMS) that need to communicate with each component of the SG to activate the energy trade between customers and producers. This chapter presents an overview of SG standards; new standardization studies, SG policies of some countries, and some important standards for SG.

17.1 Introduction

SG is influenced by many factors such as existing regulations, habits, used generation structures, etc. As the simplest definition, SG refers to modern power systems that consist of communication infrastructure, protection devices, management systems, central and distributed generators, energy storage, consumers-controlled devices, and electric vehicles (EVs). The main role of the SG concept is to monitor, control, protect, and optimize the operation of these components. This definition was published by the US National Institute of Standards and Technology (NIST) which defines the SG as a transition process from the existing power system to the future one that is based on Information and Communication Technologies (ICT) [1].

The most important point is that the implementation of SG needs the integration of ICT into the existing power system. Such a radical change needs a lot of effort on the standardization before turning the SG into reality from the prototypes. Most countries and industrial organizations have already understood the crucial role of SG standards and they have taken big steps to develop these standards. Many organizations have currently set up working groups in parallel to develop these standards. The important point is that the SG specialists have to understand and select a specific one according to the requirements of the specific

Smart Grid and Enabling Technologies, First Edition. Shady S. Refaat, Omar Ellabban, Sertac Bayhan, Haitham Abu-Rub, Frede Blaabjerg, and Miroslav M. Begovic.
© 2021 John Wiley & Sons Ltd. Published 2021 by John Wiley & Sons Ltd.
Companion website: www.wiley.com/go/ellabban/smartgrid

SG topology without going into the depth of each standard since generally these standards consist of hundreds to thousands of pages [2]. This chapter will present an overview of SG standards; new standardization studies, important standards, and SG policies of some countries.

17.2 Organizations for Smart Grid Standardization

Several national and international organizations created working groups to develop SG standards worldwide. In this section, some of the important organizations for SG standardization are introduced.

17.2.1 IEC Strategic Group on SG

The International Electro Technical Commission (IEC) publishes many well identified standards. The IEC established the Smart Grid Strategic Group, which is also referred to as IEC-SG3, in 2008 in parallel with the emerging developments in the SG domain. The main role of IEC-SG3 was to work on SG projects and develop numerous standards that include security, communication, High-Voltage Direct Current (HVDC), blackout prevention, smart substation automation, distributed energy resources (DER), advanced metering infrastructure, demand response (DR) and load management, smart home, building automation, and electric storage. The IEC-SG3 has developed the following key standards for the SG: IEC 61850, IEC 61968, IEC 61970, IEC 62351, IEC 62357 [3].

17.2.2 Technical Communities and Their Subcommittees of IEEE Power and Energy Society (PES)

The Institute of Electrical and Electronics Engineers (IEEE) is a professional association which was formed in 1963. One of the main tasks of the IEEE is developing the standards. In order to oversee the standards development process, the IEEE Standards Association (IEEE-SA) was created in IEEE. The IEEE Power and Energy Society (PES) is one of the important societies under IEEE. The mission of this society is to be the leading provider of scientific information on electric power and energy. IEEE PES technical committees provide standard, recommend practices, and guidelines. All these documents are referred as IEEE Standards.

17.2.3 National Institute of Standards and Technology

Another important organization for the standardization is the NIST, which is an agency of the US Department of Commerce. The Information Technology Laboratory (ITL) at the NIST is one of the important laboratories that offer measurement and standards infrastructure for the Nation. The Computer Security Division of ITL proposed the standard on managing risk from information systems in organizational perspective on computer security and minimum-security requirements for Federal Information and Information Systems for SG.

17.2.4 National Standard of PRC for SG

Standardization Administration is a competent department of the People's Republic of China (P.R.C.) that is responsible to develop standards, there are three types of National Standards Codes: GB, GB/T, and GB/Z. GB stands for the mandatory national standards. GB/T means voluntary national standards, while GB/Z is for the national standardization guiding technical documents. The research on the SG standards started by the Grid Corporation of P.R.C in 2009. In the same year, the Smart Grid Standards System Plan was introduced by China Electric Power Research Institute (CEPRI). This institute proposed standards on the high/ultra-high voltage direct current transmission, the EV charging and discharging, user interface, and SG scheduling.

17.3 Smart Grid Policies for Standard Developments

17.3.1 United States

The main purpose of the US government's energy policy is to provide energy supply at low cost and to protect the environment while doing so [4]. To achieve this goal, the role of renewable energy sources is becoming more prominent in the existing power system. Especially in recent years, the US has invested in renewable energy-based power generation systems and modernization of its power infrastructure. Nevertheless, the US didn't sign Kyoto Protocol, which is an international treaty that extends the 1992 United Nations Framework Convention on Climate Change (UNFCCC) that commits state parties to reduce greenhouse gas emissions. Furthermore, the US does have environment regulations to reduce carbon emissions. According to the Global Smart Grid Federation report, the US has a target of around 17% reduction, below 2005 levels, by year 2020 under the Copenhagen Accord.

On other side, the installation of smart metering infrastructure, which one of the important steps of SG deployment, is rapidly spreading in the US. Figure 17.1 shows the smart meters installations deployment in the US between 2007 and 2017. It is clearly seen that the increase in the smart meter installation grows exponentially. As of year-end 2017,

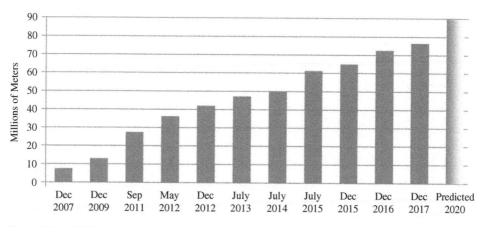

Figure 17.1 US Smart Meter Installations.

approximately 75 million smart meters were installed in houses or commercial buildings by the electric companies. This number is covering almost 60% of US households. On the other hand, according to approved plans, this number is expected to reach around 90 million by 2020 [4].

To ensure cooperation among consumers, utilities, and technology providers, and to manage the benefits of SG, the Smart Energy Consumer Collaborative (SECC) was established. Briefly, the SECC is a nonprofit organization and its mission is to provide trusted information to industry stakeholders seeking a wide understanding of consumers' views about grid modernization, energy delivery and usage. Furthermore, the SECC is aiming to assist consumers to better understand the advantages and benefits of smart energy [4]. Furthermore, the Office of Electricity Delivery and Energy Reliability, which was formed by the US Department of Energy, has produced "GRID 2030" which articulates a national vision for electricity's second 100 years. The vision of "GRID 2030" is that it energizes a competitive North American marketplace for electricity. It connects everyone to abundant, affordable, clean, efficient, and reliable electric power anytime and anywhere. It provides the best and most secure electric services available in the world [5].

17.3.2 Germany

German Commission for Electrical, Electronic & Information Technologies, abbreviated DKE, is the non-profit organization responsible for the development and adoption of standards and safety specifications in the areas of electrical engineering, electronics and information technologies [6]. German SG roadmap focuses on the development of SGs ICT infrastructure, taking national and international standards into account. Figure 17.2 presents a general overview of the active committees in the SG environment [6]. It can be seen that DKE works closely with international standardization committees and European

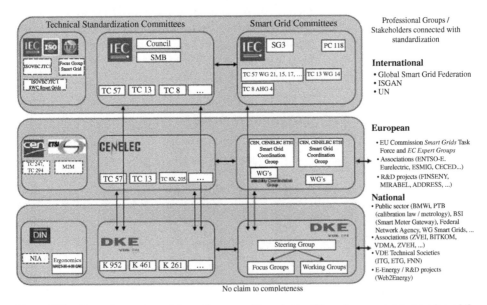

Figure 17.2 General overview of the active committees in the SG environment. Adapted from [6].

standardization committees to develop national standards for SG. The existing national standards have been improved and some of these existing standards have been used for the basic of new standards development. Furthermore, standards for HVDC, Flexible Alternating Current Transmission System (FACTS), market communications have been developed.

17.3.3 Europe

The European Union (EU) operates through supranational institutions and intergovernmental agreements. The main aim of the EU energy policy is to get 20% of its energy from renewable sources by 2020 while cutting down greenhouse emissions. The EU aims to cut down emissions as well as imported energy dependence. The EU announced 20-20-20 climate and energy package in 2007. The package focuses on emission cuts, renewables, and energy efficiency. This package targets the reduction of greenhouse gas emissions by 20%, increase renewable energy to 20% and make 20% energy efficiency improvement by 2020 [7]. Furthermore, the EU announced Electricity Directive 2009/752/EC, which requires the union member countries to install smart metering in at least 80% of households by 2020. The European Electricity Grid Initiative (EEGI), which is an R&D program for SGs, was also established to accelerate the innovation and the development of the electricity networks of the future in EU. The main role of this program is to increase the penetration of DER as well as generate and supply electricity with high power quality. In addition to this, the EEGI aims to encourage customers to be more active in the energy market as prosumers, and to ensure the economically efficient deployment of future power grid for the benefit of all players [7].

17.3.4 South Korea

South Korea was one of the first countries to implement SGs and to add SG development to its national strategy plan. For this purpose, South Korea announced the first SG demonstration project on Jeju Island in 2009. The following year, in 2010, South Korea announced the national roadmap for the SG, then the world's first law related to support for SGs was enacted in 2011 [8]. Similar to the US and the EU, South Korea prepared a road map for installing smart meters in households and they aim to replace all old meters by 2020. Moreover, 2% total electricity from renewable energy in 2012 is expected to reach 10% in 2022, which is stated under the renewable portfolio standard for South Korea. The South Korean Smart Grid Promotion Act provides a framework for sustainable SG projects, their development, deployment, and commercialization [9].

17.3.5 Australia

Australia is one of the countries which developed their energy policy that is connected to sustainable development. Australia has passed laws related to energy policies that promote renewable energy sources to reduce the country's imported energy dependence and carbon emissions. Australia aims to integrate 20% renewable energy-based power supplies in the power system by 2020. The Council of Australian Governments (COAG) establishes the framework for their energy policy. The installation of smart meters was accelerated, especially after the energy shortages in 2006 and 2007. In addition to this, Australia announced a couple of incentive packages to support SG deployment and its enabling technologies such as demand-side management (DSM), etc. The reports show that Australia and New Zealand are expected to invest around $6 billion in SG infrastructure between 2017 and 2027 [10] and [11].

17.3.6 Canada

The Canadian government supports the development of SG through various programs. For example, the decision-makers announced a multi-million fund call in January 2018 to support and accelerate the SG deployment in the country. Investments resulting from this call for proposals mainly supported large-scale SG demonstration projects to verify the benefits of SG technologies that better utilize existing electricity assets while fostering innovation and creating new industries and markets [12]. Furthermore, SmartGrid Canada, which is a not-for-profit national organization, was formed to promote a more modern and efficient electricity grid for the benefit of all Canadians. This organization is comprised of public and private organizations, including utilities, vendors, technology and service providers, academia, and other industry associations [9].

17.3.7 Japan

Japan domestic fossil fuel volume is rather low, and its vulnerability depends on imported fossil fuel. This causes Japan to have an energy structure that is easily affected from domestic/international situation of energy. To handle this challenge, Japan announced four Strategic Energy Plans in 2002, 2003, 2010, and 2014. The current Strategic Energy Plan as the fourth indicates Japan's new direction of energy policy. Some of the impressive targets from this strategic plan can be itemized as follows; (i) accelerating the penetration of wind and geothermal power, (ii) promotion of renewable energy based distributed energy systems, (iii) forming Fukishima as a renewable energy center, and (iv) leveraging DR that promotes efficient energy supply [13]. To accelerate DSM, Tokyo Electric Power Co. (TEPCO), Japan's largest utility company, has adopted smart metering in the power system. To enable remote metering and to provide users with detailed data on power use, services using smart meters have been rolled up since 2015. Currently, there are several SG pilot projects are running in Japan. For example, the project in Yokohama aims to integrate photovoltaic (PV) systems, EVs, and EMS into the existing power infrastructure.

17.3.8 China

SG development is a national priority in the energy sector of China. For that reason, State Grid Corporation of China (SGCC), China's dominant utility company, formulated a detailed strategy and plan toward the implementation of strong and smart power grids in China by 2020 [14]. To coordinate the SG research and development activities, a SG department was established in SGCC. The first version of SGCC framework consists of eight domains that include planning, generation, transmission, substation equipment and communication, distribution, utilization, dispatching and ICT. Furthermore, 26 technical fields and 92 series of standards were presented as part of this framework, which has been benefited from existing standardization roadmaps for the initial development. For instance, German DKE Roadmap, CEN/CENELEC/ETSI Working Groups are taken into consideration. The SGCC divided standardization process into three phases. Phase 1, which is mainly focused on the technical and management specification formulation, and pilot programs, was completed between 2009 and 2010 [14]. At first, nine projects had been developed so as to integrate wind-solar based renewable energy sources, energy storage technologies, smart substations, and test centers for SG. After that, 12 new projects were developed, among which nine focused on distributed generation and the micro-grid, transmission line inspection, flexible HVDC, smart house/building, advanced information platform, and security surveillance. After that, Phase 2 was started in 2011. The construction of ultra-high-voltage grid and rural distribution network were handled in this phase. In 2015, Phase 2 was completed with technical breakthroughs and extensive application of key technology and equipment. Phase 3 from 2016 to 2022 is the last phase of the SGCC standardization process and it is the leading and improving phase. Therefore, the SG with full functionalities is expected to be completed by the end of this phase. In addition to these projects, SGCC has worked actively on standardization for SGs. SGCC has worked closely with the IEC for SG standards and also joined the IEEE P2030 working group for SG interoperation [14]. At the initial stage of SG standards, the SGCC identified 22 standards overall that includes 10 national standards and 12 international standards. An overview of the SG technical standard architecture by SGCC is illustrated in Figure 17.3. Some of the international standards the SGCC identified are IEC62559, IEC 62559, IEC 61850, OGC Open GIS, IEC 61851, IEC 61970, IEC 60870, IEC 62357. The descriptions of these standards are given in Table 17.1.

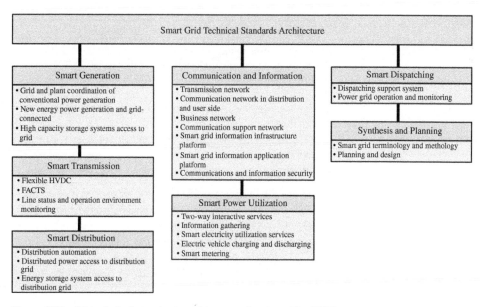

Figure 17.3 SG technical standards architecture developed by SGCC.

Table 17.1 Overview of standards.

Standard	Description
AMI-SEC System Security Requirements	Advanced metering infrastructure (AMI) and SG end-to-end security
ANSI C12.19/MC1219	Revenue metering information model
BACnet ANSI ASH-RAE 135–2008/ISO 16484-5	Building automation
Digital meter / Home gateway	Mandate M/441 of CEN/CENELEC
DNP3	Substation and feeder device automation
EDIXML	Market communication for Germany
IEC 60870	Communications protocol
IEC 60870–5	Telecontrol, EMS, DMS, DA, SA
IEC 60870–6 / TASE.2	Inter-control center communications TASE.2 Inter Control Center Communication EMS, DMS
IEC 61334	DLMS
IEC 61400–25	Wind Power Communication EMS, DMS, DER
IEC 61499	SPS and Automation, Profile for IEC 61850
IEC 61850	Substation automation and protection, Distributed generation, Wind parks, Hydro, E-Mobility
IEC 61850–7-410	Hydro Energy Communication EMS, DMS, DA, SA, DER
IEC 61850–7-420	Distributed Energy Communication DMS, DA, SA, DER, EMS
IEC 61851	EV-Communication Smart Home, e-Mobility
IEC 61968	Distribution Management, System Interfaces for Distribution Management Systems, DCIM (CIM for Distribution)
IEC 61968/61970	Application-level energy management system interfaces, CIM (Common Information Model), Domain Ontology, Interface, Exchange format, Profile, Process blueprints, CIM (Common Information Model) EMS, DMS, DA, SA, DER, AMI, DR, E-Storage
IEC 61970	Energy Management, Application-level energy management system interfaces, Core CIM
IEC 62051–54/58–59	Metering Standards DMS, DER, AMI, DR, Smart Home, E-Storage, EMobility
IEC 62056	COSEM DMS, DER, AMI, DR, Smart Home, E-Storage, EMobility
IEC 62325	Market communications using CIM
IEC 62351	Power System Management and Associated Information Exchange - Data and Communications Security
IEC 62351 Parts 1–8	Information security for power system control operations
IEC 62357	Service oriented Architecture, EMS, DMS, Metering, Security, Energy Management Systems, Distribution management Systems
IEC 62443	General Security
IEC 62559	Terminology and modeling of smart grid

Table 17.1 (Continued)

Standard	Description
IEC 62541	OPC UA (Automation)
IEC TR 61334	DLMS, Distribution Line Message Service
IEEE 1547	Physical and electrical inter-connections between utility and distributed generation (DG)
IEEE 1686–2007	Security for intelligent electronic devices (IEDs)
IEEE C37.118	Phasor measurement unit (PMU) communications
ISO / IEC 14543	KNX, BUS
NERC CIP 002–009	Cyber security standards for the bulk power system
NIST Special Publication (SP) 800–53, NIST SP 800–82	Cyber security standards and guidelines for federal information systems, in-clouding those for the bulk power system
OGC Open GIS	Specifications on Open Geographical Data Interoperability
Open Automated Demand Response (Open ADR)	Price responsive and direct load control
OpenHAN	Home Area Network communication, measurement, and control
ZigBee/Home Plug Smart Energy Profile	Home Area Network (HAN) Communications and Information Model

17.4 Smart Grid Standards

Although tremendous applications, techniques, and technological solutions have been developed for SG systems, the critical shortcomings are waiting to be solved. The major challenge is that the current SG applications lack widely accepted standards. This lack of standards limits the technology adaptation processes in SG systems. The needed standards aim to accelerate the integration of advanced applications, smart meters, smart devices, and renewable energy sources in the SG systems. Many national and international standards were developed and have been developed by various organization as explained in the previous section. Some of the recognized standard development organizations are the NIST, the American National Standards Institute (ANSI), the IEC, the IEEE, the International Organization for Standardization (ISO), the International Telecommunication Union (ITU), the Third Generation Partnership Project (3 GPP) and on the regional level, the Korean Agency for Technology and Standards (KATS), and Joint Information Systems Committee (JISC) [15, 16]. The developed standards are summarized in Table 17.1. In the following, the details of these standards are explained.

17.4.1 Revenue Metering Information Model

ANSI C12.19, which is one of the ANSI standards, identifies a table structure that is used between the end-device and the computer for the utility application. For example, the end-device can be the smart meter whereas the computer can be a personal digital assistant

(PDA) that is carried by a meter reader to read data from the smart meter. Other standards ANSI C12.21 and ANSI C12.22 define the communication requirements of ANSI C12.19 over a modem and network, respectively. In addition to these standards, to ensure two-way communications between the smart meter and client, ANSI C12.18 is specifically designed for meter communications through an optical port [16].

17.4.2 Building Automation

BACnet is a communication protocol for Building Automation and Control (BAC) networks. This communication protocol is designed to guarantee building communication and automation systems for various applications. Especially, heating, ventilating, and air-conditioning (HVAC) systems, lighting control, access control, and fire detection systems and their associated equipment can be counted as examples of these applications. The important point in this protocol is that it offers processes for computerized building automation devices to exchange information regardless of the building service they perform [17].

17.4.3 Substation Automation

To define the communication protocols among smart devices in the electrical substations, an international standard IEC 61850 was released. The abstract data models defined in IEC 61850 can be mapped to several protocols that can run over TCP/IP networks or substation LANs using high speed switched ethernet to obtain the necessary response times below four milliseconds for protective relaying [18]. This standard is used in various modern devices such as distribution automation nodes, grid measurement devices, etc. to ensure a reliable communication medium.

17.4.4 Powerline Networking

Almost all communication technologies have some significant challenges; however, the concern in wireless communications is higher due to the existence of many vulnerabilities. On the other hand, wired media such as power line communication (PLC) makes use of existing infrastructures for transmitting information between the devices and the energy management controller. The data rate of PLC technologies lies in the range of 4–10 Mbps when the used frequency band is high enough, whereas their cost of deployment is comparable with that of wireless counterparts. However, their main drawback is the quality of the transmitted information due to noise issues. For supporting and extending the use of PLCs, the HomePlug Alliance has been developed to provide standards and make PLCs affordable for SG applications. One of the versions of HomePlug is HomePlug Green PHY that is intended for use in the SG.

PRIME is an acronym for "PoweRline Intelligent Metering Evolution" and is a specification for narrow band powerline communication that provides multivendor interoperability. According to PRIME Alliance more than 5 million meters in nine countries are deployed. Some of companies have good experience in PLC technology are STMicroelectronics, Advanced Digital Design, and ZIV Medida [19].

17.4.5 Energy Management Systems

The IEC 61970 series of standards deals with the application program interfaces for EMS. Furthermore, these standards provide common information model that can be employed to exchange data between devices and networks.

The Open Automated Demand Response (OpenADR) was created to standardize, automate, and simplify DR and DER to enable utilities and aggregators to cost-effectively manage growing energy demand and decentralized energy production, and customers to control their energy future. OpenADR was created to automate and simplify DR and DER for the power industry with dynamic price and reliability signals that allow end users to modify their usage patterns to save money and optimize energy efficiency, while enhancing the effectiveness of power delivery across the SG [20].

17.4.6 Interoperability Center Communications

The IEEE American National Standards project P2030 (Guide for Smart Grid Interoperability of Energy Technology and Information Technology Operation with the Electric Power System [EPS], End-Use Applications, and Loads) provides a knowledge base addressing terminology, characteristics, functional performance and evaluation criteria, and the application of engineering principles for SG interoperability of the EPS with end use applications and loads [21].

ANSI C12.22 is the designation of a new standard that is being developed to allow the transport of ANSI C12.19 table data over networked connections. This standard uses the Advanced Encryption Standard (AES) encryption technology to ensure strong and secure communication. More information about ANSI C12.22 can be found in [22].

The International Society of Automation (ISA) developed a standard ISA100.11a, which is an open standard for wireless systems for industrial automation. This standard focuses on robustness, security and network management requirements of wireless infrastructure and low-power consuming devices. The ISA100.11a standard is simple to use and deploy and provides multivendor device interoperability [15].

17.4.7 Cyber Security

Another important aspect of the SG is cyber security since the communication medium is a crucial layer in the SG and this layer is vulnerable to cyber-attacks. To get truly effective end-to-end security, secure protocols must create protected connections based on trusted private key of the actors. To secure power system communications, IEC 62351 family of standards were developed by Working Group 15 (WG15) of IEC Technical Committee 57. These developed standards depict the architecture of a secure power system and standardizes its protocols and components [23]. The different security objectives include authentication of data transfer through digital signatures, ensuring only authenticated access, prevention of eavesdropping, prevention of playback and spoofing, and intrusion detection [15].

17.4.8 Electric Vehicles

EVs are considered game-changers in the future power grid since EVs can be counted as the battery bank of the future and the energy and information flow can be bidirectional.

The transition toward electric mobility needs well defined standards to ensure the functional interoperability of an EV and EVSE of the same physical system architecture. Some of the standards are as follows.

SAE J2293 establishes requirements for EVs and the off-board EV supply equipment used to transfer electrical energy to an EV from a power system in North America [24].

SAE J2836 establishes use cases for communication between plug-in EVs and the electric power grid, for energy transfer and other applications.

SAE J2847 recommended practice establishes requirements and specifications for communication between plug-in EVs and the electric power grid, for energy transfer and other applications [24].

17.5 Conclusion

National and International Standards are needed to accelerate the deployment of SG technologies in the existing and future power grid. To prepare the necessary standards for SG, many working groups under various organizations have conducted standardization studies. These organizations and their studies are summarized in this chapter. Furthermore, SG policies for enhancing the SG deployment in some countries are presented. Finally, the overview of SG standards is presented.

References

1 Uslar, M., Rohjans, S., Bleiker, R. et al. (2010). Survey of Smart Grid standardization studies and recommendations — Part 2. *2010 IEEE PES Innovative Smart Grid Technologies Conference Europe*, Gothenberg, Sweden (11–13 October 2010). IEEE.
2 Sato, T., Kammen, D.M., Duan, B. et al. (April 2015). *Smart Grid Standards: Specifications, Requirements, and Technologies*. Wiley, ISBN:978-1-118-65369-2.
3 IEC (2010). IEC Smart Grid Standardization Roadmap. http://www.itrco.jp/libraries/ IEC-SmartgridStandardizationRoadmap.pdf (accessed 12 February 2021).
4 Cooper, A. (2017). Electric Company Smart Meter Deployments: Foundation for a Smart Grid. http://smartenergycc.org/wp-content/uploads/2017/02/Final-Electric-Company-Smart-Meter-Deployments-Foundation-for-A-Smart-Energy-Grid.pdf (accessed 12 February 2021).
5 US Department of Energy (2003). Grid 2030, A National Vision for Electricity's Second 100 Years. https://www.energy.gov/sites/prod/files/oeprod/DocumentsandMedia/Electric_ Vision_Document.pdf (accessed 12 February 2021).
6 DKE (2013). The German RoadMap, E-Energy/Smart Grid, German Commission for Electrical, Electronic & Information Technologies of DIN and VDE. http://docplayer. net/2470907-The-german-roadmap-e-energy-smart-grid-german-commission-for-electrical-electronic-information-technologies-of-din-and-vde-in-cooperation-with.html (accessed 12 February 2021).
7 EEGI (2013). European Electricity Grid Initiative, Research and Innovation Roadmap 2013-2022. https://www.edsoforsmartgrids.eu//wp-content/uploads/public/20130228_ EEGI-Roadmap-2013-2022_to-print.pdf (accessed 12 February 2021).

8 Smart Grid Demonstration Project on Jeju Island, South Korea, Available on-line: https://www.gsma.com/iot/wp-content/uploads/2012/09/cl_jeju_09_121.pdf

9 Tuballa, M.L. and Abundo, M.L. (2016). A review of the development of smart grid technologies. 59, June 2016: 710–725.

10 Bayar, T. (2017). Australia and New Zealand to invest $6.1bn in smart grid technologies. https://www.powerengineeringint.com/smart-grid-td/australia-and-new-zealand-to-invest-6-1bn-in-smart-grid-technologies/ (accessed 12 February 2021).

11 Lapthorn, A. (2012). Smart Grid, In a New Zealand Context. University of Canterbury. https://ir.canterbury.ac.nz/bitstream/handle/10092/10716/12644631_SmartGrids%20in%20a%20New%20Zealand%20Context.pdf?isAllowed=y&sequence=1 (accessed 12 February 2021).

12 Natural Resource Canada (2018). Smart Grid Program. https://www.canada.ca/en/natural-resources-canada/news/2018/01/smart_grid_program.html (accessed 12 February 2021).

13 Japan Agency for Natural Resources and Energy (2018). Strategic Energy Plan. http://www.enecho.meti.go.jp/en/category/others/basic_plan/5th/pdf/strategic_energy_plan.pdf (accessed 12 February 2021).

14 Xu, Z., Xue, Y., and Wong, K.P. (2014). Recent advancements on smart grids in China. *Electric Power Components and Systems* 42: 251–261, 2014.

15 Gungor, V.C., Sahin, D., Kocak, T. et al. (Nov. 2011). Smart grid technologies: communication technologies and standards. *IEEE Transactions on Industrial Informatics* 7 (4): 529–539.

16 Singh, A.K. (2012). Standards for smart grid. *International Journal of Computational Engineering Research* 2 (7) http://ijceronline.com.

17 Privat, G., John, D., Socorro, R. et al. (2013). *Smart Building Functional Architecture.* Finseny.

18 Diu, A. (2014). Terms of Reference for the Trials. http://essence.ceris.cnr.it/images/documenti/RT_51.pdf (accessed 12 February 2021).

19 Archer, D., Caruso, G., Jimeno, J. et al. (2015). Integrating Active, Flexible and Responsive Tertiary Prosumers into a Smart Distribution Grid. Seventh Framework Programme, Smart Energy Grids.

20 OpenADR Alliance (2020). https://www.openadr.org/overview (accessed 12 February 2021).

21 Basso, T., Hambrick, J. and DeBlasio, D. (2012). Update and review of IEEE P2030 Smart Grid Interoperability and IEEE 1547 interconnection standards. *2012 IEEE PES Innovative Smart Grid Technologies (ISGT)*, Washington, DC, USA (16–20 January 2012). IEEE.

22 Beroset, E. (2004). An Overview of ANSI C12.22. https://electricenergyonline.com/energy/magazine/138/article/An-Overview-of-ANSI-C12-22.htm (accessed 12 February 2021).

23 Carullo, M. (2019). IEC 62351 Standards for Securing Power System Communications. https://www.nozominetworks.com/blog/iec-62351-standards-for-securing-power-system-communications (accessed 12 February 2021).

24 California ISO (2013). California Vehicle-Grid Integration Roadmap: Enabling vehicle-based grid services. http://www.fiedlergroup.com/wp-content/uploads/2014/01/Vehicle-GridIntegrationRoadmap.pdf (accessed 12 February 2021).

18

Smart Grid Challenges and Barriers

Saeed Peyghami and Frede Blaabjerg

Energy is the first concern of the world for the next upcoming decades. Fossil fuel depletion and global warming have been necessitating governments shifting to the carbon-free energy technologies. The micro-grid concept has been presented to integrate the renewable energies and storages in the vicinity of loads, not only to employ environmentally friendly energy resources, but also to prevent development of present power system infrastructures. Control and operation of interconnected microgrids as well as distributed generations (DG) require appropriate coordination among generation and consumption. The smart-grid (SG) concept is introduced to facilitate the optimal, reliable and available operation of such integrated systems. Meanwhile, the centralized, top-down generation and control systems are moving toward a distributed, bottom-up paradigm. As a result, the concepts of DG, micro-grid, SG and distributed operation pose more complexity and challenges to the modern power systems. This chapter presents the challenges and barriers that the modern SGs are facing from different perspectives.

18.1 Introduction

Decarbonization and economization are shifting the energy paradigm toward net-zero-carbon technologies in the society. Electrical energy is the most common form of energy carrier with increasing use among other forms of energy carriers worldwide. Hence, it plays a major role in decarbonization, and hence, in sustainable development. Considering the electric demand growth, the present power systems require expansion of both generation and transmission systems introducing a huge amount of investment. Moreover, the conventional power systems with top-down structure are not an optimal network due to the power loss on the long transmission lines as well as reactive power flow. Hence, the DGs has been installed in low/medium voltage distribution systems in vicinity of loads. This fact can improve the overall system efficiency and availability of electricity. Moreover, most of the DGs are renewable which are environmentally friendly energy resources aiding decarbonization process.

Energizing the distributed networks enables the operation of the distribution systems in the absence of utility grid as an islanded grid. Micro-grid has been used to call the islanded

Smart Grid and Enabling Technologies, First Edition. Shady S. Refaat, Omar Ellabban, Sertac Bayhan, Haitham Abu-Rub, Frede Blaabjerg, and Miroslav M. Begovic.
© 2021 John Wiley & Sons Ltd. Published 2021 by John Wiley & Sons Ltd.
Companion website: www.wiley.com/go/ellabban/smartgrid

distribution grids where the DGs can supply the critical loads of the islanded grid. Microgrid operation presents many advantages including higher reliability, efficiency, and power quality as well as lower operation costs. However, optimal and reliable operation of different DGs in micro-grids as well as operation of cluster of micro-grids requires wide area monitoring system. This fact can be performed by employing communication technologies to manage and operate the system as smart as possible to optimize the operational costs, emissions, and power loss with a certain level of reliability. Thus, the SG technologies have been introduced in order to appropriately monitor and control the modern power systems.

Despite the advantages introduced by the SG technologies, they may pose issues and challenges to the planning and operation of future electric energy systems. Control and communication systems together with the DGs may be the key components of SGs. Interconnected communication systems are of high importance to monitor the demands, generation, energy storage levels, grid topology and islands as well as to control the power and energy flow at among different sub-systems. However, the communication systems expose to the cyber-attacks which can affect the overall system availability. On the other hand, the DGs are mostly renewable based with uncertain prime energy sources such as PV and wind systems. These types of sources are not dispatchable and do not have enough physical inertia, hence, it may cause security issues in the power systems. Furthermore, Most of DGs are interfaced to the grid through a power electronic converter. Increasing use of power converters, while introducing flexibility and controllability, may pose stability and security issues to the power systems, and finally, affecting overall performance of power delivery. Moreover, operation and planning of distribution systems in the microgrid modes in the presence of renewable energies needs minimum reliability and adequacy requirements to supply the critical loads of system.

This chapter introduces the development of structure of modern power systems in Section 18.2. Afterwards, the general concept of reliability is explained in Section 18.3. Then, Section 18.4 discusses the challenges which the modern SGs are facing. Moreover, the new paradigm of reliability in modern SGs considering the possible challenges is presented in Section 18.5. Finally, Section 18.6 summarizes the chapter.

18.2 Structure of Modern Smart Grids

Electric power has conventionally been generated, transmitted and distributed in a centralized, top-down structure and controlled in a centralized, top-down manner. The power and energy flow from large-scale power generation units to the consumers through transmission and distribution power networks. In conventional systems, long- and short-term planning activities such as energy management, unit commitment, etc. are performed centrally. Hence, strong facilities in transmission systems such as lines and transformers are required to guarantee adequate power delivery.

The environmental issues and the limited fossil fuel resources have intensified the importance of decarbonization in energy generation systems using renewable technologies. Hence, the small-scale renewable generation units were installed in distribution systems [1]. Increasing use of these types of resources has changed the electricity market players ultimately enforcing deregulation in electric power systems. Hence, the concept of

DG has been increasing adopted not only to supply the load but also to develop to power system structures. Afterwards, the large-scale renewable generation units such as wind farms and PV power plants have been installed in transmission and medium voltage distribution networks. Consequently, the distribution networks have been energized by DGs, consequently restructuring the electric networks as shown in Figure 18.1. This figure shows a typical structure of future power networks. Compared to the traditional power networks, the large-scale renewable generation units and utility-scale energy storage systems are integrated into the transmission and medium voltage distribution networks. Furthermore, Medium Voltage DC (MVDC) transmission systems are introduced in order to interlink the medium voltage distribution networks and interconnect the High Voltage DC (HVDC) and medium voltage distribution networks [2]. This will introduce an opportunity to form hybrid electric grids as shown in Figure 18.1. Moreover, the MVDC transmission systems will change the market players in the liberalized environment.

In addition, energizing distribution networks as active distribution systems will remarkably enhance the energy efficiency, availability and security of the overall power delivery systems. The microgrid technology facilitates active networks to operate during any contingency occurrence which may cause serious issues for power delivery system. A microgrid is an active electric network having sufficient energy sources to support critical loads in the islanded mode of operation. Since the renewable energies have high uncertainty and dynamic output, utilizing an energy storage is a must. Thereby, the modern distribution networks will be integrated sub-grids forming multi-terminal hybrid Alternative Current (AC)-Direct Current (DC) networks as shown in Figure 18.2 [3, 4], and thereby, form a SG.

The microgrid structures in distribution systems can be classified based on the type of current, location and size of microgrid. Generally, the microgrid structures are divided into AC, DC and hybrid AC/DC grids [5]. Considering the location of islanding protection system over distribution feeders, it is classified as substation microgrid, full-feeder microgrid, partial-feeder microgrid, and single-customer microgrid as shown in Figure 18.2 [4]. Furthermore, following the size or rated power of the microgrid, it is subdivided into pico-, nano-, micro-, milli- and inter-grid, all can be AC or DC [3]. Conceptually, microgrids are the dynamically decoupled sub-grids where the islanded zone can form a network operating autonomously [3]. Basically, a microgrid on a distribution network is interlinked to neighboring sub-grids which guarantees adequate generation and security of supply during operation. In practice, neighboring microgrids or the main utility grid is the strong back-up. Hence, a suitable management of any probable contingency/outage, which may cause the microgrid islanding, can aid adequate and secure operation of the system. Recent advances in communication technologies make it possible to monitor and control the power systems. Adding communication layers to the multi-terminal AC-DC networks introduces SGs, which are the future of power systems.

Operation of such a complex network with a top-down, central control strategy – like traditional power networks – requires strong communication systems and central energy/power management units, which is not a cost-effective and reliable solution. Hence, distributed energy management and power control techniques, equipped with local communication networks, can facilitate appropriate operation and control of modern power systems [6, 7]. As a result, the future power systems structure and control will be as distributed as possible [8]. Moving toward distributed structures will enhance overall system

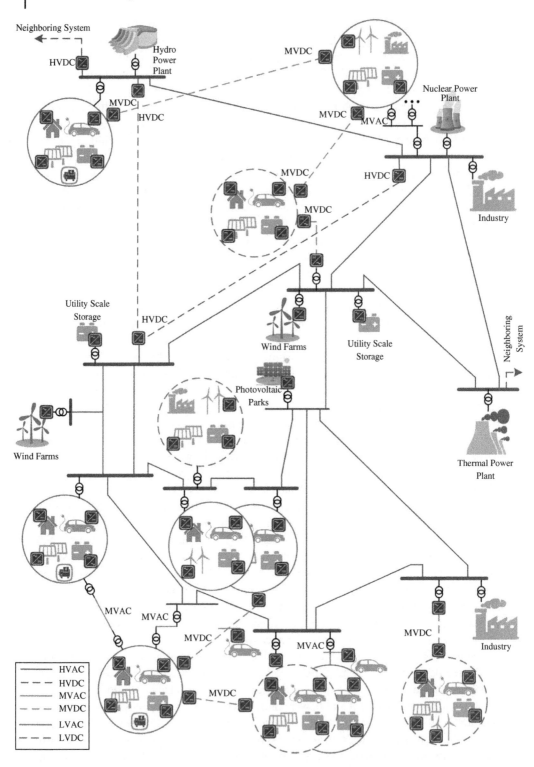

Figure 18.1 Structure of future power networks with hybrid AC and DC sub-grids.

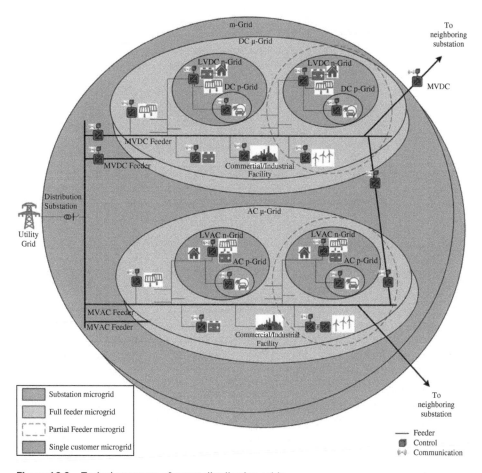

Figure 18.2 Typical structure of smart distribution grids.

efficiency and security of supply. Moreover, utilizing distributed control strategies will strengthen system resiliency subjected to the unpredictable uncertainties and malicious attacks.

18.3 Concept of Reliability in Power Systems

Planning and operation of the conventional power system have been managed hierarchically. The term planning could be divided into three main time periods including long-term facility planning, short-term planning, and operational planning [9, 10]. Facility planning is in charge of developing the power system hardware – generators, transmission lines, distribution facilities – to supply the consumers considering the energy consumption growth over short and long periods of time (5–30 years). Moreover, operational planning is related to utilization of the existing facilities in power systems by weekly and seasonal maintenance and generation scheduling activities and day-ahead, intra-day economic dispatch. The main goal of operational planning is to optimize the power system operation

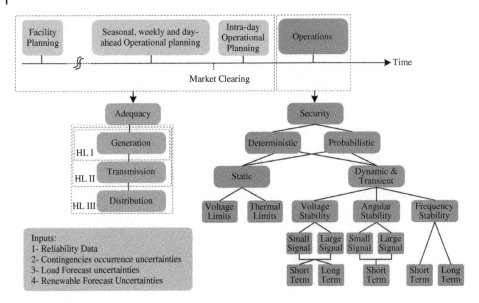

Figure 18.3 Reliability concepts in electric power systems.

attributed to the generation and delivery costs, delivery loss, environmental impacts and reliability. The term adequacy measures the system reliability during planning [11].

Adequacy represents the measure of the ability of a power system to supply customers at all times, considering the planned and unplanned contingencies in the system. In traditional electric networks with centralized, top-down structures, adequacy is evaluated in three hierarchical levels as shown in Figure 18.3, where the first level (HL I) measures the generation capacity adequacy, second level (HL 2) evaluates the composite generation-transmission, i.e. bulk power system, adequacy and third level (HL 3) evaluates the distribution system reliability.

Once power system planning is accomplished, the scheduled or planned power references of different generation units are assigned to the system operators. The operator operates the system considering any uncertainties in the load and generation forecasting, and contingencies occurrence. Actual and scheduled load-generation mismatches can be compensated for by the synchronous machines' inertia, i.e. primary reserve and frequency responsive spinning reserve, i.e. secondary reserve. Nevertheless, sudden outages or contingencies such as loss of a large generation unit, outage of busbar, transmission lines and transformers may cause serious challenges for the system. In such cases, the transmission system capacity is adequate enough and the system remains stable after a contingency, the spinning, supplemental and back-up reserves compensate the generation shortage respectively. However, a contingency may introduce voltage and frequency excursion, overloading lines and stability problems. Hence, the operator should manage and ensure the overall system security subjected to any likely scheduled and unscheduled contingency is adequate. Security management includes two main parts which are security assessment and security enhancement.

Security in electric power systems is defined as its ability to sustain/respond to sudden events and transients such as electric faults or non-anticipated component outages [12, 13].

As shown in Figure 18.3, the security of power systems can be assessed and assured by using deterministic approaches, e.g. N-1, N-2 criterion or probabilistic methods using the likelihood of outages and corresponding impacts on system performance. The operator should assess different contingencies in order to figure out that if any outage/disturbance happens, the system will remain stable or not, and if the voltages and thermal limits are fulfilled. Otherwise, remedial actions either corrective or preventive, or both, are required to maintain system security. Therefore, the security assessment in power systems is divided into two classes which are dynamic/transient and static security.

Static security assessment is associated with maintaining the steady-state values of system variables especially voltage and thermal limits of equipment after contingency occurrence. In this case, it is important that the protection system isolates the faulty zone or equipment. If the system variables violates acceptable values, and the protection system cannot prevent it, static security is not guaranteed. On the other hand, the system must be stable when encountering any contingency. Power system stability could be categorized as voltage stability, angular stability and frequency stability according to the causes of instabilities [13, 14].

Angular stability is related to the ability of electric power systems to maintain synchronism of synchronous machines after any contingencies. In steady-state conditions, there is a balance between mechanical and electromagnetic torques in all operating generators. Thus, the speed of all generators' rotors is constant, and they are synchronized together at a constant frequency. Any disturbance in the system such as loss of generation unit, load change, short circuit fault and line outage results in acceleration or deceleration of rotors of generators. Hence, the relative rotor angles will vary. The rotor angles may be settled in a new equilibrium point or become unstable depending on the disturbance severity and national operation conditions. In the case of instability, loss of synchronism can happen between one generator and the rest of power system (like a single machine infinite bus) or between two groups of generators (interarea).

A change in electromagnetic torque of generators is decomposed into two components including damping torque in phase with rotor speed deviation and synchronizing torque in phase with rotor angle deviation. The rotor angle instability may occur because of either inadequate synchronizing torque causing aperiodic (non-oscillatory) instability, or deficient damping torque resulting in oscillatory instability. According to the severity of the disturbance, the angular stability can be studied in the presence of small or large disturbances. Small disturbance (small signal) stability or dynamic stability is attributed to the synchronous machines ability to stay in synchronism with the elective power system within small disturbances. The oscillatory instability is the common form of small signal instability, while the non-oscillatory instability is compensated for by the generator Automatic Voltage Regulator (AVR). Small signal stability may occur between a small region of electric network with the rest of system and it is called local mode oscillations. Furthermore, the oscillations between a large group of generators with a global effect is called interarea mode oscillations.

Large-disturbance (large-signal) stability, i.e. transient stability, is attributed to the electric power system ability in maintaining system synchronism after severe disturbances. Transient instability is a consequence of sufficient synchronizing torque causing aperiodic rotor angle oscillations. Rotor angle oscillation and excursion finally trigger the protection relay separating the unstable machines. Angular stability is a short-term phenomenon where the transient stability occurs within a few seconds and might expand to a few tens of seconds in very large power systems. Moreover, the dynamic instability may occur within a few tens of seconds.

Voltage stability is defined as the electric power system ability in maintaining/restoring the system voltages at all busses after any disturbance in the system. Voltage instability can occur when the progressive voltage excursion cannot be prevented by the operators or automatic controllers. The consequences will be loss of generation, loss of load, outage of transmission line, and finally cascaded outages. Outspreading the voltage instability can cause voltage collapse and ultimately lead to a blackout.

Voltage stability is categorized as small- and large-signal stability. The concept of small-signal stability is associated with the power-system ability and its voltages under small disturbances such as small-load variations. This form of stability is induced by continuous and discontinuous control systems as well as load characteristics. Moreover, large signal-stability is concerned with the ability of electric power systems to maintain its voltage when subjected to large disturbances such as loss of generation and contingencies. This ability depends on characteristics of load and system, continuous controls such as reactive power controllers, and discontinuous controls such as transformer tap changers, as well as protection systems.

The dynamics of voltage stability will have a longer time frame starting from a few seconds up to several minutes. Hence, it has short- and long-term behavior. Short-term stability is associated with the fast dynamics of loads such as electronically controlled loads, induction motors, converter fed generation units and converters of HVDC and MVDC systems. Its dynamics may last a few seconds. Furthermore, long-term stability is related to the slow dynamics coming from transformer tap changers, thermostatically controlled loads and current limiters of generator with the time frame in the order of several minutes.

Frequency stability is associated with the power system ability to retain system frequency subjected to a severe contingency causing mismatch between demand and generation. Frequency instability can be in the form of frequency fluctuations which may cause generator trip or load disconnection. Severe disturbances introduce large digressions in the power system variables such as voltage, frequency and power flow. Controllers, protection systems and processes will react to these severe violations in order to sustain overall system stability. The reaction of processes and controls could be very fast (short-term) in the order of fractions of seconds such as under-frequency load shedding, generator controls and protections as well as very slow (long-term) in the order of several minutes such as prime mover response. Power system stability can be guaranteed by proper coordination of control units and protection systems as well as sufficient equipment responses. For instance, the existence of adequate spinning and non-spinning reserves, non-critical load curtailment, generation trip, network splitting and, etc. could appropriately facilitate maintaining system stability. The system instability may occur due to the insufficient under-frequency load shedding, lack of adequate reserve, turbine overspeed controls, boiler protections and so on.

18.4 Smart Grid Challenges and Barriers

This section presents the challenges and barriers SGs face in terms of control, operation, planning, marketing, and reliability.

18.4.1 Low Inertia Issues – Frequency Support

In synchronous machine-based power systems, the load-generation balance is immediately achieved by inertial response of generators, which is supported by the kinetic energy of the rotating mass inertia in large-scale generator shafts. The frequency changes caused by the load variations will be measured by the governor system. Thus, the frequency will be changed according to the droop characteristics of governor systems. Finally, it will be compensated for by the frequency secondary controller. The droop control employs the primary reserve in order to balance demand-generation during a short-time period. Furthermore, the secondary frequency controller employs the spinning reserve and hence releases primary reserve. Over a longer time period, the load increment is supplied by the planned generation units during unit commitment at the tertiary control level.

By increasing the integration of the renewable generations, especially wind and solar PV power plants, the inertial response of the system will become weak due to lack of rotating mass in the system. This will cause a high rate of change of frequency (ROCOF), and thus, will result in protection relay reaction. In full power electronic based power systems, it may lead to cascaded outages and finally blackout in the system. In order to maintain system reliability and stability, these units should support the grid frequency. To do so, the inertial response can be provided by (i) control of energy storage units such as dc link capacitor or battery to mimicking the synchronous generator swing equation behavior (ii) de-loading the renewable units, and (iii) wind turbine blades. Furthermore, the secondary control compensates the frequency variation in the system by the storage systems or de-loaded renewable units.

18.4.2 Moving Toward Full/More Renewable Energies

Full/more renewable energy-based grids have gained increasing interest due to the decarbonization, fossil fuel depletion and decreasing costs of renewable technologies. However, this brings more challenges to the grid operation and security because of the variability in the generated power by renewable resources. The first challenge is associated with the system adequacy issues raised by load and generation uncertainties, where the demand and generation mismatch must be supported. Large-scale energy storages and hydro-power plants can enable full renewable energy grids. The main challenges raised by energy storages include the number of required storages, their sizing and placement. Different factors may affect the decision-making on the sizing and placement. System security, especially the frequency stability, might be an important factor which should be taken into account in planning of storages. Moreover, the system adequacy especially in the case of microgrid operation may affect the location and number of energy storages in medium-voltage and low-voltage distribution grids. In order to optimize sizing and placement of the storages, new regulations and technical solutions should be provided especially in terms of protection system and islanding management.

On the other hand, inter-connecting the power systems together with HVDC systems might enable full renewable grid operation, where the neighboring grids can support the security and adequacy of such systems. In this case, a new market might be developed for supporting overall system security. Operating at a maximum power point can enable incorporation of full renewable operations into the electric grids, while it may not be an optimal and cost-effective approach from an energy standpoint. Therefore, there should be a compromise between loss of renewable energy and security of supply provided by de-loading renewable generations which will, in turn, introduce market challenges.

Moreover, most of the renewable resources are connected to the grid through power converters. The power converters may pose stability issues to the system. Thereby, smarter solutions for their control systems compatible with the entire power system should be developed [15]. Otherwise, a small violation in the system voltage/frequency may cause loss of a large part of the system [16] and maybe lead to blackout.

18.4.3 Protection Challenges

Thermal limit: the thermal limits of electro-mechanical equipment during fault is around 10–20 times of its rated power for a few seconds. Therefore, the conventional protection system can appropriately detect and clear the fault in a timely manner without affecting system operation and continuity. However, semiconductor-based devices have less thermal limit compared to electro-mechanical facilities. It is around 1.5–2 times that of rated current for a few seconds in Silicon (Si), and 2–3 times for 3–5 seconds in Gallium-Nitride (GaN) and Silicon-Carbide (SiC) technologies. Thereby, advanced devices and approaches should be developed to facilitate the protection system in order to timely and properly identify, and clear the fault or isolate faulty zones. Inappropriate response of a protection system may lead to loss of power electronic based units, which in turn, can cause cascaded outages and ultimately cause blackout.

Operation mode: The fault current is supported by the grid and DGs during grid connection mode, while the DGs inject the fault current during islanding. Protection system coordination and settings should be adapted for operation modes with different short-circuit current levels. Furthermore, in systems with low inertia, such as islanded AC microgrids, the transient instability is expected, which requires a fast protection strategy to separate the faulty region.

18.4.4 Control Dynamic Interactions

The control unit of power converters are equipped with various controllers with different functionalities. These are inner current regulator, inner voltage regulator, grid synchronization unit usually using Phase Locked Loop (PLL), active/reactive power controllers, and a frequency controller. In order to fulfill its functionalities, each controller is designed in a specific frequency range which can vary from a few hertz to several kilohertz. Therefore, these controllers may have electro-mechanic and electro-magnetic dynamic interaction with other units in the electric network. It may happen across a wide range of frequency from a few hertz up to the switching frequency of converter [9, 17, 18].

The most likely and dangerous interactions in traditional power systems happen between different elements with torsional modes of turbine-generators at sub- and super-synchronous frequencies. Sub-synchronous Resonance (SSR) is the result of interaction between series compensated lines with torsional modes of turbo-generator train in some compensation factors. Moreover, interaction between control systems of HVDC and variable speed drives and power system stabilizers with the torsional modes of turbo-generator may cause sub-synchronous oscillations. Sub-synchronous oscillations may occur among power converters with other converters or series capacitive compensated networks. Furthermore, oscillations at higher frequencies can happen between converters (control of hardware) with other converters (control or hardware), torsional modes of generators, and other passive components.

Most of the interactions occur between the control system of at least one unit with the control system of other units or power system hardware. The interactions can lead to instability in power girds with poor/negative damping ratio. This type of stability has been discussed in [9, 17, 18]. It will become severe by increasing the integration of power converters into electric networks, and it may interrupt power delivery system [17, 19–22].

18.4.5 Reliability Issues

In a liberalized environment, conventional reliability measures may not ensure an acceptable performance of the system. This is due to the fact that conventional concepts are based on centralized, top-down paradigm, where in the distributed systems, introducing new definitions are required. The Cigre C1.27 working group has modified the reliability definition in modern power systems [23]. This new definition includes the terms of adequacy and security considering integrating renewable resources, DGs, and storages. In this new definition, not only the end-consumers should be supplied with a required level of reliability but also the power system should ensure reliability of both consumers and suppliers including private generation units. However, this re-definition is still in its first step and much more effort should be made to carry out and provide comprehensive reliability definitions in modern power systems. The first challenge might be defining a local reliability concept covering the dynamic structure of microgrids. New reliability indices are required [24] not only applicable for grid connected operation, but also for measuring the islanded microgrids reliability. Furthermore, like load point reliability indices, new reliability metrics are of importance for DG points in order to guarantee a reliable grid for DG owners. The next challenge will be reliability evaluation methods for dynamic microgrids considering the complexity of the system such as islanded points, accessibility of private sources/storages, liberalization and interoperability. Due to the complexity of the system, local reliability concepts may reduce the analysis and calculation burdens. Furthermore, the cyber-security concept should be included in the reliability evaluation, where the operation and control of most of the units rely on communication and software technologies. New reliability concepts should also provide reliability indices considering the power quality issues. For instance, a low-quality electric power may not be reliable for some hospital equipment while it might be reliable for street lighting. Especially, in the case of microgrids, one microgrid may not require higher power quality while another one may have sensitive loads. Considering reliability worth analysis, this fact may introduce more complexity.

High proliferation of power electronics in modern electric power networks also pose new reliability challenges. The main issue caused by power electronic converters is the control interactions among converter control and other sub-systems. This will bring security challenges in case of increasing use of power converters [17]. This challenge will be even greater once the operator wants to evaluate the security level of system in terms of various contingency considering different causes of power system stability. The first issue is the feasibility of the operator access to the control system of each DG, e.g. PV converters, wind converters and so on. Since the control system interactions in wide range of frequency depend on the converter characteristics and corresponding control system specifications, assessing the overall system security can be a difficult task considering the high penetration of power electronics. The next challenge, in terms of security, is the dynamic and static security assessment approaches and the time consumption during stability analysis, providing that the system operator may not be able to assess the system security associated with various forms of stability at an expected time period. Therefore, some aggregated and smart approaches should be developed to assess and manage system security.

The second challenge caused by power converters is the reliability of converter components which depends on the operating conditions [25–33]. Therefore, an accurate reliability model of converters is required for system-level analysis especially taking into account that power converters are also exposed to wear-out failures [26, 32, 33]. On the other hand, power converter technologies are developing every day and the reliability characteristics of modern technologies, such as new semiconductor devices, will be different from current technologies. As a result, any converter-level and system-level reliability and risk analysis following the historical reliability data may cause non-optimal decision making. Therefore, reliability modeling of converters is of paramount importance for modern power system reliability assessment. Moreover, an item or a component reliability is associated with the applied mission profile [25–33]. Taking into account the mission profiles, especially for wind and PV applications with high fluctuations in climate conditions, brings more complexity in terms of reliability prediction. This will further increase the analysis and calculation burdens in future power systems adequacy assessment especially in the case of moving toward 100% renewable energies. Therefore, simplified and aggregated models for reliability of power converters could be developed to facilitate analysis of the reliability of large-scale modern power electronic based power systems.

18.4.6 Marketing

Microgrid technology in distribution systems may pose some marketing issues to operation and planning of modern electric power networks. At the low-voltage microgrid level, the frequency and voltage support by the DGs and energy storages should be performed by the available and assessable units in an optimal scheme. From the stability and security point of view, the available units must maintain the microgrid security in the islanded mode. However, the private DG/storage owners may not support the microgrid. As a result, operational planning may be a complex task for operators, and hence, some regulations are required to highlight security in operational planning and optimization to convince private owners to support the microgrid.

Moreover, the MVDC transmission systems among different sub-grids will affect the marketing of future systems. Transmission/Distribution Systems, Transmission System Operator (TSO) and Distribution System Operator (DSO) unbundling as well as generation and operation unbundling together with DGs in a liberalized environment poses new challenges for power system marketing, which requires technical solutions to socio-economically manage the SGs.

Furthermore, cyber security will introduce high costs for power systems, which should be considered in the planning phase of SGs. However, the source of the cyber security induced costs should be defined by the planners. These costs may be supported by the utility and/or customers. It should be identified which parts have the greatest benefits from having a secure grid, i.e. the large-scale dispatchable units, the large-scale renewable generations, low- /medium-scale DGs, customers (residential, industrial and commercial end-consumers), independent system operator, aggregators, retailers and so on. An optimization program should be employed to assign the cyber-security costs to the different parts. [34].

18.5 New Reliability Paradigm in Smart Grids

The paradigm shift from the top-down structure to the distributed structure in modern power systems poses new challenges in terms of reliability (Figure 18.4). Microgrids are becoming active and dynamic elements of power systems which should be able to operate in the islanded mode being subjected to any contingency in the system. Therefore, facility and operational planning must fulfill minimum adequacy requirements taking into account the islanded operation mode. Moreover, the overall system security should be guaranteed considering the issues raised by modern technologies.

18.5.1 Adequacy

The distribution systems as microgrids are energized by DGs which are able to supply the critical demand of corresponding zone within islanding. The majority of DGs rely on renewable technologies such as wind and PV, which are facing considerable uncertainties in production. Hence, long-term energy storage systems are required to compensate the generation system uncertainties, consequently improving system reliability. Hence, the

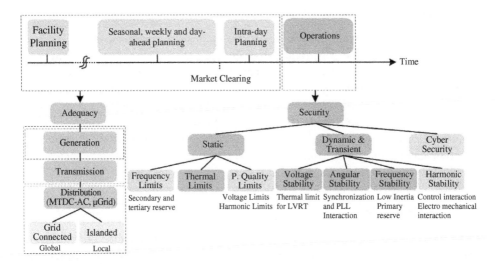

Figure 18.4 Reliability concepts in modern SGs.

storage systems and DGs are crucial elements of microgrids. Moreover, the medium-voltage DC grids are the enabling technology to efficiently interconnect the high voltage ac-dc grids to the distribution systems as well as efficiently integrate renewable technologies to the grid. During islanding or other outage in the system, the storage and generation system capacity in the distribution system should be adequate enough to ensure supply of the system demand with a certain level of reliability. As a result, the conventional hierarchy of adequacy assessment in three levels may not be a suitable approach for modern power system reliability evaluation and enhancement.

Top-down reliability assessment approaches sound to be changed into the distributed or decentralized manners. The distribution active systems with DGs and storages should be adequate enough to supply their local demand. This could be called "local adequacy" meaning the local system has enough capacity to supply its load during islanding operation. While the entire system must have global adequacy implying that the power system should have enough capacity to supply overall demand.

Islanding and separation have accrued in conventional power systems due to the frequency of instability [13]. However, traditional power networks have large-scale thermal power generation units, which are dispatchable and may be able to support primary and secondary reserves (spinning reserve). In modern electric power networks, with high levels of renewable energy penetration, the primary and secondary reserves are limited due to their un-dispatchability. This may be supported by energy storage in small-scale microgrids while it is challenging for large-scale islanded grids. Therefore, sufficient storage capacities, and their location, together with DG placement should be planned considering the overall system reliability.

Liberalization should also pose new challenges to system planners. New markets will be introduced by MVDC transmission systems while it will be the main link between distribution systems. This may affect system adequacy and may be a new player in the electric market.

18.5.2 Security

Modern power system security in the top-level sounds similar to conventional systems. However, by increasing the penetration of renewable energies and connecting them by HVDC or converters to the balk power system i.e. inertia will decrease. Consequently, unlike conventual power plants, the system will suffer from poor static and dynamic security. Moreover, the distribution systems in the islanded mode also suffer from poor inertia. The power electronic systems tend to instability due to the interaction between high-speed control systems of converters, implying poor dynamic security. Moreover, the communication system among different operators, as well as between operators and components, will open the system to cyber-attacks and cyber security will be of importance.

18.5.3 Static Security

Besides the conventional static security assessment ensuring the retention of voltage and thermal limits in the corresponding boundaries, power quality limits and frequency support capability should be taken into consideration for modern power systems security.

Power quality becomes an important issue in some of the applications especially for the electronic loads in, e.g. hospital, data center and, etc. Therefore, in case of a contingency, the islanded sections should meet the minimum power quality requirements. In the conventional approach only, voltage limits were important, however, quality of voltage is also of importance for modern power systems. For instance, the quality of power feeding street lighting might not be a reliable power for hospital. Moreover, the lower power quality may damage the home appliance and commercial machines. It will also increase system power loss and decrease equipment lifetime. Therefore, power quality should be ensured during static security assessment.

18.5.4 Dynamic/Transient Security

Dynamic security in modern power systems is of paramount importance due to the vulnerability of the system in terms of frequency, harmonic and voltage stability. Lack of sufficient inertial response exposes the system to frequency instability. Emulating the inertia by primary droop control, e.g. in wind turbines, to be operated under maximum allowable power as a primary or a secondary reserve could improve frequency stability, while subjected to high degree of uncertainty. Hence, in the case of insufficient input wind or solar energy, the total amount of this kind of reserve will be very small and might be unable to respond to load violations. This becomes severe in power systems with high penetration of renewable energy resources. Energy storages should then support the system for frequency stability. Even if they have enough adequacy to enhance the frequency response of the system, the private energy storages may not be accessible. Thereby, the system operator should take preventive or corrective actions to prevent system failure by load shedding, network splitting to islands, etc. The regions with private DGs and storages, load shedding may not be possible and the corresponding operator may prefer to become islanded. This will result in even more frequency violations in the system and can finally lead to blackout of the main grid.

Another issue which may affect the system security is the harmonic instability, particularly in systems with high penetration of power electronic converters. The system operator should retain the system as securely as possible in terms of any contingency leading to harmonic stability. The conventional security assessment activities in power system operation consist of voltage, frequency and angular stability analysis. However, the harmonic stability is becoming a serious issue of modern power systems. The interest time for harmonic stability is around micro-seconds to seconds. Hence, strong assessment tools and techniques are required to check the system security for numerous likely contingencies. Since the system controls and components have to be accurately modeled for harmonic stability analysis, the operator may not have enough time to assess all contingencies. Therefore, advanced harmonic stability techniques should be developed for large power systems. These techniques can rely on sensitivity analysis, stability margin prediction, frequency domain analysis, etc.

18.5.5 Cyber Security

Modern power systems reliability, performance and manageability have remarkably intensified the importance of bidirectional communications among various domains such as generation units, transmission and distribution systems, consumers, operators, market players and service providers. The time frame of interest in the aforementioned domains is from fractions of seconds, associated with the operation, to a few hours in the marketing domain. From a power system operation standpoint, any unbalance between generation and demand might cause any types of instability depending on disturbance characteristics, i.e. its size and location, leading to the generation-demand mismatch and ultimately over-all black out. Hence, communication is used in various domains to exchange information for monitoring, control, protection, planning, etc. Home Area Networks (HAN) are used to interface end-consumers to the SG for supporting their energy and facilitating demand-side management functionality. The Neighboring Area Network (NAN) is used to interface the HAN and Wide Area Networks (WAN) in distribution systems, and the WAN is used for monitoring and control of generation systems and transmission networks. These communication networks exposed to attacks from different sources can finally damage and make an area insecure which, in turn, can infect other areas.

Modern power systems require strong cyber infrastructure including measurements, communications, controls, computers and software in order to efficiently and reliably manage the power system with new market players including DGs, electric vehichles (EVs) and storages. On the other hand, digitalization and liberalization make the system move toward distributed control strategies and grid structures, and hence, communication is becoming more important. The communication system adds flexibility and reliability for the modern power systems, while being exposed to cyber-attacks. Hence, cyber security is of paramount importance for future power systems. An insecure cyber layer will cause dynamic and static instabilities leading to islanding and ultimately blackout. Cyber security should be considered in both planning and operation phases of modern power systems.

18.6 Summary

Future power systems are revolutionizing to be as distributed as possible not only in gen-eration, but also in control and operation. Modern technologies are facilitating the power system modernization in order to aid decarbonization and sustainable development. Such technologies introduce various challenges in terms of control, operation, planning, and marketing, which can affect overall system performance. Socio-economical and technical solutions are required to facilitate smarter power generation and delivery in future power systems. These solutions can include aggregated and simplified models for reliability, secu-rity, stability, power management, energy management analyses as well as technical approaches to overcome marketing challenges regarding DSO/TSO unbundling, genera-tion/operation unbundling, cyber security costs, DG. Thereby, the conventional reliability concepts should be modified not only in definition but also in introducing new metrics and solutions to assess and enhance the reliability and availability of the re-structured power systems. Reliability worth analysis is of paramount importance in the liberalized environ-ment taking into account modern technologies such as DGs, distributed storages, micro-grids, communication and SG technicities. Moreover, new marketing and operation regulations are required for distribution systems in the islanded mode to guarantee system security as well as optimal and stable operations.

References

1 Blaabjerg, F., Yang, Y., Yang, D., and Wang, X. (Jul. 2017). Distributed power-generation systems and protection. *Proceedings of the IEEE* 105 (7): 1311–1331.

2 Petz, I. (2017). Distributed Power Generation: Future Energy. https://www.siemens-energy.com/global/en/news/magazine/2019/distributed-power-grids.html (accessed 12 February 2021).

3 Boroyevich, D., Cvetkovic, I., Burgos, R., and Dong, D. (2013). Intergrid: a future electronic energy network? *IEEE Journal of Emerging and Selected Topics in Power Electronics* 1 (3): 127–138.

4 Bower, W.I., Ton, D.T., Guttromson, R. et al. (2014). The advanced microgrid. Integration and interoperability. https://www.energy.gov/sites/prod/files/2014/12/f19/AdvancedMicrogrid_Integration-Interoperability_March2014.pdf (accessed 12 February 2021).

5 Peyghami, S., Mokhtari, H., and Blaabjerg, F. (2018). Autonomous operation of a hybrid AC/DC microgrid with multiple interlinking converters. *IEEE Transactions on Smart Grid* 9 (6): 6480–6488.

6 Peyghami, S., Davari, P., Mokhtari, H., and Blaabjerg, F. (2019). Decentralized droop control in DC microgrids based on a frequency injection approach. *IEEE Transactions on Smart Grid* 99, no. To be published/DOI: 10.1109/TSG.2019.2911213: 1–11.

7 Azizi, A., Peyghami, S., Mokhtari, H., and Blaabjerg, F. (Dec. 2019). Autonomous and decentralized load sharing and energy management approach for DC microgrids. *Electric Power Systems Research* 177: 1–11.

8 Peyghami, S., Mokhtari, H., and Blaabjerg, F. (2018). Distributed and decentralized control of DC microgrids. In: *DC Distribution Systems and Microgrids* (eds. T. Dragičević, P. Wheeler and F. Blaabjerg), 23–42. IET.

9 Kundur, P., Balu, N., Lauby, M. et al. (2007). *Power System Stability and Control*, 2e, vol. 20073061. New York: Taylor & Francis Group, LLC.

10 Sheblé, G.B. (2001). *Power System Planning (Reliability)*. Bosa Roca, United States: Taylor & Francis Inc.

11 Billinton, R. and Allan, R.N. (1984). *Reliability Evaluation of Power Systems,"* First. New York: Plenum Press.

12 Morison, K., Wang, L., and Kundur, P. (2004). Power system security assessment. *IEEE Power and Energy Magazine* 2 (October): 30–39.

13 Kundur, P., Paserba, J., Ajjarapu, V. et al. (2004). Definition and classification of power system stability IEEE/CIGRE joint task force on stability terms and definitions. *IEEE Transactions on Power Apparatus and Systems* 19 (2): 1387–1401.

14 Kundur, P., Balu, N., and Lauby, M. (1994). *Power System Stability and Control*. New York: McGraw-hill.

15 PES-TR67 (2018). Impact of IEEE 1547 Standard on Smart Inverters. https://resourcecenter.ieee-pes.org/publications/technical-reports/PES_TR0067_5-18.html (accessed 12 February 2021).

16 Vartanian, C., Bauer, R., Casey, L. et al. (Nov. 2018). Ensuring system reliability: distributed energy resources and bulk power system considerations. *IEEE Power and Energy Magazine* 16 (6): 52–63.

17 Wang, X. and Blaabjerg, F. (2019). Harmonic stability in power electronic based power systems: concept, modeling, and analysis. *IEEE Transactions on Smart Grid* 10 (3): 2858–2870.

18 ENTSOE (2018). Interactions between HVDC Systems and Other Connections. ENTSO-E Guidance Document for National Implementation for Network Codes on Grid Connection. https://eepublicdownloads.entsoe.eu/clean-documents/Network%20codes%20documents/ NC%20RfG/IGD-Interactions_between_HVDC_Controllers_final.pdf (accessed 12 February 2021).

19 Buchhagen, C., Greve, M., Menze, A. and Jung, J. (2016). Harmonic Stability-Practical Experience of a TSO. *15th Wind Integration Workshop*, Vienna, Austria (15–17 November 2016).

20 Li, C. (2018). Unstable operation of photovoltaic inverter from field experiences. *IEEE Transactions on Power Delivery* 33 (2): 1013–1015.

21 Mollerstedt, E. and Bernhardsson, B. (2017). Out of control because of harmonics – an analysis of the harmonic response of an inverter locomotive. *IEEE Transactions on Power Electronics* 32 (11): 8922–8935.

22 Peyghami, S., Azizi, A., Mokhtari, H. and Blaabjerg, F. (2018). Active Damping of Torsional Vibrations Due to the Sub-Harmonic Instability on a Synchronous Generator. *IEEE ECCE EUROPE*, Riga, Latvia (17–21 September 2018). IEEE.

23 Cigre Working Group C1.27 (2018). The Future of Reliability – Definition of Reliability in Light of New Developments in Various Devices and Services Which Offer Customers and System Operators New Levels of Flexibility. Technical Brochure.

24 Wang, S., Li, Z., Wu, L. et al. (Aug. 2013). New metrics for assessing the reliability and economics of microgrids in distribution system. *IEEE Transactions on Power Apparatus and Systems* 28 (3): 2852–2861.

25 Peyghami, S., Davari, P., and Blaabjerg, F. (2019). System-level reliability-oriented power sharing strategy for DC power systems. *IEEE Transactions on Industry Applications* 55 (5): 4865–4875.

26 Peyghami, S., Wang, H., Davari, P., and Blaabjerg, F. (2019). Mission profile based system-level reliability analysis in DC microgrids. *IEEE Transactions on Industry Applications* 55 (5): 5055–5067.

27 Hahn, F., Andresen, M., Buticchi, G. and Liserre, M. (2017). Mission Profile Based Reliability Evaluation of Building Blocks for Modular Power Converters. *International Exhibition & Conference for Power Electronics, Intelligent Motion, Renewable Energy and Energy Management,* Nuremberg, Germany (7–9 May 2019). Curran Associates, Inc.

28 De Léon-Aldaco, S.E., Calleja, H., Chan, F., and Jiménez-Grajales, H.R. (2013). Effect of the Mission profile on the reliability of a power converter aimed at photovoltaic applications-a case study. *IEEE Transactions on Power Electronics* 28 (6): 2998–3007.

29 Ma, K., Liserre, M., Blaabjerg, F., and Kerekes, T. (Feb. 2015). Thermal loading and lifetime estimation for power device considering mission profiles in wind power converter. *IEEE Transactions on Power Electronics* 30 (2): 590–602.

30 Peyghami, S., Wang, Z. and Blaabjerg, F. (2019). Reliability Modeling of Power Electronic Converters: A General Approach. *IEEE COMPEL*, Toronto, Canada (17–20 June 2019). IEEE.

31 IEC 61709 (2017). Electric Components – Reliability – Reference Conditions for Failure Rates and Stress Models for Conversion. https://standards.iteh.ai/catalog/standards/iec/42393689-38ca-4537-90a4-5bc80655a7dc/iec-61709-2017 (accessed 12 February 2021).

32 International Electrotechnical Commission. "Electric components– Reliability– Reference conditions for failure rates and stress models for conversion: IEC 61709, 2017."

33 FIDES (2010). FIDES Guide 2009 Edition A. Reliability Methodology for Electronic Systems. http://www.embedded.agh.edu.pl/www/fpga/dydaktyka/MPiMS/Data/UTE_FIDES_Guide_2009_-_Edition_A%20-%20September%202010_english_version.pdf (accessed 12 February 2021).

34 Peyghami, S., Palensky, P., and Blaabjerg, F. (2020). An overview on the reliability of modern power electronic based power systems. *IEEE Open Journal of Power Electronics* 1: 34–50.

Index

Smart Grid and Enabling Technologies, First Edition. Shady S. Refaat, Omar Ellabban, Sertac Bayhan, Haitham Abu-Rub, Frede Blaabjerg, and Miroslav M. Begovic.
© 2021 John Wiley & Sons Ltd. Published 2021 by John Wiley & Sons Ltd.
Companion website: www.wiley.com/go/ellabban/smartgrid